Periodic Table

Legend:
- Atomic number — Atomic weight
- Element symbol
- Element name

Example: 1 / 1.008 / **H** / Hydrogen

1 IA	2 IIA	3 IIIB	4 IVB	5 VB	6 VIB	7 VIIB	8	9 VIIIB	10	11 IB	12 IIB	13 IIIA	14 IVA	15 VA	16 VIA	17 VIIA	18 VIIIA
1 1.008 **H** Hydrogen																	2 4.003 **He** Helium
3 6.941 **Li** Lithium	4 9.012 **Be** Beryllium											5 10.811 **B** Boron	6 12.011 **C** Carbon	7 14.007 **N** Nitrogen	8 15.999 **O** Oxygen	9 18.998 **F** Fluorine	10 20.180 **Ne** Neon
11 22.990 **Na** Sodium	12 24.305 **Mg** Magnesium											13 26.982 **Al** Aluminum	14 28.086 **Si** Silicon	15 30.974 **P** Phosphorus	16 32.066 **S** Sulfur	17 35.453 **Cl** Chlorine	18 39.948 **Ar** Argon
19 39.098 **K** Potassium	20 40.078 **Ca** Calcium	21 44.956 **Sc** Scandium	22 47.88 **Ti** Titanium	23 50.942 **V** Vanadium	24 51.996 **Cr** Chromium	25 54.938 **Mn** Manganese	26 55.847 **Fe** Iron	27 58.933 **Co** Cobalt	28 58.69 **Ni** Nickel	29 63.546 **Cu** Copper	30 65.39 **Zn** Zinc	31 69.723 **Ga** Gallium	32 72.61 **Ge** Germanium	33 74.922 **As** Arsenic	34 78.96 **Se** Selenium	35 79.904 **Br** Bromine	36 83.80 **Kr** Krypton
37 85.468 **Rb** Rubidium	38 87.62 **Sr** Strontium	39 88.906 **Y** Yttrium	40 91.224 **Zr** Zirconium	41 92.906 **Nb** Niobium	42 95.94 **Mo** Molybdenum	43 (98) **Tc** Technicium	44 101.07 **Ru** Ruthenium	45 102.91 **Rh** Rhodium	46 106.42 **Pd** Palladium	47 107.87 **Ag** Silver	48 112.41 **Cd** Cadmium	49 114.82 **In** Indium	50 118.71 **Sn** Tin	51 121.75 **Sb** Antimony	52 127.60 **Te** Tellurium	53 126.91 **I** Iodine	54 131.29 **Xe** Xenon
55 132.91 **Cs** Cesium	56 137.33 **Ba** Barium	71 174.97 **Lu** Lutetium	72 178.49 **Hf** Hafnium	73 180.95 **Ta** Tantalum	74 183.85 **W** Tungsten	75 186.21 **Re** Rhenium	76 190.2 **Os** Osmium	77 192.22 **Ir** Iridium	78 195.08 **Pt** Platinum	79 196.97 **Au** Gold	80 200.59 **Hg** Mercury	81 204.38 **Tl** Thallium	82 207.2 **Pb** Lead	83 208.98 **Bi** Bismuth	84 (209) **Po** Polonium	85 (210) **At** Astatine	86 (222) **Rn** Radon
87 (223) **Fr** Francium	88 226.03 **Ra** Radium	103 (260) **Lr** Lawrencium	104 (261) **Rf** Rutherfordium	105 (262) **Ha** Hahnium	106 (263) **Sg** Seaborgium	107 (264) **Ns** Nielsbohrium	108 (265) **Hs** Hassium	109 (266) **Mt** Meitnerium	110 (281) **Ds** Darmstadtium	111 (272) **Rg** Roentgnenium							

Lanthanides:

57 138.91 **La** Lanthanum	58 140.11 **Ce** Cerium	59 140.91 **Pr** Praseodynium	60 144.24 **Nd** Neodynium	61 (145) **Pm** Promethium	62 150.36 **Sm** Samarium	63 151.96 **Eu** Europium	64 157.25 **Gd** Gadolinium	65 158.93 **Tb** Terbium	66 162.50 **Dy** Dysprosium	67 164.93 **Ho** Holmium	68 167.26 **Er** Erbium	69 168.93 **Tm** Thulium	70 173.04 **Yb** Ytterbium

Actinides:

89 227.03 **Ac** Actinium	90 232.04 **Th** Thorium	91 231.04 **Pa** Protactinium	92 238.03 **U** Uranium	93 237.05 **Np** Neptunium	94 (244) **Pu** Plutonium	95 (243) **Am** Americium	96 (247) **Cm** Curium	97 (247) **Bk** Berkelium	98 (251) **Cf** Californium	99 (252) **Es** Einsteinium	100 (257) **Fm** Fermium	101 (258) **Md** Mendelevium	102 (259) **No** Nobelium

International Atomic Weights[a] Based on $^{12}C=12$

Element	Symbol	Atomic Number	Atomic Weight	Element	Symbol	Atomic Number	Atomic Weight
Actinium	Ac	89	(227)	Meitnerium	Mt	109	(268)
Aluminum	Al	13	26.9815	Mercury	Hg	80	200.59
Americium	Am	95	(243)	Molybdenum	Mo	42	95.94
Antimony	Sb	51	121.760	Neodymium	Nd	60	144.24
Argon	Ar	18	39.948	Neon	Ne	10	20.1797
Arsenic	As	33	74.9216	Neptunium	Np	93	(237)
Astatine	At	85	(210)	Nickel	Ni	28	58.6934
Barium	Ba	56	137.327	Niobium	Nb	41	92.906
Berkelium	Bk	97	(247)	Nitrogen	N	7	14.0067
Beryllium	Be	4	9.0122	Nobelium	No	102	(259)
Bismuth	Bi	83	208.980	Osmium	Os	76	190.23
Bohrium	Bh	107	(264)	Oxygen	O	8	15.9994
Boron	B	5	10.811	Palladium	Pd	46	106.42
Bromine	Br	35	79.904	Phosphorus	P	15	30.9738
Cadmium	Cd	48	112.411	Platinum	Pt	78	195.078
Calcium	Ca	20	40.078	Plutonium	Pu	94	(244)
Californium	Cf	98	(251)	Polonium	Po	84	(209)
Carbon	C	6	12.0107	Potassium	K	19	39.0983
Cerium	Ce	58	140.116	Praseodymium	Pr	59	140.908
Cesium	Cs	55	132.905	Promethium	Pm	61	(145)
Chlorine	Cl	17	35.4527	Protactinium	Pa	91	231.036
Chromium	Cr	24	51.9961	Radium	Ra	88	(226)
Cobalt	Co	27	58.9332	Radon	Rn	86	(222)
Copper	Cu	29	63.546	Rhenium	Re	75	186.207
Curium	Cm	96	(247)	Rhodium	Rh	45	102.9055
Dubnium	Db	105	(262)	Rubidium	Rb	37	85.4678
Dysprosium	Dy	66	162.50	Ruthenium	Ru	44	101.07
Einsteinium	Es	99	(252)	Rutherfordium	Rf	104	(261)
Erbium	Er	68	167.26	Samarium	Sm	62	150.36
Europium	Eu	63	151.964	Scandium	Sc	21	44.956
Fermium	Fm	100	(257)	Seagborgium	Sg	106	(266)
Fluorine	F	9	18.9984	Selenium	Se	34	78.96
Francium	Fr	87	(223)	Silicon	Si	14	28.0855
Gadolinium	Gd	64	157.25	Silver	Ag	47	107.8682
Gallium	Ga	31	69.723	Sodium	Na	11	22.9898
Germanium	Ge	32	72.61	Strontium	Sr	38	87.62
Gold	Au	79	196.967	Sulfur	S	16	32.066
Hafnium	Hf	72	178.49	Tantalum	Ta	73	180.948
Hassium	Hs	108	(269)	Technetium	Tc	43	(98)
Helium	He	2	4.0026	Tellurium	Te	52	127.60
Holmium	Ho	67	164.930	Terbium	Tb	65	158.925
Hydrogen	H	1	1.00794	Thallium	Tl	81	204.3833
Indium	In	49	114.818	Thorium	Th	90	232.0381
Iodine	I	53	126.9045	Thulium	Tm	69	168.934
Iridium	Ir	77	192.217	Tin	Sn	50	118.710
Iron	Fe	26	55.845	Titanium	Ti	22	47.867
Krypton	Kr	36	83.80	Tungsten	W	74	183.84
Lanthanum	La	57	138.9055	Uranium	U	92	238.0289
Lawrencium	Lw	103	(262)	Vanadium	V	23	50.9415
Lead	Pb	82	207.2	Xenon	Xe	54	131.29
Lithium	Li	3	6.941	Ytterbium	Yb	70	173.04
Lutetium	Lu	71	174.967	Yttrium	Y	39	88.906
Magnesium	Mg	12	24.3050	Zinc	Zn	30	65.39
Manganese	Mn	25	54.9380	Zirconium	Zr	40	91.224
Mendelevium	Md	101	(258)				

[a]Parentheses indicate the atomic weight of the most stable isotope.

Analytical Chemistry for Technicians

Fourth Edition

Analytical Chemistry for Technicians

Fourth Edition

John Kenkel

CRC Press
Taylor & Francis Group
Boca Raton   London   New York

CRC Press is an imprint of the
Taylor & Francis Group, an **informa** business

CRC Press
Taylor & Francis Group
6000 Broken Sound Parkway NW, Suite 300
Boca Raton, FL 33487-2742

© 2014 by Taylor & Francis Group, LLC
CRC Press is an imprint of Taylor & Francis Group, an Informa business

No claim to original U.S. Government works

Printed on acid-free paper
Version Date: 20130422

International Standard Book Number-13: 978-1-4398-8105-7 (Hardback)

Library of Congress Cataloging-in-Publication Data

Kenkel, John.
 Analytical chemistry for technicians / John Kenkel. -- Fourth edition.
 pages cm
 Includes bibliographical references and index.
 ISBN 978-1-4398-8105-7 (hardback)
 1. Chemistry, Analytic. I. Title.

QD75.22.K445 2013
543--dc23
                                                                    2013011752

Visit the Taylor & Francis Web site at
http://www.taylorandfrancis.com

and the CRC Press Web site at
http://www.crcpress.com

*This book is dedicated to the hundreds of hardworking students
that have passed through my classroom and laboratory over
the past 36 years. Without them, the wonderful career that has
defined my professional life would have been a mere dream.*

*This book is also dedicated to the precious women in my personal life that
I dearly love—my wife, Lois, and my daughters, Sister Emily, Jeanie, and
Laura. May God bless you and keep you forever in the palm of His hand.*

Contents

List of Experiments...xvii
Preface...xix
Acknowledgments...xxi
Author xxiii
Introduction to Laboratory Work xxv

Chapter 1   Introduction to Analytical Science...1

    1.1   Analytical Science Defined 1
    1.2   Classifications of Analysis..2
    1.3   The Sample 3
    1.4   The Analytical Process..3
    1.5   Analytical Technique and Skills 4
    1.6   Elementary Statistics 5
           1.6.1   Errors...5
           1.6.2   Definitions 6
           1.6.3   Distribution of Measurements 8
           1.6.4   Student's t 10
           1.6.5   Rejection of Data 12
           1.6.6   Final Comments on Statistics...13
    1.7   Precision, Accuracy, and Calibration 13

Chapter 2   Sampling and Sample Preparation 19

    2.1   Introduction 19
    2.2   Obtaining the Sample..19
    2.3   Statistics of Sampling 20
    2.4   Sample Handling 21
           2.4.1   Chain of Custody..21
           2.4.2   Maintaining Sample Integrity 22
    2.5   Sample Preparation—Solid Materials...23
           2.5.1   Particle Size Reduction 23
           2.5.2   Sample Homogenization and Division 23
           2.5.3   Solid–Liquid Extraction 24
           2.5.4   Other Extractions from Solids..24
    2.6   Water Purification and Use..25
           2.6.1   Purifying Water by Distillation.......................................25
           2.6.2   Purifying Water by Deionization 26
    2.7   Total Sample Dissolution and Other Considerations......................26
           2.7.1   Hydrochloric Acid 27
           2.7.2   Sulfuric Acid 27
           2.7.3   Nitric Acid 28
           2.7.4   Hydrofluoric Acid...28
           2.7.5   Perchloric Acid 28
           2.7.6   "Aqua Regia"...28
           2.7.7   Acetic Acid 28
           2.7.8   Ammonium Hydroxide 29

	2.8	Fusion 30
	2.9	Sample Preparation: Liquid Samples, Extracts, and Solutions of Solids 30
		2.9.1 Extraction from Liquid Solutions 30
		2.9.2 Dilution, Concentration, and Solvent Exchange 32
		2.9.3 Sample Stability 32
	2.10	Liquid–Liquid Extraction 32
		2.10.1 Introduction 32
		2.10.2 The Separatory Funnel 33
		2.10.3 Theory 34
		2.10.4 Calculations Involving Equation 2.2 35
		2.10.5 Calculations Involving Equation 2.3 36
		2.10.6 Calculations Involving a Combination of Equations 2.3 (or 2.7) and 2.4 37
		2.10.7 Calculation of Percent Extracted (Equation 2.5) 37
		2.10.8 Evaporators 38
	2.11	Solid–Liquid Extraction 38
	2.12	Distillation of a Mixture of Liquids 39
	2.13	Reagents Used in Sample Preparation 41
	2.14	Labeling and Record Keeping 41

Chapter 3   Gravimetric Analysis 49

	3.1	Introduction 49
	3.2	Weight vs. Mass 49
	3.3	The Balance 49
	3.4	The Desiccator 51
	3.5	Calibration and Care of Balances 52
	3.6	When to Use Which Balance 52
	3.7	Details of Gravimetric Methods 53
		3.7.1 Physical Separation Methods and Calculations 53
			3.7.1.1 Loss on Drying 55
			3.7.1.2 Loss on Ignition 55
			3.7.1.3 Residue on Ignition 56
			3.7.1.4 Insoluble Matter in Reagents 56
			3.7.1.5 Solids in Water and Wastewater 56
			3.7.1.6 Particle Size by Analytical Sieving 57
		3.7.2 Chemical Alteration/Separation of the Analyte 58
		3.7.3 Gravimetric Factors 59
		3.7.4 Using Gravimetric Factors 61
	3.8	Experimental Considerations 63
		3.8.1 Weighing Bottles 63
		3.8.2 Weighing by Difference 63
		3.8.3 Isolating and Weighing Precipitates 64

Chapter 4   Introduction to Titrimetric Analysis 73

	4.1	Introduction 73
	4.2	Terminology 73
	4.3	Review of Solution Concentration 75
		4.3.1 Molarity 75
		4.3.2 Normality 77

	4.4	Review of Solution Preparation	79
		4.4.1 Solid Solute and Molarity	80
		4.4.2 Solid Solute and Normality	81
		4.4.3 Solution Preparation by Dilution	82
	4.5	Stoichiometry of Titration Reactions	82
	4.6	Standardization	84
		4.6.1 Standardization Using a Standard Solution	84
		4.6.2 Standardization Using a Primary Standard	86
		4.6.3 Titer	88
	4.7	Percentage Analyte Calculations	88
	4.8	Volumetric Glassware	91
		4.8.1 Volumetric Flask	91
		4.8.2 Pipet	94
		4.8.3 Buret	98
		4.8.4 Cleaning and Storing Procedures	100
	4.9	Pipetters, Automatic Titrators, and Other Devices	100
		4.9.1 Pipetters	100
		4.9.2 Bottle-Top Dispensers	102
		4.9.3 Digital Burets and Automatic Titrators	102
	4.10	Calibration of Glassware and Devices	103
	4.11	Analytical Technique	103

Chapter 5 Applications of Titrimetric Analysis | 113

	5.1	Introduction	113
	5.2	Acid–Base Titrations and Titration Curves	113
		5.2.1 Titration of Hydrochloric Acid	113
		5.2.2 Titration of Weak Monoprotic Acids	115
		5.2.3 Titration of Monobasic Strong and Weak Bases	116
		5.2.4 Equivalence Point Detection	116
		5.2.5 Titration of Polyprotic Acids: Sulfuric Acid and Phosphoric Acid	118
		5.2.6 Titration of Potassium Biphthalate	121
		5.2.7 Titration of Tris-(Hydroxymethyl)Amino Methane	122
		5.2.8 Titration of Sodium Carbonate	122
	5.3	Examples of Acid/Base Determinations	123
		5.3.1 Alkalinity of Water or Wastewater	124
		5.3.2 Back Titration Applications	124
		5.3.3 Indirect Titration Applications	126
	5.4	Other Acid/Base Applications	127
	5.5	Buffer Solution Applications	127
		5.5.1 Conjugate Acids and Bases	128
		5.5.2 Henderson–Hasselbalch Equation	129
	5.6	Complex Ion Formation Reactions	133
		5.6.1 Introduction	133
		5.6.2 Complex Ion Terminology	133
		5.6.3 EDTA and Water Hardness	135
		5.6.4 Expressing Concentration Using Parts per Million	138
		5.6.4.1 Solution Preparation	139
		5.6.5 Water Hardness Calculations	141
		5.6.6 Other Uses of EDTA Titrations	143

5.7   Oxidation–Reduction Reactions 144
        5.7.1   Review of Basic Concepts and Terminology 144
        5.7.2   The Ion-Electron Method for Balancing Equations 147
        5.7.3   Analytical Calculations 148
        5.7.4   Applications 150
                  5.7.4.1   Potassium Permanganate 150
                  5.7.4.2   Iodometry: An Indirect Method 150
                  5.7.4.3   Prereduction and Preoxidation 152
5.8   Other Examples 152

Chapter 6   Introduction to Instrumental Analysis 165

6.1   Review of the Analytical Process 165
6.2   Instrumental Analysis Methods 166
6.3   Basics of Instrumental Measurement 167
        6.3.1   Sensors, Signal Processors, Readouts, and Power Supplies 168
        6.3.2   Calibration of an Analytical Instrument 168
        6.3.3   Mathematics of Linear Relationships 170
        6.3.4   Method of Least Squares 171
        6.3.5   The Correlation Coefficient 172
6.4   Preparation of Standards 172
6.5   Blanks and Controls 173
        6.5.1   Reagent Blanks 173
        6.5.2   Sample Blanks 174
        6.5.3   Controls 174
6.6   Post-Run Calculations in Instrumental Analysis 174
        6.6.1   Calculation of ppm Analyte in a Solution Given Mass and
                  Volume Data 175
        6.6.2   Calculation of ppm Analyte in a Solid Sample Given Mass Data 175
        6.6.3   Calculation of the Mass of Analyte Found in an Extract 175
        6.6.4   Calculation of ppm Analyte in a Liquid or Solid That Was
                  Extracted 176
        6.6.5   Calculation When a Dilution Is Involved 176
6.7   Laboratory Data Acquisition and Information Management 178
        6.7.1   Data Acquisition 178
        6.7.2   Laboratory Information Management 179

Chapter 7   Introduction to Spectrochemical Methods 185

7.1   Introduction 185
7.2   Characterizing Light 185
        7.2.1   Wavelength, Speed, Frequency, Energy, and Wavenumber 186
7.3   The Electromagnetic Spectrum 189
7.4   Refractometry 190
7.5   Absorption and Emission of Light 193
        7.5.1   Brief Summary 193
        7.5.2   Atoms vs. Molecules and Complex Ions 196
        7.5.3   Absorption Spectra 197
        7.5.4   Light Emission 201
7.6   Absorbance, Transmittance, and Beer's Law 202
7.7   Effect of Concentration on Spectra 207

Chapter 8   UV-Vis and IR Molecular Spectrometry 215

　　　8.1   Review 215
　　　8.2   UV-Vis Instrumentation 215
　　　　　8.2.1   Sources 215
　　　　　　8.2.1.1   Tungsten Filament Lamp 215
　　　　　　8.2.1.2   Deuterium Lamp 216
　　　　　　8.2.1.3   Xenon Arc Lamp 216
　　　　　8.2.2   Wavelength Selection 216
　　　　　　8.2.2.1   Absorption Filters 217
　　　　　　8.2.2.2   Monochromators 217
　　　　　8.2.3   Sample Compartment 220
　　　　　　8.2.3.1   Single-Beam Spectrophotometer 220
　　　　　　8.2.3.2   Beam Splitting and Chopping 221
　　　　　　8.2.3.3   Double-Beam Designs 222
　　　　　　8.2.3.4   Diode Array Design 223
　　　　　　8.2.3.5   Summary 225
　　　　　8.2.4   Detectors 226
　　　　　　8.2.4.1   Photomultiplier Tube 226
　　　　　　8.2.4.2   Photodiodes 228
　　　8.3   Cuvette Selection and Handling 228
　　　8.4   Interferences, Deviations, Maintenance, and Troubleshooting 229
　　　　　8.4.1   Interferences 229
　　　　　8.4.2   Deviations 229
　　　　　8.4.3   Maintenance 230
　　　　　8.4.4   Troubleshooting 230
　　　8.5   Fluorometry 231
　　　8.6   Introduction to IR Spectrometry 233
　　　8.7   IR Instrumentation 234
　　　8.8   Sampling 235
　　　　　8.8.1   Liquid Sampling 235
　　　8.9   Solid Sampling 240
　　　　　8.9.1   Solution Prepared and Placed in a Liquid Sampling Cell 240
　　　　　8.9.2   Thin Film Formed by Solvent Evaporation 240
　　　　　8.9.3   KBr Pellet 240
　　　　　8.9.4   Nujol Mull 242
　　　　　8.9.5   Reflectance Methods 242
　　　　　　8.9.5.1   Specular Reflectance 242
　　　　　　8.9.5.2   Internal Reflectance 242
　　　　　　8.9.5.3   Diffuse Reflectance 244
　　　　　8.9.6   Gas Sampling 244
　　　8.10   Basic IR Spectra Interpretation 244
　　　8.11   Quantitative Analysis 247

Chapter 9   Atomic Spectroscopy 259

　　　9.1   Review and Comparisons 259
　　　9.2   Brief Summary of Techniques and Instrument Designs 260
　　　9.3   Flame Atomic Absorption 262
　　　　　9.3.1   Flames and Flame Processes 262
　　　　　9.3.2   Spectral Line Sources 263

            9.3.2.1   Hollow Cathode Lamp 264

            9.3.2.2   Electrodeless Discharge Lamp 265

      9.3.3   Premix Burner 265

      9.3.4   Optical Path 267

      9.3.5   Practical Matters and Applications 268

            9.3.5.1   Slits and Spectral Lines 268

            9.3.5.2   Linear and Nonlinear Standard Curves 269

            9.3.5.3   Hollow Cathode Lamp Current 271

            9.3.5.4   Lamp Alignment 271

            9.3.5.5   Aspiration Rate 271

            9.3.5.6   Burner Head Position 271

            9.3.5.7   Fuel and Oxidant Sources and Flow Rates 271

      9.3.6   Interferences 271

            9.3.6.1   Chemical Interferences 272

            9.3.6.2   Spectral Interferences 273

      9.3.7   Safety and Maintenance 274

  9.4   Graphite Furnace Atomic Absorption 275

      9.4.1   General Description 275

      9.4.2   Advantages and Disadvantages 277

  9.5   Inductively Coupled Plasma 278

  9.6   Miscellaneous Atomic Techniques 281

      9.6.1   Flame Photometry 281

      9.6.2   Cold Vapor Mercury 282

      9.6.3   Hydride Generation 282

      9.6.4   Spark Emission 282

      9.6.5   Atomic Fluorescence 282

  9.7   Summary of Atomic Techniques 282

Chapter 10 Introduction to Chromatography 291

  10.1   Introduction 291

  10.2   Chromatography 291

  10.3   "Types" of Chromatography 292

      10.3.1   Partition Chromatography 292

      10.3.2   Adsorption Chromatography 293

      10.3.3   Ion-Exchange Chromatography 294

      10.3.4   Size Exclusion Chromatography 295

  10.4   Chromatography Configurations 295

      10.4.1   Paper and Thin-Layer Chromatography 296

      10.4.2   Classical Open-Column Chromatography 298

      10.4.3   Instrumental Chromatography 301

      10.4.4   Instrumental Chromatogram 301

      10.4.5   Quantitative Analysis with GC and HPLC 305

  10.5   Electrophoresis 306

Chapter 11 Gas Chromatography 311

  11.1   Overview 311

  11.2   Vapor Pressure and Solubility 311

  11.3   Instrument Components 312

  11.4   Sample Injection 314

11.5   Column Details..316
    11.5.1   Instrument Logistics..316
    11.5.2   Packed, Open Tubular, and Preparative Columns....................317
    11.5.3   The Nature and Selection of the Stationary Phase..................318
    11.5.4   Column Temperature...319
    11.5.5   Carrier Gas Flow Rate 320
11.6   Detectors..321
    11.6.1   Flame Ionization Detector (FID) 321
    11.6.2   Thermal Conductivity Detector (TCD)..................................322
    11.6.3   Electron Capture Detector (ECD)..323
    11.6.4   Nitrogen/Phosphorus Detector (NPD) 324
    11.6.5   Flame Photometric Detector (FPD) 324
    11.6.6   Electrolytic Conductivity (Hall Detector)..............................324
    11.6.7   Gas Chromatography–Mass Spectrometry (GC-MS)..............324
    11.6.8   Photoionization Detector (PID)...325
11.7   Qualitative Analysis 325
11.8   Quantitative Analysis 326
    11.8.1   Quantitation Methods...326
    11.8.2   Response Factor Method..326
    11.8.3   Internal Standard Method...327
    11.8.4   Standard Additions Method..328
11.9   Troubleshooting..329
    11.9.1   Diminished Peak Size...329
    11.9.2   Unsymmetrical Peak Shapes...329
    11.9.3   Altered Retention Times 330
    11.9.4   Baseline Drift...330
    11.9.5   Baseline Perturbations..330
    11.9.6   Appearance of Unexpected Peaks..330

Chapter 12 High-Performance Liquid Chromatography and Electrophoresis 341

12.1   Introduction 341
    12.1.1   Summary of Method 341
    12.1.2   Comparisons with GC..341
12.2   Mobile Phase Considerations 342
12.3   Solvent Delivery 344
    12.3.1   Pumps...344
    12.3.2   Gradient vs. Isocratic Elution..345
12.4   Sample Injection 346
12.5   Column Selection 348
    12.5.1   Normal Phase Columns..348
    12.5.2   Reverse-Phase Columns 348
    12.5.3   Adsorption Columns 349
    12.5.4   Ion Exchange and Size Exclusion Columns 349
    12.5.5   The Size of the Stationary Phase Particles.............................349
    12.5.6   Column Selection 349
12.6   Detectors..350
    12.6.1   UV Absorption...350
    12.6.2   Diode Array..351
    12.6.3   Fluorescence...351
    12.6.4   Refractive Index 353

           12.6.5 Electrochemical...354
                   12.6.5.1 Conductivity..354
                   12.6.5.2 Amperometric...354
    12.7   Qualitative and Quantitative Analysis...................................355
    12.8   Troubleshooting 356
           12.8.1 Unusually High Pressure..356
           12.8.2 Unusually Low Pressure...356
                   12.8.2.1 System Leaks..356
                   12.8.2.2 Air Bubbles...357
                   12.8.2.3 Column "Channeling" 357
                   12.8.2.4 Decreased Retention Time..............................357
                   12.8.2.5 Baseline Drift 357
    12.9   Electrophoresis 357
           12.9.1 Introduction 357
           12.9.2 Capillary Electrophoresis...359
                   12.9.2.1 Electroosmotic Flow 361
                   12.9.2.2 Sample Introduction.......................................361
                   12.9.2.3 Analyte Detection...361

Chapter 13 Mass Spectrometry...371

    13.1   Basic Principles 371
    13.2   Sample Inlet Systems and Ion Sources...................................372
    13.3   Mass Analyzers 373
    13.4   The Ion Detector..376
    13.5   Mass Spectra...377
    13.6   ICP-MS..378
    13.7   GC-MS...378
    13.8   LC-MS..380
    13.9   Tandem Mass Spectrometry 381

Chapter 14 Electroanalytical Methods 387

    14.1   Introduction 387
    14.2   Transfer Tendencies: Standard Reduction Potentials 391
    14.3   Determination of Overall Redox Reaction Tendency: E°_{cell} 393
    14.4   The Nernst Equation..394
    14.5   Potentiometry 396
           14.5.1 Reference Electrodes..396
                   14.5.1.1 The Saturated Calomel Reference Electrode (SCE).........396
                   14.5.1.2 The Silver–Silver Chloride Electrode..............398
           14.5.2 Indicator Electrodes 399
                   14.5.2.1 The pH Electrode...399
           14.5.3 Combination Electrodes 400
                   14.5.3.1 The Combination pH Electrode.....................400
                   14.5.3.2 Ion-Selective Electrodes 401
           14.5.4 Other Details of Electrode Design 403
           14.5.5 Care and Maintenance of Electrodes 403
           14.5.6 Potentiometric Titrations...404
    14.6   Voltammetry and Amperometry 405
           14.6.1 Voltammetry...405

14.6.2 Amperometry 406
14.7 Karl Fischer Titration 406
14.7.1 End Point Detection...406
14.7.2 Elimination of Extraneous Water..407
14.7.3 The Volumetric Method 407
14.7.4 The Coulometric Method 409

Chapter 15 Miscellaneous Instrumental Techniques... 417

15.1 X-Ray Methods.. 417
15.1.1 Introduction 417
15.1.2 X-Ray Diffraction Spectroscopy 418
15.1.3 X-Ray Fluorescence Spectroscopy 421
15.1.4 Applications.. 421
15.1.5 Safety Issues Concerning X-Rays 422
15.2 Nuclear Magnetic Resonance Spectroscopy 422
15.2.1 Introduction 422
15.2.2 Instrumentation 423
15.2.3 The NMR Spectrum..425
15.2.3.1 Chemical Shifts 425
15.2.3.2 Peak Splitting and Integration 427
15.2.4 Solvents and Solution Concentration.......................................428
15.2.5 Analytical Uses 428
15.3 Viscosity 428
15.3.1 Introduction 428
15.3.2 Definitions 429
15.3.3 Temperature Dependence...430
15.3.4 Capillary Viscometry 430
15.3.5 Rotational Viscometry 433
15.4 Thermal Analysis 434
15.4.1 Introduction 434
15.4.2 DTA and DSC 434
15.4.3 DSC Instrumentation...436
15.4.4 Applications of DSC...437
15.5 Optical Rotation 437

Appendix 1: Formulas for Solution Concentration and Preparation Calculations...............443

Appendix 2: The Language of Quality Assurance and Good Laboratory Practice (GLP) Laws: A Glossary 447

Appendix 3: Significant Figure Rules 451

Appendix 4: Answers to Questions and Problems..453

Index...501

List of Experiments

Experiment 1: Assuring the Quality of Weight Measurements 14

Experiment 2: Weight Uniformity of Dosing Units 15

Experiment 3: A Study of the Dissolving Properties of Water, Some Common Organic
Liquids, and Laboratory Acids 42

Experiment 4: Loss on Drying 64

Experiment 5: Particle Sizing 65

Experiment 6: The Determination of Salt in a Salt–Sand Mixture 65

Experiment 7: The Gravimetric Determination of Sulfate in a Commercial Unknown 65

Experiment 8: High-Precision Glassware: A Calibration Experiment 104

Experiment 9: Preparation and Standardization of HCl and NaOH Solutions 106

Experiment 10: Titrimetric Analysis of a Commercial Soda Ash Unknown for Sodium
Carbonate 152

Experiment 11: Titrimetric Analysis of a Commercial KHP Unknown for KHP 152

Experiment 12: EDTA Titrations 153

Experiment 13: Plotting a Standard Curve Using Excel Spreadsheet Software (Version 2010) 179

Experiment 14: Percentage of Sugar in Soft Drinks by Refractometry 207

Experiment 15: Colorimetric Analysis of Prepared and/or Real Water Samples for Iron 207

Experiment 16: Design an Experiment: A Study of the Effect of pH on the Analysis of
Water Samples for Iron 208

Experiment 17: Design an Experiment: Determining the Concentration at Which a Beer's
Law Plot Becomes Nonlinear 208

Experiment 18: The Determination of Phosphorus in Environmental Water 208

Experiment 19: Spectrophotometric Analysis of a Prepared Sample for Toluene 248

Experiment 20: Extraction of Iodine with Cyclohexane 248

Experiment 21: Determination of Nitrate in Drinking Water by UV Spectrophotometry 249

Experiment 22: Fluorometric Analysis of a Prepared Sample for Riboflavin 249

Experiment 23: Design an Experiment: Determination of Riboflavin in a Vitamin Tablet
Using Fluorometry 250

Experiment 24: Quantitative Infrared Analysis of Isopropyl Alcohol in Toluene 250

Experiment 25: Quantitative Flame Atomic Absorption Analysis of a Prepared Sample 284

Experiment 26: The Analysis of Soil Samples for Iron Using Atomic Absorption 284

Experiment 27: Design an Experiment: A Study to Determine the Optimum pH for the
Extraction Solution for Experiment 26 285

Experiment 28: The Determination of Sodium in Soda Pop 285

Experiment 29: Design an Experiment: Sodium in Soda Pop by the Standard Additions
Method 286

Experiment 30: Design an Experiment: Analysis of Fertilizer for Potassium 286

Experiment 31: Thin-Layer Chromatography Analysis of Cough Syrups for Dyes 306

Experiment 32: Thin-Layer Chromatography Analysis of Jelly Beans for Food Coloring 307

Experiment 33: A Qualitative Gas Chromatography Analysis of a Prepared Sample 330

Experiment 34: The Quantitative Gas Chromatography Analysis of a Prepared Sample
for Toluene by the Internal Standard Method 331

Experiment 35: The Determination of Ethanol in Wine by Gas Chromatography and the
Internal Standard Method 332

Experiment 36: Design an Experiment: Determination of Ethanol in Cough Medicine or Other Pharmaceutical Preparation..333

Experiment 37: A Study of the Effect of the Changing of GC Instrument Parameters on Resolution...333

Experiment 38: The Quantitative Determination of Methyl Paraben in a Prepared Sample by HPLC...362

Experiment 39: HPLC Determination of Caffeine in Soda Pop 363

Experiment 40: Design an Experiment: Dependence of Caffeine Analysis on the pH of the Mobile Phase..364

Experiment 41: The Analysis of Mouthwash by HPLC: A Research Experiment 364

Experiment 42: The Determination of 4-Hydroxyacetophenone by Capillary Electrophoresis...365

Experiment 43: The Quantitative GC-MS Analysis of a Prepared Sample for Chlorobenzene by the Internal Standard Method...382

Experiment 44: GC-MS Determination of Ethylbenzene in Gasoline by Combined Internal Standard and Standard Additions Methods 382

Experiment 45: Determination of the pH of Soil Samples 409

Experiment 46: Analysis of a Prepared Unknown for Fluoride Using an Ion-Selective Electrode...409

Experiment 47: Design an Experiment: Determination of Fluoride in Toothpaste Using an Ion-Selective Electrode 410

Experiment 48: Operation of Metrohm Model 701 Karl Fischer Titrator (for Liquid Samples)...410

Preface

The primary purpose of this text continues to be as a training manual for chemistry-based laboratory technicians. It is designed to emphasize the practical rather than the theorical. The practical begins with classical quantitative analysis, because this is what instills the mindset of analytical skill and technique into the students' psyche, and very practicing scientists should agree that this is a very important part of a technician's training. But equally important is what follows this in the book: a very practical approach to the complex world of the sophisticated electronic instrumentation that a technician will find in common use in the real-world laboratory. My hope is that laboratory supervisors will find that technicians who have utilized this textbook will have an analytical mindset and a basic understanding of the analytical instrumentation needed for success on the job.

It has been 10 years since the publication of the third edition of *Analytical Chemistry for Technicians*. I have taught the sequence of courses that use this text over 20 times since then. Each time I teach these courses, it seems I have fresh ideas and new developments to discuss with my students. These have resulted in new drawings for my PowerPoint slides and new (and I hope better) explanations of what can be complex instrument designs and functions. It is true that my students have a limited background in chemistry when they come to me, and so I strive to write, draw, and speak with this in mind.

And now the time has come for the fourth edition. What is new and what has changed? Following is the list:

- Over 150 new photographs and either new or reworked drawings spanning every chapter to assist the visual learner.
- A new chapter on mass spectrometry.
- Thirteen new laboratory experiments, including nine "design an experiment" exercises.
- An "Introduction to Laboratory Work" section before Chapter 1 to give students a preview of general laboratory considerations, safety, laboratory notebooks, and instrumental analysis.
- "Application Notes" in each chapter. A few of these are carryovers of the "Workplace Scenes" in the third edition, but most are new. The intent is to give students a larger hint of how specific techniques may be routinely used in the laboratory.
- Calculation summary tables in Chapters 3 and 4.
- Relevant section headings in the end-of-chapter Questions and Problems section to help better organize the material for students.
- An appendix providing a glossary of quality assurance and good laboratory practice (GLP) terms.
- More end-of-chapter problems in Chapters 1, 2, 3, 6, and 14 covering statistics, liquid–liquid extraction, gravimetric analysis, post-run instrumental analysis, and electrode potential.
- More examples of calculations covering liquid–liquid extraction theory and post-run instrumental analysis in Chapters 2 and 6.
- The topic of refractometry has been moved to Chapter 7 from Chapter 15 since it represents a basic technique involving light.
- The topics of liquid–liquid extraction, solid–liquid extraction, and distillation–deionization have been moved to Chapter 2 from Chapter 11 as sample and solution preparation techniques rather than analytical separation techniques. Chapter 11 has become Chapter 10 and is now dedicated to chromatography, and this is reflected in the chapter title "Introduction to Chromatography."

- The topic of capillary electrophoresis has been moved from the former Chapter 11 to the new Chapter 12 and expanded.
- Chapter 1 has been expanded to include more information on statistics.
- Expanded end-of-chapter "report"-type questions have been included in Chapters 5, 7, 8, 9, 11, 12, and 14 to give improved guidance to students as to how to answer these questions.
- De-emphasis of the Kjeldahl method in Chapter 5.
- Appendix 1 has been reworked to include the percent unit and deleting standardization and percentages of analyte formulas.
- The material from the former Chapter 10 ("Other Spectroscopic Methods") has been moved to the new mass spectrometry chapter and to a new Chapter 15, "Miscellaneous Instrumental Technique."
- Coverage of the mathematical relationships involving wavelength, frequency, energy, and wavenumber have been retained, but the calculations involving these relationships have been deleted.
- The first few sections of Chapter 11, "Gas Chromatography," have been rewritten to reflect the popularity of capillary columns and to include more information about injectors. GC-MS coverage has been moved to Chapter 13, "Mass Spectrometry."
- The chapters on bioanalysis and physical testing methods have been deleted. Approximately 25% of this latter chapter is in the new Chapter 15, "Miscellaneous Instrumental Technique."
- Coverage of recrystallization has been deleted.

One of the biggest changes is the introduction of an entirely new chapter on mass spectrometry, Chapter 13, which includes coverage of GC-MS, LC-MS, LC-MS-MS, and ICP-MS. This chapter follows other chapters on basic quantitative analysis, atomic and molecular spectroscopy, and gas and liquid chromatography. It comes in the third course in the analytical chemistry sequence at Southeast Community College, and while the students' background is far from advanced at that point, I find that they are well prepared for an illustrated but limited discussion of this important topic.

What do I mean by illustrated? Reading this text, students will see dozens of new illustrative drawings and photographs throughout, not just in Chapter 13. I find that most of my students learn and I can teach best from illustrations. Whether the topic is laboratory glassware, spectrophotometers, gas chromatographs, mass spectrometers, or electrodes, students are visual learners, and all the drawings and photographs were created by me. I had no outside artist or photographer for this text. Over the years, I have learned to use the drawing tools in PowerPoint to great advantage. I have also learned how to use digital photography to great advantage. The drawings and photographs are exactly as I intended them to appear because I created them. Some were done over the last 10 years of teaching and others were done as I sat at my desk or laboratory bench the past 2 years and contemplated how best to convey the information in this fourth edition.

From monochromators to capillary columns, from absorption spectra to standard curves, from titration curves to electrodes, from double-beam spectrophotometers to quadrupole mass analyzers, and from flame atomic absorption to ICP-MS, there are many new and, I trust, improved illustrations and photographs for students to ponder.

Of course, I am anxious for students to get their hands on this fourth edition and to begin to peruse its contents. It comes to you after 36 years of practice, and it has evolved considerably since the first edition appeared in 1988. Good luck to you!

John Kenkel
Southeast Community College

Acknowledgments

I am very grateful to Southeast Community College for the professional niche they provided me starting in 1977 and continuing for a full 36 years. In all that time, I have come to know many students who are hungry to learn about science and to have a career in the field. They were my primary inspiration year after year.

As always, I am indebted to a number of people who assisted in some direct way during the development of the manuscript. My editor at CRC Press, Barbara Glunn, offered me the opportunity and was very friendly and encouraging at every step along the way, especially when I asked if I could delay the project six months due to a serious family illness.

A number of my colleagues at institutions across the United States served as reviewers on this project and I appreciate their work very much. They are:

Tracy Holbrook, Cape Fear Community College
Cynthia Peck, Delta College
Michele Mangels, Cincinnati State Technical and Community College
Susan Marine, Miami University Middletown
Charlie Newman, Mount San Antonio College
Onofrio Gaglione, College of Southern Nevada
Kazumasa Lindley, College of Southern Nevada
Renee Madyun, Los Angeles Trade Tech College

I am indebted to my wife, Lois, for her superior proofreading skills. One thing I have learned during my various writing projects over the years is that many errors seem to creep in unnoticed despite a seemingly thorough review by the author. For this project, Lois' help was greatly appreciated.

John Kenkel
Southeast Community College

Author

John Kenkel is a chemistry instructor at Southeast Community College (SCC) in Lincoln, Nebraska. Throughout his 36-year career at SCC, he has been directly involved in the education of chemistry-based laboratory technicians in a vocational program. He has also been heavily involved in chemistry-based laboratory technician education on a national level, having served on a number of American Chemical Society (ACS) committees, including the Committee on Technician Activities and the Coordinating Committee for the Voluntary Industry Standards project. In addition to these, he has served a 5-year term on the ACS Committee on Chemistry in the Two-Year College, the committee that administers the Two-Year College Chemistry Consortium (2YC$_3$) conferences. He was the general chair of this committee in 1996. He later served a 3-year term on this committee as the Industrial Sponsor Chair (2006–2008).

Mr. Kenkel has authored several popular textbooks for chemistry-based technician education. Three editions of *Analytical Chemistry for Technicians* preceded the current edition, the first published in 1988, the second in 1994, and the third in 2003. In addition, he has authored five other books: *Basic Chemistry Concepts and Exercises*, published in 2011, *Chemistry: An Industry-Based Introduction* and *Chemistry: An Industry-Based Laboratory Manual*, both published in 2000–2001, *Analytical Chemistry Refresher Manual*, published in 1992, and *A Primer on Quality in the Analytical Laboratory*, which was published in 2000. All were published through CRC Press/Lewis Publishers.

Mr. Kenkel has been the principal investigator for a series of curriculum development project grants funded by the National Science Foundation's Advanced Technological Education Program, from which four of his nine books evolved. He has also authored or coauthored four articles on the curriculum work in the *Journal of Chemical Education* and has presented this work at more than 20 conferences.

In 1996, Mr. Kenkel won the prestigious National Responsible Care Catalyst Award for excellence in chemistry teaching sponsored by the Chemical Manufacturer's Association. He has a master's degree in chemistry from the University of Texas at Austin (1972) and a bachelor's degree in chemistry (1970) from Iowa State University. His research at the University of Texas was directed by Professor Allen Bard. He was employed as a chemist from 1973 to 1977 at Rockwell International's Science Center in Thousand Oaks, California.

Introduction to Laboratory Work

GENERAL CONSIDERATIONS

There is laboratory work, and then there is laboratory work! No doubt you have had a general chemistry course before enrolling in the course for which this textbook is used. And no doubt there was a laboratory section included in this general chemistry course. In this general chemistry laboratory section, you paid attention to detail to the extent that you were able to pass the class with a satisfactory grade. Perhaps you worked as part of a group of other students and never were quite on your own for any of the tasks involved. You probably used ordinary laboratory balances, prepared solutions in Erlenmeyer flasks or beakers, and used graduated cylinders to transfer volumes of solutions from one container to another. In short, you did the work, but it was not important to strive for highly precise measurements or results nor perhaps even highly accurate measurements or results. All of that is about to change.

Perhaps you wondered about the nature of laboratory work in the real world. Perhaps you thought about the analysis of pharmaceutical, food, or beverage samples, that is, samples that represent products consumed by the general public. Or perhaps you thought about the analysis of a hospital patient's blood or urine. Would a supervisor demand more careful attention to detail in these situations? Indeed, the answer is a resounding "yes." In this textbook, high precision and high accuracy are the name of the game. You will usually be working by yourself on a given analysis in the laboratory and striving to meet the demands of your instructor to develop your analytical skill and technique so you will be qualified to work in a real-world analytical laboratory. After all, you will be using balances where the weight of your fingerprint is measureable and where you strive to keep containers to be weighed free of fingerprints. You will be using glassware where a fraction of a drop will make a difference in the position of a meniscus on a calibration line and make a difference in the results. In short, when you complete the class for which this textbook is used and you enter the real world, you will have the confidence to be successful no matter what demands are thrown at you. You will be an analytical laboratory technician who works at a high level of analytical skill to complete the tasks assigned.

SAFETY IN THE ANALYTICAL LABORATORY

The analytical chemistry laboratory is a very safe place to work. However, that is not to say that the laboratory is free of hazards. The dangers associated with contact with hazardous chemicals, flames, etc. are very well documented, and as a result, laboratories are constructed and procedures are carried out with these dangers in mind. Hazardous chemical fumes are, for example, vented into the outdoor atmosphere with the use of fume hoods. Safety showers for diluting spills of concentrated acids on clothing are now commonplace. Eyewash stations are strategically located for immediately washing one's eyes in the event of accidental contact with a hazardous chemical. Fire blankets, extinguishers, and sprinkler systems are also located in and around analytical laboratories for immediately extinguishing flames and fires. Also, a variety of safety gear, such as safety glasses, aprons, and shields, is available. There is never a good excuse for personal injury in a well-equipped laboratory where well-informed analysts are working.

Although the pieces of equipment mentioned above are now commonplace, it remains for the analysts to be well-informed of potential dangers and of appropriate safety measures. To this end, we list below some safety tips of which any laboratory worker must be aware. This list should be studied carefully by all students who have chosen to enroll in an analytical chemistry course. This is not intended to be a complete list, however. Students should consult with their instructor in order to establish safety ground rules for the particular laboratory in which they will be working. Total awareness of hazards and dangers and what to do in case of an accident is the responsibility of each student and the instructor.

1. Safety glasses must be worn at all times by students and instructors. Visitors to the laboratory must be appropriately warned and safety glasses made available to them.
2. Fume hoods must be used when working with chemicals that may produce hazardous fumes.
3. The location of fire extinguishers, safety showers, and eyewash stations must be known.
4. All laboratory workers must know how and when to use the items listed in number 3.
5. There must be no unsupervised or unauthorized work in the laboratory.
6. A laboratory is never a place for practical jokes or pranks.
7. The toxicity of all the chemicals you will be working with must be known. Consult the instructor, material safety data sheets (MSDSs), safety charts, and container labels for safety information about specific chemicals. Recently, many common organic chemicals, such as benzene, carbon tetrachloride, and chloroform, have been deemed unsafe. Eating, drinking, or smoking in the laboratory is never allowed.
8. Never use laboratory containers (beakers or flasks) to drink beverages.
9. Shoes (not open-toed) must always be worn; hazardous chemicals may be spilled on the floor or feet.
10. Long hair should always be tied back.
11. Mouth pipetting is never allowed.
12. Cuts and burns must be immediately treated. Use ice on new burns and consult a doctor for serious cuts.
13. In the event of acid spills on one's person, flush thoroughly with water immediately. Be aware that acid–water mixtures will produce heat. Removing clothing from the affected area while water flushing may be important so as to not trap hot acid–water mixtures against the skin. Acids or acid–water mixtures can cause very serious burns if left in contact with skin, even if only for a very short time.
14. Weak acids (such as citric acid) should be used to neutralize base spills, and weak bases (such as sodium carbonate) should be used to neutralize acid spills. Solutions of these should be readily available in the laboratory in case of emergency.
15. Dispose of all waste chemicals from the experiments according to your instructor's directions.
16. In the event of an accident, report immediately to your instructor, regardless of how minor you perceive it to be.
17. Always be watchful and considerate of others working in the laboratory. It is important not to jeopardize their safety or yours.
18. Always use equipment that is in good condition. Any piece of glassware that is cracked or chipped should be discarded and replaced.

It is impossible to foresee all possible hazards that may manifest themselves in an analytical laboratory. Therefore, it is very important for all students to listen closely to their instructor and obey the rules of their particular laboratory in order to avoid injury. Neither the author of this text nor its publisher assumes any responsibility whatsoever in the event of injury.

THE LABORATORY NOTEBOOK

Another item basic to all analytical laboratory work is proper record keeping. In the real world, laboratory notebooks are considered legal documents. Indeed, good record keeping is central to good analytical science. Not only must data obtained from samples and analytes be recorded, but it must be recorded with diligence and with considerable thought being given to integrity and purpose.

Accordingly, an analytical laboratory will usually have strict guidelines with respect to laboratory notebooks. The following typifies what these guidelines might be.

I. General Guidelines
A. All notebooks must begin with a table of contents. All pages must be numbered, and these numbers must be referenced in the table of contents. The table of contents must be updated as projects are completed and new projects begun.
B. All notebook entries must be made in ink. Use of graphite pencils or other erasable writing instruments is strictly prohibited.
C. No data entries will be erased or made illegible. If an error was made, a single line is drawn through the entry. Do not use correction fluid. Initial and date corrections and indicate why the correction was necessary.
D. Under no circumstances will the notebook be taken or otherwise leave the laboratory unless there are data to be recorded at a remote site, such as a remote sampling site, or unless special permission is granted by the supervisor.
E. The following notebook format should be maintained for each project undertaken: (1) title and date; (2) purpose or objectives statement; (3) data entries; (4) results; (5) conclusions. Each of these are explained below. In each case, write out and underline the words "title and date," "objective," "data," etc. to clearly identify the beginning of each section.
F. Make notebook entries for a given project on consecutive pages where practical. Begin a new project on the front side of a new page. You may skip pages only in order to comply with this guideline.
G. Draw a single diagonal line through blank spaces that consist of four or more lines (including any pages skipped according to guideline "F" above). These spaces should be initialed and dated.
H. Never use a highlighter in a notebook.
I. Each notebook page must be signed, dated, and possibly witnessed.

II. Title and Date
A. All new experiments will begin with the title of the work and the date the work is performed. If the work was continued on another date, that date must be indicated at the point the work was restarted.
B. The title will reflect the nature of the work or shall be the title given to the project by the study director.

III. Purpose or Objectives Statement
A. Following the title and date, a statement of the purpose or objective of the work will be written. This statement should be brief and to the point.
B. If appropriate, the standard operating procedure (SOP) will be referenced in this statement.

IV. Data Entries
A. Enter data into the notebook as the work is being performed. This means that loose pieces of paper used for intermediate recordings are prohibited. Entries should be made in ink only.
B. If there is any deviation from the SOP, permission must be obtained from the study director, and this must be thoroughly documented by indicating exactly what the deviation is and why it occurred.
C. The samples analyzed must be described in detail. Such descriptions may include the source of the sample, what steps were taken to ensure that the sample represents the whole (reference SOP if appropriate), and what special coding may be assigned and

what the codes mean. If the codes are recorded in a separate notebook (such as a field notebook), this notebook must be cross-referenced.

D. Show the mathematical formulas utilized for all calculations and include a sample calculation.

E. Construct data tables whenever useful and appropriate.

F. Both numerical data and important observations should be recorded.

G. Limit attachments (chart recordings, computer printouts, etc.) to one per page. Clear tape or glue may be used. Do not use staples. Only one fold in attachments is allowed. Do not cover any notebook entries with attachments.

V. Results

A. The results of the project, such as numerical values representing analysis results, should be reported in the notebook in table form if appropriate. Otherwise, a statement of the outcome is written, or if a single numerical value is the outcome, it is reported here. In order to identify what is to be reported as results, consider what the client wants and needs to know.

VI. Conclusions

A. After results are reported, the experiment is drawn to a close with a brief concluding statement indicating whether the objective was achieved.

Two sample pages from the notebook of a student performing Experiment 7 are shown in Figure I.1. However, the student did not follow some of the guidelines (as indicated in the caption). Notice that each page is signed with the student's name and is also dated, as suggested by Part I under General Guidelines above. This is often required in an industry for patent protection.

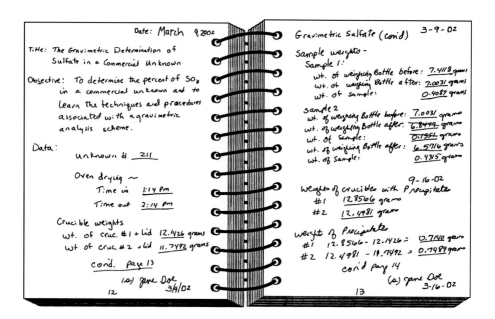

FIGURE I.1 Sample pages from a laboratory notebook that a student is using for Experiment 7 in this text. However, some of the guidelines in this section were not followed. How many errors can you find?

INSTRUMENTAL ANALYSIS

Technology has "exploded" in recent years to the point that we carry in our pockets amazing devices that do amazing things. In the analytical laboratory, there are equally amazing devices that do many more amazing things. Scientists have combined the known principles of chemistry, physics, and electronics with high technology to produce devices that are, in short, unbelievable. You get a taste of this in Chapters 6–15 in this textbook.

An example is the inductively coupled plasma instrument discussed in Chapter 9. There are several definitions of "plasma" in the dictionary. You may think of the fluid component of blood when you hear the word, but the chemistry/physics definition is a partially ionized stream of gas containing positive ions and electrons and is capable of conducting electricity and interacting with a magnetic field. Research has brought this into the routine workings in an analytical laboratory, and using this and associated hardware and software, we now are able to quickly and easily determine ultrasmall concentrations of metals in a wide variety of materials.

Another example is the mass spectrometry instrument to which Chapter 13 is dedicated. Once again, scientists have harnessed a combination of chemistry and physics principles to invent and use such devices as the electron impact ion source, the quadrupole mass analyzer, and ion detectors that generate electrical signals based on the impact of a collision between ions in the gas phase with dynodes. With these devices, scientists are able to analyze a variety of materials for organic and inorganic constituents, including such things as illegal drugs in an athlete's urine and ultrasmall concentrations of pesticide residues in apples.

These are but two examples. The chapters ahead are going to confuse you, enlighten you, excite you, and make you hungry for more. Get ready!

1 Introduction to Analytical Science

1.1 ANALYTICAL SCIENCE DEFINED

Imagine yourself strolling down the aisle in your local grocery store to select your favorite foods for your lunch. You pick up a jar of peanut butter, look at the label, and read that there are 190 mg of sodium in one serving. You think to yourself: "I wish I knew how they knew that for sure." After picking up the lunch items you want, you proceed to the personal hygiene aisle to look for your toothpaste. Again you look at the label and notice that the fluoride content is 0.15% w/v. "How do they know that?," you again say to yourself. Finally, you stop by the pharmaceutical shelves and pick up a bottle of your favorite vitamin. Looking at the label, you see that there are 1.7 mg of riboflavin in every tablet and marvel how the manufacturer can know that that is really the case.

There is a seemingly endless list of examples of scenarios like the above that one can think of without even leaving the grocery store! We could also visit a hardware store and look at the labels of cleaning fluids, adhesives, paint or varnish formulations, paint removers, garden fertilizers, and insecticides and make similar statements. Although you may question how the manufacturers of these products know precisely the content of their products in such a quantitative way, you yourself may have undertaken exactly that kind of work at some point in your life right in your own home. If you have an aquarium, you may have come to know that it is important to not let the ammonia level in the tank get too high and may have purchased a kit to allow you to monitor the ammonia level. Or, you may have purchased a water test kit to determine the pH, the hardness, or even the nitrate concentration in the water that comes from your tap. You may have a soil test kit to determine the nitrate, phosphate, and potassium levels of the soil in your garden. Then you think: "Gee, it's actually pretty easy." But when you sit down and read the paper or watch the evening news, you are baffled again by how a forensic scientist determines that a criminal's DNA was present on a murder weapon, or how someone determined the ammonia content in the atmosphere of the planet Jupiter without even being there, or how it can be possible to determine the ozone level high above the North Pole.

The science that deals with the identification and/or quantification of the components of material systems such as these is called **analytical science**. It is called that because the process of determining the level of any or all components in a material system is called **analysis**. It can involve both physical and chemical processes. If it involves chemical processes, it is called **chemical analysis** or, more broadly, **analytical chemistry**. The sodium in the peanut butter, the nitrate in the water, and the ozone in the air in the above scenarios are the substances that are the objects of analysis. The word for such substances is **analyte**, and the word for the material in which the analyte is found is called the **matrix** of the analyte.

Another word often used in a similar context is the word "assay." If a material is known by a particular name and an analysis is carried out to determine the level of that named substance in the material, the analysis is called an **assay** for that named substance. For example, if an analysis is being carried out to determine what percent of the material in a bottle labeled "aspirin" is aspirin, the analysis is called an assay for aspirin. In contrast, an *analysis* of the aspirin would imply the determination of other minor ingredients as well as the aspirin itself.

The purpose of this book is to discuss in a systematic way the techniques, the methods, the equipment, and the processes of this important, all-encompassing science.

1.2   CLASSIFICATIONS OF ANALYSIS

Analytical procedures can be classified in two ways: first, in terms of the goal of the analysis, and second, in terms of the nature of the method used. In terms of the goal of the analysis, classification can be based on whether the analysis is "qualitative" or "quantitative." **Qualitative analysis** is identification. In other words, it is an analysis carried out to determine only the identity of a pure analyte, the identity of an analyte in a matrix, or the identity of several or all components of a mixture. Stated another way, it is an analysis to determine *what* a material is or what the components of a mixture are. Such an analysis does not report the amount of the substance. If a chemical analysis is carried out and it is reported that there is mercury present in the water in a lake and the quantity of the mercury is not reported, then the analysis was a qualitative analysis. **Quantitative analysis**, on the other hand, is the analysis of a material for *how much* of one or more components is present. Such an analysis is undertaken when the identity of the components is already known and when it is important to also know the quantities of these components. It is the determination of the quantities of one or more components present per some quantity of the matrix. For example, the analysis of the soil in your garden that reports the potassium level as 342 parts per million would be classified as a quantitative analysis. The major emphasis of this text is on quantitative analysis, although some qualitative applications will be discussed for some techniques (see also Application Note 1.1).

Analysis procedures can be additionally classified into procedures that involve physical properties, wet chemical analysis procedures, and instrumental chemical analysis procedures. **Analysis using physical properties** involves no chemical reactions and at times relatively simple devices (although possibly computerized) to facilitate the measurement. Physical properties are especially useful for identification, but may also be useful for quantitative analysis in cases where the value of a property, such as specific gravity or refractive index (Chapter 7), varies with the quantity of an analyte in a mixture.

Wet chemical analysis usually involves chemical reactions and/or classical reaction stoichiometry, but no electronic instrumentation beyond a weighing device. Wet chemical analysis techniques are classical techniques, meaning they have been in use in analytical laboratory for many years, before electronic devices came on the scene. If executed properly, they have a high degree of inherent accuracy and precision, but they take more time to execute.

Instrumental analysis can also involve chemical reactions, but always involves modern sophisticated electronic instrumentation. Instrumental analysis techniques are "high-technology" techniques, often utilizing the ultimate in complex hardware and software. Although sometimes not as precise as a carefully executed wet chemical method, instrumental analysis methods are fast and can offer a much greater scope and practicality to the analysis. In addition, instrumental methods are generally used to determine the minor constituents, or constituents that are present in low levels, rather than the major constituents of a sample. We discuss wet chemical methods in Chapters 3 and 5 of this text. Chapter 6 is concerned with instrumental methods in general, and Chapters 7–15 involve specific instrumental methods.

APPLICATION NOTE 1.1: "CHARACTERIZING" A MATERIAL

A combination of the qualitative analysis and quantitative analysis of a material or matrix is sometimes called **characterizing** the material. A "total" analysis such as this might involve a complete reporting of the properties of a material as well as the identity and quantity of component substances. For example, a company that manufactures a perfume might "characterize" their product as having a particular fragrance, a particular staying power, and a particular feel on the skin but may also report the identity of the ingredients and the quantity of each. The characterization of the raw materials used to make a product as well as the final product itself, and even the package in which the product is contained, is often considered a very important part of a manufacturing effort because of the need to assure the product's quality.

1.3 THE SAMPLE

A term for the material under investigation is **bulk system**. The bulk system in the case of analyzing toothpaste for fluoride is the toothpaste in the tube. The bulk system in the case of determining the ammonia level in the water in an aquarium is all the water in the aquarium.

When we want to analyze a bulk system such as these in an analytical laboratory, it is usually not practical to literally place the entire system under scrutiny in the laboratory. We cannot, for example, bring all the soil found in a garden into the laboratory to determine the phosphate content. We therefore collect a representative *portion* of the bulk system and bring this portion into the laboratory for analysis. Hence, a portion of the water in a lake is analyzed for mercury and a portion of the peanut butter in a jar of peanut butter is analyzed for sodium. This portion is called a **sample**. The analytical laboratory technician analyzes these samples by subjecting them to certain rigorous laboratory operations that ultimately result in the identity or quantity of the analyte in question. The key is that the sample must possess all the characteristics of the entire bulk system with respect to the analyte and the analyte concentration in the system. In other words, it must be a **representative sample**—it must truly represent the bulk system. There is much to discuss with respect to the collection and preparation of samples and we will do that in Chapter 2.

1.4 THE ANALYTICAL PROCESS

The process by which an analyte's identity and/or concentration level in a sample is determined in the laboratory may involve many individual steps. For us to have a coherent approach to the subject, we will group the steps into major parts and study each part individually. In general, these parts vary in specifics according to what the analyte and analyte matrix are and what methods have been chosen for the analysis. In this section, we present a general organizational framework for these parts and we will then proceed in later chapters to build upon this framework for each major method of analysis to be encountered. Let us call this framework the **analytical process**.

There are five parts to the analytical process. These are the following: (1) obtain the sample, (2) prepare the sample, (3) carry out the analysis method, (4) work up the data, and (5) calculate and report the results. These are expressed in the flowchart in Figure 1.1. The terminology used in

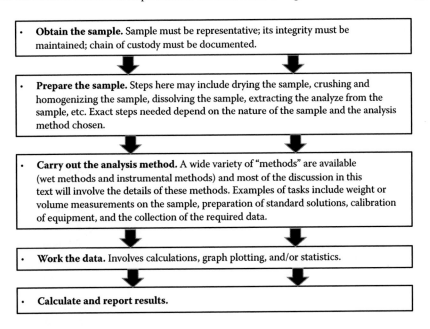

FIGURE 1.1 Flowchart of the analytical process described in the text.

Figure 1.1 and the various steps in the carrying out of the method may be foreign to you now, but these will be discussed as we progress through the coming chapters.

1.5 ANALYTICAL TECHNIQUE AND SKILLS

If the label on a box of Cheerios™ cereal states that there are 22 g of carbohydrates in each serving, how does the manufacturer know with certainty that it is 22 g and not 20 g or 25 g? If the label on a bottle of rubbing alcohol says that it is 70% isopropyl alcohol, how does the manufacturer know that it is 70% and not 65% or 75%? The answers have to do to with the quality of the manufacturing process and also with how accurately the companies' quality assurance laboratories can measure these ingredients. However, much of it also has to do with the skills of the technicians performing the analyses.

An analytical laboratory technician is a person with a special mindset and special skills. He or she must be a thinking person—a person who pays close attention to detail and never waivers in his/her pursuit of high quality data and results, even in simple things performed in the laboratory. We say that he/she possesses **good analytical technique** or **good analytical skills**.

Quality is emphasized because of the value and importance that are usually riding on the results of an analysis. Great care must be exercised in the laboratory when handling the sample and all associated materials. Contamination and/or loss of sample through avoidable accidental means cannot be tolerated. The results of a chemical analysis could affect such ominous decisions as the freedom or incarceration of a prisoner on trial; whether to proceed with an action that could mean the loss of a million dollars for an industrial company or the life or death of a hospital patient.

Students should develop a kind of psychology for functioning in an analytical laboratory—a psychology that facilitates the good technique and skills. One must always stop and think before proceeding with a new step in the procedure. What might happen in this step that would cause contamination or loss of the sample? A simple example would be when stirring a solution in a beaker with a stirring rod. You may wish to remove the stirring rod from the beaker when going on to the next step. However, if you stop and think in advance, you would recognize that you need to rinse wetness adhering to the rod back into the beaker as you remove it. This would prevent the loss of that part of the solution adhering to the rod. Such a loss might result in a significant error in the determination.

Buzzwords in an analytical laboratory include **precision** and **accuracy**. To the novice, these words may be synonymous, but to the analytical laboratory technician, each has a specific meaning. As we will discuss later, precision has to do with reproducibility or a measurement, whereas accuracy relates to whether the measurement is correct. In the pages that follow, you will learn about high precision weighing devices where fingerprints make a difference and high-precision glassware where a meniscus that is only slightly out of position will also make a difference.

Quality of technique in an analytical laboratory is so important that there are laws governing the actions of the scientists in such a laboratory. These laws, passed by Congress and placed into the Federal Registry in the late 1980s, are known as **Good Laboratory Practices**, or **GLP**. The GLP laws address such things as labeling, record keeping and storage, documentation, and updating of laboratory procedures (known as **standard operating procedures**, or **SOPs**), and the laboratory **protocols**, which are formal written documents defining and governing the work a laboratory is performing. They also address such things as who has authority over various aspects of the work, who has authority to change SOPs and the processes by which they are changed. The GLP also allow for **audits**, or regular inspections of a laboratory by outside personnel to ensure compliance with the regulations. GLP regulations receive much attention in analytical laboratories (see Appendix 2).

1.6 ELEMENTARY STATISTICS

1.6.1 ERRORS

Errors in the analytical laboratory are basically of two types: determinate errors and indeterminate errors. **Determinate errors**, also called **systematic errors**, are errors that were known to have occurred, or at least were determined later to have occurred, in the course of the laboratory work. They may arise from avoidable sources, such as contamination, wrongly calibrated instruments, reagent impurities, instrumental malfunctions, poor sampling techniques, errors in calculations, etc. Results from laboratory work in which avoidable determinate errors are known to have occurred must be rejected or, if the error was a calculation error, recalculated.

Determinate errors may also arise from unavoidable sources. An error that is known to have occurred but was unavoidable is called a **bias**. Such an error occurs each time a procedure is executed and thus its effect is usually known and a correction factor can be applied. For example, if it is determined that a calibration mark on a piece of high-precision glassware, for example, a pipette, is slightly out of place, the same error is repeated each time this pipette is used. In this case, one can determine the exact error and then apply a correction factor to all results in which the pipette was used.

Indeterminate errors, also called **random errors**, on the other hand, are errors that are not specifically identified and are therefore impossible to avoid. Since the errors cannot be specifically identified, results arising from such errors cannot be immediately rejected or compensated for as in the case of determinate errors. Rather, a statistical analysis must be performed to determine whether the results are far enough "off-track" so as to merit rejection.

Accuracy in the laboratory is obviously an important issue. If the analysis results reported by a laboratory are not accurate, everything a company or government agency strives for may be in jeopardy. If the customer discovers the error, especially through painful means, the trust the public has placed in the entire enterprise is lost. For example, if a baby dies due to nitrate contamination in drinking water that a city's health department had determined to be safe, that department, indeed the entire city government, is liable. In this "worst-case scenario," some employees would likely lose their jobs and perhaps even be brought to justice in a court of law.

The correct answer to an analysis is usually not known in advance. So the key question becomes, "How can a laboratory be absolutely certain that the result it is reporting is accurate?" There are several items that must be considered to answer this question. First, all instruments and measuring devices must be properly calibrated and functioning properly. Second, the bias, if any, of a method must be determined and accounted for. Third, the method must be valid method for the analysis being performed. In other words, it must be proven that the chosen method does indeed provide the answer that is sought. Fourth, the sample that has been acquired must truly represent the entire bulk system that is under investigation. In other words, all possibilities for avoidable determinate errors for a given analysis must have been investigated and eliminated.

Assuming that this is the case, it then becomes a question of indeterminate errors. In other words, if it is known that the method is valid, that all instruments and measuring devices are functioning as expected and are calibrated properly, that any bias has been accounted for, and that the sample being tested is representative of the bulk system under investigation, then it comes down to the treatment of possible indeterminate errors. Such treatment is a matter of statistics. The analyst relies on the repeatability of a measurement or analysis results to be the indicator of accuracy. If a series of tests all provide the same, or nearly the same answer in this scenario, then it is taken to be an accurate answer. Obviously, the degree of precision required, or what "nearly the same answer" means in terms of percent error, and how to deal with the data in order to have the confidence that is needed or wanted are important questions. Statistical methods take a look at the data, provide some mathematical indication of the precision, and reject or retain suspect data values, based on predetermined limits.

1.6.2 DEFINITIONS

We now provide some definitions that are fundamental to statistical analysis. Please refer to Example 1.1 that follows for an example of how each is calculated.

1. **Mean:** In the case in which a given measurement on a sample is repeated a number of times, the average of all measurements is an important number and is called the "mean." It is calculated by adding together the numerical values of all measurements and dividing this sum by the number of measurements. It is used in all statistical methods in one way or another, as we will see.

2. **Median:** For this same series of identical measurements on a sample, the "middle" value is sometimes important and is called the median. If the total number of measurements is an even number, there is no single "middle" value. In this case, the median is the average of two "middle" values. For a large number of measurements, the mean and the median should be the same number.

3. **Mode:** The value that occurs most frequently in the series is called the "mode." Ideally, for a large number of identical measurements, the mean, median, and mode should be the same. However, this rarely occurs in practice. If there is no value that occurs more than once, or if there are two values that equally occur most frequently, then there is no mode.

4. **Deviation:** How much each measurement differs from the mean is an important number and is called the deviation. A deviation is associated with each measurement, and if a given deviation is large compared with others in a series of identical measurements, this may signal a potentially rejectable measurement (outlier) that will be tested by the statistical methods. Mathematically, the deviation is calculated as follows:

$$d = (m - e) \tag{1.1}$$

in which d is the deviation, m is the mean, and e represents the individual experimental measurement.

5. **Sample Standard Deviation:** The most common measure of the dispersion of data around the mean for a limited number of samples (<20) is the sample standard deviation:

$$s = \sqrt{\frac{(d_1^2 + d_2^2 + d_3^2 \ldots\ldots)}{(n-1)}} \tag{1.2}$$

in which n is the number of measurements. The term (n − 1) is referred to as the number of **degrees of freedom**, and s represents the standard deviation.

Example 1.1

The percent of moisture in a powdered pharmaceutical sample is determined by six repetitions of the Karl Fisher method to be 3.048%, 3.035%, 3.053%, 3.044%, 3.049%, and 3.046%. What are the mean, median, mode, and sample standard deviation for these data?

Solution 1.1

The mean is the average of all measurements. Thus, we have

$$\text{mean} = \frac{(3.048 + 3.035 + 3.053 + 3.044 + 3.049 + 3.046)}{6} = 3.045833 = 3.046\%$$

The median is the "middle" value of an odd number of values. If there is an even number, the median is the average of the two "middle" values. Thus, we have

$$\text{median} = \frac{(3.048 + 3.046)}{2} = 3.047\%$$

There is no mode in this case because there is no value appearing more than once.

The sample standard deviation is calculated according to Equation 1.2, in which the "d" values are deviations calculated by subtracting each individual percent value from the mean according to Equation 1.1. The deviations (absolute values) are 0.002, 0.011, 0.007, 0.002, 0.003, and 0.000. To substitute into Equation 1.2, we must square the deviations. The squares of the deviations are 0.000004, 0.000121, 0.000049, 0.000004, 0.000009, 0.000000. Substituting into Equation 1.2, we have

$$s = \sqrt{\frac{(0.002)^2 + (0.011)^2 + (0.007)^2 + (0.002)^2 + (0.003)^2 + (0.000)^2}{5}}$$

$$s = \sqrt{\frac{(0.000004 + 0.000121 + 0.000049 + 0.000004 + 0.000009 + 0.000000)}{5}}$$

$$= 0.0061155 = 0.006$$

The significance of the sample standard deviation is that the smaller it is numerically, the more precise the data and thus presumably (if free from bias and determinate error) the more accurate the data.

6. **Population Standard Deviation:** The dispersion of data around the mean for the entire population of possible samples (an infinite number of samples), which is approximated by upwards of 20 to 30 samples, is called the population standard deviation and is given the symbol σ (Greek letter sigma).

7. **Variance:** A more statistically meaningful quantity for expressing data quality is the variance. For a finite number of samples, it is defined as

$$v = s^2 \qquad (1.3)$$

It is considered to be more statistically meaningful because if the variation in the measurements is due to two or more causes, the overall variance is the sum of the individual variances. The variance for Example 1.1 is $(0.0061155)^2$, or 0.0000037, or 0.000004. For an entire population of samples, s is replaced by σ.

8. **Relative Standard Deviation:** One final deviation parameter is the relative standard deviation, RSD. It is obtained by dividing s by the mean.

$$\text{RSD} = \frac{s}{m} \qquad (1.4)$$

Multiplying the RSD by 100 gives the percentage of RSD:

$$\% \text{ Relative Standard} = \text{RSD} \times 100 \qquad (1.5)$$

Muliplying the RSD by 1000 gives the relative parts per thousand (ppt) standard deviation:

$$\text{Relative ppt Standard Deviation} = \text{RSD} \times 1000 \tag{1.6}$$

The RSD (Equation 1.5) is also called the **coefficient of variance**, c.v.

RSD relates the standard deviation to the value of the mean and represents a practical and popular expression of data quality. Often, a laboratory will use a value of the percentage of RSD, such as ≤10%, as a measure of acceptability. Again, for an entire population of samples, s is replaced by σ.

Example 1.2

What is the percentage of RSD for the data in Example 1.1?

Solution 1.2

The percentage of RSD is calculated according to Equation 1.5 or

$$\%RSD = \frac{s}{m} \times 100 = \frac{0.0061125}{3.045833} \times 100 = 0.201\%$$

1.6.3 Distribution of Measurements

If the entire universe of data (the **population**), as opposed to just a small number of samples, were graphically displayed in a plot of frequency of occurrence vs. individual measurement values, a bell-shaped curve would result in which the peak of the curve would coincide with the mean, as shown in Figure 1.2. This graph is called the **normal distribution curve**. It shows that, for an entire population, the measurements are dispersed around the mean with an equal "drop-off" from the mean in each direction. This mean is recognized as the **true mean** because the entire population was analyzed. The true mean is designated as μ. The **population standard deviation** is associated with μ and, as indicated in Section 1.6.2, is designated as σ. A bias can be depicted on a normal

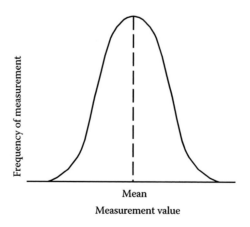

FIGURE 1.2   Normal distribution curve and the definition of the mean (see text for discussion).

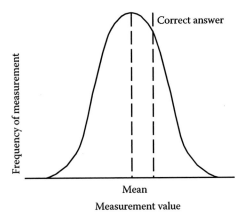

FIGURE 1.3   Normal distribution curve in which the mean is different from the right answer due to the presence of a bias (see text for discussion).

distribution curve by drawing a vertical line at the position of the correct answer to an analysis (see Figure 1.3). In addition, the concept of precision may also be depicted. The more precise the data, the tighter the data points are bunched around the mean and the smaller the σ. The less precise the data are, the less tight the data and the larger the σ (see Figure 1.4). On a normal distribution curve, 68.3% of the data fall within 1σ on either side of the mean, 95.5% of the data fall within 2σ on either side of the mean, and 99.7% of the data fall within 3σ on either side of the mean. Figure 1.5 shows the 2σ limits on either side of the mean.

For a small number of samples, it is sometimes useful to plot a histogram of the data to show the distribution pictorially. A **histogram** is a bar graph that plots ranges of values on the x-axis and frequency on the y-axis. An example is shown in Figure 1.6. Each vertical bar represents a range of measurement values on the x-axis. The height of each bar represents the number of values falling within each range.

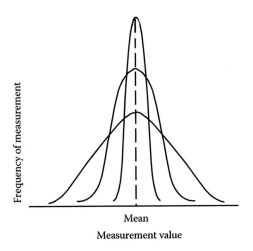

FIGURE 1.4   Several normal distribution curves with different standard deviations. The smaller the standard deviation, the more tightly bunched the curve is around the mean.

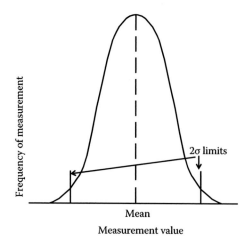

FIGURE 1.5 Normal distribution curve with markers for ±2σ.

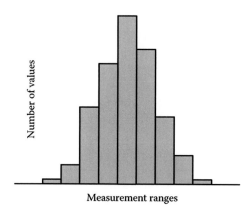

FIGURE 1.6 Example of a histogram.

1.6.4 STUDENT'S T

To express a certain degree of confidence that the mean determined in a real data set is the true mean, **confidence limits** are established based on the degree of confidence, or **confidence level**, that the analyst wishes to have for the analysis. The confidence limit is the interval around the mean that probably contains the true mean, μ. The confidence level is the probability (in percent) that the mean occurs in a given interval. A 95% confidence level means that the analyst is confident that for 95% of the tests run, the sample will fall within the set limits.

The more measurements made on a given bulk system, the more the histogram in Figure 1.6 would begin to look like the normal distribution curve in Figure 1.2. In other words, the more measurements made, the closer the value of the mean will be to the true mean and the more we can rely on the mean to be the correct answer in the absence of bias. Similarly, the more measurements made, the closer the value of the standard deviation is to the true standard deviation. From a practical point of view, however, we only run an experiment enough times to provide us with the confidence interval desired. The confidence interval represents the range from the lower confidence limit to the upper confidence limit. For example, for a mean of 23.54, if the confidence limits are 23.27 and 23.81 (0.27 on either side of the mean), the confidence interval would be ±0.27. To express the degree of confidence in the mean, the answer to the analysis, or what could be called the "true mean," could then be expressed as 23.54 ± 0.27.

For a small number of measurements (for which s is the symbol for the standard deviation), a statistically appropriate way of determining the confidence interval for a desired confidence level so that the mean can be expressed as the mean for an entire population, is the Student's t method. This method expresses the true mean for an entire population (μ) as follows:

$$\mu = m \pm \frac{ts}{\sqrt{n}} \tag{1.7}$$

in which t is a constant depending on the confidence level and n is the number of measurements. The values of t required for a desired confidence level and for a given number of measurements are given in Table 1.1.

Example 1.3

What is the true mean (with confidence interval) for the data in Example 1.1 using Student's t for the 95% confidence level?

Solution 1.3

Using Equation 1.7 and the value of t from Table 1.1 for a confidence level of 95%, we get

$$\mu = m \pm \frac{ts}{\sqrt{n}}$$

$$\mu = 3.046 \pm \frac{2.57 \times 0.0061155}{\sqrt{6}}$$

$$= 3.046 \pm 0.006$$

TABLE 1.1

Values of t Required for a Desired Confidence Level and for a Given Number of Measurements

	Confidence Level		
n − 1	**90%**	**95%**	**99%**
1	6.314	12.706	63.657
2	2.920	4.303	9.925
3	2.353	3.182	5.841
4	2.132	2.776	4.604
5	2.015	2.571	4.032
6	1.943	2.447	3.707
7	1.895	2.365	3.500
8	1.860	2.306	3.355
9	1.833	2.262	3.250
10	1.812	2.228	3.169
∞	1.645	1.960	2.576

1.6.5   REJECTION OF DATA

It may at times appear that a single measurement (an outlier) is so different from the others that the analyst wonders if there was some determinate error that was not detected. In that case, a decision must be made as to whether this measurement should be "rejected," meaning not included in the calculation of the mean. This measurement should not be immediately rejected as being "bad" because, in the absence of a full investigation to determine a cause, it may, in fact, be legitimate. If a legitimate measurement is rejected, then a bias is introduced, and the mean, while assumed to be the correct answer, is actually flawed. There must be some criterion adopted for the rejection or retention of such data.

The first course of action would be for the analyst to inspect his/her technique, chemicals, notebook records, and perhaps the equipment used, to try to detect a determinate error. If a cause is found, then the measurement should be rejected and the reasons for such rejection documented. If a cause is not found, and if time is not a factor, it would be advisable to repeat the measurement, perhaps many times, to see if the anomaly appears again. If it does, the situation is not resolved unless a cause is established in the course of the repetition. If it does not appear again, its seriousness has diminished because there are more measurements from which the mean is calculated.

For small data sets ($3 \leq n \leq 10$), which are often encountered in chemical analysis, a simple method to determine if an outlier is rejectable is the **Q test**. In this test, a value for Q is calculated and compared with a table of Q values that represent a certain percentage of confidence that the proposed rejection is valid. If the calculated Q value is greater than the value from the table, then the suspect value is rejected and the mean recalculated. If the Q value is less than or equal to the value from the table than the calculated mean is reported. Q is defined as follows:

$$Q = \frac{\text{gap}}{\text{range}} \tag{1.8}$$

where the "gap" is the difference between the suspect value and it nearest neighbor and the "range" is the difference between the lowest and highest values (see Figure 1.7). A table of Q values for the 90% confidence level is given in Table 1.2.

Example 1.4

Determine if any of the values in Example 1.1 should be rejected based on the Q test at the 90% confidence level.

Solution 1.4

The six values realigned from lowest to highest are 3.035%, 3.044%, 3.046%, 3.048%, 3.049%, and 3.053%. The outlier would be 3.035%, since it has a larger gap (0.009) from its nearest neighbor (3.044) than has 3.053. Thus, we have

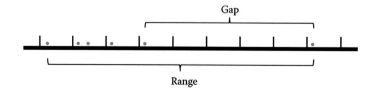

FIGURE 1.7   Illustration of the "gap" and the "range" as used in the Q test.

TABLE 1.2
**Q Values for the 90% Confidence
Level**

Number of Measurements	Q (at 90% Confidence Level)
3	0.94
4	0.76
5	0.64
6	0.56
7	0.51
8	0.47
9	0.44
10	0.41

$$Q = \frac{gap}{range} = \frac{0.009}{0.018} = 0.5$$

Since the Q value for 6 measurements is 0.56 (Table 1.2) and since the calculated value (0.5) is less, the suspect value cannot be rejected.

1.6.6 FINAL COMMENTS ON STATISTICS

When chemists talk about an analytical method or when instrument vendors tout their products, they often quote the standard deviation that is achievable with the method or instrument as a measure of quality. For example, the manufacturer of an HPLC pump may declare that the digital flow control for the pump, with flow rates from 0.01 to 9.99 mL/min has an RSD <0.5%, or a chemist declares that her atomic absorption instrument gives results within 0.5% RSD. The most fundamental point about standard deviation is that the smaller it is, the better, because the smaller it is, the more precise the data (the more tightly bunched the data are around the mean) and, if free of bias, the greater the chance that the data are more accurate. Chemists have come to know through experience that a 0.5% RSD for the flow controller and, under the best of circumstances, a 0.5% RSD for atomic absorption results, are favorable RSD values compared with other comparable instruments or methods.

1.7 PRECISION, ACCURACY, AND CALIBRATION

We have made references in the foregoing discussion to the "precision" of data, or how "precise" the data are. We have also made reference to the "accuracy" of data. **Precision** refers to the repeatability of a measurement. If you repeat a given measurement over and over and these measurements deviate only slightly from one another, within the limits of the number of significant figures obtainable, then we say that the data are precise, or that the results exhibit a high degree of precision. The mean of such data may or may not represent the real value of that parameter. In other words, it may not be accurate. **Accuracy** refers to the correctness of a measurement or how close it comes to the correct value of a parameter.

For example, if an analyst has an object that he/she knows weighs exactly 1.0000 g, the accuracy of a laboratory balance (weight measuring device) can be determined.* The object can be weighed

---

* Standard weights certified by the National Institute of Standards and Technology (NIST) are available.

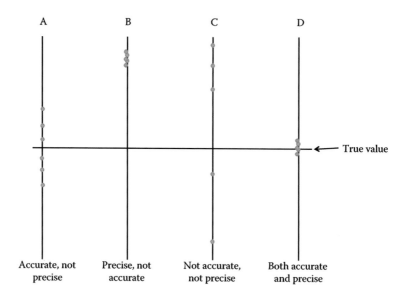

A — Accurate, not precise
B — Precise, not accurate
C — Not accurate, not precise
D — Both accurate and precise

True value

FIGURE 1.8 Illustration of the difference between accuracy and precision. In this illustration, "accurate" describes the data set when the mean of the measurements fall on or near the true value line. "Precise" describes the data set when the measurements are bunched together.

on the balance to see if the balance will read 1.0000 g. Suppose this balance is reputed to have a repeatability of ±0.0002 g. If a series of repeated weight measurements using this balance are all between 0.9998 and 1.0002 g, we say the balance is both precise and accurate. If, on the other hand, a series of repeated weight measurements using this balance are all between 0.9983 and 0.9987 g, we say that the balance is precise, but not accurate. If repeated weight measurements using this balance are all between 0.9956 and 0.9991 g, the data are neither precise nor accurate. Finally, if repeated weight measurements using this balance are all between 0.9956 and 1.0042 g, such that the mean is 1.0000 g, then the balance is not precise, but it does "appear" to be accurate. These facts on accuracy and precision are illustrated further in Figure 1.8.

If it is established that a measuring device provides a value for a known sample that is in agreement with the known value to within established limits of precision, that device is said to be "calibrated." Thus, **calibration** refers to a procedure that checks the device to confirm that it provides the known value. An example is an analytical balance as discussed above. Sometimes the device can be electronically adjusted to give the known value, such as in the case of a pH meter that is calibrated with solutions of known pH. However, calibration can also refer to the procedure by which the measurement value obtained on a device for a known sample becomes known. An example of this is a spectrophotometer in which the absorbance values for known concentrations of solutions are measured. We will encounter all of these calibration types in our studies.

EXPERIMENTS

EXPERIMENT 1: ASSURING THE QUALITY OF WEIGHT MEASUREMENTS

Note: This experiment assumes that a permanent log and a quality control chart are constantly maintained for each analytical balance in use in the laboratory. Each day you use a given analytical balance, log in with your name and date. The following calibration check should be performed weekly on all balances. If, according to the log, the calibration of the balance you want to use has not been checked in over a week, perform this procedure. Review Chapter 3, Section 3.3, and Appendix 1 for basic information concerning the analytical balance.

1. Obtain a certified 500-mg standard weight or other weight as suggested by your instructor. Do not touch the weight, but handle it with tweezers and never allow any water or other foreign material to touch it.
2. Check the calibration of the analytical balance you have chosen to use by weighing this standard weight on this balance. When finished, store the standard weight in the specified protected location.
3. Along with your name and date in the log book, also record the measured weight of the standard weight.
4. Plot your measured weight on the control chart. If any irregularity is observed, report to your instructor.

EXPERIMENT 2: WEIGHT UNIFORMITY OF DOSING UNITS

1. Randomly collect 10 ibuprofen (or other) tablets from a bottle of tablets available from a pharmacy.
2. Handling the tablets with tweezers, carefully weigh each on an analytical balance (see Section 3.3).
3. Calculate the mean, standard deviation, and percentage of RSD of this data set.
4. Evaluate the results from step 3. Comment on the uniformity of the tablet weight. Also, note the milligrams of ibuprofen (or other) per tablet found on the label and compare this with your results. If the label value is lower than the mean you calculated, give some possible reasons for this.

QUESTIONS AND PROBLEMS

Analytical Science Defined

1. Define: analytical science, analysis, chemical analysis, analyte, matrix, assay.
2. When is an analysis an "assay" and when is it not? Give examples of both.

Classification of Analysis

3. Distinguish between qualitative analysis and quantitative analysis. Give examples.
4. Imagine yourself as an analytical chemist and someone brings you an oily rag to analyze in order to identify the material on the rag. Is this a qualitative analysis or a quantitative analysis?
5. Distinguish among wet chemical analysis, instrumental analysis, and analysis using physical properties.
6. Imagine taking a tour of an industrial facility and having a particular laboratory being described to you as the "wet lab." What do you suppose is the kind of activity going on in such a laboratory?
7. When would you choose a wet chemical analysis procedure over an instrumental analysis procedure? When would you choose an instrumental analysis procedure over a wet chemical procedure?

The Sample

8. Define "representative sample" without using any form of the word "represent" in your answer.

The Analytical Process

9. What are the five steps in the analytical process?
10. What does it mean to "obtain a sample?" What is a "sample?"
11. What does it mean to "prepare" a sample?

12. What is an "analytical method" and how does it fit into the total analysis process?
13. What does it mean to "carry out the analytical method?"
14. What happens after a chemist acquires data from an analytical method?

Analytical Technique and Skill

15. What does it mean to say that a laboratory worker has "good analytical technique?"
16. Why should a stirring rod, which is being removed from a beaker containing a solution of the sample being analyzed, be rinsed back into that solution with distilled water?
17. Explain "GLP" and "SOP."
18. What sort of things do the GLP regulations address?

Elementary Statistics

19. Given the care with which laboratory equipment (balances, burets, instruments, etc.) is calibrated at the factory, why should the chemical analyst worry about errors?
20. Distinguish between determinate errors and indeterminate errors.
21. Define "bias."
22. An analyst determines that the analytical balance he used in a given analytical test is wrongly calibrated. Is this a determinate error or an indeterminate error? Explain.
23. A student determines the weight of an object on an analytical balance to be 12.2843 g. The actual weight, unknown to the student or to anyone else, is 12.2845 g. Is the error in the student's measurement determinate or indeterminate? Explain.
24. Can a person really trust the results of a laboratory analysis to be accurate? Explain.
25. Distinguish clearly between "accuracy" and "precision."
26. A given analytical test was performed five times. The results of the analysis are represented by the following values: 6.738%, 6.738%, 6.737%, 6.739%, and 6.738%. Would you say that these results are precise? Can you say that they are accurate? Explain both answers.
27. A given analytical test was performed five times. The results of the analysis are represented by the following values: 37.23%, 32.91%, 45.38%, 35.22%, and 41.81%. Would you say that these results are precise? Can you say that they are accurate? Explain both answers.
28. Suppose the correct answer to the analysis represented in question 26 above is 6.923%. What can you say now about the precision and accuracy?
29. Calculate the standard deviation and the RSD for the following data.

Measurement	Value (g)
1	16.7724
2	16.7735
3	16.7722
4	16.7756
5	16.7729
6	16.7716
7	16.7720
8	16.7733

30. A series of eight absorbance measurements using an atomic absorption spectrophotometer are as follows: 0.855, 0.836, 0.848, 0.870, 0.859, 0.841, 0.861, and 0.852. According to the instrument manufacturer, the precision of the absorbance measurements using this instrument should not exceed 1% RSD. Does it in this case?

31. Why is the RSD considered a popular and practical expression of data quality?
32. Explain this statement: "An RSD of 1% can be achieved in this experiment".
33. What laboratory analysis results might cause a batch of raw material at a manufacturing plant to be rejected from potential use in the plant process? Explain.
34. How can a quality control chart signal a problem with a routine laboratory procedure?
35. Using the information given in each row of the table below, calculate the item marked with a question mark (?). Follow significant figures rules when reporting the answers to any calculations.

	Value 1	Value 2	Value 3	Value 4	Value 5	Mean	Median	Mode
(a)	5.004	5.185	4.941	5.066	5.085	?	?	?
(b)	0.5503	0.5612	0.5678	0.5612	0.5587	?	?	?
(c)	43.49	43.49	43.89	43.02	43.49	?	?	?
(d)	12.559	12.300	12.695	12.428	12.717	?	?	?
(e)	0.1190	0.1159	0.1176	0.1190	0.1212	?	?	?
(f)	0.0988	0.0969	0.0980	0.0961	xxxxx	?	?	?
(g)	32.85	32.99	32.85	32.71	xxxxx	?	?	?

36. Using the information given in each row of the table below, calculate the items marked with a question mark (?).

	Value 1	Value 2	Value 3	Value 4	Value 5	Mean	s	RSD	% RSD	ppt RSD
(a)	xxxxx	xxxxx	xxxxx	xxxxx	xxxxx	xxxxx	xxxxx	0.093	?	?
(b)	xxxxx	xxxxx	xxxxx	xxxxx	xxxxx	xxxxx	xxxxx	0.331	?	?
(c)	xxxxx	xxxxx	xxxxx	xxxxx	xxxxx	3.3992	0.0488	?	?	?
(d)	xxxxx	xxxxx	xxxxx	xxxxx	xxxxx	16.38	0.16	?	?	?
(e)	0.375	0.366	0.371	xxxxx	xxxxx	?	?	?	?	?
(f)	0.1102	0.1096	0.1109	xxxxx	xxxxx	?	?	?	?	?
(g)	1.4490	1.4356	1.4402	1.4391	xxxxx	?	?	?	?	?
(h)	56.70	55.95	56.54	56.29	xxxxx	?	?	?	?	?
(i)	0.409	0.402	0.429	0.405	0.420	?	?	?	?	?
(j)	23.69	23.55	23.76	23.88	23.47	?	?	?	?	?

37. Using the information given in each row of the table below, determine the item marked with a question mark (?). The true mean (μ) is the mean with the confidence interval at the confidence level indicated as determined by the Student's t method, for example 27.48 ± 0.14.

	Mean	n − 1	Confidence Level (%)	s	t (from Table 1.1)	μ
(a)	16.37	5	90	0.28	?	?
(b)	0.238	6	95	0.019	?	?
(c)	5.20	4	99	0.10	?	?
(d)	67.30	7	95	0.09	?	?
(e)	1.265	5	90	0.045	?	?
(f)	38.034	6	99	0.104	?	?

38. Fill in the "?" blanks in the following table using the Q value from Table 1.2 at the 90% confidence level to judge whether the suspect value is rejectable.

	Gap	Range	Q	n	Suspect Value Rejected? (Yes/No)
(a)	0.058	1.238	?	5	?
(b)	0.18	3.95	?	6	?
(c)	1.24	23.09	?	4	?
(d)	0.48	12.82	?	5	?
(e)	0.86	67.10	?	6	?
(f)	0.12	5.88	?	4	?

39. Fill in the "?" blanks in the following table using the Q value from Table 1.2 at the 90% confidence level to judge whether the suspect value is rejectable.

	Value 1	Value 2	Value 3	Value 4	Value 5	Value 6	Gap	Range	Q	Suspect Value	Rejected? (Yes/No)
(a)	14.89	14.95	14.78	14.82	14.50	xxxxx	?	?	?	?	?
(b)	9.336	9.015	9.328	9.341	9.327	9.348	?	?	?	?	?
(c)	48.25	48.60	48.33	47.51	xxxxx	xxxxx	?	?	?	?	?
(d)	2.447	2.567	2.560	2.558	2.553	xxxxx	?	?	?	?	?
(e)	29.04	29.11	29.25	29.09	29.00	29.19	?	?	?	?	?
(f)	0.283	0.290	0.259	0.288	xxxxx	xxxxx	?	?	?	?	?

2 Sampling and Sample Preparation

2.1 INTRODUCTION

The process for analyzing a bulk system for the identity and/or quantity of an analyte involves the five steps that were summarized in Figure 1.1. These were (1) obtain a sample of the bulk system, (2) prepare this sample for the analytical method to be used, (3) execute the method chosen for the analysis, (4) work the data, and (5) calculate and report the results. A great deal of detail is presented in this book, and in other books, concerning Step 3. The reason for this is the large variety of methods that are available and routinely used. Each of these methods must be studied separately and in detail if we are to understand specifically how each method produces the result that is sought.

However, no analytical method, no matter how simple or sophisticated, no matter how specialized or routine, no matter how easy or difficult, and no matter how costly, will produce the correct result if the sample is not correctly obtained and prepared. The first two steps of the analytical process are therefore on at least equal footing with the analytical method in terms of importance to the end result. Thus, while the topics of sampling and sample preparation are given the space of only one chapter in this book, their critical importance should not go unnoticed. Quality sampling and sample preparation is crucial to the success of an analysis.

2.2 OBTAINING THE SAMPLE

As stated in Chapter 1, a laboratory analysis is almost always meant to give a result that is indicative of a concentration in a large system. For example, a farmer wants an analysis result to represent fertilizer needs in an entire 40-acre field. A pharmaceutical manufacturer wants an analysis result to represent the concentration of an active ingredient in each tablet in 80 cases of its product, each case containing 3 dozen bottles of 100 tablets each. A governmental environmental control agency wants a single laboratory analysis to represent the concentration of a toxic chemical in every cubic inch of soil within 5 miles of a hazardous waste dumpsite.

The critical part of any sampling task is to obtain a sample that represents the bulk system as well as possible. As stated in Chapter 1, the sample must possess all the characteristics of the entire bulk system with respect to the analyte and the analyte concentration in the system. In other words, it must be a **representative sample**—it must truly represent the bulk system. Whatever concentration level is found for a given component of a sample is also then taken to be the concentration level in the entire system. For example, in order to analyze the water in a lake for mercury, a bottle is filled with the water and is then taken into the laboratory for analysis. If the mercury level in this sample is determined to be 12 ppm, then, assuming the sample is representative of the entire lake, the entire lake is assumed to be 12 ppm in mercury.

Obviously, there are different degrees of difficulty and different sampling modes involved with obtaining samples for analysis, depending on the type of sample to be gathered, whether the source of the sample is homogeneous, the location of and access to the system, etc. For example, obtaining a sample of blood from a hospital patient is completely different from obtaining a sample of coal from a train car full of coal.

As far as blood is concerned, the time of day, along with a knowledge of the patient's recent dietary habits, is important. Or perhaps the patient is on some sort of medication that could affect the analysis.

With the coal sample it is important to recognize that the coal held in a train car may not be homogeneous and a sampling scheme that takes this into account must be implemented, so that a proper sample preparation scheme can be planned.

The key word in any case is "representative." A laboratory analysis sample must be representative of the whole so that the result of the chemical analysis represents the entire system that it is intended to represent. If there are variations in composition, such as with the coal example above, or at least suspected variations, small samples must be taken from all suspect locations. If results for the entire system are to be reported, these small samples are then mixed and made homogeneous to give the final sample to be tested. Such a sample is called a **composite sample**. In some cases, analysis on the individual samples may be more appropriate. Such samples are called **selective samples**.

Consider the analysis of soil from a farmer's field. The farmer wants to know whether he needs to apply a nitrogen-containing fertilizer to his field. It is conceivable that different parts of the field could provide different types of samples in terms of nitrogen content. Suppose there is a cattle feed lot nearby, perhaps uphill from part of the field and downhill from another part of the field such that runoff from the feed lot affects part of the field but not the other part. If the farmer wishes to have an analysis report for the field as a whole, then the sample taken should include combined portions from all parts of the field that may be different (a composite sample) so that it will truly represent the field as a whole. Alternatively, two selective samples could be taken, one from above the feed lot and one from below the feed lot, so that two analyses are performed and reported to the farmer. These would be referred to as selective samples. At any rate, one wants the results of the chemical analysis to be correct for the entire area for which the analysis is intended.

Consider the analysis of the leaves on a tree for pesticide residue. The tree grower wants to know if the level of pesticide residue on the leaves indicates whether the tree needs another pesticide application. Again, the analyst must consider all parts of the tree that might be different. Leaves at the top, in the middle, and at the bottom should be sampled (one can imagine differences in application rates at the different heights); leaves on the outside and leaves close to the trunk should be sampled; and perhaps there should also be difference between the shady side and the sunny side of the tree. All leaf samples are then combined for a composite sample.

Consider the analysis of a blood sample for alcohol content (imagine that a police officer suspects a motorist to be intoxicated). The problem here is not sampling different locations within a system, but rather a time factor. The blood must be sampled within a particular time frame, which would demonstrate intoxication at the time the motorist was stopped.

If it is thought that a bulk system is homogeneous for a particular component, then a **random sample** is taken. This would be just one sample taken from one location at random in the bulk system. An example would be when determining the level of active ingredient in a pharmaceutical product stored in boxes of individual bottles in a warehouse. Having no reason to assume a greater concentration level in one bottle or box compared with another, a sample is chosen at random.

Other designations for samples are **bulk sample**, **primary sample**, **secondary sample**, **subsample**, **laboratory sample**, and **test sample**. These terms are used when a sample of a bulk system is divided, possibly a number of times, before actually being used in an analysis. For example, a water sample from a well may be collected in a large bottle (bulk sample or primary sample) from which a smaller sample is acquired by pouring it into a vial to be taken into the laboratory (secondary sample, subsample, or laboratory sample), then pouring it into a beaker (another secondary sample or subsample), before a portion is finally carefully measured into a flask (test sample) and diluted to make the sample solution.

The problems associated with sampling are unique to every individual situation. The analyst simply needs to take all possible variations into account when obtaining the sample, so that the sample taken to the laboratory truly does represent what it is intended to represent.

2.3 STATISTICS OF SAMPLING

A consideration of statistics is required in a discussion of sampling because of the randomness with which samples are acquired. Just as random, indeterminate errors associated with laboratory work

(Chapter 1) are dealt with by statistics because there is no other way to deal with them, so also with sampling errors. We may take pains to see that a sample randomly acquired is representative, but the compositions of a series of such samples taken from the same system always vary to some unknown extent.

A novice might think that a laboratory analyst obtains a single sample from a bulk system, analyzes it one time in the laboratory, and reports the answer to this one analysis as the analysis results. If variances in the sampling and laboratory work are both insignificant, these results may be valid. However, due to possible large variances in both the sampling and the laboratory work, such a result cannot be considered reliable. In Chapter 1, we indicated that the correct procedure is to perform the analysis many times and deal with the variances with statistics.

The fact that sampling introduces a second statistics problem means that we must also consider taking a large number of samples and deal with the results with statistics just as we perform a laboratory analysis a large number of times and deal with those results with statistics.

Chemists want to have as low a variance (or standard deviation) as possible for the greatest accuracy. If it is not possible to have a low enough standard deviation to suit the need, then the number of measurements (either the number of samples, the number of laboratory analyses, or both) must be increased. If increasing the number of measurements is not desirable (due to an increased work load or expense, etc.), then the analyst must live with a larger error.

2.4 SAMPLE HANDLING

The importance of a high quality representative sample has already been noted. How to obtain the sample and what to do with it once it reaches the laboratory are obviously important factors. But the handling of the sample between the sampling site and the laboratory is often something that is given less than adequate consideration. The key concept is that the sample's integrity must be strictly maintained and preserved. If a preservative needs to be added, then someone needs to be sure to add it. If maintaining integrity means to refrigerate the sample, then it should be adequately refrigerated. If there is a specified holding time, then this should be accurately documented. If preventing contamination means that we must use a particular material for the storage container (e.g., glass vs. plastic), then that material should be used.

2.4.1 CHAIN OF CUSTODY

It is very important to document who has handled the sample, what responsibility each handler has at various junctures between the sampling site and the laboratory and what actions the sample handler has taken relating to sample integrity while the sample was in his/her custody. In other words, the **chain of custody** must be maintained and documented. A sample can have a number of custodians along the way to the laboratory. A sample of lake water may be taken by a sampling technician at the site. The sampling technician may give it to a driver who transports it to the analysis site. A shipping/receiving clerk may log in the sample and give it to a subordinate who takes it to the laboratory. Along the way, this sample is in the hands of five different handlers, the sampling technician, the driver, the shipping/receiving clerk, the subordinate, and the laboratory technician. Each should maintain documentation of his/her activity and duties and copies of the chain of custody should be filed (see Figure 2.1).

FIGURE 2.1   Representation of the chain of custody for a water sample from a remote site as described in the text. The point is that this chain of custody needs to be documented. (From Kenkel, J., *A Primer on Quality in the Analytical Laboratory*, CRC Press/Lewis Publishers, 2000.)

2.4.2 MAINTAINING SAMPLE INTEGRITY

It is important for all sample custodians to maintain the sample in its original physical and chemical condition so that it remains representative of the bulk system in terms of the analyte identity and concentration. Possible changes to avoid are (1) loss of sample matrix or solvent through evaporation or other means, (2) loss of analyte through evaporation, chemical reaction, temperature effects, bacterial effects, etc., (3) contamination with additional analyte (or another chemical that would interfere with the analysis) through contact with other matrices or chemical systems, and (4) moisture absorption or adsorption by exposure to humid air.

To avoid loss by evaporation, the sample should be sealed in its container. For liquid samples, the container should be filled to the top with no headspace. If any matrix component is volatile, it may need to be stored at a lower temperature than room temperature and care should taken when opening the container in the laboratory. Care should be taken not to lose any sample when transferring to another container or to a filter, etc. It may be important to not leave any portion of the sample behind (such as on the container walls) when transferring. It is also important to consider the cleanliness of the container or the laboratory equipment that comes into contact with the sample.

To avoid loss of analyte by chemical reaction or by temperature or bacterial effects, it may be necessary to take extraordinary precautions specific to the analyte itself. Such precautions may include adding a preservative to the sample, maintaining specific conditions of temperature, humidity, or other environmental conditions, and avoiding sunlight or oxygen, etc. In addition, laboratory equipment that comes into contact with the sample in the preparation process should be clean and free of material that would remove analyte or add contaminant.

APPLICATION NOTE 2.1: AN EXAMPLE OF SAMPLE PRESERVATION: ENVIRONMENTAL WATER

Environmental water samples to be analyzed for metals are best stored in quartz or Teflon containers. However, because these containers are expensive, polypropylene containers are often used. Borosilicate glass may also be used, but soft glass should be avoided because it can leach traces of metals into the water. If silver is to be determined, the containers should be light absorbing (dark-colored). Samples should be preserved by adding concentrated nitric acid so that the pH of the water is less than 2. The iron in well water samples, for example, will precipitate as iron oxide upon exposure to air and would be lost to the analysis if not for this acidification.

Environmental water samples to be analyzed for phosphate are not stored in plastic bottles unless kept frozen because phosphates can be absorbed onto the walls of plastic bottles. Mercuric chloride is used as a preservative and acid (such as the nitric acid suggested for metals above) should *not* be used unless total phosphorus is determined. All containers used for water samples to be used for phosphate analysis should be acid rinsed and commercial detergents containing phosphates should not be used to clean sample containers or laboratory glassware.

Environmental water samples analyzed for nitrate should be analyzed immediately after sampling. If storage is necessary, samples may be stored for up to 24 h at 4°C or indefinitely if sulfuric acid is added (2 mL of concentrated acid per liter) and they are stored at 4°C. In such samples, however, nitrate and nitrite cannot be differentiated.

(*Reference: Standard Methods for the Examination of Water and Wastewater*, 19th edition, APHA, Washington, DC, AWWA, Denver, CO [2000].)

Other matrices or chemical systems that come into contact with a sample include the sample container, spatulas and scoops, grinders and mixers, and filters, etc. Precautions should be taken to assure that these do not unintentionally affect or alter the composition of the sample in any way.

A good sample custodian is one who has all the analytical laboratory skills and utilizes them in executing any analytical laboratory operation. A conscientious and careful laboratory technician will succeed in maintaining sample integrity (see also Application Note 2.1).

2.5 SAMPLE PREPARATION—SOLID MATERIALS

Except for the moon and for the planet Mars, no extraterrestrial body has had pieces of its mass directly examined by scientists in an earthly laboratory. This means that there has been no laboratory sample preparation scheme performed on samples of solid matter from any of the other planets, their moons, the comets, or the asteroids. And yet, we read repeatedly about how scientists have been able to surmise the chemical composition of the solid materials on the surfaces of these celestial bodies following, for example, the close approach of or the landing of a spacecraft loaded with special scientific equipment.

Obviously, the technology exists for obtaining analytical results without special preparation and analysis in a laboratory. However, at the present time there is no acceptable substitute for direct laboratory examination of laboratory samples if we want the kind of accuracy and confidence we have come to expect. All conventional methods for analysis of solid materials require one or more of the following preparation activities before an analytical method can be properly executed: (1) particle size reduction, (2) homogenization and division, (3) partial dissolution, and (4) total dissolution. Let us briefly discuss each of these individually.

2.5.1 PARTICLE SIZE REDUCTION

If a solid sample is to be efficiently dissolved, either partially or totally, the solvent must have intimate contact with small particles, since without contact, there can be no dissolution. The reason is simply that a solvent will only have contact with the outer surface of a particle and not the entire particle. Thus, it would be ideal if a solid's particle size can be reduced to that of a fine powder, if possible, so that the solvent contact can be as complete as possible. Thus, sample preparation activities such as crushing, milling, grinding, or pulverizing are common. If the solid sample is of a state that reducing to a powder is not possible (such as with a polymer film or piece of organic tissue), then particle size reduction may take the form of cutting, chopping, or blending, etc. Thus, crushers, ball mills, mortars and pestles, scissors, and blenders are common items used in sample preparation in analytical laboratories.

2.5.2 SAMPLE HOMOGENIZATION AND DIVISION

The amount of solid sample called for in an analysis procedure may be less, sometimes considerably less, than the amount of sample at hand. In that case, a portion of the larger sample must be taken. Obtaining a portion that is representative of the larger sample (and therefore representative of the entire bulk system) requires that the larger sample be homogeneous. Particle size reduction procedures may not leave a sample homogeneous. For example, large dirt clods that may be part of a dried soil sample may have been reduced to powder through crushing procedures but this powder, from which a test sample must be taken, may not be sufficiently homogeneous without first undergoing a mixing procedure. Thus, a mixing procedure and a procedure by which a sample is divided to create the required portions may follow particle size reduction (see Figure 2.2). In addition, to ensure that the solid particles are of uniform size, they may be passed through a sieve (see Chapter 3 for a description of sieves).

FIGURE 2.2   Photograph of a sample of fertilizer that is obviously not homogeneous. Notice the various light- and dark-shaded granules. To ensure a homogeneous sample when the amount to be tested is small, such a sample must have the particle size reduced and the smaller particles must then be thoroughly mixed.

2.5.3   SOLID–LIQUID EXTRACTION

An analyte may be present in one material phase (either a solid or liquid sample) and, as part of the sample preparation scheme, be required to be separated from the sample matrix and placed in another phase (a liquid). Such a separation is known as an **extraction**—the analyte is "extracted" from the initial phase by the liquid and is deposited (dissolved) in the liquid while other sample components are insoluble and remain in the initial phase. If the sample is a solid, the extraction is referred to as a **solid–liquid extraction**. In other words, a solid sample is placed in the same container as the liquid and the analyte is separated from the solid because it dissolves in the liquid, whereas other sample components do not.

The process can take one of two forms. In one, the sample and liquid are shaken (or otherwise agitated) together in the same container, the resultant mixture filtered, and the filtrate, which then contains the analyte, collected. In the other, fresh liquid is continuously cycled through the solid sample via a continuous evaporation/condensation process over a period of hours (that usually does not require an extra filtration step) and the liquid collected. This latter method is known as a **Soxhlet extraction**. Soxhlet extraction will be discussed in more detail in Section 2.11.

2.5.4   OTHER EXTRACTIONS FROM SOLIDS

Extractions from solid samples may also be performed with supercritical fluids. A **supercritical fluid** is the physical state of a substance that is neither a liquid or gas but something in between, meaning its properties change to a point different from either the liquid phase or the gas phase of that substance. For example, the solvent properties of supercritical fluids are better than that of their corresponding gas phase and their viscosities (resistance to flow) are considerably less than their corresponding liquid phase. The supercritical fluid state is achieved at a high temperature and extremely high pressure, exceeding the so-called critical temperature and pressure for that substance.

Because their solvent properties are very good, and because their viscosities are very low, supercritical fluids can be used for very efficient extraction of analytes from solid phase samples. The solid phase sample is held in a tube or cartridge and the supercritical fluid made to flow through (minimal pressure required). The fluid with the analyte is then made to flow through a trap solvent. The analyte dissolves in this solvent and the fluid reverts back to the gas phase.

Finally, extractions from solids can be performed by heating followed by solvent trapping. Such a procedure is known as **thermal extraction**.

2.6 WATER PURIFICATION AND USE

Before we continue to discuss issues involving sample preparation, it is appropriate that we bring water into the discussion. First, ordinary water can be an excellent solvent for many samples. Due to its extremely polar nature, water will dissolve most substances of likewise polar or ionic nature. Obviously, then, when samples are composed solely of ionic salts or polar substances, water would be an excellent choice. An example might be the analysis of a commercial iodized table salt for sodium iodide content. A list of "solubility rules" for ionic compounds in water can be found in Table 2.1.

But besides its possible use for dissolving samples, we also recognize that water is widely used in many other ways as part of the overall analytical strategy. For this reason, chemists have come to rely on laboratory methods to obtain a high purity in order to avoid contamination from the minerals dissolved in ordinary tap water. The most important of these is probably distillation.

2.6.1 PURIFYING WATER BY DISTILLATION

Distillation is a method of purification of liquids contaminated with either dissolved solids or miscible liquids. In the case of water, dissolved solids are usually the focus. The dissolved solids are metal carbonates, sulfates, chlorides, etc. The method consists of boiling and evaporating the water followed by condensation of the vapors in a "condenser." A condenser is a tube usually cooled by isolated, cold tap water (see Figure 2.3). The theory is that the water vapor (and thus condensed water emerging from the condenser) will be purer than the original water. The separation is based on the fact that the contaminants have different boiling points and vapor pressures than the liquid to be purified.

All liquids have a tendency to evaporate. This means that their atoms or molecules have a tendency to escape the liquid phase and enter the gas phase even if the liquid is not boiling. For

TABLE 2.1
Solubility Rules for Ionic Compounds in Water

Compound Class	Soluble?	Exceptions
Nitrates	Yes	None
Acetates	Yes	Silver acetate is sparingly soluble
Chlorides/Bromides	Yes	Chlorides/bromides of Ag, Pb, and Hg are insoluble
Sulfates	Yes	Sulfates/bromides of Pb are soluble in hot water
Carbonates	No	Carbonates of Na, K, and NH_4 are soluble
Phosphates	No	Phosphates of Na, K, and NH_4 are soluble
Chromates	No	Chromates of Na, K, NH_4, and Mg are soluble
Hydroxides	No	Hydroxides of Na, K, and NH_4 are soluble; hydroxides of Ba, Ca, and Sr are slightly soluble
Sulfides	No	Sulfides of Na, K, NH_4, Ca, Mg, and Ba are soluble
Sodium Salts	Yes	Some rare exceptions
Potassium Salts	Yes	Some rare exceptions
Ammonium Salts	Yes	Some rare exceptions
Silver Salts	No	Silver nitrate and perchlorate are soluble; Silver acetate and sulfate are sparingly soluble

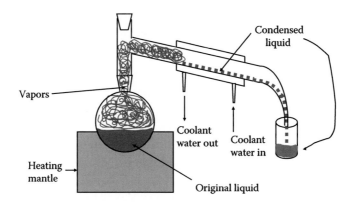

FIGURE 2.3 Drawing of a simple distillation apparatus as described in the text.

example, liquid water has a noticeable such tendency—wet laundry dries when hung out. Some liquids have a strong tendency to do this, whereas other liquids have a weaker tendency. This tendency is called the **vapor pressure**. The vapor pressure increases when the temperature increases and is a maximum at the temperature at which the liquid boils (the **boiling point**). Different substances have different vapor pressures and boiling points. Distillation separates substances from each other based on these facts.

When the liquid is boiled and evaporated, the vapors (and recondensed liquids) created have a composition different from the original liquid. The substances with lower boiling points and higher vapor pressures are therefore separated from substances that have high boiling points and low vapor pressures, i.e., the dissolved minerals in tap water.

Distillation of water to remove hardness minerals is an example and probably the most common application in an analytical laboratory. Of all the applications of distillation, it is one of the easiest to perform. Although water is known to have a relatively high boiling point and low vapor pressure compared with other liquids, the dissolved minerals are ionic solids that generally have extremely high boiling points (indeed, extremely high *melting* points) and extremely low vapor pressures. Thus, a simple distilling apparatus and a single distillation, or, at most, two (doubly distilled) or three (triply distilled) distillations, will produce very pure water.

2.6.2 PURIFYING WATER BY DEIONIZATION

Another method for obtaining pure water is deionization. Water is often "deionized" by passing it through a cartridge containing a mixed-bed ion-exchange resin (Section 10.3.3). Ion-exchange resins are polymers that have ionic bonding sites in their structures. Initially, hydrogen ions and hydroxides ions are bound to these sites. When water containing the dissolved minerals pass through, the metals ions change places with the hydrogen ions and the anions change places with the hydroxide ions. The released hydrogen ions, and hydroxides then combine to form more water. The water emerging from this cartridge then is "deionized." The ions have been removed. Deionization of water is often done in conjunction with distillation, such that the water is both deionized and distilled prior to use. Also, if the water is contaminated with organics, or other low boiling substances, a charcoal filter cartridge is often used as well.

2.7 TOTAL SAMPLE DISSOLUTION AND OTHER CONSIDERATIONS

In the event that the extractions discussed in Section 2.5 are not useful or feasible, a *total* dissolution of a sample may be required. Total dissolution involves the proper choice of an acid—one that will indeed dissolve the total solid sample. This will usually involve the use of acid–water mixtures, but sometimes fairly concentrated acid solutions, any of which may accompanied by heating. For this reason, it is useful to summarize the application of the most common laboratory acids.

2.7.1 HYDROCHLORIC ACID

Strong acids are used frequently for the purpose of sample dissolution when water would not do the job. One of these is hydrochloric acid, HCl. Concentrated HCl is actually a saturated solution of hydrogen chloride gas, fumes of which are very pungent. Such a solution is 38.0% HCl and about 12 M. Hydrochloric acid solutions are used especially for dissolving metals, metal oxides, and carbonates not ordinarily dissolved by water. Examples are iron and zinc metals, iron oxide ore, and the metal carbonates of which the scales in boilers and humidifiers are composed. Being a strong acid, it is very toxic and must be handled with care. It is stored in a blue color-coded container (see Figure 2.4).

2.7.2 SULFURIC ACID

An acid that is considered a stronger acid than HCl in many respects is sulfuric acid, H_2SO_4. When sulfuric acid contacts clothing, paper, etc., one can see an almost instantaneous reaction—paper towels turn black and disintegrate, clothing fibers become weak and holes readily form. Concentrated sulfuric acid is about 96% H_2SO_4 (the remainder being water), or about 18 M, and is a clear, colorless, syrupy dense liquid. It reacts violently with water, evolving much heat, and so water solutions

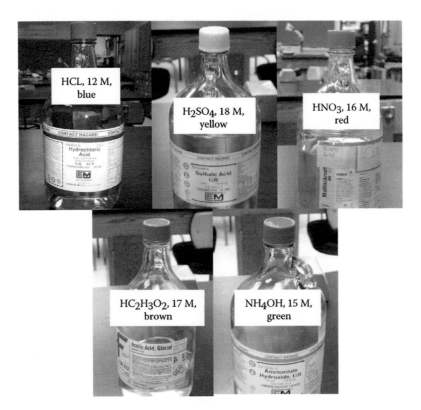

FIGURE 2.4 Photographs of the containers of four common concentrated laboratory acids and one common concentrated laboratory base as they are purchased from chemical vendors. The formulas of the acids, their concentrations, and the color codes used on the bottles and bottle caps are indicated for each. Clockwise from upper left: hydrochloric acid, sulfuric acid, nitric acid, ammonium hydroxide, and acetic acid.

of sulfuric acid must be prepared cautiously, often to include a means of cooling the container. Its sample dissolution application is limited mostly to organic material, such as vegetable plants. It is not as useful for metals because many metals form insoluble sulfates. It is the solvent of choice for the Kjeldahl analysis (Chapter 5) for such materials as grains, and products of grain processing. It is also used to dissolve aluminum and titanium oxides on airplane parts. It is stored in a yellow color-coded container (see Figure 2.4).

2.7.3   NITRIC ACID

Another acid that has significant application is nitric acid, HNO_3. This acid is also very danger-ous and corrosive and is aptly referred to as an oxidizing acid. This means that a reaction other than hydrogen gas displacement (as with HCl) occurs when it contacts metals. Frequently, oxides of nitrogen form in such a reaction, and noxious brown, white, and colorless gases are evolved. Concentrated HNO_3 is 70% HNO_3 (16 M) and is used for applications where a strong acid with addi-tional oxidizing power is needed. These include metals such as silver and copper, as well as organic materials such as in a wastewater sample. Nitric acid will turn skin yellow after only a few seconds of contact. It is stored in a red color-coded container (see Figure 2.4).

2.7.4   HYDROFLUORIC ACID

An acid that has some very useful and specific applications, but is also very dangerous, is hydro-fluoric acid, HF. This acid reacts with skin in a way that is not noticeable at first, but becomes quite serious if left in contact for a period of time. It has been known to be especially serious if trapped against the skin or after diffusing under fingernails. Treatment of this is difficult and pain-ful. Concentrated HF is about 50% HF (26 M). It is an excellent solvent for silica (SiO_2)-based mate-rials such as sand, rocks, and glass. It can also be used for stainless steel alloys. Since it dissolves glass, it must be stored in plastic containers. This is also true for low pH solutions of fluoride salts.

2.7.5   PERCHLORIC ACID

Another important acid for sample preparation and dissolution is perchloric acid, $HClO_4$. It is an oxidizing acid like HNO_3, but is considered to be even more powerful in that regard when hot and concentrated. It can be used for metals, since metal perchlorates are highly soluble, but it is most useful for more difficult organic samples, such as leathers and rubbers, often in combination with nitric acid. It can also be used for stainless steels and other more stable alloys. Commercial $HClO_4$ is 72% (12 M). It is a very dangerous acid, especially when hot and in contact with organic matter, and should only be used in a fume hood designed for the collection of its vapors. Contact with alcohols, including polymeric alcohols, such as cellulose, and other oxidizable materials, should be avoided due to the potential for explosions.

2.7.6   "AQUA REGIA"

An acid mixture that is prepared by mixing one part concentrated HNO_3 with three parts con-centrated HCl is called "aqua regia." This mixture is among the most powerful dissolving agents known. It will dissolve the very noble metals (gold and platinum) as well as the most stable of alloys.

2.7.7   ACETIC ACID

A weak acid that has some important applications is acetic acid, an organic acid with the formula $HC_2H_3O_2$. The main application is in the preparation of buffer solutions (Chapter 5). Although it is weak, it is still a rather dangerous acid with serious strong vinegar fumes. Concentrated acetic acid has a 99.5% concentration (17 M). It is often referred to as "glacial" acetic acid because it has a relatively high freezing point and in a cooler laboratory environment will turn to a solid. It has a brown color code (see Figure 2.4).

2.7.8 AMMONIUM HYDROXIDE

As we will see in Chapter 5, weak bases are also useful for preparing buffer solutions. Ammonium hydroxide, NH_4OH, is a very common weak base. Although it has limited sample dissolution application, it is sold as a concentrated solution like the others in this list. Like hydrochloric acid and acetic acid, it is very dangerous because of fumes, in this case, stifling ammonia fumes. Concentrated ammonium hydroxide has a concentration of 58% (15 M). It has a green color code (see Figure 2.4).

Table 2.2 summarizes the above discussion. Occasionally, organic liquids are also used for total sample dissolution. Common laboratory organic solvents are described in Section 2.9.

TABLE 2.2
Various Laboratory Acids and Other Chemicals

Name	Formula	Description	Example Uses
Water	H_2O	Clear, colorless liquid with relatively high vapor pressure, highly polar	Dissolving polar and ionic compounds
Hydrochloric acid	HCl	Commercially available concentrated solution is 38% (12 M) HCl Evolves pungent fumes and must be handled in fume hood	Dissolving some metals and metal ores
Sulfuric acid	H_2SO_4	Commercially available concentrated solution is 96% (18 M) H_2SO_4 A dense, syrupy liquid Reacts on contact with skin and clothing Evolves much heat when mixed with water	Organic samples, such as for Kjeldahl analysis (see Chapter 5) Also oxides of Al and Ti
Nitric acid	HNO_3	Commercially available concentrated solution is 70% (16 M) HNO_3 Reacts with clothing and skin—turns skin yellow Evolves thick brown and white fumes when in contact with most metals	Dissolving more noble metals (e.g., copper, silver) and also some organic samples
Hydrofluoric acid	HF	Commercially available concentrated acid is 50% (26 M) Must be stored in plastic containers, since it attacks glass Very damaging to skin	Dissolving silica-based and stainless steel
Perchloric acid	$HClO_4$	Commercially available concentrated solution is 72% (12 M)	Dissolving difficult organic samples and stable metal alloys
Aqua regia		A mixture of concentrated HNO_3 and HCl in the ratio of 1:3 HNO_3/HCl	Dissolving highly unreactive metals, such as gold
Acetic acid	$HC_2H_3O_2$	Commercially available concentrated acid is 99.5% (17 M) Has strong vinegar odor and should be handled in a fume hood	Buffer solutions
Ammonium hydroxide	NH_4OH	A weak base Commercially available concentrated solution is 58% (15 M) Has a very strong ammonia odor and must be used in a fume hood	Buffer solutions

2.8 FUSION

For extremely difficult samples, a method called "fusion" may be employed. Fusion is the dissolving of a sample using a molten inorganic salt, generally called a "flux," as the solvent. This flux dissolves the sample and, upon cooling, results in a solid mass that is then soluble in a liquid reagent. The dissolving power of the flux is mostly due to the extremely high temperatures (usually 300°C to 1000°C) required to render most inorganic salts molten.

Additional problems arise within fusion methods, however. One is the fact that the flux must be present in a fairly large quantity in order to be successful. The measurement of the analyte must not be affected by this large quantity. Also, while a flux may be an excellent solvent for difficult samples, it will also dissolve the container to some extent, creating contamination problems. Platinum crucibles are commonly used, but nickel, gold, and porcelain have been successfully used for some applications.

Probably the most common fluxes are sodium carbonate, Na_2CO_3, lithium tetraborate, $Li_2B_4O_7$, and lithium metaborate, $LiBO_2$. Fluxes may be used by themselves or in combination with other compounds, such as oxidizing agents (nitrates, chlorates, and peroxides). Applications include silicates and silica-based samples and metal oxides.

For dissolving particularly difficult metal oxides, the acidic flux potassium pyrosulfate ($K_2S_2O_7$) may be used.

2.9 SAMPLE PREPARATION: LIQUID SAMPLES, EXTRACTS, AND SOLUTIONS OF SOLIDS

Even though a sample may be dissolved in a liquid, it may not be ready for the method chosen. There may be other sample components present that may interfere or the solvent may not be suitable for the chosen method. Thus, original liquid solutions and the extracts and solutions resulting from the procedures described in Section 2.5 may require separation or dilution/concentration procedures.

2.9.1 EXTRACTION FROM LIQUID SOLUTIONS

In Section 2.5, we described separation procedures in which analytes are extracted from solid samples via contact with liquid solvents that selectively dissolve the analyte and leave other components undissolved or unextracted. There are several methods by which analytes can be extracted from liquid matrices as well.

In **liquid–liquid extraction**, the liquid solution containing the analyte (usually a water solution) is brought into contact with a liquid solvent (usually a nonpolar organic solvent) that is immiscible with the first solvent. The container is usually a separatory funnel (Figure 2.5). Since the two solvents are immiscible, there are two separate liquid phases (layers) in the separatory funnel. Shaking the separatory funnel brings the two solvents into intimate contact such that the analyte then moves from the original (first) solvent to the new (second) solvent. Being immiscible, the two layers can then be separated from each other by allowing the two layers to drain one at a time through the valve ("stopcock") at the bottom of the funnel. The desired solution is then carried forward to the next step. We will build upon this brief discussion in Section 2.10. Box 2.1 presents a list of some nonpolar organic solvents that are most often used as the extracting solvent for aqueous solutions.

In **solid phase extraction**, the liquid solution containing the analyte is passed through a cartridge containing a solid sorbent. The sorbent either retains the analyte or its matrix. If it is the analyte that is retained by the sorbent, the cartridge is first washed with a solvent that will sweep away interferences. The analyte is then eluted with a stronger solvent and collected. If it is the matrix that is retained by the sorbent, the analyte is eluted immediately and collected.

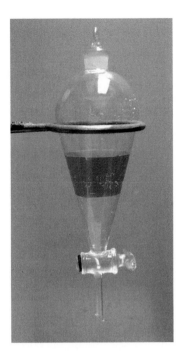

FIGURE 2.5   Photograph of a separatory funnel containing two immiscible liquid layers.

BOX 2.1   SOME NONPOLAR ORGANIC SOLVENTS THAT ARE OFTEN USED FOR EXTRACTING ANALYTES FROM AQUEOUS SOLUTIONS

Aliphatic hydrocarbons n-hexane, cyclohexane, and n-heptane: Aliphatic hydrocarbons are nonpolar. Their solubility in water is virtually nil. They are less dense than water, and thus would be the top layer in a separatory funnel with a water solution. They are obviously poor solvents for polar compounds, but are very good for extracting traces of nonpolar solutes from water solutions. They are highly flammable and have a low toxicity level.

Methylene chloride: This solvent is a slightly polar solvent also known as dichloromethane, CH_2Cl_2. Its solubility in water is 1.32 g/100 mL. It is denser than water (density = 1.33 g/mL); thus it would be the bottom layer when used with a water solution in a separatory funnel. It may form an emulsion when shaken in a separatory funnel with water solutions. It is not flammable and is considered to have a low toxicity level.

Toluene: Toluene ($C_6H_5–CH_3$) is a nonpolar aromatic liquid. It is flammable and has a density of 0.87 g/mL and thus would be the top layer in the separatory funnel with water. It is only slightly soluble in water (0.47 g/L). Its acute toxicity is low.

Diethyl ether: Diethyl ether ($CH_3CH_2–O–CH_2CH_3$), also frequently referred to as ethyl ether and as ether, is a mostly nonpolar organic liquid that is highly volatile and extremely flammable, a dangerous combination. Also, explosive peroxides form with time. Precautions regarding storage to discourage peroxide formation include storage in metal containers in explosion-proof refrigerators. It should be disposed of after about 9 months of storage. Ether is only slightly soluble (approximately 60 g/L at 25°C) in water. Its density is 0.71 g/mL; therefore, it would be the top layer in a separatory funnel with water. Since it is volatile, it is easily evaporated from extraction fractions. Its acute toxicity is low.

Chloroform: Chloroform (trichloromethane, $CHCl_3$) is a nonflammable organic liquid with low miscibility with water (approximately 7.5 g/L at 25°C). It shows carcinogenic effects in animal studies and should be avoided when another solvent would do as well. Vapors should not be inhaled, and contact with the skin should be avoided.

In a **purge-and-trap** procedure, volatile analytes can be purged from a liquid sample by helium sparging, vigorous bubbling of helium through the sample. The helium/analyte mixture is guided through an adsorption cartridge in which the analytes are "trapped" on the adsorbent in a manner similar to solid phase extraction. The analytes are then desorbed by heating the cartridge. The gases that come off the cartridge are typically analyzed by gas chromatography (Chapter 11).

2.9.2 Dilution, Concentration, and Solvent Exchange

The analyte in a sample may be too concentrated or too dilute for the chosen method. If it is too concentrated, a dilution with a compatible solvent may be performed. The dilution should be performed with volumetric glassware (Chapter 4) and with good analytical technique so that the dilution factor is known and so that accuracy is not diminished.

If the analyte is too dilute for the chosen method, its concentration may be increased in any number of ways. One is a controlled evaporation of the solvent (such that the factor by which its concentration is increased is known). Another is to perform an extraction that results in a smaller solution volume for the same quantity of analyte. Another is to evaporate the analyte solution to dryness and then reconstitute (i.e., redissolve) with a smaller volume of solvent.

The reconstitution can occur with a different solvent, a process known as **solvent exchange**. A different solvent may be more compatible with the chosen method.

2.9.3 Sample Stability

In the event that the sample cannot be carried forward through the analytical method procedure because the analyte is not stable (e.g., it might be thermally unstable and would decompose in the higher temperature used in the method procedure), it may be important to protect the sample from decomposition in some way or to derivatize the analyte. Protection from decomposition might mean storing the sample (e.g., a biological sample) in a refrigerator, protecting it from light, or protecting it from exposure to air or humidity (i.e., store in a desiccator), etc. Derivatizing the analyte means to chemically convert it to a form (a derivative) that is stable so that the quantity of the analyte can be determined indirectly by analyzing for the derivative.

2.10 LIQUID–LIQUID EXTRACTION

2.10.1 Introduction

One popular method of separating an analyte species from a complicated liquid sample is the technique known as **liquid–liquid extraction** or **solvent extraction**, first mentioned in Section 2.9.1. In this method, the sample containing the analyte is a liquid solution, typically a water solution, that also contains other solutes. The need for the separation usually arises from the fact that the other solutes, or perhaps the original solvent, interfere in some way with the analysis technique chosen. An example is a water sample that is being analyzed for a pesticide residue. The water may not be a desirable solvent and there may be other solutes that may interfere. It is a "selective dissolution"

**APPLICATION NOTE 2.2: EXTRACTING ACETAMINOPHEN FROM
HUMAN PLASMA**

The analysis of human plasma for acetaminophen, the active ingredient in some pain reliev-
ers, involves a unique extraction procedure. Small-volume samples (approximately 200 µL)
of heparinized plasma, which is plasma that is treated with heparin, a natural anticoagulant
found in biological tissue, are first placed in centrifuge tubes and treated with 1 N HCl to
adjust the pH. Ethyl acetate is then added to extract the acetaminophen from the samples.
The tubes are vortexed, and after being allowed to separate, the ethyl acetate layer containing the acetaminophen is decanted. The resulting solutions are evaporated to dryness and
then reconstituted with an 18% methanol solution, which is the final sample preparation step
before testing with an HPLC instrument (Chapter 12). The procedure is a challenge because
the initial sample size is so small.

method—a method in which the analyte is removed from the original solvent and subsequently
dissolved in a different solvent (extracted) while most of the remainder of the sample remains unextracted, i.e., remains behind in the original solution.

The technique obviously involves two liquid phases—one the original solution and the other the
extracting solvent. The important criteria for a successful separation of the analyte are (1) that these
two liquids be immiscible and (2) that the analyte be more soluble in the extracting solvent than the
original solvent (see Application Note 2.2).

2.10.2 THE SEPARATORY FUNNEL

As mentioned in Section 2.9.1, the extraction takes place in a specialized piece of glassware known
as a separatory funnel. The **separatory funnel** is manufactured especially for solvent extraction. It
has a "teardrop" shape with a stopper at the top and a stopcock at the bottom (see again Figure 2.5).
The sample and solvent are placed together in the funnel, the funnel is tightly stoppered and, while
holding the stopper in with the index finger, shaken vigorously for a moment.

Following this, the funnel may need to be vented, since one of the liquids is likely to be a volatile organic solvent, such as methylene chloride. Venting is accomplished by opening the stopcock
while the funnel is inverted. This shaking/venting step is then repeated several times such that
the two liquids have plenty of opportunity for the intimate contact required for the analyte to pass
into the extracting solvent to the maximum possible extent (see Figure 2.6a for a shaking/venting
illustration).

Following this procedure, the funnel is positioned in a padded ring in a ring stand and left undis-
turbed for a period of time to allow the two immiscible layers to once again separate. The purpose
of the specific design of the separatory funnel is mostly to provide for easy separation of the two
immiscible liquid layers after the extraction takes place. All one needs to do is remove the stopper, open the stopcock, allow the bottom layer to drain, then close the stopcock when the interface
between the two layers disappears from sight in the stopcock (see Figure 2.6b). The denser of the
two liquids is the bottom layer and will be drained through the stopcock first. The entire process
may need to be repeated several times, since the extraction is likely not to be quantitative. This
means that another quantity of fresh extracting solvent may need to be introduced into the separatory funnel with the sample and the shaking procedure repeated. Even so, the experiment may never
be completely quantitative (see the next section for the theory of extraction and a more in-depth
discussion of this problem).

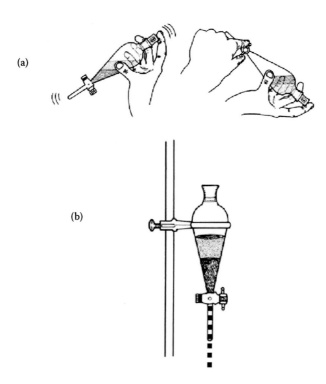

(a)

(b)

FIGURE 2.6 (a) Drawings of the shaking/venting process described in the text. (b) Drawing of the draining of the lower denser layer of liquid in the separatory funnel. The stopper is removed and the stopcock opened.

2.10.3 Theory

The process of a solute dissolved in one solvent being "pulled out," or "extracted," into a new solvent actually involves an equilibrium process. At the time of initial contact, the solute will move from the original solvent to the extracting solvent at a particular rate, but, after a time, it will begin to move back to the original solvent at a particular rate. When the two rates are equal, we have equilibrium. We can thus write the following:

$$A_{orig} \leftrightarrows A_{ext} \tag{2.1}$$

where A refers to "analyte" and orig and ext refer to "original solvent" and "extracting solvent," respectively. If the analyte is more soluble in the extracting solvent than the original solvent, then, at equilibrium, assuming equal volumes of the two solvents, a greater percentage will be found in the extracting solvent and less in the original solvent. If the analyte is more soluble in the original solvent, then the greater percentage of analyte will be found in the original solvent. Thus, the amount that gets extracted depends on the relative distribution between the two layers, which, in turn, depends on the solubilities in the two layers. A **distribution coefficient** analogous to an equilibrium constant (also called the **partition coefficient**) can be defined as follows:

$$K = \frac{[A]_{ext} \text{ (after ext.)}}{[A]_{orig} \text{ (after ext.)}} \tag{2.2}$$

where $[A]_{ext}$ (after ext.) is the concentration of the analyte in the extracting solvent after extraction and $[A]_{orig}$ (after ext.) is the concentration of the analyte in the original solvent after extraction.

The distribution coefficient is dimensionless. The concentrations in the above equation can have any units (molarity, grams/L, etc.), as long as both have the same unit so that the units cancel. Often, the value of K is approximately equal to the ratio of the solubilities of "A" in the two solvents. If the value of K is very large, the transfer of solute to the extracting solvent is considered to be quantitative. A value around 1.0 would indicate equal distribution, and a small value would indicate very little transfer.

The following expansion of Equation 2.2 can also be useful:

$$K = \frac{W_{ext}\ (\text{after ext.})/V_{ext}}{W_{orig}\ (\text{after ext.})/V_{orig}} \tag{2.3}$$

in which W_{ext} (after ext.) is the weight of the analyte found in the extracting solvent after extraction, V_{ext} is the volume of the extracting solvent used, W_{orig} (after ext.) is the weight of the analyte remaining in the original solvent after extraction, and V_{orig} is the volume of the original solvent used.

There are multiple examples of possible calculations using these relationships, and these can be concisely summarized as follows: (1) Given any two of the parameters seen in Equation 2.2, what is the value of the third? (2) Given any four of the parameters seen in Equation 2.3, what is the fifth? One can imagine, for example, performing an experiment to determine the distribution coefficient, or a performing a quantitative analysis in which the weight of the analyte in the original solvent before extraction is to be determined. This latter parameter is the sum of the weight in the original solvent after extraction and the weight of the analyte in the extracting solvent after extraction.

$$W_{orig}\ (\text{before ext.}) = W_{orig}\ (\text{after ext.}) + W_{ext}\ (\text{after ext.}) \tag{2.4}$$

where W_{orig} (before ext.) is the weight of the analyte in the original solvent before extraction, the total amount of analyte involved. One can also imagine wanting to determine the percent of analyte extracted. This would be defined as follows:

$$\%\ \text{extracted} = \frac{W_{ext}\,(\text{after ext.})}{W_{orig}\ (\text{before ext.})} \times 100 \tag{2.5}$$

2.10.4   CALCULATIONS INVOLVING EQUATION 2.2

In this case, if K is to be calculated and the two concentrations are given, it is a matter of dividing $[A]_{ext}$ by $[A]_{orig}$. If K is given and one of the two concentrations is to be calculated, we simply solve the equation for the unknown and substitute the values, as seen in Example 2.1 below.

Example 2.1

The distribution coefficient for a given extraction experiment is 98.0. After extraction, if the concentration in the extracting solvent is 0.0127 M, what is the concentration in the original solvent?

Solution 2.1

$$K = \frac{[A]_{ext}\,(\text{after ext.})}{[A]_{orig}\ (\text{after ext.})}$$

$$98 = \frac{0.0127}{[A]_{orig} \ (after \ ext.)}$$

$$98.0 \times [A]_{orig} \ (after \ ext.) = 0.0127$$

$$[A]_{orig} \ (after \ ext.) = \frac{0.0127}{98}$$

$$[A]_{orig} = 0.000130 \ M$$

2.10.5   CALCULATIONS INVOLVING EQUATION 2.3

If K is to be calculated, the calculation would involve substituting the known values into Equation 2.3. However, if one of the other parameters is to be calculated, then a more complex "solving for x" step would be involved. For this, it may be helpful to use the general Equation 2.6 below.

$$\frac{a/b}{c/d} = \frac{a \times d}{b \times c} \tag{2.6}$$

Equation 2.3 then becomes the following:

$$K = \frac{W_{ext} \ (after \ ext.) \times V_{orig}}{W_{orig} \ (after \ ext.) \times V_{ext}} \tag{2.7}$$

Notice that when the volume of original solvent and the volume of the extracting solvent are the same, these two volumes cancel out and their values need not be known.

Example 2.2

What weight of analyte is found in 50.0 mL of an extracting solvent when the distribution coefficient is 231, and the weight of analyte found in 75.0 mL of the original solvent after extraction was 0.00723 g?

Solution 2.2

Substituting into Equation 2.7, we have

$$231 = \frac{W_{ext} \ (after \ ext.) \times 75.0}{0.00723 \times 50.0}$$

$$W_{ext} \ (after \ ext.) \times 75.0 = 231 \times 0.00723 \times 50.0$$

$$W_{ext} \ (after \ ext.) = \frac{231 \times 0.00723 \times 50.0}{75.0}$$

$$= 1.11 \, g$$

2.10.6 CALCULATIONS INVOLVING A COMBINATION OF EQUATIONS 2.3 (OR 2.7) AND 2.4

A combination of Equations 2.3 and 2.4 would be necessary when the total weight of analyte either needs to be calculated or is given and one of either W_{ext} (after ext.) or W_{orig} (after ext.) needs to be calculated (see Example 2.3 below).

Example 2.3

If the distribution coefficient is 12.9 and the weight of analyte found in 20.0 mL of the extraction solvent after extraction is 0.0016 g, how many grams of analyte are in the original solvent before and after extraction when 15.0 mL of the original solvent are used?

Solution 2.3

Once again, utilizing Equation 2.7, we have the following:

$$12.9 = \frac{0.0016 \times 15.0}{W_{orig} \text{ (after ext.)} \times 20.0}$$

$$0.0016 \times 15.0 = 12.9 \times W_{orig} \text{ (after ext.)} \times 20.0$$

$$W_{orig} \text{ (after ext.)} = \frac{0.0016 \times 15.0}{12.9 \times 20.0} = 0.000093023 \text{ g}$$

$$= 0.000093 \text{ g}$$

Then, utilizing Equation 2.4, we have the following:

$$W_{orig} \text{ (before ext.)} = 0.0016 + 0.000093023$$

$$= 0.0017 \text{ g}$$

2.10.7 CALCULATION OF PERCENT EXTRACTED (EQUATION 2.5)

The fraction extracted, also known as the **extraction efficiency**, is a measure of the success of an extraction. As stated in Equation 2.5, it is defined as the weight of the analyte found in the extracting solvent after extraction divided by the total weight in the original solvent before extraction.

Example 2.4

What is the percentage extracted in Example 2.3?

Solution 2.4

Substituting into Equation 2.5, we have the following:

$$\% \text{ extracted} = \frac{0.0016}{(0.0016 + 0.000093023)} \times 100$$

$$= 94\%$$

It may be possible to evaluate the percent extracted by a separate experiment. A solution of the analyte in the original solvent may be prepared such that W_{orig} (before extraction) is known. Following this, an extraction is performed on this solution using a particular volume of extracting solvent (V_{ext}). This volume of extract is then analyzed quantitatively for the analyte by some appropriate analysis technique. Knowing the concentration of the analyte, then, and the volume of extract converted to liters (L_{ext}), one can calculate the percent extracted.

$$\% \text{ extracted} = \frac{[A]_{ext} \text{ (after ext.)} \times L_{ext}}{W_{orig} \text{(before ext.)}} \qquad (2.8)$$

Experiment 20 in Chapter 8 is such an experiment.

2.10.8   EVAPORATORS

Following an extraction procedure, the analyte concentration may not be high enough to be detected with the method chosen, e.g., gas chromatography (Chapter 11). In that case, it is necessary to evaporate some solvent so as to concentrate the extracted analyte in a small volume of solution, thus dramatically increasing its concentration. It is not uncommon for several hundred milliliters of solution to be reduced to just a few milliliters in this process. The modern procedure is to use a commercial evaporator that gently heats the solution while blowing a stream of nitrogen gas toward the surface. If the solvent is volatile, the required evaporation takes place in a rather short period of time and the solution is transferred to a small vial for the analysis.

2.11   SOLID–LIQUID EXTRACTION

There are instances in which the analyte needs to be extracted from a solid material sample rather than a liquid (see also Section 2.5.3). As in the above discussion for liquid samples, such an experiment is performed either because it is not possible or necessary to dissolve the entire sample or because it is undesirable to do so because of interferences that may also be present. In these cases, the weighed solid sample, preferably finely divided, is brought into contact with the extracting liquid in an appropriate container (not a separatory funnel) and usually shaken or stirred for a period of time, sometimes at an elevated temperature, such that the analyte species is removed from the sample and dissolved in the liquid. The time required for this shaking is determined by the rate of the dissolving. A separatory funnel is not used since two liquid phases are not present, but rather a liquid and a solid phase. A simple beaker, flask, or test tube usually suffices.

Following the extraction, the undissolved solid material is then filtered out and the filtrate analyzed. Examples of this would be soil samples to be analyzed for metals, such as potassium or iron, and polymer film or insulation samples to be analyzed for formaldehyde residue. The extracting liquid may or may not be aqueous. Soil samples being analyzed for metals, for example, utilize aqueous solutions of appropriate inorganic compounds, sometimes acids, while soil samples, or the polymer films or insulation samples referred to above, that are being analyzed for organic compounds utilize organic solvents for the extraction. As with the liquid–liquid examples, the extract is then analyzed by whatever analytical technique is appropriate—atomic absorption for metals and spectrophotometry or gas or liquid chromatography for organics. Methods for increasing the concentration of the analyte in the extract may also be required, especially for organics.

It may be desirable to try to keep the sample exposed to *fresh* extracting solvent as much as possible during the extraction in order to maximize the transfer to the liquid phase. This may be accomplished by pouring off the filtrate and reintroducing fresh solvent periodically during the extraction and then combining the solvent extracts at the end. There is a special technique and apparatus, however, that has been developed, called the **Soxhlet extraction**, which accomplishes

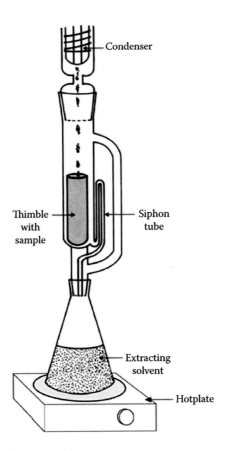

Condenser

Thimble
with
sample

Siphon
tube

Extracting
solvent

Hotplate

FIGURE 2.7   Drawing of a Soxhlet extraction apparatus.

this automatically (previously described in Section 2.5.3). The Soxhlet apparatus is shown in Figure 2.7. The extracting solvent is placed in the flask at the bottom, whereas the weighed solid sample is placed in the solvent-permeable thimble in the compartment directly above the flask. A condenser is situated directly above the thimble. The thimble compartment is a sort of cup that fills with solvent when the solvent in the flask is boiled, evaporated, and condensed on the condenser. The sample is thus exposed to freshly distilled solvent as the cup fills. When the cup is full, the glass tube next to the cup is also full, and when it (the tube) begins to overflow, the entire contents of the cup are siphoned back to the lower chamber and the process repeated. The advantages of such an apparatus are (1) fresh solvent is continuously in contact with the sample (without having to introduce more solvent, which would dilute the extract) and (2) the experiment takes place unattended and can conveniently occur overnight if desired.

2.12   DISTILLATION OF A MIXTURE OF LIQUIDS

We previously discussed the distillation of water being useful to purify water so as to avoid interference from the dissolved minerals in tap water (see Section 2.6.1). Organic liquids that are contaminated with other organic liquids usually constitute a much more difficult situation. Such liquids probably have such similar boiling points and vapor pressures that a distillation of a mixture of two or more would result in all being present in the distillate (the condensed vapors)—an unsuccessful separation. However, the liquid that has the highest vapor pressure and/or lowest boiling point, while not being completely separated, would be present in the initial distillate at a higher concentration level than before the distillation. It follows that if this distillate were then to be redistilled, perhaps over and over again, further enrichment of this component would take place such that an acceptable separation would eventually take place. However, the time involved in such a procedure would be prohibitive. A procedure known as fractional distillation solves the problem.

Fractional distillation involves repeated evaporation/condensation steps before the distillate is actually collected. These repeated steps occur in a "fractionating column" (tube) above the original heated container—a column that contains a high surface area of inert material for condensing the vapors. As the vapors condense on this material, the material itself heats up and the condensate re-evaporates. The re-evaporated liquid then moves further up the column, contacts more cold inert material and the process occurs again—and again and again as the liquid makes its way up the column. If a fractionating column were used that is long enough and contains a sufficient quantity of the high surface area material, any purification based on differences in boiling point and vapor pressure can be effected. An illustration of a distillation apparatus fitted with a fractionating column is shown in Figure 2.8. The high surface area packing material in a fractionating column typically consists of glass beads, glass helices, or glass wool.

Each time a single evaporation/condensation step occurs in a fractionating column, the condensate has passed through what has been called a theoretical plate. A **theoretical plate** is thus that segment of fractionating column in which one evaporation/condensation step occurs. The name is derived from the image that the condensate is captured on small "plates" inside the fractionating column from which it is again boiled and evaporated. A fractionating column used for a given liquid mixture is then identified as having a certain number of theoretical plates and given liquid mixtures are known to require a certain number of theoretical plates in order to achieve a given purity. The **height equivalent to a theoretical plate**, or HETP, is the length of fractionating column corresponding to one theoretical plate. If the number of theoretical plates required is known, then the analyst can select a height of column that would contain the proper number of plates according to manufacturer's specifications, or according to his own measurements of a homemade column. Height selection is not entirely experimental, however. The use of liquid–vapor composition diagrams to predict the theoretical plates required can help. These diagrams are based on boiling point and vapor pressure differences in a pair of liquids. Further discussion of the use of these diagrams is beyond the scope of this book.

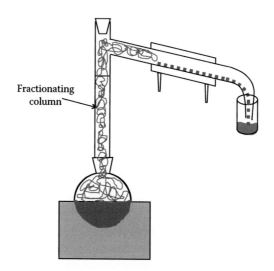

FIGURE 2.8 Drawing of a distillation apparatus with the fractionating column in place for the separation of two liquids. Compare to Figure 2.3.

The concept of theoretical plates is applicable to instrumental chromatography as well. This will be discussed in Chapter 10.

2.13 REAGENTS USED IN SAMPLE PREPARATION

A **reagent** is a substance used in a chemical reaction in an analytical laboratory because of its specific applicability to a given system or procedure. Reagents used in sample preparation must be wholly applicable to the preparation undertaken in terms of identity and purity. In other words, if a given procedure calls for certified ACS grade methyl alcohol, for example, then certified ACS grade methyl alcohol should be used and not methyl alcohol of a lesser grade (lower purity). In addition, this purity must not have diminished since the reagent was purchased. If a given item is designated with a particular **shelf life**, then the analyst must be aware of the date beyond which the composition of a reagent can no longer be trusted.

There are a number of different purity grades of chemicals that are available. Some of the more common are listed in Box 2.2.

2.14 LABELING AND RECORD KEEPING

Samples gathered and solutions prepared by laboratory personnel must be properly labeled at the time of sampling or preparation. In addition, a complete record of the sampling or preparation should be maintained. Sound quality assurance practices include a notebook record where one can find the source and concentration of the material used, the identity and concentration of the standard being prepared, the name of the analyst who prepared it, the specific procedure used, the date it was sampled or prepared, and the expiration date for any stored solutions. The reagent label should have a clear connection to the notebook record. A good label includes an ID number that would match the notebook record, the name of the material and its concentration, the date, the name of the analyst, and the expiration date.

BOX 2.2   EXAMPLES OF PURITY GRADES OF CHEMICALS

Primary standard: A specially manufactured analytical reagent of exceptional purity for standardizing solutions and preparing reference standards.

ACS certified: A reagent that meets or exceeds the specifications of purity put forth by the American Chemical Society. The certificate of analysis is on the label.

Certified reagent: A reagent that meets the standards of purity established by the manufacturer. The certificate of analysis is on the label.

USP/NF: Reagents that meet the purity requirements of the U.S. Pharmacopeia (USP) and the National Formulary (NF). Generally of interest to the pharmaceutical profession, these specifications may not be adequate for reagent use.

Spectro grade or spectranalyzed: Solvents of suitable purity for use in spectrophotometric procedures. A certificate of analysis is on the label.

High-performance liquid chromatography (HPLC) grade: Solvents of suitable purity for use in liquid chromatography procedures.

Practical: Chemicals of sufficiently high quality to be suitable for use in some syntheses. Organic chemicals of practical grade may contain small amounts of intermediates, isomers, or homologues.

Technical: Chemicals of reasonable purity for applications that have no official standard for purity.

EXPERIMENTS

EXPERIMENT 3: A STUDY OF THE DISSOLVING PROPERTIES OF WATER, SOME COMMON ORGANIC LIQUIDS, AND LABORATORY ACIDS

INTRODUCTION

In this experiment, the miscibility of water with some common organic solvents and the dissolving power of water, hydrochloric acid, sulfuric acid, and nitric acid will be studied. First, small amounts of water will be mixed with roughly equal amounts of the organic solvents and the miscibilities observed. Then, small volumes of water and the acids will be allowed to contact granules of some selected metals and other compounds and their dissolving power noted. The objective is to confirm some of the statements made in Tables 2.1, 2.2, and Box 2.1 and also to discover the behavior of some solvents and acids not listed.

PART A: A STUDY OF THE MISCIBILITY OF WATER WITH SELECTED ORGANIC SOLVENTS

1. Obtain a test tube rack and nine small test tubes. Add distilled water to each tube such that each is about a fourth to a third full. Place a piece of labeling tape on each.
2. Obtain samples of the following nine organic liquids contained in individual small dropper bottles: n-hexane (or other alkane), acetonitrile, methylene chloride, acetone, toluene, methanol, diethyl ether, ethyl acetate, ethylbenzene, ethanol, and chloroform. Then label each of the test tubes from step 1 with the names, or an abbreviation of the names, of these liquids.
3. Add small amounts of each liquid indicated in Step 2 to the water in the test tubes with the corresponding label. Stopper, shake, and observe the miscibility. Record the miscibility ("miscible" or "immiscible") and the polarity ("polar" or "nonpolar"), in a table, such as the following, in the DATA section of your notebook.

Solvent	Miscibility	Polarity
n-Hexane		
Acetonitrile		
Methylene chloride		
Acetone		
Toluene		
Methanol		
Diethyl ether		
Chloroform		
Ethyl acetate		
Ethylbenzene		
Ethanol		

Compare the results in your table with what was stated in Box 2.1 for miscibility with water and polarity for some of the liquids.

4. Optional: Check out the liquids in the above table to see if they are miscible with each other rather than with water. You might expect those that are polar to be immiscible with those that are nonpolar. Check out, for example, the miscibility of acetone and methanol with those you identified as nonpolar. Place your observations in a table and draw some appropriate conclusions regarding polarities.
5. Dispose of the contents of each test tube according to the directions of your instructor.

PART B: A STUDY OF THE SOLUBILITY OF SOME SELECTED INORGANIC MATERIALS IN WATER, ACETIC ACID, HYDROCHLORIC ACID, SULFURIC ACID, AND NITRIC ACID

1. Obtain small samples of NaCl, $CaCO_3$, Fe_2O_3, Al_2O_3, and the metals aluminum, zinc, iron, copper, and silver. Obtain around 20 small test tubes and a test tube rack. Place a piece of labeling tape on each tube.
2. Make a table or "grid" in the DATA section of your notebook as follows:

Acid	NaCl CaCO₃ Fe₂O₃ Al₂O₃ Al Zn Fe Cu Ag
Water	
Acetic	
Hydrochloric	
Sulfuric	
Nitric	

3. Place very small granules of each of the nine materials indicated in step 1 individually into test tubes. Label each of the nine tubes with the chemical symbol/formula of the contents.
4. Half-fill each of the test tubes in step 2 with distilled water. Now also place a symbol for water (such as a "W" or "H_2O)" on each label. After 2 min, carefully observe the contents of each tube to determine whether there has been a reaction. If there has been no visible reaction, stopper and shake, then observe whether the material has dissolved. Place a "S" (for "soluble") in the grid if the material has either reacted or dissolved (or both), and an "I" (for "insoluble") if nothing happened.
5. Obtain dropper bottles containing 50% (by volume) solutions of acetic acid, hydrochloric acid, sulfuric acid, and nitric acid. Repeat Steps 2 and 3 using the acetic acid in place of the water, testing all those for which water gave an "I." Then repeat with hydrochloric acid and sulfuric acid. Repeat finally with nitric acid, but only for those that acetic, hydrochloric, and sulfuric gave an "I."
6. Compare the results with statements of uses of these acids in Tables 2.1 and 2.2.

QUESTIONS AND PROBLEMS

Introduction

1. What general tasks relating to the chemical analysis of a material sample must be performed prior to executing the chosen method of analysis?
2. Why are sampling and sample preparation procedures as crucial to the success of an analysis as the analytical method chosen?

Obtaining the Sample

3. What is a representative sample? Do not use the words "represent" or "representative" in your answer.
4. Under what circumstances would a "selective sample" be more appropriate than a "composite sample"?
5. How would you obtain a representative sample of each of the following
   (a) The water in a creek?
   (b) The polymer film manufactured by a chemical company?
   (c) The aspirin in a bottle of aspirin tablets?

(d) The soil in your yard?
(e) Old paint on a building?
(f) Animal tissue being analyzed for pesticide residue?
6. Differentiate between "bulk sample," "primary sample," "secondary sample," "subsample," "laboratory sample," and "test sample."

Statistics of Sampling

7. Why is statistics important in sampling?

Sampling Handling

8. Why should one be concerned about who has custody of a sample between the sampling site and the laboratory?
9. If a water sample is to be analyzed for trace levels of metals, why is a **glass** container for sampling and storage inappropriate?
10. If a sample to be analyzed in a laboratory is subject to alteration due to the action of bacteria, what can be done to preserve its integrity?
11. Look up in a reference book exactly what is involved in taking and preserving a sample of soil to be analyzed for nitrate.

Sample Preparation—Solid Materials

12. Why is it important to reduce the size of solid particles present in a sample?
13. Why is it important for a sample to be homogeneous before it is divided to create the test sample?
14. Define "extraction" as it is used in a chemical analysis laboratory.
15. Why is the separation of an analyte from a sample matrix called an "extraction"?
16. What two forms may a solid–liquid extraction take? Why does one require filtering, whereas the other may not?
17. What is a "supercritical fluid"? Why are supercritical fluids useful in solid–liquid extraction?

Water Purification and Use

18. What does vapor pressure have to do with purification by distillation?
19. Water has a very high vapor pressure compared with the hardness minerals that may be dissolved in it. Why is that a good thing when wanting to purify tap water by distillation?
20. Why does one simple distillation remove most of the dissolved hardness minerals from tap water while many distillations (or a fractionating column) are required to separate a mixture of two liquids?
21. How is distilled water different from deionized water?

Total Sample Dissolution and Other Considerations

22. Identify water or a specific acid as the least drastic dissolving agent for each of the following:
(a) a mixture of sodium chloride and sodium sulfate.
(b) the material from a silver mine.

    (c) a sample of stainless steel.

    (d) a sample of iron carbonate ore.

    (e) the material from a gold mine.

    (f) a sample of sand from a beach.

    (g) the oxide on the surface of an airplane part.

23. What concentrated acid is correctly described by each of the following?

    (a) A very dense, syrupy liquid.

    (b) A liquid that can be absorbed through the skin.

    (c) A liquid that turns skin yellow.

    (d) A liquid that gives off unpleasant fumes.

    (e) A liquid that is mixed with HCl to give "aqua regia".

    (f) A solvent for silver metal.

    (g) A solvent for corrosion products on an airplane part.

    (h) A solvent for sand.

    (i) A solvent for stainless steel.

    (j) A solvent for iron ore.

    (k) A solvent for leather.

    (l) A liquid that gets especially hot when mixed with water.

    (m) An acid that turns paper towels black on contact.

24. What is "aqua regia"?

25. Fill in the blanks with a specific example of the kind of sample dissolved by the acid indicated.

    (a) Hydrofluoric acid _____

    (b) Nitric acid _____

    (c) Sulfuric acid _____

    (d) Hydrochloric acid _____

    (e) Aqua regia _____

    (f) Perchloric acid _____

26. Fill in the blanks with the name of the acid described by the statement.

    (a) The bottle containing this acid has a red color code. _____

    (b) This acid, when mixed with nitric acid in certain proportions, gives aqua regia. _____

    (c) When an open bottle of this acid sits next to an open bottle of ammonium hydroxide, thick white fumes form. _____

    (d) This acid gets especially hot when mixed with water. _____

    (e) This acid diffuses through skin and is especially bad when it gets under the fingernails. _____

    (f) The bottle containing this acid has a yellow color code. _____

    (g) This acid is used to dissolve the sample for the Kjeldahl analysis. _____

27. From the following list, choose those samples that would be best dissolved by HCl and those that would be best dissolved by HNO_3. Not all samples will be used.

    (a) NaCl

    (b) Iron ore

    (c) Gold metal

    (d) Lettuce leaves

    (e) Copper metal

    (f) Sand

    (g) Aluminum oxide

Fusion

28. What is meant by "fusion" as a method for sample dissolution?
29. What is meant by the "flux" in a fusion procedure?
30. What materials must be used as containers for the flux/sample mixture in a fusion procedure? Explain.

Sample Preparation—Liquid Samples, Extracts, and Solutions of Solids

31. Differentiate between solid–liquid extraction, liquid–liquid extraction, and solid phase extraction.
32. Give two examples of when an "extraction" is useful rather than total sample dissolution.
33. Why are the solvents listed in Box 2.1 useful for extracting analytes from water samples?
34. What are the hazards associated with the use of diethyl ether and what specific precautions are taken?
35. Of the solvents listed in Box 2.1, which would be the top layer in a separatory funnel containing an aqueous solution? Explain what the solvents' density has to do with your answer.
36. List the organic solvents listed in Box 2.1 that (a) are toxic and (b) are flammable.
37. What is a "purge-and-trap" procedure?
38. What must be done to a sample if is too concentrated for the chosen method? What must be done if it is too dilute?
39. What is meant by "solvent exchange"?

Liquid–Liquid Extraction

40. Describe in words what happens in a liquid–liquid extraction experiment.
41. What are the two important criteria for a successful separation by solvent extraction?
42. Describe the glassware article called the "separatory funnel" and tell for what purpose and how it is used?
43. Iodine dissolved in water gives the water a brown color. Iodine dissolved in hexane gives the hexane a pink color. Hexane and water are immiscible. How can these facts be used to demonstrate a liquid–liquid extraction?
44. If the distribution coefficient, K, for a given solvent extraction is 16.9,
    (a) What is the molar concentration of the analyte found in the extracting solvent if the concentration in the original solvent **after** the extraction is 0.027 M?
    (b) What is the molar concentration of the analyte found in the extracting solvent if the concentration in the orginal solvent **before** the extraction was 0.045 M?
45. How many moles of analyte are extracted if 50.0 mL of extracting solvent is brought into contact with 50.0 mL of original solvent, the concentration of analyte in the original solvent is 0.060 M, and the distribution coefficient is 23.8?
46. In an extraction experiment, it is found that 0.0376 g of an analyte are extracted into 50.0 mL of solvent from 150.0 mL of a water sample. If there were originally 0.192 g of analyte in this volume of the water sample, what is the distribution coefficient? What percent of analyte was extracted?
47. The distribution coefficient for a given extraction experiment is 5.27. If 0.037 g of analyte are found in 75 mL of the extracting solvent after extraction and 100 mL of original solvent was used, how many grams of analyte were in this volume of original solvent before extraction?

48. Using the information given in each row of the table below, calculate the items marked with a question mark (?).

	K	[A]$_{orig}$ (after ext.)	[A]$_{ext}$ (after ext.)	V$_{orig}$	V$_{ext}$	[A]$_{orig}$ (before ext.)
(a)	?	0.0104 M	0.372 M	xxxxx	xxxxx	xxxxx
(b)	?	0.0183 M	0.0398 M	xxxxx	xxxxx	xxxxx
(c)	87.4	0.00367	?	xxxxx	xxxxx	xxxxx
(d)	62.6	?	0.482 M	xxxxx	xxxxx	xxxxx
(e)	?	0.0822 M	?	25.00 mL	25.00 mL	0.102 M
(f)	?	?	0.330 M	25.00 mL	25.00 mL	0.389 M
(g)	?	?	0.144 M	25.00 mL	5.00 mL	0.160 M
(h)	?	?	0.0017 M	25.00 mL	10.00 mL	0.0029 M
(i)	29.3	?	0.229 M	50.00 mL	10.00 mL	?
(j)	35.2	?	0.0386 M	25.00 mL	25.00 mL	?

49. Using the information given in each row of the table below, calculate the items marked with a question mark (?).

	Grams Extracted	Grams Not Extracted	Total Grams	K	V$_{orig}$ (mL)	V$_{ext}$ (mL)	% Extracted
(a)	0.0023	xxxxx	0.0026	xxx	xxxx	xxxx	?
(b)	0.0029	0.0035	?	xxx	xxxx	xxxx	?
(c)	0.0040	0.0011	?	?	50.0	50.0	?
(d)	0.0036	0.0011	?	?	50.0	10.0	?
(e)	0.0031	?	0.0037	?	25.0	10.0	?
(f)	0.0019	?	?	18.2	25.0	10.0	?
(g)	?	?	0.0045	?	50.0	5.00	67.2%

50. How does the distribution "coefficient" differ from the distribution "ratio"?

51. What is the purpose of an "evaporator"? Describe a modern way to evaporate excess solvent.

Solid–Liquid Extraction

52. What is "solid–liquid extraction"? Without going into great detail, tell what are two different ways to accomplish a solid–liquid extraction?

53. Is a separatory funnel appropriate for a solid–liquid extraction? Explain.

54. Give some examples of the types of samples and analytes to which solid–liquid extraction would be applicable.

Distillation of Mixtures of Liquids

55. What special problems exist when trying to separate two organic liquids by distillation?

56. What is a theoretical plate and the height equivalent to a theoretical plate (HETP) as they pertain to distillation?

Reagents Used in Sample Preparation

57. Why must the quality of all reagents used in the laboratory be assured before use?

58. Perhaps the most common grade of chemical used in the laboratory is "ACS certified." Explain what "ACS certified" refers to.

59. Can chemicals labeled as "technical" grade be used as standards in analytical procedures? Explain.
60. What designation of purity is used for solvents that are appropriate for spectrophotometric analysis? Liquid chromatography analysis?

Labeling and Record Keeping

61. How important is the label on a sample or reagent? What information should appear on the label of a sample to be analyzed? What information should appear in the notebook record?

3 Gravimetric Analysis

3.1 INTRODUCTION

We now begin to discuss the parts of the analytical strategy designated "carry out the analytical method" and "work the data" (refer again to the analytical strategy flowchart given in Figure 1.1). As stated in Chapter 1, there is a wide variety of analytical methods, but they can all be classified in one of two categories: wet chemical methods and instrumental methods. We will present a modified analytical strategy flowchart at the beginning of each discussion with the parts designated "carry out the analytical method" and "work the data" highlighted.

The first of the two wet chemical methods we will consider is the gravimetric method of analysis. **Gravimetric analysis** is characterized by the fact that the measurement of weight is the primary measurement, usually the only measurement made on the sample, its components, and/or its reaction products. Thus, weight measurements are used in the calculation of the results and are often the only measurements used in the calculation of the results. The analytical strategy flowchart specific to gravimetric analysis is given in Figure 3.1.

3.2 WEIGHT VS. MASS

The most fundamental, and possibly the most frequent, measurement made in an analysis laboratory is that of weight (or mass). While we speak of mass and weight often in the same breath, it is of some importance to recognize that they are not the same. Mass is the "quantity" or "amount" of a substance being measured. This quantity is the same no matter where the measurement is made—on the surface of the earth, in a spaceship speeding toward the moon, or on the surface of Mars. Weight is a measure of the earth's gravitational force exerted on a quantity of matter. Weight is one way to measure mass. In other words, we can measure the quantity of a substance (mass) by measuring the earth's gravitational effect on it (weight).

Since nearly 100% of all weight measurements made in any analysis laboratory are made on the surface of the earth where the gravitational effect is nearly constant, weighing has become the normal method of measuring mass. Weighing devices are calibrated in grams, which is defined as the basic unit of mass in the metric system. Technically, however, analytical quantities are mass quantities and not weight quantities. Despite this, because the term "weight" is used almost universally to describe amount of matter as a matter of common language, this is the term we will use throughout this text to describe the amount of a material.

3.3 THE BALANCE

The laboratory instrument built for measuring weight is called the **balance**. The name is derived from obsolete mechanical devices that utilized known weights to "balance" the object to be weighed across a fulcrum like a teeter-totter. Most balances in use today are "electronic" balances rather than "mechanical" balances. An **electronic balance** is one that uses an electromagnet to "balance" the object to be weighed on a single pan. The older mechanical balances also utilize a single pan concept, but these also use the teeter-totter design, so they are mechanical and not electronic.

Different examples of laboratory work require different degrees of precision, i.e., different numbers of significant figures to the right of the decimal point. Thus, there are a variety of balance designs available, and this variety reflects this requirement. Some balances are read to the nearest

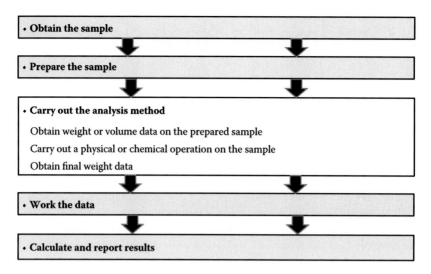

FIGURE 3.1   Steps characteristic of gravimetric analysis methodology.

gram; some are read to the nearest tenth or hundredth of a gram; some are read to the nearest milligram; and still others are read to the nearest tenth and hundredth of a milligram (see Figure 3.2).

Balances with few such significant figures to the right of the decimal point (zero to three) are often referred to as "ordinary" balances or "top-loading" balances (precision is ±100 mg to ±1 mg). A **top-loading balance** is an electronic ordinary balance with a pan on the top as shown in Figure 3.3a. The electronic top loaders often have a "tare" feature. A chemical can be conveniently weighed on a piece of weighing paper, for example, without having to determine the weight of the paper. **Taring** means that the balance is simply zeroed with the weighing paper on the pan.

A balance that is used to obtain four or five digits to the right of the decimal point in the analytical laboratory is called an **analytical balance** (precision is ±0.1 mg or ±0.01 mg). The modern laboratory utilizes single pan electronic analytical balances almost exclusively for this kind of precision. Figure 3.3b shows a typical modern electronic analytical balance. Notice that it is a single pan balance with the pan enclosed. The chamber housing the pan has transparent walls for easy viewing. Sliding doors on the right and left sides and on the top make the pan accessible for loading and handling samples. The reason for enclosing the pan in this manner is to avoid the effect of air currents on the weight. Most modern analytical balances also have the tare feature mentioned above.

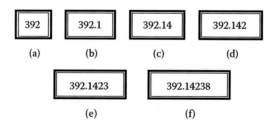

FIGURE 3.2   Depiction of the displays of balances read (a) to the nearest gram, (b) to the nearest tenth of a gram, (c) to the nearest hundredth of a gram, (d) to the nearest milligram, (e) to the nearest tenth of a milligram, and (f) to the nearest hundredth of a milligram.

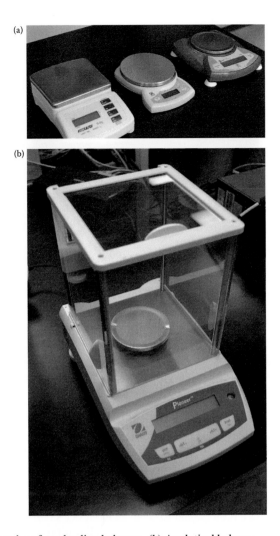

FIGURE 3.3   (a) Three styles of top-loading balances. (b) Analytical balance.

The operator of an analytical balance should keep in mind that this instrument is extremely sensitive and, therefore, must be handled carefully and correctly. Also, this discussion presumes that, if necessary, the sample has been dried (such as in a laboratory drying oven) and has been kept dry (such as by storing in a desiccator; Section 3.4) prior to weighing.

3.4   THE DESICCATOR

A desiccator is a storage container used either to dry samples, or, more commonly, to keep samples and other materials dry and protected from the laboratory environment once they have been dried by other means. A typical laboratory desiccator is shown in Figure 3.4. A quantity of water-absorbing material, called the desiccant, is placed in the bottom of the container. The desiccant will absorb all the moisture inside the sealed vessel, thus providing a dry environment. A good commercial desiccant is called Drierite™ (anhydrous $CaSO_4$). This substance can be purchased in the "indicating" form, in which case a color change (from blue to pink) will be observed when the material is saturated with moisture and can no longer function. When a desiccant is saturated, it must be replaced. Drierite can be recharged by heating in a vacuum oven.

FIGURE 3.4 Typical laboratory desiccator. It is made of heavy glass and comes in a variety of sizes. The purpose of the stopcock protruding from the lid is to connect to a vacuum pump.

3.5 CALIBRATION AND CARE OF BALANCES

It is important to keep the balances, especially analytical balances, and the immediate area clean. Spilled chemicals can react over time with the various finished surfaces on the balances and may inhibit their sensitive and accurate operation at some point. In addition, spills and debris can contaminate articles to be weighed if they are allowed to remain in the enclosed space or the immediate area. It is a good idea to hire a professional to clean a laboratory's balances on a regular basis.

Balances should also be calibrated on a regular basis. **Calibration** refers to the process by which a balance is checked to see if the weight obtained for an object is the correct weight. To calibrate a balance, a known weight is used to see if that weight is the weight displayed by the balance within an acceptable error range for that balance. If it is, the balance is said to be calibrated for that weight. If it is not, it is said to be "out of calibration." Such known weights are available from the federal agency charged with providing standard reference materials, the National Institute of Standards and Technology (NIST), and must be stored in a dry, protected environment in the laboratory. The NIST can provide an entire set of weights so that the calibration of a balance can be checked over a wide range of weights. If it is determined that a balance is out of calibration, it must be taken out of service and repaired or replaced.

3.6 WHEN TO USE WHICH BALANCE

The question of which balance to use under given circumstances, an ordinary balance or an analytical balance, remains. An analyst must be able to recognize when a weight measurement with four or five decimal places is needed (analytical balance) or when one with only one to three decimal places is needed (ordinary balance).

First, an analytical balance should not be used if it is not necessary to have the number of significant figures that this balance provides. It does not make sense to use a highly sensitive instrument to measure a weight that does not require such sensitivity. This includes a weight measurement when the overall objective is strictly qualitative or when quantitative results are to be reported to two significant figures or less. Whether the weight measurements are for preparing solutions to be used in such a procedure, or for obtaining an appropriate weight of sample for such an analysis, etc., they need not be made on an analytical balance since the outcome will be either only qualitative or does not necessarily require more than one or two significant figures.

Second, if the results of a quantitative analysis are to be reported to three or more significant figures, then weight measurements that enter directly into the calculation of the results should be made on an analytical balance so that the results of the analysis can be correctly reported after the calculation is performed.

Third, if a weight is only incidental to the overall result of an accurate quantitative analysis, then it need not be made on an analytical balance. This means that if the weight measurement to be performed has no bearing whatsoever on the quantity of the analyte tested, but is only needed to support the chemistry or other factor of the experiment, then it need not be measured on an analytical balance.

Fourth, if a weight measurement does directly affect the numerical result of an accurate quantitative analysis (in a way other than entering directly into the calculation of the results, which is the second case mentioned above), then it must be performed on an analytical balance.

These last two points require that the analyst carefully consider the purpose of the weight measurement and whether it will directly affect the quantity of analyte tested or the numerical value to be reported. While it is often required that weight measurements be made on an analytical balance so that the result can have the appropriate degree of precision, it is also sometimes true that weight measurements taken during such a procedure need not be made with an analytical balance. It depends on whether the degree of precision desired in the results will be satisfactorily maintained by a weight that has fewer significant figures than an analytical balance provides.

3.7 DETAILS OF GRAVIMETRIC METHODS

Gravimetric analysis methods proceed with the following steps: (1) the weight or volume of the prepared sample is obtained; (2) the analyte is either physically separated from the sample matrix or chemically altered and its derivative separated from the sample matrix; and (3) the weight of the separated analyte or its derivative is obtained. The data thus obtained are then used to calculate the desired results.

3.7.1 PHYSICAL SEPARATION METHODS AND CALCULATIONS

The method by which an analyte is physically separated from the matrix varies depending on the nature of the sample and what form the analyte is in relative to its matrix. For example, the analyte or its matrix may be sufficiently volatile so that one or the other can be separated by evaporation at a temperature attainable by laboratory ovens or burners. In that case, the analyte weight is measured either by sample weight loss if the analyte has been evaporated or directly if the matrix has been evaporated. In either case, the weight of a container may also be involved.

As another example, consider that the analyte may be able to be filtered or sieved (separated according to particle size) from the matrix. If the sample is a liquid suspension, its volume is first measured. This is followed by a filtering or wet sieving step to capture the desired solid material. The filtering or sieving medium or unit is weighed first and thus the weight of the analyte is determined from the weight gain that occurs as a result of the filtering or sieving. If the sample is a solid, its weight is measured and the sample either dry-sieved (through a stack of "nested" metal sieves with different sized holes) or dissolved such that insoluble matter is filtered. In the case of sieving, the weight gain of each sieve is the analyte weight for that particle size. In the case of insoluble matter, the weight gain of the filter is the analyte weight.

In each of these cases, the percentage of the analyte is often calculated. The weight percent of an analyte in a sample is calculated using the definition of weight percent:

$$\text{Weight } \% = \frac{\text{Weight of part}}{\text{Weight of whole}} = \times 100 \tag{3.1}$$

or

$$\% \text{ analyte} = \frac{\text{Weight of analyte}}{\text{Weight of sample (S)}} = \times 100 \tag{3.2}$$

Table 3.1 gives examples of gravimetric analysis when the analyte is physically separated from the matrix. Some additional points about these are given below (see also Application Note 3.1).

TABLE 3.1

Summary of the Gravimetric Methods That Involve a Physical Separation of the Analyte as Discussed in the Text

Named Method	Analyte(s)	Calculation
Loss on drying	Material evaporated from a sample at oven-drying temperatures	Percent Numerator: weight loss upon heating Denominator: weight of sample before heating
Loss on ignition	Material evaporated from a sample at ignition temperatures	Percent Numerator: weight loss upon ignition Denominator: weight of sample before ignition
Residue on ignition	Material remaining after applying ignition temperatures to samples	Percent Numerator: weight of sample (residue) after ignition Denominator: weight of sample before ignition
Insoluble matter in reagents	Material captured on filter after solubilizing a solid sample	Percent Numerator: weight of solid captured on the filter after parent solid is solubilized Denominator: weight of solid material before solubilizing
Solids in water and wastewater	Solids that are (1) captured on a filter (**suspended solids**); (2) remaining after evaporating the water (**total solids**); (3) remaining after both filtering out the suspended solids and evaporating the water (**total dissolved solids**); (4) remaining after applying ignition temperatures to dried samples (**total volatile solids** and **volatile suspended solids**); (5) allowed to settle out after a period of time (**settleable** solids), etc.	Milligrams per liter Numerator: milligrams of solid Denominator: liters of water or wastewater sample used
Particle size analysis	Solids of a particular particle size that are captured on a sieve that has the appropriate mesh size for allowing smaller particle sizes to pass through	Percent Numerator: weight of particles captured on sieve Denominator: weight of sample used

**APPLICATION NOTE 3.1: THE DETERMINATION OF
MOISTURE, TOTAL SOLIDS, AND ASH IN FOOD**

The moisture, total solids, and ash in foods can be determined by exactly those methods discussed in this chapter. Dry (e.g., cereals, uncooked pasta) and moist foods (e.g., meats, vegetables) are first prepared in a grinder and mixed. Wet foods (e.g., sauces, canned products) are prepared in a high-speed blender. Although there are some additional variations depending on the type of food, a small quantity of the food thus prepared is weighed into a preheated and preweighed crucible. The crucible is then placed in a drying oven set at 110°C and heated to constant weight. The weight loss is considered to be the moisture and the residue the total solids. The percentage of each is calculated.

The procedure for ash (residue on ignition) is also as discussed in this chapter. Again, with some slight variations depending on the type of food, the sample is prepared by grinding or blending as described above and then weighed into a preignited and preweighed crucible. A Bunsen burner can be used to char the sample, but the ignition takes place in a muffle furnace set at 550°C. The percent of ash is identical to the percent residue on ignition calculation in Table 3.1.

(*Reference:* James, C., *Analytical Chemistry of Foods*, Chapman & Hall (1995), Boca Raton, FL, pp. 72–75.)

3.7.1.1 Loss on Drying

This is an example of a loss through volatilization (evaporation) under temperature conditions at which water would volatilize, hence the word "drying." The loss can occur by elevating the temperature of the sample to just above the boiling point of water (although a different temperature may be specified), or through desiccation. This is not necessarily limited to water, however, as any sample component that volatilizes in the case of elevated temperature, for example, would be included in the weight loss. Thus, it is called "loss on drying" rather than the "determination of water or moisture by drying." If it is known that only water is lost in the drying step, then the latter would accurately describe it. For example, only water would be lost if the loss occurs through desiccation (placing the sample in a desiccator) and not by heating. In any case, the sample is usually contained in either a predried weighing bottle (see Section 3.8.1) or an evaporating dish and placed in an oven or desiccator for a specified period or to a constant weight. **Constant weight** refers to repeated drying steps until two consecutive weights agree to within a specified precision, such as having two consecutive weights that do not differ by more than 0.25%. The results are calculated as either the percent loss in weight or the percent moisture or water.

Loss on drying is a routine analytical procedure in the pharmaceutical, food, and agricultural industries. Special loss-on-drying equipment is often used (see Figure 3.5).

3.7.1.2 Loss on Ignition

This is similar to loss on drying, except that an extremely high temperature, such as that of a Meker burner or a muffle furnace, is used such that changes other than just moisture evaporation occur, a process known as **ignition**. The changes can still be just the simple evaporation of water and other components that would volatilize at the temperature used, although some procedures require that the sample be dried first. The changes may also include chemical changes, such as carbon dioxide evolution from carbonate materials, such as in cements or limestone, as in the following:

$$CaCO_3 + heat \rightarrow CO_2 \,(\uparrow) + CaO$$

FIGURE 3.5 Automated loss-on-drying unit that includes a heating element, a built-in balance, and a computerized results calculator. When the lid is closed, the heating element in the lid heats the sample such that the moisture and other volatile materials are lost. When constant weight is achieved, the percent of loss on drying is calculated and displayed.

In any case, the sample is usually contained in a crucible (prepared in advance by ignition without a sample) and placed in the oven or over the Meker burner for a specified period of time or to a constant weight, as in the loss on drying.

Applications of loss on ignition are found in pharmaceutical, agricultural, cement, and rock and mineral industries.

3.7.1.3 Residue on Ignition

The mineral particles and ash that remain after the ignition of a sample as described above may be important to calculate and report. The mineral and ash are known as the "residue." In pharmaceutical applications, the sample is pretreated with sulfuric acid prior to ignition in order to consume the carbon in the sample; these are sometimes referred to as sulfated ash residues. The process usually must be repeated to obtain a constant weight.

3.7.1.4 Insoluble Matter in Reagents

As part of a general analysis report, the labels on certified chemicals that are soluble in water indicate the percent of insoluble matter present (see example in Figure 3.6). This insoluble matter is determined by preparing a solution of the chemical, which sometimes involves heating to boiling in a covered beaker and maintaining a higher temperature for a specified time to effect dissolution. A certain weight of the chemical is used, and this is the sample weight for the calculation. The solution is filtered through a tared or preweighed, fine-porosity sintered glass crucible. After rinsing thoroughly with water this crucible is then dried and weighed again.

3.7.1.5 Solids in Water and Wastewater

When analyzing liquid samples for solids, such as water and wastewater samples, the analysis is based on a volume of the sample rather than a weight. Thus, although this is considered an example of a gravimetric analysis because the weight of the solids is determined, a volume of the sample is measured rather than a weight. There are several categories of solids that may be determined. Refer again to Table 3.1 (see Figure 3.7).

MEETS A. C. S. SPECIFICATIONS
Maximum Limits of Impurities
Ammonium hydroxide Ppt. 0.010%
Arsenic (As) 0.0001%
Calcium and
    magnesium Ppt. 0.010%
Chloride (Cl) 0.002%
Heavy metals (as Pb) 0.0005%
Insoluble matter 0.010%
Iron (Fe) 0.0005%
Lost on heating at 285°C. 1.0%
Nitrogen compounds
    (as N) 0.001%
Phosphate (PO_4) 0.001%
Potassium (K) 0.005%
Silica (SiO_2) 0.005%

FIGURE 3.6 Partial view of a general analysis report on a bottle of a certified chemical showing "insoluble matter" as 0.010%. Notice also the "loss on heating" report.

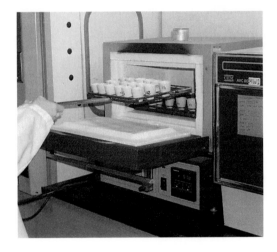

FIGURE 3.7 Muffle furnace in use for a volatile suspended solids analysis.

3.7.1.6 Particle Size by Analytical Sieving

Solid materials that have been crushed or that otherwise exist in variable particle sizes can be analyzed gravimetrically to determine the percentages of the various particle sizes, i.e., the particle size distribution. For this to occur, the various sized particles must be separated from each other so that their individual weights can be measured. This is accomplished by sieving. A sieve is a brass or stainless steel cylinder approximately 2 in. tall and 5–12 in. in diameter, open on one end and having a plate with holes or wire mesh on the other end (see Figure 3.8). Often these sieves are "nested" or stacked as in Figure 3.8, in order to analyze for a number of different particle sizes at once or to obtain a particle size distribution profile. The individual sieves in the nest are weighed in advance and then are stacked in the order of descending aperture size from top to bottom. The sample is introduced to the top sieve. The nested sieves are then shaken using a sieve shaker so that the particles fall through the stack until they reach a sieve through which they cannot pass due to the small size of the apertures in the sieve. Thus, particles of a particular range of sizes are found on

FIGURE 3.8 Nested sieves in position on a sieve shaker.

a particular sieve. The final weight of each sieve is then determined and the percentage calculated. Often, the percentages are expressed as a cumulative percentage (see Table 3.1).

Analytical sieving may also be done with the solids in a liquid suspension. In this case, it is called "wet sieving" and the suspension is transferred to the nested sieves and rinsed through with solvent. The suspended particles are captured on the sieves in the same manner as described above.

Particle size analysis by analytical sieving is important in the agricultural industry for soil analysis and in the pharmaceutical industry for various raw materials (such as sugar spheres) and liquid suspensions.

3.7.2 CHEMICAL ALTERATION/SEPARATION OF THE ANALYTE

The physical separation of an analyte from a sample so that its weight can be measured, which the above examples typify, can be difficult, if not impossible, to accomplish. It is important to recognize that in many cases there are no physical means by which such a separation can take place. For example, if you wished to determine the amount of sodium sulfate present in a mixture with sodium chloride, it would not be possible to separate it from the sodium chloride by physical means so as to determine its weight.

However, a gravimetric procedure might still be used if a chemical reaction is employed to convert the analyte to another chemical form that is both able to be separated cleanly and able to be weighed accurately. In the example of determining sodium sulfate in the presence of sodium chloride, one can dissolve the mixture in water and precipitate the sulfate with barium chloride to form barium sulfate. Sodium chloride would not react.

$$Na_2SO_4 + BaCl_2 \rightarrow BaSO_4\ (\downarrow) + NaCl$$

$$NaCl + BaCl_2 \rightarrow no\ reaction$$

The weight of the precipitate after filtering and drying can then be measured free of any influence from the NaCl and converted back to the weight of the analyte with the use of a gravimetric factor (see the next section) and its percent in the sample calculated. Examples will be given in Section 3.7.4.

3.7.3 GRAVIMETRIC FACTORS

A **gravimetric factor** is a number used to convert, by multiplication, the weight of one chemical to the weight of another. Such a conversion can be very useful in an analytical laboratory. For example, if a recipe for a solution of iron calls for 55 g of $FeCl_3$ but a technician finds only iron wire on the chemical shelf, he/she would want to know how much iron metal is equivalent to the 55 g of $FeCl_3$ so that he/she could prepare the solution with the iron wire instead and have the same weight of iron in either case. In one formula unit of $FeCl_3$, there is one atom of Fe, so the fraction of iron(III) chloride that is iron metal is calculated as follows:

$$\text{Fraction of iron(III) chloride that is iron} = \frac{\text{atomic weight of Fe}}{\text{formula weight of } FeCl_3} \tag{3.3}$$

$$= 0.34429 \ Fe/FeCl_3$$

The weight of iron metal that is equivalent to 55 g of iron(III) chloride can be calculated by multiplying the weight of iron(III) chloride by this fraction:

$$\text{weight of iron equivalent to 55 g of iron(III) chloride} = 55 \ g \times 0.34429 = 19 \ g$$

This fraction is therefore a gravimetric factor because it is used to convert the weight of $FeCl_3$ to the weight of Fe as shown.

In general, many gravimetric factors are calculated by simply dividing the atomic or formula weight of the substance whose weight we are seeking (substance sought) by the atomic or formula weight of the substance whose weight we already know (substance known), as we did in the above example. However, the calculation can be further complicated when the number of atoms in the symbol of the particular element or compound in the numerator does not match the number of atoms of the same element in the symbol or formula of the compound in the denominator. If we symbolize the atomic weight and formula weight used in the above by the corresponding chemical symbols and formulas, we can write the gravimetric factor for our example as follows:

$$\text{gravimetric factor} = \frac{\text{sought}}{\text{known}} = \frac{Fe}{FeCl_3} \tag{3.4}$$

The one element that the numerator has in common with the denominator (the "common element") is iron, and there is one atom (no subscripts) of iron in each and so there is no such complication in this case. When the number of atoms of a common element are not the same, we must multiply by coefficients to make them the same, as is done in balancing equations. To illustrate, consider an example in which we want to know the weight of NaCl that is equivalent to a given weight of $PbCl_2$. The NaCl is the compound sought and the $PbCl_2$ is the compound known. The common element is Cl, and the numerator must be multiplied by 2 in order to equalize the number of Cl in the two formulas. In one formula unit of $PbCl_2$, there are two atoms of Cl:

$$\text{gravimetric factor} = \frac{\text{sought}}{\text{known}} = \frac{NaCl \times 2}{PbCl_2} \tag{3.5}$$

The general formula for a gravimetric factor is

$$\text{Gravimetric factor} = \frac{\text{atomic or formula weight of chemical sought} \times Q_s}{\text{atomic or formula weight of chemical known} \times Q_k} \quad (3.6)$$

where Q_s is the balancing coefficient for the common element in the formula of the substance sought and Q_k is the balancing coefficient for the common element in the formula of the substance known.

Example 3.1

What is the gravimetric factor for converting a weight of AgCl to a weight of Cl?

Solution 3.1

substance sought   Cl

substance known   AgCl

$$\text{gravimetric factor} = \frac{\text{Cl} \times 1}{\text{AgCl}} = \frac{35.453 \times 1}{143.32 \times 1}$$
$$= 0.24737 \text{ Cl/AgCl}$$

Example 3.2

What is the gravimetric factor if a weight of Cl is to be determined from a weight of $PbCl_2$?

Solution 3.2

substance sought   Cl

substance known $PbCl_2$

$$\text{gravimetric factor} = \frac{\text{Cl} \times 2}{PbCl_2 \times 1} = \frac{35.453 \times 2}{278.1 \times 1} = 0.2550 \text{ Cl/PbCl}_2$$

Example 3.3

If you have a weight of $AlCl_3$ and you want to know the weight of Cl_2 that is equivalent to this, what gravimetric factor is required?

Solution 3.3

substance sought Cl_2

substance known $AlCl_3$

$$\text{gravimetric factor} = \frac{Cl_2 \times 3}{AlCl_3 \times 2} = \frac{70.906 \times 3}{133.34 \times 2} = 0.79765 \; Cl_2/AlCl_3$$

In cases in which there are two common elements, one of which is oxygen, the oxygen is ignored and the other element is balanced.

Example 3.4

What gravimetric factor is required to calculate the weight of SO_3 from a weight of $BaSO_4$?

Solution 3.4

$$\text{substance sought} \quad SO_3$$

$$\text{substance known} \quad BaSO_4$$

Besides oxygen, sulfur is the common element.

$$\text{gravimetric factor} = \frac{SO_3 \times 1}{BaSO_4 \times 1} = \frac{80.06 \times 1}{233.40 \times 1} = 0.3430 \; SO_3/BaSO_4$$

3.7.4 Using Gravimetric Factors

As stated earlier, gravimetric factors are used to convert the weight of one chemical to the weight of another, as in the example we cited at the beginning of Section 3.7.3. Below is another example of such a conversion.

Example 3.5

How many grams of copper(II) sulfate hexahydrate are required to prepare a solution that has the equivalent of 0.339 g of copper dissolved?

Solution 3.5

The formula of copper(II) sulfate hexahydrate is $CuSO_4 \cdot 6H_2O$ and its formula weight is 267.68 g/mol.

$$\text{substance sought} \; CuSO_4 \cdot 6H_2O$$

$$\text{substance known} \; Cu$$

$$\text{gravimetric factor} = \frac{CuSO_4 \cdot 6H_2O \times 1}{Cu \times 1} = \frac{267.68 \times 1}{63.54 \times 1} = 4.21278 = 4.213$$

$$\text{grams of } CuSO_4 \cdot 6H_2O = \text{grams of copper} \times \text{gravimetric factor}$$
$$= 0.339 \times 4.21278 = 1.43 \text{ g } CuSO_4 \cdot 6H_2O$$

**APPLICATION NOTE 3.2: THE DETERMINATION
OF SULFUR IN DRY FERTILIZERS**

It can be important to determine the amount of sulfur in a fertilizer. A common procedure
for determining the sulfur that originates as sulfate is similar to that of Experiment 6 in this
textbook. This procedure first involves weighing a test portion of the dry fertilizer into a beaker and extracting the sulfate with 200 mL of a 7% HCl solution. The extraction mixture is
gently boiled for 10 min. After this, there will be some undissolved fertilizer present, and so
a filtration step is required. The filtrate is then carried through the Experiment 7 procedure
to determine the amount of sulfate in the extract and ultimately to determine the percent of
sulfate in the fertilizer. If the answer is to be reported as %S, then the gravimetric factor to
be used in the calculation would be the atomic weight of sulfur divided by the formula weight
of barium sulfate.

   (*Reference:* Horowitz, W., editor, *Official Methods of Analysis of AOAC International*,
17th edition, vol. 1, para. 2.6.28, Gaithersberg, MD, USA (2000).)

In the case in which the analyte participates in a chemical reaction, the product of which
is weighed, the weight of this product must be converted to the weight of the analyte before
the percent can be calculated. This can be done with the use of a gravimetric factor. Examples
include the sulfate determination as in Experiment 7 in this chapter (see also Application
Note 3.2).

Example 3.6

A sample that weighed 0.8112 g is analyzed for phosphorus (P) content by precipitating the phosphorus as $Mg_2P_2O_7$. If the precipitate weighs 0.5261 g, what is the % P in the sample?

Solution 3.6

$$\% \text{ analyte} = \frac{\text{weight of product} \times \text{gravimetric factor}}{\text{weight of sample}} \times 100$$

Substance sought P

Substance known $Mg_2P_2O_7$

$$\text{Gravimetric factor} = \frac{P \times 2}{MgP_2O_7 \times 1} = \frac{30.9738 \times 2}{222.57} = 0.278329 = 0.27833$$

$$\% P = \frac{0.5261 \times 0.278329}{0.8112} = \times 100 = 18.05\%$$

3.8 EXPERIMENTAL CONSIDERATIONS

3.8.1 WEIGHING BOTTLES

A weighing bottle is a small glass bottle with a ground-glass top and stopper. It is used for containing, drying, weighing, and dispensing finely divided solids for all types of analyses, as suggested by their use in the laboratory experiments in this chapter and those to follow. For gravimetric analysis, weighing bottles can be used for containing samples in loss-on-drying experiments and in other instances in which a sample must be dried before proceeding and then conveniently weighed and dispensed later.

The drying takes place in a drying oven, an oven that can be set to temperatures up to 100°C and beyond. The weighing bottle, with sample contained, is usually placed in the oven, with lid ajar, in a small beaker and the beaker covered with a watch glass, as in Figure 3.9.

3.8.2 WEIGHING BY DIFFERENCE

While dispensing and weighing a small amount of the dried sample contained in the weighing bottle, it is prudent to avoid contacting the sample with laboratory utensils or weighing paper in order to avoid loss of a portion of the sample that might adhere to the utensil or to the paper. In such instances, weighing by difference is recommended.

Weighing by difference is the act of determining the weight of a sample by obtaining two weighing bottle weights, one before and one after dispensing the sample, and then subtracting the two weights. In the process, no utensil or weighing paper contacts the sample because the sample is shaken into the receiving vessel directly from the weighing bottle. If the two weights are determined on an analytical balance, it is desirable to avoid touching the weighing bottle with fingers from the time the first weight is obtained until the second weight is obtained. This avoids putting fingerprints on the weighing bottle that would add measurable weight to the weighing before the second weight

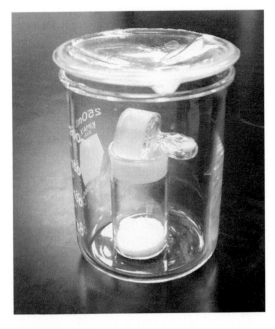

FIGURE 3.9 Weighing bottle with sample, lid ajar, in a small beaker, ready to be placed in a drying oven for the purpose of drying the contained sample.

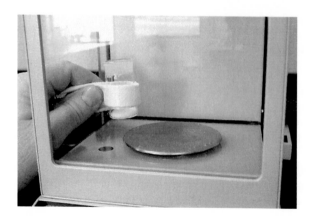

FIGURE 3.10   Handling a weighing bottle with a rolled-up piece of paper towel to avoid fingerprints.

is determined. One can conveniently avoid fingerprints by wearing clean gloves, or by the use of a rolled up paper towel, as shown in Figure 3.10.

3.8.3   Isolating and Weighing Precipitates

In a gravimetric analysis experiment in which a chemical reaction is used to obtain a precipitate, the manner in which the precipitate is isolated and weighed involves a filtration step, including a full rinse with distilled water, and two weight measurements. The filtration can involve a filtering crucible (connected to a vacuum system) or simply a piece of filter paper and a funnel. If a filtering crucible is used, its weight is measured before filtering and again after filtering, after the water solvent has been evaporated in a drying oven. The difference in the two weights is the weight of the precipitate.

The distilled water rinse mentioned in this discussion is necessary to rinse the filtering crucible or filter paper and precipitate free of any other soluble chemicals that might be present. The rinse is accomplished with the use of streams of distilled water from a squeeze bottle directed over the entire crucible or filter paper.

EXPERIMENTS

EXPERIMENT 4: LOSS ON DRYING

INTRODUCTION

A number of experiments in this text require that a sample be dried before proceeding with the chosen method. Examples in this chapter include the sulfate unknown (see Experiment 7, Step 1). An example in Chapter 4 is the KHP standard in Experiment 9 (see Experiment 9, Step 1). In this experiment, the percent loss on drying of one of these materials will be determined. Remember to wear safety glasses.

1. Clean a shallow glass weighing bottle and lid and dry it with lid ajar (Figure 3.9) in a drying oven for 30 min at 105°C. After drying, avoid fingerprints. Allow to cool to room temperature and weigh on an analytical balance. Record the weight.
2. Add the sample to be dried to the weighing bottle. Record the exact weight, weighing the bottle and sample.
3. Place the weighing bottle with the sample, lid ajar, back in the drying oven and dry for the time specified in Experiments 7 or 9.
4. Remove from the drying oven, place the lid on the bottle, allow to cool to room temperature in your desiccator and then weigh.

5. This step is optional. Repeat steps 3 and 4 until you have constant weight (successive weights agreeing to within 0.001 g).
6. Calculate and report the loss on drying as a percentage. Store your sample in your desiccator if the experiment that utilizes this material (Experiment 7 or 9) is to be performed at a later time.

EXPERIMENT 5: PARTICLE SIZING

Remember to wear safety glasses.

1. Obtain some sand, such as from a bag of playground sand from a building materials supplier, or from a beach, etc. Obtain test sieves with apertures of 2 and 1 mm and also a bottom collection pan.
2. Clean the test sieves to make sure there are no particles stuck in the apertures. Weigh each empty test sieve to the nearest gram and then stack the sieves.
3. Tare a beaker or other container and weigh, to the nearest gram, a sample of the sand into the beaker.
4. Pour the sample into the top sieve, place the sieve stack on a sieve shaker and shake for 5 min. Obtain the new weight of each test sieve to the nearest gram.
5. Reassemble and shake for an additional 5 min and weigh again. If the weights have changed appreciably, reassemble and shake again. Repeat until you have a constant weight (within one gram) for each test sieve.
6. Calculate the percentage of each particle size range (greater than 2 mm, between 1 and 2 mm, and less than 1 mm). Also report the cumulative percentages.

EXPERIMENT 6: THE DETERMINATION OF SALT IN A SALT–SAND MIXTURE

1. Obtain a beaker containing the salt/sand mixture from your instructor. Record the letter on the beaker in your notebook or on your data sheet. This is your unknown identification. Next, weigh the beaker with contents on an analytical balance and record the weight in your notebook or on your data sheet.
2. Add approximately 50 mL of distilled water to the beaker and swirl for a few minutes to dissolve the salt.
3. Let settle for 5 min and then carefully decant the supernatant into another container. Take care not to lose any of the sand while decanting.
4. Repeat Steps 2 and 3 at least five more times so that you are reasonably sure that all of the salt has been dissolved and separated from the sand.
5. Place the beaker with the wet sand on a hot plate set on a low setting and heat for 30 min to evaporate to dryness. After 30 min, remove the beaker from the hot plate, set it on the bench top, and allow to cool to room temperature.
6. Weigh the beaker with the dry sand on the analytical balance and record the weight in your notebook or on your data sheet. Next dump the sand in the trash, tapping it while inverted to make reasonably sure that there is no longer any sand in the beaker. Now, weigh the empty beaker and record the weight in your notebook or on your data sheet.
7. Using the three weights recorded in the above steps, calculate the percent salt in the salt/sand mixture. Record the calculation in your notebook or on your data sheet along with the resulting percent salt. If required, fill out a results card and turn in to your instructor.

EXPERIMENT 7: THE GRAVIMETRIC DETERMINATION OF SULFATE IN A COMMERCIAL UNKNOWN

INTRODUCTION

This experiment represents an example of a gravimetric analysis based on the stoichiometry of a precipitation reaction (as discussed in Sections 3.7.2 through 3.7.4) rather than a simple physical separation, such as heating. A sample containing a soluble sulfate compound, such as Na_2SO_4 or K_2SO_4, is dissolved and the sulfate precipitated with $BaCl_2$ as $BaSO_4$. The percent of SO_3 is determined using a gravimetric factor as discussed in the text. Read the entire experiment completely before beginning the lab work. The procedure calls for you to run the experiment in duplicate. Your instructor may want you to run it in triplicate. Remember to wear safety glasses.

PROCEDURE
Session 1

1. Obtain a sample of unknown sulfate from your instructor. Record the unknown number in your notebook. If not already done in Experiment 4, dry the sample using a weighing bottle and drying oven set at 110°C for 1 h.

2. While waiting for your sample to dry, prepare two porcelain Gooch crucibles by washing with soapy water and a brush and drying with a paper towel. Label each as "1" and "2" using a permanent lab marker. Place a filter paper on the bottom of each. Dry both crucibles in a drying oven set at 110°C until Step 9. Since others will be using the drying oven for this, be sure you are able to identify yours. You may use two clean small beakers, each labeled with your name, to hold the crucibles in the oven.

3. Thoroughly clean, using soapy water and a brush, two 400-mL beakers, two watch glasses (for use as covers for the beakers), and two glass stirring rods that are fitted with a rubber policeman. The beakers should be free of any water hardness residue and scratches. If there is water hardness residue in a beaker, rinse with a 6-M HCl solution and then rinse with distilled water to remove the acid. Also locate a hot plate and acquire beaker tongs for handling the beakers when they are hot. Label the beakers "1" and "2" using labeling tape and a permanent lab marker.

4. Obtain and clean a 150-mL beaker and place in it about 130 mL of barium chloride stock solution (26 g/L concentration) prepared by your instructor. Place a watch glass on this beaker also. Let this beaker with the solution sit on your lab bench for now. Obtain and clean a 100-mL graduated cylinder and a 10-mL graduated cylinder.

5. After the 1-h drying time for the sample has expired, remove the unknown from the oven, place the lid on the weighing bottle and place the weighing bottle in your desiccator to cool for 15 min. After cooling, weigh accurately by difference, on an analytical balance, two samples of your unknown, each weighing around 0.4 g, into the two 400-mL beakers. The weights can be recorded in the DATA section of your notebook as below. Sample 1 should go into the beaker labeled "1" and so forth.

    Sample 1   Weight of weighing bottle before: _____ g
                    Weight of weighing bottle after: _____ g
    Sample 2   Weight of weighing bottle before: _____ g
                    Weight of weighing bottle after: _____ g

6. Using the graduated cylinders, add to each of the two beakers 125 mL of water and 3 mL of concentrated HCl. Place one glass stirring rod, policeman end pointing up, in each beaker. Stir the mixtures until the unknowns have dissolved. Leaving the stirring rods in the beakers, place the watch glasses on the beakers and place both beakers on a hot plate. Place also the barium chloride beaker on the hot plate.

7. Carefully heat all solutions to boiling, then turn off the hot plate, remove the beakers from the hot plate and, while stirring, add 55 mL (use the 100-mL graduated cylinder) of the barium chloride to each of the unknowns. Use beaker tongs for handling hot beakers. The fine, white precipitate, barium sulfate, forms almost immediately. The amount of barium chloride used is in excess of the amount required. This is to ensure complete precipitation of the sulfate in your unknown. Continue to stir both solutions for at least 20 s. Again, leave the stirring rods in the beakers.

8. Place the two sample beakers back on the hot plate and turn it on to a low setting. Adjust the setting on the hotplate so that the two remaining solutions remain very hot (but not quite boiling). If either sample boils, remove from the hot plate to cool it for a moment and then place it back on the hotplate. Allow the precipitate to "digest" for 1 h. **Digestion** refers to a procedure in which a sample is heated or stored for a period of time to allow a particular chemical or physical process to occur. In this case, a digestion step is required in order to change the small, finely divided particles of the barium sulfate precipitate into larger,

filterable particles. Heating for a period of 1 h accomplishes this goal. While the digestion takes place, proceed to Step 9.

9. Using crucible tongs, remove the crucibles from the drying oven and place in your desiccator. Let cool in your desiccator for 15 min and then weigh each crucible on the analytical balance, recording the weights in your notebook. Make sure you have the weights labeled in your notebook as 1 and 2, corresponding to the labels on the crucibles. Following weighing, using tongs, place the crucibles back into your desiccator to store until Session 2.

10. After the 1-h digestion is completed, take the beakers off the hot plate and allow to cool on your bench for 10 min. Rinse the watch glasses into the beakers using a stream of distilled water from a squeeze bottle. Then, place the watch glasses back on the beakers and store until Session 2.

Session 2

11. Warm approximately 500 mL of distilled water on a hot plate and then place it in a squeeze bottle. Warm water is needed for rinsing when transferring the precipitate from the beakers to the crucible for filtration and also for rinsing the precipitate in the crucible.

12. Read this entire step before beginning. Set up a vacuum system using a vacuum pump, a 500-mL vacuum flask, and a crucible holder. Place crucible 1 in the crucible holder, and after moistening the filter paper with distilled water, turn on the vacuum pump to draw the water through the filter paper. This step will "seat" the filter paper in the crucible. Now filter solution 1 through Gooch crucible 1. Do so initially without disturbing the precipitate at the bottom of the beaker—transfer the supernatant first. Use the stirring rod as a guide (ask instructor to demonstrate) so that the solution does not get on the outside of the beaker. When 10–20 mL remain in the beaker, swirl the beaker to mix the precipitate with the supernatant before resuming the transfer. In this process, the transfer must be a **quantitative transfer**, meaning that every last trace of solution and precipitate in the beaker must be transferred to the Gooch crucible. All of the precipitate must end up in the crucible so that its total weight can be determined via the analytical balance. You may freely utilize the warm water from the squeeze bottle to rinse down the wall of the beaker and swirling again to help accomplish this. For precipitate particles that adhere to the walls of the beaker, use the rubber policeman to scrub the walls and then rinse the policeman into the crucible. Rinse the glass end of the stirring rod into the crucible first before inverting it. Be careful not to spill any solution, or lose any precipitate, on the outside of the crucible or anywhere else. Once the transfer is complete, rinse the interior of the crucible and the precipitate with approximately 20 mL (estimated) of the warm water from the squeeze bottle. Repeat with solution 2 and crucible 2.

13. The precipitate must be dried before the weight of the precipitate can be accurately determined. Place the crucibles in the drying oven as before and dry for at least 1 h. Then, allow them to cool in your desiccator for 15 min and weigh.

14. Calculate the %SO$_3$ for each of the samples and report the results to your instructor.

QUESTIONS AND PROBLEMS

Introduction

1. Define gravimetric analysis.
2. Compare Figure 3.1 with Figure 1.1. What is special about the gravimetric method of analysis?

Weight vs. Mass

3. What is the difference between "weight" and "mass"?

The Balance

4. Why is a weighing device called a "balance"?
5. What is a "single-pan balance"?
6. What does it mean to say that a technician measures a weight to the nearest 0.01 g? What does it mean to say that the precision of a given balance is ±0.1 mg?
7. What is a "top-loading balance"?
8. What does it mean to say that a balance has a "tare" feature?
9. What is an "analytical balance"?
10. The analytical balance is a much more sensitive weighing device than an "ordinary" balance. Name three things that are important to remember when using the analytical balance that are not important when using an ordinary balance. (Hint: Refer to Appendix 2.)

The Desiccator

11. What is a desiccator? What is Drierite™? What is an "indicating desiccant"?

Calibration and Care of Balances

12. What does it mean to check the calibration of a balance?

When to Use Which Balance

13. When a weight measurement enters directly into the calculation of a quantitative analysis result that is to be reported to four significant figures, is an ordinary top-loading balance satisfactory for this measurement? Explain.

Details of Gravimetric Methods

14. What is a physical separation of an analyte as opposed to a chemical separation?
15. Using the information given in each row of the table below, calculate all the items marked with a question mark (?). Weight units are grams.

	Weight of Container + Sample	Weight of Empty Sample Container	Weight of Container + Analyte	Weight of Empty Analyte Container	Weight of Sample	Weight of Analyte	% Analyte
(a)	xxxxx	xxxxx	xxxxx	xxxxx	3.5992	1.0449	?
(b)	15.6603	14.0338	xxxxx	xxxxx	?	0.3059	?
(c)	xxxxx	xxxxx	10.3834	9.9192	0.9481	?	?
(d)	29.0038	20.4819	6.0127	5.1304	?	?	?
(e)	21.9477	19.5820	11.0382	10.8383	xxxxx	xxxxx	?
(f)	18.7049	18.0288	xxxxx	xxxxx	xxxxx	0.2174	?
(g)	xxxxx	xxxxx	15.3938	14.9300	0.6729	xxxxx	?

16. (a) Define "loss on drying."
    (b) A soil sample, as received by a laboratory, weighed 5.6165 g. After drying in an oven, this same sample weighed 2.7749 g. What is the percent loss on drying?

17. In an experiment in which the percent loss on drying in a sample was determined, the following data were obtained
    Weight of crucible + sample before drying: 11.9276 g
    Weight of crucible + sample after drying: 10.7742 g
    Weight of empty crucible: 7.6933 g
    What is the percent loss on drying in the sample?

18. What is meant by a step in a procedure that states "heat to constant weight"?

19. Define "loss on ignition" and "residue on ignition."

20. A sample of grain was analyzed for loss on ignition. Initially, a sample of this grain in an evaporating dish weighed 29.6464 g. After heating in an oven at 500°C for 2 h, this same sample in the dish weighed 20.9601 g. If the evaporating dish alone weighed 11.6626 g, what is the percent loss on ignition and the percent residue on ignition?

21. An analyst performs an experiment to determine the percent loss on ignition of a grain sample. The grain is placed in a preweighed evaporating dish, the dish is weighed again, the volatiles are driven off by heating in a muffle furnace, and the dish weighed a third time. Calculate the percent of volatile organics and the percent residue given the following data:
    Weight of empty evaporating dish: 28.3015 g
    Weight of evaporating dish with the grain: 39.4183 g
    Weight of dish with grain after heating: 33.1938 g

22. Using the information given in each row of the table below, calculate all the items marked with a question mark (?). Weight units are grams.

	Weight of Empty Container	Weight of Container + Sample	Weight of Container + Residue	Weight of Sample	Weight of Residue	Weight Loss	% Loss
(a)	xxxxx	xxxxx	xxxxx	0.9875	xxxxx	0.1265	?
(b)	xxxxx	xxxxx	xxxxx	1.2445	1.1739	?	?
(c)	13.8837	14.2842	xxxxx	?	0.3382	?	?
(d)	14.9047	xxxxx	16.2284	1.5378	?	?	?
(e)	11.7835	12.3256	xxxxx	xxxxx	0.1760	xxxxx	?
(f)	20.4659	xxxxx	20.9117	1.5557	xxxxx	xxxxx	?
(g)	15.9257	17.4072	16.9993	?	xxxxx	?	?
(h)	12.3590	13.3039	12.8902	xxxxx	xxxxx	xxxxx	?

23. Using the information given in each row of the table below, calculate all the items marked with a question mark (?). Weight units are grams.

	Weight of Empty Container	Weight of Container + Sample	Weight of Container + Residue	Weight of Sample	Weight of Residue	Weight Loss	% Residue
(a)	xxxxx	xxxxx	xxxxx	4.1944	3.8776	xxxxx	?
(b)	xxxxx	xxxxx	xxxxx	1.5000	xxxxx	0.0882	?
(c)	21.6423	24.4287	xxxxx	?	2.3771	xxxxx	?
(d)	13.3555	xxxxx	13.8103	0.9228	?	xxxxx	?
(e)	17.0042	18.5237	xxxxx	?	?	0.2564	?
(f)	14.2294	15.6688	xxxxx	xxxxx	1.1373	xxxxx	?
(g)	19.0025	xxxxx	20.6131	1.9304	xxxxx	xxxxx	?
(h)	25.5904	27.2049	26.8202	?	?	xxxxx	?
(i)	18.4586	18.9583	18.8902	xxxxx	xxxxx	xxxxx	?

24. What does "insoluble matter" in the general analysis report on the label of a chemical bottle refer to?

25. Consider the analysis of a salt-sand mixture. If the mixture contained in a beaker is treated with sufficient water to dissolve the salt, what is the percent of both salt and sand in the mixture given the following data:
Weight of mixture: 5.3502 g
Weight of sand isolated from mixture after filtering and drying: 4.2034 g

26. The following data are obtained from a particle size analysis. What percent of the sample has a particle size greater than 2 mm diameter?
Weight of sample: 12.5 g
Weight of 2 mm sieve before sieving: 382.1 g
Weight of 2 mm sieve after sieving: 389.3 g

27. Using the information given in each row of the table below, calculate all the items marked with a question mark (?). Weight units are grams.

	Weight of Container + Sample	Weight of Empty Container	Weight of Empty Filter or Sieve	Weight of Filter or Sieve with Analyte	Weight of Sample	Weight of Captured Matter or Particles	% Insoluble Matter or Captured Particles
(a)	xxxxx	xxxxx	xxxxx	xxxxx	76.22	54.92	?
(b)	xxxxx	xxxxx	210.3	294.2	132.8	?	?
(c)	439.1	291.4	xxxxx	xxxxx	?	90.1	?
(d)	xxxxx	xxxxx	209.46	289.45	150.39	xxxxx	?
(e)	628.4	402.5	xxxxx	xxxxx	xxxxx	105.4	?
(f)	345.03	300.48	158.20	149.48	?	?	?
(g)	198.7	139.6	60.3	89.2	xxxxx	xxxxx	?

28. What is the gravimetric factor in each of the following gravimetric analysis examples?

	Substance Sought	Substance Weighed
(a)	Ag	$AgBr$
(b)	SO_3	$BaSO_4$
(c)	Ag_2O	$AgCl$
(d)	Na_3PO_4	$Mg_2P_2O_7$
(e)	Pb_3O_4	$PbCrO_4$
(f)	SiF_6	CaF_2
(g)	Co_3O_4	$Co_2P_2O_7$
(h)	Bi_2S_3	Bi_2O_3

29. What is the gravimetric factor
   (a) For obtaining the weight of Ag_2CrO_4 from the weight of $AgCl$?
   (b) If one is calculating the percent of Na_2SO_4 in a mixture when the weight of Na_3PO_4 is measured?
   (c) When converting the weight of $HgCl_2$ to the weight of Hg_2Cl_2?

30. What is the gravimetric factor that must be used in each of the following experiments?
   (a) The weight of $Mg_2P_2O_7$ is known and the weight of MgO is to be calculated.
   (b) The weight of Fe_3O_4 is to be converted to the weight of FeO.
   (c) The weight of Mn_3O_4 is to be determined from the weight of Mn_2O_3.

31. If a technician wishes to prepare a solution containing 55.3 mg of barium, how many grams of barium chloride dihydrate does she need to weigh?

32. Using the information given in each row of the table below, calculate the items marked with a question mark (?).

	Known Weight	Substance to Be Weighed	Grams of Substance to Be Weighed Equivalent to Known Weight
(a)	0.45 g Na	NaCl	?
(b)	1.48 g Ag	$AgNO_3$	?
(c)	0.80 g K	K_2SO_4	?
(d)	10.0 mg N	KNO_3	?
(e)	50.0 mg P	KH_2PO_4	?
(f)	75.0 mg Cl	$MgCl_2$	?

33. What weight of K_2SO_4 is equivalent to 0.6603 g of K_3PO_4?
34. What weight of P_2O_5 is equivalent to 0.6603 g of P?
35. What is the percent of K_2CrO_4 in a sample that weighed 0.7193 g if the weight of the Cr_2O_3 precipitate derived from the sample was 0.1384 g?
36. The gravimetric factor for converting the weight of $BaCO_3$ to Ba is 0.6959. If the weight of $BaCO_3$ derived from a sample was 0.2644 g, what weight of Ba was in this sample?
37. If 0.9110 g of a sample of silver ore yielded 0.4162 g of AgCl in a gravimetric experiment, what is the percentage of Ag in the ore?
38. Given the following data, what is the % S in the sample?
    Weight of weighing bottle before dispensing sample: 5.3403 g
    Weight of weighing bottle after dispensing sample: 4.8661 g
    Weight of crucible with $BaSO_4$ precipitate: 19.3428 g
    Weight of crucible empty: 18.7155 g
39. Given the following data, what is the % Fe in the sample?
    Weight of weighing bottle before dispensing sample: 3.5719 g
    Weight of weighing bottle after dispensing sample: 3.3110 g
    Weight of crucible with Fe_2O_3 precipitate: 18.1636 g
    Weight of empty crucible: 18.0021 g
40. Nickel can be precipitated with dimethylglyoxime (DMG) according to the following reaction:

$$Ni^{2+} + HDMG \rightarrow Ni(DMG)_2 + 2H^+$$

   If 2.0116 g of a nickel-containing substance is dissolved and the nickel precipitated as above so that the $Ni(DMG)_2$ weighs 2.6642 g, what is the percentage of nickel in the substance? The formula weight of $Ni(DMG)_2 = 288.92$ g/mol.
41. Imagine an experiment in which the percentage of manganese, Mn, in a manganese ore is to be determined by gravimetric analysis. If 0.8423 g of the ore yielded 0.3077 g of Mn_3O_4 precipitate, what is the %Mn in the ore?

Experimental Considerations

42. What is the advantage to weighing by difference as opposed to weighing using a spatula and weighing paper?
43. Why is the barium sulfate precipitate in Experiment 7 "digested"?
44. During what period of the "weighing by difference" process must fingerprints on the item being weighed be avoided? Explain.

45. In Experiment 7, why is the desiccator not used to store the crucible and filter paper immediately after the filtration and before placing in the drying oven?

46. In Experiment 7, consider an error in which a student accidentally poked a hole in the filter paper with the stirring rod as he or she was filtering and some precipitate was observed with the filtrate. If the student takes no steps to correct the error, would the %SO_3 calculated at the end be higher or lower than the true percentage? Explain.

4 Introduction to Titrimetric Analysis

4.1 INTRODUCTION

The gravimetric analysis methods introduced in Chapter 3 represent one of two major subcategories of wet chemical analysis. As we discussed, gravimetric analysis methods primarily utilize the measurement of weight and that is why this subcategory is called "gravimetric analysis." In this chapter, we introduce **titrimetric analysis**, the other major subcategory. Titrimetric analysis methods utilize solution chemistry heavily, and therefore it should not be surprising that volumes of solutions are prepared, measured, transferred, and analyzed frequently in this type of analysis. In fact, titrimetric analysis is often referred to as **volumetric analysis** for that reason. Solution reaction stoichiometry lies at the heart of these methods. Thus, methods of solution preparation, methods of measuring and transferring liquid volumes, and methods of utilizing solution reaction stoichiometry in analyzing solutions are all topics with which we will be concerned in our discussion of titrimetric analysis (see Figure 4.1).

4.2 TERMINOLOGY

Let us begin with the definition of standard solution. A **standard solution** is a solution that has a concentration of solute known to some high degree of precision, such as a molarity known to four decimal places. For example, a solution of HCl with a concentration of 0.1025 M is a standard solution of HCl. The concentration can be known, in some cases, directly through the preparation of this solution. It may also become known by performing an experiment.

If the concentration of a standard solution is to be known directly through its preparation, then the amount of solute present must be accurately measured with a high degree of precision. In addition, a **volumetric flask** (Figure 4.2), which is an accurate, high-precision but very common type of flask, must be used as the container so that the total volume of solution can also be accurately measured with good precision. If the solute is a pure solid material capable of being weighed accurately, then an analytical balance must be used to weigh it. If the solute is already in solution, then the solution can be prepared by diluting this solution. However, the solution to be diluted must be a standard solution with a concentration known with equal or better precision (same or more significant figures) than the solution being prepared. In addition, the volume of the solution to be diluted must be precisely measured. A **volumetric transfer pipet** (Figure 4.2), a high-precision type of transfer pipet, is often the piece of glassware used for this latter measurement, although other devices are available. Flasks, pipets, and pipetting devices will be discussed in detail in Sections 4.8 and 4.9.

If the concentration of a standard solution is to become known by experiment, the experiment must be carried out with a high degree of accuracy and precision. Such an experiment is called **standardization**. Similarly, to **standardize** a solution means to determine its concentration to three or more significant figures. For example, if a solution of NaOH is made up to be approximately 0.1 M, a standardization experiment may be performed and the concentration determined to be 0.1012 M. We will discuss the details of such an experiment in Section 4.6.

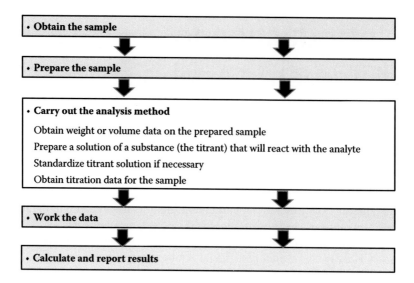

FIGURE 4.1   Steps characteristic of titrimetric analysis methodology.

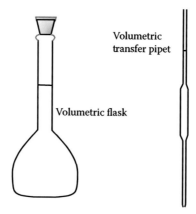

FIGURE 4.2   Drawings of the volumetric flask and the volumetric transfer pipet used in titrimetric analysis and other types of analysis as well.

Another piece of glassware often used in titrimetric analysis (for both standardization experiments and sample analyses) is the buret. A traditional glass **buret** (Figure 4.3) is a long and narrow graduated cylinder with a dispensing valve at the bottom. The dispensing valve is called a **stopcock**. The solution in the buret is dispensed by turning the stopcock to the open position. The experiment used for titrimetric analysis and solution standardization is called a titration. A **titration** is an experiment in which a solution of a reactant is added to a reaction flask with the use of a buret. The solution being dispensed via the buret (or the substance dissolved in the solution dispensed via the buret) is called the **titrant**. The solution held in the reaction flask before the titration (or the substance dissolved in this solution) is called the **substance titrated**. For example, a given titration experiment may have a solution of NaOH in the buret and a solution of HCl in the reaction flask. In that case, NaOH is the titrant and HCl is the substance titrated.

To effectively utilize the stoichiometry of the reaction involved in a titration, both the titrant and the substance titrated need to be measured exactly. The reason is that one is the known quantity and the other is the unknown quantity in the stoichiometry calculation. The buret is

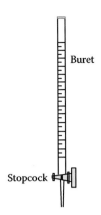

FIGURE 4.3   Drawing of a buret.

an accurate (if carefully calibrated) and relatively high-precision device because it is long and narrow. If a meniscus is read in a narrow graduated tube, it can be read with higher precision (more significant figures) than in a wider tube. Thus, a buret provides the required precise measurement of the titrant.

The addition of titrant from the buret must be stopped at precisely the correct moment—the moment at which the last trace of substance titrated is consumed by a fraction of a drop of titrant added, so that the correct volume can be read on the buret. That exact moment is called the **equivalence point** of the titration. In order to detect the equivalence point, an indicator is often used. An **indicator** is a substance added to the reaction flask ahead of time in order to cause a color change at or near the equivalence point, i.e., to provide a visual indication of the equivalence point. For example, the use of a chemical named phenolphthalein as an indicator for a titration in which a strong base is used as the titrant and an acid as the substance titrated would give a color change of colorless to pink in the reaction flask near the equivalence point. Although the color change may occur "near" the equivalence point and not exactly at the equivalence point, it is usually not a concern. The reason will become clear in a later discussion. The point of a titration at which an indicator changes color, the visual indication of the equivalence point, is called the **end point** of the titration. As we will see, equivalence points can be determined in other ways as well.

Electronic devices such as **automatic titrators** and **digital burets** may be used in place of the traditional glass buret and manual titration. Such devices provide electronic control over the addition of titrant and so, with proper calibration, are accurate, high-precision devices. These will be discussed in Section 4.9.

Solution preparation, standardization, and sample analysis activities all involve solution concentration. Let us review molarity and normality as methods of expressing solution concentration.

4.3   REVIEW OF SOLUTION CONCENTRATION

4.3.1   MOLARITY

Molarity is defined as the number of moles of solute dissolved per liter of solution and may be calculated by dividing the number of moles dissolved by the number of liters of solution (see Equation M.1 in Box 4.1). Solutions are referred to as being, for example, 2.0 molar, or 2.0 M. The "M" refers to "molar," and the solution is said to have a molarity of 2.0, or 2.0 mol dissolved per liter of solution. It is important to recognize that, for molarity, it is the number of moles dissolved per liter of *solution* and not per liter of solvent.

BOX 4.1   CALCULATION OF THE MOLARITY OF A SOLUTION FROM SOLUTION PREPARATION DATA

$$\text{molarity} = \frac{\text{moles of solute}}{\text{liters of solution}} \tag{M.1}$$

$$\text{molarity} = \frac{\text{grams of solute/FW}}{\text{liters of solution}} \tag{M.2}$$

$$C_A = \frac{C_B \times V_B}{V_A} \tag{M.3}$$

These are formulas for calculating the molarity of a solution from solution preparation data as derived in the text. FW = formula weight of the solute; C_A = concentration after dilution; C_B = concentration before dilution; V_B = volume before dilution; V_A = volume after dilution.

Example 4.1

What is the molarity of a solution that has 4.5 mol of solute dissolved in 300.0 mL of solution?

Solution 4.1

Equation M.1 in Box 4.1 defines molarity. The number of milliliters given must first be converted to liters, and so 0.3000 L is used in the denominator.

$$\text{molarity} = \frac{4.5\ \text{mol}}{0.3000\ \text{L}} = 15\,M = 15\ M$$

If the number of grams of the solute used is given rather than moles, these grams must be converted to moles by dividing by the formula weight (FW) as shown in Equation M.2 in Box 4.1.

Example 4.2

What is the molarity of a solution of NaOH that has 0.491 g dissolved in 400.0 mL of solution?

Solution 4.2

Equation M.2 is used and, once again, the milliliters are first converted to L.

$$\text{molarity} = \frac{(0.491/39.997)\ \text{moles}}{0.4000\ \text{L}} = 0.0307\ M = 0.0307\ M$$

When solutions are prepared such that their concentration is known directly through their preparation by weighing a pure solid into the container as discussed in Section 4.2, the calculation performed in Example 4.2 is typical.

If a solution is prepared by diluting another solution whose concentration is precisely known (another possibility mentioned in Section 4.2), the following **dilution equation** may be used to calculate molarity.

$$C_B \times V_B = C_A \times V_A \tag{4.1}$$

Here "C" refers to concentration (any unit, but the same on both sides), "V" refers to volume (any unit, but also the same on both sides), "B" refers to "before dilution," and "A" refers to "after dilution." In this case, C_A is to be calculated and the dilution equation is rearranged so as to solve it for C_A. The result is Equation M.3 in Box 4.1.

Example 4.3

What is the molarity of a solution prepared by diluting 10.00 mL of a 4.281-M solution to 50.00 mL?

Solution 4.3

Using Equation M.3, we have the following.

$$C_A = \frac{4.281 \times 10.00}{50.00} = 0.8562 \text{ M}$$

4.3.2 NORMALITY

Normality is similar to molarity except that it utilizes a quantity of chemical called the **equivalent** rather than the mole, and is defined as **the number of equivalents per liter** rather than the number of moles per liter. It is dependent on the specific chemistry of the analysis. It may be calculated by dividing the equivalents of solute by the liters of solution in which that number of equivalents is dissolved, as shown in Equation N.1 in Box 4.2. If there are 2.0 equivalents dissolved per liter, a solution would be referred to as being 2.0 normal, or 2.0 N. The equivalent is either the same as the mole, or it is some fraction of the mole, depending on the reaction involved, and the **equivalent weight**, or **the weight of one equivalent**, is either the same as the formula weight, or some fraction of the formula weight. Normality is either the same as molarity, or some multiple of molarity. Let us illustrate with acids and bases in acid/base neutralization reactions.

**BOX 4.2   CALCULATION OF THE NORMALITY OF A
SOLUTION FROM SOLUTION PREPARATION DATA**

$$\text{normality} = \frac{\text{equivalents of solute}}{\text{liters of solution}} \tag{N.1}$$

$$\text{normality} = \text{molarity} \times \text{equivalents per mole} \tag{N.2}$$

$$\text{normality} = \frac{\text{grams of solute/EW}}{\text{liters of solution}} \tag{N.3}$$

$$C_A = \frac{C_B \times V_B}{V_A} \tag{N.4}$$

These are formulas for calculating the normality of a solution from solution preparation data as derived in the text. EW = equivalent weight; C_A = concentration after dilution; C_B = concentration before dilution; V_B = volume before dilution; V_A = volume after dilution.

The **equivalent weight of an acid in an acid/base neutralization reaction** is defined as the formula weight divided by the number of hydrogens lost per formula of the acid in the reaction. Acids may lose one or more hydrogens (per formula) when reacting with a base.

$$HCl + NaOH \rightarrow NaCl + H_2O \text{ (one hydrogen lost by HCl)} \qquad (4.2)$$

$$H_2SO_4 + 2\,NaOH \rightarrow Na_2SO_4 + 2\,H_2O \text{ (two hydrogens lost by } H_2SO_4) \qquad (4.3)$$

The equivalent weight of HCl is the same as its formula weight, and the equivalent weight of H_2SO_4 is half the formula weight. The equivalent weight is analogous to the formula weight in our molarity discussions. In the case of the HCl, there is one hydrogen lost per formula, one equivalent is therefore the same as the mole, and the equivalent weight is the same as the formula weight. In the case of H_2SO_4 in the reaction in Equation 4.3, there are two hydrogens lost per formula, 2 equivalents per mole when used as in the above reaction, and the equivalent weight is half the formula weight.

The **equivalent weight of a base in an acid/base neutralization reaction** is defined as the formula weight divided by the number of hydrogens accepted per formula of the base in the reaction. This definition is based on the Bronsted–Lowry concept of a base (a compound that accepts hydrogens when reacting with an acid). Thus, all hydroxides, carbonates, ammonia, etc. are Bronsted–Lowry bases and accept hydrogens when reacting with acids. Bases may accept one or more hydrogens (per formula) when reacting with an acid. In both reactions above (Equations 4.2 and 4.3), sodium hydroxide is accepting one hydrogen per formula unit. Indeed, there is only one hydroxide group in one formula of NaOH to combine with one hydrogen to form water and therefore sodium hydroxide can only accept one hydrogen per formula.

Meanwhile, sodium carbonate is capable of accepting either one or two hydrogens per formula.

$$Na_2CO_3 + HCl \rightarrow NaCl + NaHCO_3 \text{ (one hydrogen accepted by } Na_2CO_3) \qquad (4.4)$$

$$Na_2CO_3 + 2\,HCl \rightarrow 2\,NaCl + H_2CO_3 \text{ (two hydrogens accepted by } Na_2CO_3) \qquad (4.5)$$

Thus, the equivalent weight of sodium carbonate may either be equal to the formula weight divided by 1 (105.989 g per equivalent; Equation 4.4) or the formula weight divided by 2 (52.9945 g per equivalent; Equation 4.5) depending on which reaction is involved.

Example 4.4

What is the normality of a solution of sulfuric acid if it is used as in Reaction 4.3 and there is 0.248 mol dissolved in 250.0 mL of solution?

Solution 4.4

Normality is calculated here by combining Equations N.1 and N.2 in Box 4.2:

$$\begin{aligned}
\text{normality} &= \frac{\text{equivalents of solute}}{\text{liters of solution}} \\[6pt]
&= \frac{\text{moles of solute} \times \text{equivalents per mole}}{\text{liters of solution}} \\[6pt]
&= \frac{0.248 \text{ mol} \times 2 \text{ equivalents/mol}}{0.2500 \text{ L}} \\[6pt]
&= 1.98 \text{ N}
\end{aligned}$$

(Note that there are exactly 2 equivalents per mole and that the number 2 does not diminish the number of significant figures in the answer because it has an infinite number of significant figures. This is also true in other calculations in this section.)

If the number of grams of the solute used is given rather than moles or equivalents, these grams must be converted to equivalents by dividing by the equivalent weight (EW) as shown in Equation N.3 in Box 4.2.

Example 4.5

What is the normality of a solution of oxalic acid dihydrate ($H_2C_2O_4 \cdot 2H_2O$, formula weight 126.07 g/mol) if it is to be used as in the following reaction and 0.4920 g of it is dissolved in 250.0 mL of solution?

$$H_2C_2O_4 + 2\ NaOH \rightarrow Na_2C_2O_4 + 2\ H_2O$$

Solution 4.5

In the reaction given, two hydrogens are lost per molecule of oxalic acid. The equivalent weight is therefore the formula weight divided by 2, or 63.035 g per equivalent. The normality is calculated according to Equation N.3 in Box 4.2.

$$\text{normality} = \frac{(0.4920/63.035)\text{equivalents}}{0.2500\ L} = 0.03122\ N = 0.0307\ M$$

Example 4.6

What is the normality of a solution of sodium carbonate if 0.6003 g of it is dissolved in 500.00 mL of solution and it is to be used as in Equation 4.5?

Solution 4.6

As stated above, the equivalent weight of sodium carbonate as used in Equation 4.5 is the formula weight divided by 2 because it accepted two hydrogens in the reaction. Thus, utilizing Equation N.3 in Box 4.2, we have the following:

$$\text{normality} = \frac{(0.6003/52.9945)\ \text{equivalents}}{0.5000\ L} = 0.03122\ N = 0.0307\ M$$

Normality applies mostly to acid/base neutralization, but the concentrations of other chemicals used in other kinds of reactions, such as in oxidation–reduction reactions, may also be expressed in normality. Oxidation/reduction will be discussed in Chapter 5.

4.4 REVIEW OF SOLUTION PREPARATION

We now consider the preparation of solutions. Solutions are prepared for a wide variety of reasons. We have already discussed the use of standard solutions in titrimetric analysis and that these solutions sometimes must be prepared with high precision and accuracy so that their concentrations may be known directly through the preparation process. Even if the need for good precision and accuracy through the preparation is not necessarily important, an analyst frequently prepares other solutions with concentrations known less precisely. Thus, the familiarity with solution preparation schemes, highly precise or not, is very important.

4.4.1 SOLID SOLUTE AND MOLARITY

To prepare a given volume of a solution of a given molarity of solute when the weight of the pure, solid solute is to be measured, it is necessary to calculate the grams of solute that are required. This number of grams can be calculated from the desired molarity and the volume of solution that is desired. Grams can be calculated by multiplying moles by the formula weight.

$$\text{moles} \times \frac{\text{grams}}{\text{mole}} = \text{grams} \tag{4.6}$$

The moles that are required can be calculated from the volume (liters) and molarity (moles per liter).

$$\text{moles} = \frac{\text{moles}}{\text{liter}} \times \text{liters} \tag{4.7}$$

Combining Equations 4.6 and 4.7 gives us the formula needed.

$$\text{grams} = \text{liters} \times \frac{\text{moles}}{\text{liter}} \times \frac{\text{grams}}{\text{mole}} \tag{4.8}$$

The liters in Equation 4.8 are the liters of solution that are desired, the moles per liter is the molarity desired, and the grams per mole is the formula weight of the solute. Thus, we have

$$\text{grams to weigh} = L_D \times M_D \times FW_{SOL} \tag{4.9}$$

where L_D refers to the liters that are desired, M_D to desired molarity, and FW_{SOL} to the formula weight of solute. This is Equation PM.1 in Box 4.3. The grams of solute thus calculated is weighed, and placed in the container. Water is added to dissolve the solute and to dilute the solution to volume. Following this, the solution is shaken to make it homogeneous.

BOX 4.3 PREPARATION OF MOLAR SOLUTIONS

If the solute is to be weighed into the preparation vessel:

$$\text{grams to weigh} = L_D \times M_D \times FW_{SOL} \tag{PM.1}$$

If the solute is already dissolved, but the solution is to be diluted:

$$V_B = \frac{C_A \times V_A}{C_B} \tag{PM.2}$$

These are formulas used for preparing molar solutions as derived in the text. L_D = liters desired; M_D = molarity desired; FW_{SOL} = formula weight of the solute, C_A = concentration after dilution; C_B = concentration before dilution; V_B = volume before dilution; V_A = volume after dilution.

Example 4.7

How would you prepare 500.0 mL of a 0.20-M solution of NaOH from pure, solid NaOH?

Solution 4.7

Using Equation PM.1 in Box 4.3, we have the following.

$$\text{grams to weigh} = L_D \times M_D \times FW_{SOL}$$

$$\text{grams to weigh} = 0.5000 \text{ L} \times 0.20 \text{ mol/L} \times 39.997 \text{ g/mol} = 4.0 \text{ g}$$

The analyst would weigh 4.0 g of NaOH, place it in a container with a 500-mL calibration line, add water to dissolve the solid, and then dilute to volume, and shake, stir, or swirl.

If the solute is a liquid (somewhat rare), the weight calculated from Equation PM.1 can, but rather inconveniently, be measured on a balance. However, this weight can also be converted to milliliter using the density of the liquid. In this way, the volume of the liquid can be measured, rather than its weight, and it can be pipetted into the container.

4.4.2 Solid Solute and Normality

To determine the weight of a solid that is needed to prepare a solution of a given normality, we can derive an equation in a manner similar to that in Section 4.4.1 for molarity.

$$\text{grams} = \text{liters} \times \frac{\text{equivalents}}{\text{liter}} \times \frac{\text{grams}}{\text{equivalent}} \tag{4.10}$$

$$\text{grams to weigh} = L_D \times N_D \times EW_{SOL} \tag{4.11}$$

This is Equation PN.1 in Box 4.4, in which L_D is the liters of solution desired, N_D is the normality desired, and EW_{SOL} is the equivalent weight of the solute.

As in Section 4.3, acid/base neutralization reactions will be illustrated here. To calculate the equivalent weight of an acid, the balanced equation representing the reaction in which the solution is to be used is needed so that the number of hydrogens lost per formula in the reaction can be deter-

---

BOX 4.4   PREPARATION OF NORMAL SOLUTIONS

If the solute is to be weighed into the preparation vessel:

$$\text{grams to weigh} = L_D \times N_D \times EW_{SOL} \tag{PN.1}$$

If the solute is already dissolved, but the solution is to be diluted:

$$V_B = \frac{C_A \times V_A}{C_B} \tag{PN.2}$$

These are formulas used for preparing normal solutions as derived in the text. L_D = liters desired; N_D = normality desired; EW_{SOL} = Equivalent weight of the solute, C_A = concentration after dilution; C_B = concentration before dilution; V_B = volume before dilution; V_A = volume after dilution.

mined. The equivalent weight of an acid is the formula weight of the acid divided by the number of hydrogens lost per molecule (see Section 4.3).

Example 4.8

How many grams of KH_2PO_4 are needed to prepare 500.0 mL of a 0.200-N solution if it is to be used as in the following reaction?

$$KH_2PO_4 + 2\ KOH \rightarrow K_3PO_4 + 2\ H_2O$$

Solution 4.8

There are two hydrogens lost per formula of KH_2PO_4. Therefore the equivalent weight of KH_2PO_4 is the formula weight divided by 2. Utilizing Equation PN.1 in Box 4.4, we have the following:

$$\text{grams to weight} = 0.5000\ L \times 0.200\ \text{equiv/L} \times \frac{136.09}{2}\ \text{g/equiv} = 6.80\ \text{g}$$

4.4.3   SOLUTION PREPARATION BY DILUTION

If a solution is to be prepared by diluting another solution, whether high precision and accuracy are important or not, the **dilution equation**, Equation 4.1, is again used.

$$C_B \times V_B = C_A \times V_A \tag{4.12}$$

where C, V, B, and A have the same meaning as discussed previously. For solution preparation, V_B, the volume of the solution to be diluted (the volume before dilution), is calculated so that the analyst can know how much of this more concentrated solution to measure out to prepare the less concentrated solution. Remember that V_B and V_A can have any volume unit, but the units for both V_B" and V_A must be the same unit (for example, mL). The same is true of the concentration units. This equation, solved for V_B is shown in Box 4.3 as Equation PM.2 and again in Box 4.4 as Equation PN.2.

Example 4.9

How many milliliters of 12 M hydrochloric acid are needed to prepare 500.0 mL of a 0.60-M solution?

Solution 4.9

Using Equation PM.2 in Box 4.3, we have the following.

$$V_B = \frac{0.60\ M \times 500.0}{12.0\ M} = 25\ \text{mL}$$

4.5   STOICHIOMETRY OF TITRATION REACTIONS

At the equivalence point of a titration, exact stoichiometric amounts of the reactants have reacted, i.e., the amount of titrant added is the exact amount required to consume the amount of substance titrated in the reaction flask. If the reaction is one-to-one in terms of moles (moles of titrant equals the moles of substance titrated as is the case for the reaction represented by Equation 4.2, for example), then the moles of titrant added equals the moles of substance titrated consumed.

$$\text{moles}_T = \text{moles}_{ST} \qquad (4.13)$$

where moles_T represents the moles of titrant added to get to the equivalence point and moles_{ST} represents the moles of substance titrated in the reaction flask. If the reaction is 2:1 in terms of moles as in Equation 4.3, then the moles of titrant is twice the moles of substance titrated.

$$\text{moles}_T = \text{moles}_{ST} \times 2 \qquad (4.14)$$

and the moles of substance titrated is the half the moles of titrant.

$$\text{moles}_{ST} = \text{moles}_T \times \tfrac{1}{2} \qquad (4.15)$$

or vice versa, depending on which substance is the titrant and which is the substance titrated, And so it goes for 3:1, 3:2, etc. reactions. In general terms, then, we have the following:

$$\text{moles}_{ST} = \text{moles}_T \times \text{mole ratio (ST/T)} \qquad (4.16)$$

where mole ratio (ST/T) is the ratio of moles of substance titrated (ST) to the moles of titrant (T) as gleaned from the chemical equation representing the reaction. When working with moles as in these equations, the molarity method of expressing concentration and formula weights are needed in the calculations. For example, to convert the buret reading for the titrant in liters (L_T) to moles of titrant (which can then be used in Equation 4.16), we must multiply L_T by the titrant molarity (mol/L) as the following shows.

$$\text{liters} \times \frac{\text{moles}}{\text{liter}} = \text{moles} \qquad (4.17)$$

Thus, we have the following:

$$L_T \times M_T = \text{moles}_T \qquad (4.18)$$

Combining Equations 4.16 and 4.18, we have the following:

$$\text{moles}_{ST} = L_T \times M_T \times \text{mole ratio (ST/T)} \qquad (4.19)$$

This equation will be useful as we explore titration calculations in Sections 4.6 and 4.7.

Some analysts prefer to work with equivalents rather than with moles. In that case, the normality method of expressing concentration is used and the equivalent weight is needed rather than the formula weight. The equivalent weight of one substance reacts with the equivalent weight of the other substance. In other words, the reaction is *always* one-to-one, one equivalent of one substance *always* reacts with one equivalent of the other. Thus, we can write the following as a true statement at the equivalence point of the titration.

$$\text{equiv}_T = \text{equiv}_{ST} \qquad (4.20)$$

where equiv_T represents the equivalents of titrant used to get to the equivalence point and equiv_{ST} represents the equivalents of substance titrated that reacted in the reaction flask. Multiplying the liters of titrant (L_T) by the normality of the titrant (N_T) gives the equivalents of titrant.

$$\text{liters} \times \frac{\text{equivalents}}{\text{liters}} = \text{equivalents} \qquad (4.21)$$

Therefore, we have the following:

$$L_T \times N_T = \text{equiv}_T \qquad (4.22)$$

Combining this with Equation 4.20, we have the following.

$$\text{equiv}_{ST} = L_T \times N_T \qquad (4.23)$$

Like Equation 4.19, this equation will also be useful as we explore titration calculations in Sections 4.6 and 4.7.

4.6 STANDARDIZATION

Standardization was defined in Section 4.2 as a titration experiment in which the concentration of a solution becomes known to a high degree of precision and accuracy. In a standardization experiment, the solution being standardized is compared with a known standard. This known standard can be either a solution that is already a standard solution or it can be an accurately weighed solid material. In either case, the solute of the solution to be standardized reacts with the known standard in the titration vessel. If the solution to be standardized is the titrant, then the known standard is the substance titrated, and vice versa. We now describe these two methods and the calculations involved.

4.6.1 STANDARDIZATION USING A STANDARD SOLUTION

In the case of standardization using a standard solution, it is a volume of the substance titrated that is measured into the reaction flask. Since volume in liters multiplied by molarity gives moles (see Equation 4.17), Equation 4.19 becomes

$$L_T \times M_T \times \text{mole ratio (ST/T)} = L_{ST} \times M_{ST} \qquad (4.24)$$

Also, since volume in liters multiplied by normality gives us equivalents (see Equation 4.21), Equation 4.23 becomes

$$L_T \times N_T = L_{ST} \times N_{ST} \qquad (4.25)$$

As in the discussion accompanying Equation 4.1, the volume of titrant and the volume of substance titrated may be expressed in any volume unit as long as they are both the same unit. Thus, we have Equations 4.26 and 4.27.

$$V_T \times M_T \times \text{mole ratio (ST/T)} = V_{ST} \times M_{ST} \qquad (4.26)$$

$$V_T \times N_T = V_{ST} \times N_{ST} \qquad (4.27)$$

These equations are summarized in Box 4.5 as Equations SS.1 and SS.2. In these equations, V_T and V_{ST} represent the volumes of titrant and substance titrated, respectively.

The experiment is performed by precisely measuring the volume of the solution of substance titrated (either the solution to be standardized or the known standard solution) into the reaction flask and then titrating it with the other solution. At the end point, V_T and V_{ST} are known and one

BOX 4.5   STANDARDIZATION USING A STANDARD SOLUTION

If working with molarities, either M_T or M_{ST} is to be calculated:

$$V_T \times M_T \times \text{mole ratio (ST/T)} = V_{ST} \times M_{ST} \qquad \text{(SS.1)}$$

If working with normalities, either N_T or N_{ST} is to be calculated:

$$V_T \times N_T = V_{ST} \times N_{ST} \qquad \text{(SS.2)}$$

These are formulas, as derived in the text, used for calculating the molarity or normality of a solution when standardizing using another standard solution. V_T = volume of titrant; M_T = molarity of titrant; ST = substance titrated; T = titrant; V_{ST} = volume of substance titrated; M_{ST} = molarity of substance titrated; N_T = normality of titrant; N_{ST} = normality of substance titrated.

of the two concentrations is known (the known standard). Thus, the other concentration can then be calculated.

Example 4.10

Standardization of a solution of sulfuric acid required 29.03 mL of 0.06477 M NaOH when exactly 25.00 mL of the H_2SO_4, was used. What is the molarity of the H_2SO_4? Refer to Equation 4.3 for the reaction involved.

Solution 4.10

Equation SS.1 in Box 4.5 is used since the question has to do with molarity.

$$V_T \times M_T \times \text{mole ratio (ST/T)} = V_{ST} \times M_{ST}$$

$$mL_{H_2SO_4} \times M_{H_2SO_4} \times \text{mole ratio (NaOH/H}_2SO_4) = mL_{NaOH} \times M_{NaOH}$$

$$29.03 \text{ mL} \times 0.06477 \text{ M} \times 2 = 25.00 \text{ mL} \times M_{NaOH}$$

$$M_{NaOH} = 0.1504 \text{ M}$$

Example 4.11

Standardization of a solution of sulfuric acid required 28.50 mL of 0.1077 N NaOH when exactly 25.00 mL of the H_2SO_4, was used. What is the normality of the H_2SO_4? Refer to Equation 4.3 for the reaction involved.

Solution 4.11

Equation SS.2 in Box 4.5 is used since the question concerns normality.

$$V_T \times N_T = V_{ST} \times N_{ST}$$

$$mL_{H_2SO_4} \times N_{H_2SO_4} = mL_{NaOH} \times N_{NaOH}$$

$$25.00 \text{ mL} \times N_{H_2SO_4} = 28.50 \text{ mL} \times 0.1077 \text{ N}$$

$$N_{H_2SO_4} = \frac{28.50 \text{ mL} \times 0.1077}{25.00 \text{ mL}} = 0.1228 \text{ N}$$

4.6.2 STANDARDIZATION USING A PRIMARY STANDARD

The alternative to the above is using an accurately weighed solid material as the known standard. Such a material is called a **primary standard**. Thus, a primary standard is a material that can be weighed accurately for the purpose of either preparing a standard solution (which then does not have to be standardized) or for comparison to a solution with which it reacts for the purpose of standardizing that solution. For standardization with a primary standard, Equation 4.19 becomes

$$L_T \times M_T \times \text{mole ratio (PS/T)} = \text{moles}_{PS} \qquad (4.28)$$

where PS refers to primary standard. Knowing that grams divided by the formula weight gives moles, we have the following:

$$L_T \times M_T \times \text{mole ratio (PS/T)} = \frac{\text{grams}_{PS}}{FW_{PS}} \qquad (4.29)$$

In addition, Equation 4.23 becomes

$$L_T \times N_T = \frac{\text{grams}_{PS}}{EW_{PS}} \qquad (4.30)$$

These two equations are summarized in Box 4.6 as Equations SP.1 and SP.2, respectively. The solution to be standardized is the titrant. In this case, the volume of the titrant must be expressed in liters as shown.

The experiment consists of weighing the primary standard on an analytical balance into the titration flask, dissolving, and then titrating it with the solution to be standardized.

BOX 4.6 STANDARDIZATION USING A PRIMARY STANDARD

If working with molarities, M_T is to be calculated:

$$L_T \times M_T \times \text{mole ratio (PS/T)} = \frac{\text{grams}_{PS}}{FW_{PS}} \qquad (SP.1)$$

If working with normalities, N_T is to be calculated:

$$L_T \times N_T = \frac{\text{grams}_{PS}}{EW_{PS}} \qquad (SP.2)$$

These are formulas, as derived in the text, used for calculating the molarity or normality of a solution when standardizing using a primary standard. L_T = liters of titrant; M_T = molarity of titrant; PS = primary standard; T = titrant; FW_{PS} = formula weight of primary standard; N_T = normality of titrant; EW_{PS} = equivalent weight of primary standard.

Example 4.12

In a standardization experiment, 0.4920 g of primary standard sodium carbonate (Na_2CO_3) was exactly neutralized by 19.04 mL of hydrochloric acid solution. What is the molarity of the HCl solution? Refer to Equation 4.5 for the reaction involved.

Solution 4.12

Since the problem concerns molarity, Equation SP.1 in Box 4.6 applies.

$$L_T \times M_T \times \text{mole ratio (PS/T)} = \frac{\text{grams}_{PS}}{FW_{PS}}$$

$$L_{HCl} \times M_{HCl} \times \text{mole ratio (Na}_2\text{CO}_3/\text{HCl)} = \frac{\text{grams}_{Na_2CO_3}}{FW_{Na_2CO_3}}$$

$$0.01904 \times M_{HCl} \times \frac{1}{2} = \frac{0.4920}{105.989}$$

$$M_{HCl} = 0.4876 \text{ M}$$

Example 4.13

In a standardization experiment, 0.5067 g of primary standard sodium carbonate (Na_2CO_3) was exactly neutralized by 27.86 mL of a hydrochloric acid solution. What is the normality of the HCl solution? Refer to Equation 4.5 for the reaction involved.

Solution 4.13

Equation SP.2 in Box 4.6 is used as follows.

$$L_{HCl} \times N_{HCl} = \frac{\text{grams}_{Na_2CO_3}}{EW_{Na_2CO_3}}$$

$$0.02786 \text{ L} \times N_{HCl} = \frac{0.5067}{52.9945}$$

$$N_{HCl} = 0.3432 \text{ N}$$

Obviously, the quality of the primary standard substance is ultimately the basis for a successful standardization. This means that it must meet some special requirements with respect to purity, etc., and these are enumerated below.

1. It must be 100% pure, or at least its purity must be known.
2. If it is impure, the impurity must be inert.
3. It should be stable at drying oven temperatures.
4. It should not be hygroscopic—it should not absorb water when exposed to laboratory air.
5. The reaction in which it takes part must be quantitative and preferably fast.
6. A high formula weight is desirable, so that the number of significant figures in the calculated result is not diminished.

Most substances used as primary standards can be purchased as a primary standard grade and this is usually appropriate and sufficient for standardization experiments. However, how a primary standard grade chemical, as sold by a chemical supply company, itself becomes a standard is a legitimate question. In the United States, there is a federal agency that produces and certifies this ultimate standard. The agency is the National Institute for Standardization and Technology (NIST), the same agency that provides standard weights for balance calibration (Chapter 3). NIST manufactures and sells standards (Standard Reference Materials, or SRMs) for a wide variety of laboratory applications. Being the ultimate standards, such materials are expensive and are not used beyond essential standardizations and calibrations and are used for that purpose most often by the chemical supply companies only. Standards distributed by the supply companies are then called "certified reference materials" (CRMs). CRMs are, in modern analytical chemistry language, **traceable** to SRMs.

4.6.3 TITER

The strength (concentration) of a titrant can also be expressed as its "titer." The **titer** of a titrant is defined as the weight (mg) of substance titrated that is consumed by 1 mL of titrant. Thus, it is specific to a particular substance titrated, meaning it is expressed with respect to a specific substance titrated. For example, if the analyst is using an oxalic acid solution to titrate a solution of calcium oxide, the titer of the oxalic acid solution would be expressed as its CaO titer, or the weight of CaO that is consumed by 1 mL of the oxalic acid solution. Titer is typically used for repetitive routine work in which the same titrant is used repetitively to titrate a given analyte.

Example 4.14

What is the titer (expressed in mg/mL) of a solution of oxalic acid dihydrate with respect to calcium oxide if 21.49 mL of it were needed to titrate 0.2203 g of CaO?

Solution 4.14

$$\text{titer} = \frac{\text{milligrams of substance titrated}}{\text{milliliters of titrant}}$$

The titer of the oxalic acid solution with respect to calcium oxide is calculated as follows:

$$\text{titer of oxalic acid solution} = \frac{220.3 \text{ mg of CaO}}{21.49 \text{ mL of titrant}} = 10.25 \text{ mg/mL}$$

4.7 PERCENTAGE ANALYTE CALCULATIONS

The ultimate goal of any titrimetric analysis is to determine the amount of the analyte in a sample, and this also involves a stoichiometry calculation. This amount of analyte is often expressed as a percentage, as it was for the gravimetric analysis examples in Chapter 3. This percentage is calculated via the basic equation for percent used previously for the gravimetric analysis examples.

$$\% \text{ analyte} = \frac{\text{weight of analyte}}{\text{weight of sample}} \times 100 \tag{4.31}$$

As with gravimetric analysis, the weight of the sample (the denominator in Equation 4.31) is determined by direct measurement in the laboratory or by weighing by difference. The weight of the analyte in the sample is determined from the titration data via a stoichiometry calculation. As discussed previously, we calculate moles of substance titrated (in this case, the analyte) as in Equation 4.19.

$$\text{moles}_{\text{analyte}} = L_T \times M_T \times \text{mole ratio (ST/T)} \qquad (4.32)$$

The weight of the analyte is then calculated by multiplying by its formula weight (grams per mole).

$$\text{grams}_{\text{analyte}} = \text{moles}_{\text{analyte}} \times FW_{\text{analyte}} \qquad (4.33)$$

Combining Equations 4.32 and 4.33 we have the following.

$$\text{grams}_{\text{analyte}} = L_T \times M_T \times \text{mole ratio (ST/T)} \times FW_{\text{analyte}} \qquad (4.34)$$

Finally, combining Equations 4.31 and 4.34, we have the following.

$$\% \text{ analyte} = \frac{L_T \times M_T \times \text{mole ratio(ST/T)} \times FW_{\text{analyte}}}{\text{weight of sample}} \times 100 \qquad (4.35)$$

This equation is also given in Box 4.7 as Equation PA.1.

If normality and equivalents are used, we calculate equivalents of analyte as in Equation 4.25 (again, the substance titrated is the analyte).

$$\text{equiv}_{\text{analyte}} = L_T \times N_T \qquad (4.36)$$

The weight of the analyte (the numerator in Equation 4.31) can then be calculated by multiplying the equivalents of substance titrated by the equivalent weight (grams per equivalent).

$$L_T \times N_T \times EW_{ST} = \text{grams}_{ST} \qquad (4.37)$$

BOX 4.7   PERCENT ANALYTE CALCULATIONS

If the molarity of the titrant is to be used:

$$\% \text{ Analyte} = \frac{L_T \times M_T \times \text{mole ratio (ST / T)} \times FW_{\text{analyte}}}{\text{Weight of sample}} \times 100 \qquad (PA.1)$$

If the normality of the titrant is to be used:

$$\% \text{ Analyte} = \frac{L_T \times N_T \times EW_{\text{analyte}}}{\text{Weight of sample}} \times 100 \qquad (PA.2)$$

Back titration, if normalities are to be used:

$$\% \text{ Analyte} = \frac{L_T \times N_T - L_{BT} \times N_{BT} \times EW_{\text{analyte}}}{\text{Weight of sample}} \times 100 \qquad (PA.3)$$

These are formulas, as derived in the text, used for calculating the percent of an analyte in an unknown for three different scenarios. L_T = liters of titrant; M_T = molarity of titrant; ST = substance titrated; T = titrant, FW_{analyte} = formula weight of the analyte; EW_{analyte} = equivalent weight of the analyte; L_{BT} liters of back titrant; N_{BT} = normality of back titrant.

Thus, the percent calculation then becomes the following.

$$\% \text{ analyte} = \frac{L_T \times N_T \times EW_{analyte}}{\text{weight of sample}} \times 100 \qquad (4.38)$$

This is Equation PA.2 in Box 4.7.

Example 4.15

In the analysis of a sample for KH_2PO_4 content, a sample weighing 0.3994 g required 18.28 mL of 0.1011 M KOH for titration. The equation below represents the reaction involved. What is the % KH_2PO_4 in this sample?

$$2 \text{ KOH} + KH_2PO_4 \rightarrow K_3PO_4 + 2 \text{ } H_2O$$

Solution 4.15

Since the question involves molarity, Equation PA.1 in Box 4.7 applies and we have the following:

$$\% \text{ analyte} = \frac{L_{KOH} \times M_{KOH} \times \text{mole ratio}\left(\dfrac{KH_2PO_4}{KOH}\right) \times FW_{KH_2PO_4}}{\text{weight of sample}} \times 100$$

$$= \frac{0.01818 \times 0.1011 \times \dfrac{1}{2} \times 136.085}{0.3994} \times 100$$

$$= 31.84\%$$

Example 4.16

In the analysis of a soda ash (impure Na_2CO_3) sample for sodium carbonate content, 0.5203 g of the soda ash required 36.42 mL of 0.1167 N HCl for titration. What is the % Na_2CO_3 in this sample?

$$Na_2CO_3 + 2 \text{ HCl} \rightarrow 2 \text{ NaCl} + H_2CO_3$$

Solution 4.16

Since the question involves normality, Equation PA.2 in Box 4.7 applies and we have the following:

$$\% \text{ analyte} = \frac{L_{HCl} \times N_{HCl} \times EW_{analyte}}{\text{weight of sample}} \times 100$$

There are two hydrogens accepted by sodium carbonate, so the equivalent weight is the formula weight divided by 2, or 52.9945 g per equivalent. Thus, we have the following:

$$\% \text{ } Na_2CO_3 = \frac{0.03642 \times 0.1167 \times 52.9945}{0.5203} \times 100$$

$$= 43.29\%$$

4.8   VOLUMETRIC GLASSWARE

As indicated in Section 4.1 (and as should be apparent from the discussion thus far in this chapter), titrimetric analysis methods utilize solution chemistry heavily, and therefore volumes of solutions are prepared, measured, transferred, and analyzed with some degree of frequency in this type of analysis. It should not be surprising that analytical laboratory workers need to be well versed in the selection and proper use of the glassware and devices used for precise volume measurement.

The three volumetric glassware products we will discuss are the volumetric flask, the pipet, and the buret. Drawings of the flask and the pipet were shown in Figure 4.2. Let us study the characteristics of each type individually.

4.8.1   VOLUMETRIC FLASK

The container that is typically used for precise solution preparation is the volumetric flask. This container has a single calibration line placed in a narrow diameter neck (see Figure 4.4). The reason a narrow diameter neck is desirable is that the volume of solution can be controlled very precisely. Even a fraction of a drop of solvent gives a noticeable change in the position of the meniscus when preparing the solution. If one were to use a beaker or Erlenmeyer flask for this, it may take as much as several milliliters of solvent to cause noticeable change in the position of the meniscus. Thus, volumes of solutions measured with a volumetric flask are precise to four significant figures.

Conveniently, the volumetric flask can be purchased in a variety of sizes, from 5 mL up to several liters. Figure 4.5 shows some of the various sizes.

The calibration line is affixed so that the indicated volume is **contained** rather than delivered. Accordingly, the legend "TC" is imprinted on the base of the flask, thus marking the flask as a vessel "to contain" the volume indicated as opposed to "to deliver" a volume. The reason this imprint is important is that a contained volume is different from a delivered volume, since a small volume of solution remains adhering to the inside wall of the vessel and is not delivered when the vessel is drained. If a piece of glassware is intended to deliver a specified volume, the calibration must obviously take this small volume into account in the sense that it **will not** be part of the **delivered**

FIGURE 4.4   Left, a photograph of a 1000-mL volumetric flask. Right, a close-up view of the neck of the flask showing the single calibration line.

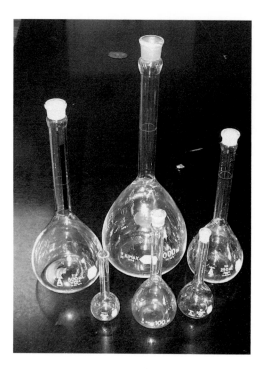

FIGURE 4.5 Volumetric flasks come in a variety of sizes.

volume. On the other hand, if a piece of glassware is not intended to deliver a specified volume, but rather to contain the volume, the calibration must take this small volume into account in the sense that it will be part of the **contained** volume. Other pieces of glassware, namely most pipets and all burets, are "TD" vessels, meaning they are calibrated "to deliver." On the right in Figure 4.6, a close-up photograph of the base of a volumetric flask clearly shows the "TC" imprint (just below the flask's capacity, 1000 mL).

Notice the other markings on the base of the flask in Figure 4.6. The imprint "20°C" indicates that the flask is calibrated to contain the indicated volume when the temperature is 20°C, which is a standard temperature of calibration. This marking is needed since the volume of liquids and liquid solutions changes slightly with temperature. For highest accuracy, the temperature of the contained fluid should be adjusted to 20°C. Notice the "22" marking visible in the left photograph in Figure 4.6. This refers to the size of the tapered top found on some flasks. The stopper that is used can be either a ground glass stopper, to match the flask opening, or it can simply be a plastic tapered

FIGURE 4.6 Close-up views of the labels found on a 1000-mL volumetric flask. On the left, the stopper size (22) is evident, and on the right, besides the capacity of the flask (1000 mL), the TC designation and the 20°C designation are evident.

stopper, which can also be used in a ground glass opening. It has an identical numerical imprint on it (22 in this case), and such number designations on these two items must match to indicate that the stopper is the correct size. Figure 4.7 shows a ground glass stopper being inserted into the ground glass opening. The opening may not necessarily be designed for a tapered stopper as in this example. It is fairly common for a flask to be designed for a "snap cap" rather than a ground glass stopper. In this case, the top of the flask is not tapered, nor is it a ground glass opening. Rather, it has an unusually large lip around the opening, over which the snap cap is designed to seal. A number designation, however, is usually used as with the tapered opening to indicate the size of the opening and the size of the cap required. This number is usually found on both the flask base and the cap as in the taper design. The analyst should always be aware of the type and size of cap required for a given volumetric flask. If any amount of solution should leak out due to an improperly sealed cap while shaking before the solution is homogeneous, its concentration cannot be trusted. Figure 4.8 shows a volumetric flask with the large lip being fitted with a snap cap and the imprint on the flask indicating the size of the cap required.

Some volumetric glassware products have a large "A" imprint on the label. This designates the item as a Class "A" item, meaning that more stringent calibration procedures were undertaken when it was manufactured. Class "A" glassware is thus more expensive, but it is most appropriate when highly precise work is important. This imprint may be found on both flasks and pipets.

One problem with the volumetric flask exists because of its unique shape, and that is the difficulty in making prepared solutions homogeneous. When the flask is inverted and shaken, the solution in the neck of the flask is not agitated. Only when the flask is set upright again is the solution drained from the neck and mixed. A good practice is to invert and shake at least a dozen times in ensure homogeneity.

Volumetric flasks should **not** be used to prepare solutions of reagents that can etch glass (such as sodium hydroxide and hydrofluoric acid), since if the glass is etched, its accurate calibration is lost. Volumetric flasks should **not** be used for storing solutions. Their purpose is to prepare solutions

FIGURE 4.7   Tapered ground glass stopper may be used with a flask with a ground glass opening. A tapered plastic stopper may also be used.

FIGURE 4.8 (Left) Top of a volumetric flask designed for a snap cap. (Right) Imprint on the base of the flask indicating the size of snap cap required.

accurately. If they are used for storage then they are not available for their intended purpose. Finally, volumetric flasks should **not** be used to contain solutions when heating or performing other tasks for which their accurate calibration serves no useful purpose. There are plenty of other glass vessels to perform these functions.

4.8.2 PIPET

As indicated previously, most pipets are pieces of glassware that are designed to deliver (TD) the indicated volume. Pipets come in a variety of sizes, and shapes. The most common is probably the **volumetric** or **transfer pipet** shown in Figure 4.9. This pipet, like the volumetric flask, has a single calibration line. It can thus be used in delivering only rather common volumes, meaning whole-number volumes. The correct use of a volumetric pipet is given in Box 4.8. See also Figure 4.10.

We summarize the steps in Box 4.8 as follows: rinse out soapy water or any other solution or wetness held inside the pipet several times with the solution to be transferred. Draw the intended

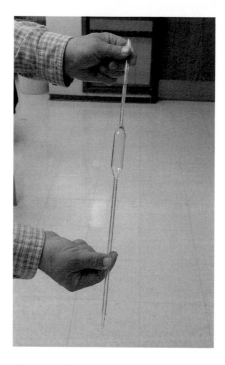

FIGURE 4.9 A 25-mL volumetric, or transfer, pipet.

BOX 4.8 STEPS INVOLVED IN TRANSFERRING A SOLUTION WITH A VOLUMETRIC OR TRANSFER PIPET

Step 1 If the pipet is wet on the inside with soapy water, first rinse thoroughly with tap or distilled water. The pipet may be inverted and drained through the top. If the pipet has been rinsed in this way, was stored filled with water, or is wet on the inside with another solution, wipe the outside dry with a paper towel and, with a rubber pipet bulb (or other device; see Figure 4.10), blow out any water on the inside that does not drain out naturally. Again wipe the area around the tip dry with a paper towel. You may have to blow and wipe again (or several times) to rid the inside of as much wetness as possible. This will help minimize the potential problem of wetness from the inside leaking into the container of the solution being transferred. If the pipet has been in dry storage, you may skip this step.

Step 2 Prepare the device of choice (Figure 4.10) and seat the opening over the top opening of the pipet.

Step 3 If the pipet is wet on the inside, such as with storage water, immerse the tip of the pipet into the solution to be transferred (held in a "rinse beaker" and not the original container) while simultaneously drawing the solution into about half the pipet's capacity. This immediate simultaneous dipping/releasing helps ensure that none of the wetness from the inside the pipet leaks into the original solution container. Rinse so that the entire inside surface becomes wet with this solution. Empty by draining into a sink (again, invert and drain through the top). Repeat this rinsing step at least two more times. If the inside of the pipet is dry, it should still be rinsed with the solution, but the danger of contaminating the solution in the original container is not a problem, so the use of a rinse beaker is not necessary. To prevent contamination or dilution, it is important that any wetness on the inside of the pipet consists of the solution to be transferred.

Step 4 Fill the pipet to well past the calibration line with the solution to be transferred using again one of the devices in Figure 4.10.

Step 5 While holding the meniscus above the calibration line, remove the tip from the solution and wipe with a towel. To avoid contamination from the towel, tilt the pipet to a 45° angle so that a small volume of air is drawn into the tip before wiping.

Step 6 With the pipet held vertically once again carefully release the meniscus to move it to the calibration line.

Step 7 When the meniscus is at the calibration line, stop it there and touch the tip to the *outside* of the receiving vessel so as to remove any solution that may be suspended there.

Step 8 Touch the tip to the *inside* of the receiving vessel and release the solution. The tip should stay in contact with the inside wall of the receiving vessel until the end of the delivery. Be careful not to shake the pipet so that some solution is lost and an air bubble appears in the tip before releasing the solution.

Step 9 When draining is complete, give the pipet a half-twist and remove the pipet from the receiving vessel.

Top

Tip

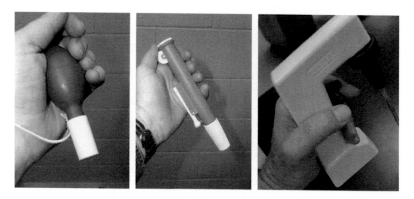

FIGURE 4.10   Left to right: a rubber pipette bulb, a plastic pipette filler, and a motorized, battery-powered pipette filler.

solution up past the calibration line and hold it there while the outside of the pipet tip is wiped dry. Allow the meniscus to be lowered to the calibration line by draining into a waste container or sink. Touch the tip to the outside of the container and allow to drain into the receiving vessel. Touch the tip to the inside of the receiving vessel and remove.

For precise work, volumetric pipets that are labeled as Class A have a certain time in seconds imprinted near the top (see Figure 4.11), which is the time that should be allowed to elapse from the time the finger is released until the pipet is completely drained. The reason for this is that the film of solution adhering to the inner walls after delivery will continue to slowly run down with time, and the length of time one waits to terminate the delivery thus becomes important. The intent with Class A pipets, then, is to take this "run-down" time into account by terminating the delivery in the specified time. After this specified time has elapsed, the delivery is terminated.

Several additional styles of pipets other than the volumetric pipet are available. Two of these are shown in Figure 4.12. These are pipets that have many graduation lines, much like a buret, and are called **measuring pipets**. They are used whenever odd volumes are needed. There are two types of measuring pipets—**the Mohr pipet** (left in Figure 4.12) and the **serological pipet** (right). The difference is whether or not the calibration lines stop short of the tip (Mohr pipet) or go all the way to the tip (serological pipet). The serological pipet is better in the sense that the meniscus needs to be read only once since the solution can be allowed to drain completely out. In this case, the last

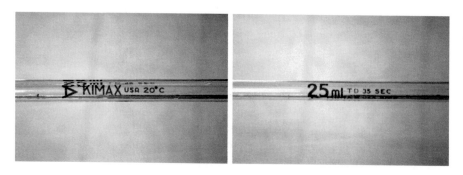

FIGURE 4.11   Top portion of a 25-mL volumetric pipet showing the labels indicating that it is class A (left) and that it has a rundown time of 35 s (right).

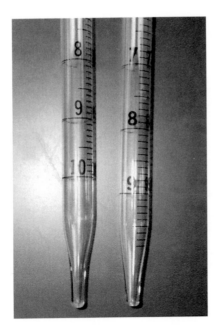

FIGURE 4.12   Tips of two styles of measuring pipets. The Mohr pipet is shown on the left and the serological pipet on the right. The graduation lines on the Mohr pipet stop short of the tip, but on the serological pipet, pass through the tip.

drop of solution is blown out with the pipetting device. With the Mohr pipet, however, the meniscus must be read twice—once before the delivery (such as on the top graduation line) and again after the delivery is complete. The solution flow out of the pipet must be halted at the correct calibration line and the error associated with reading a meniscus is doubled since the meniscus must be read twice. The delivery of 4.62 mL, for example, is done as in Figure 4.13a with a Mohr pipet but as in Figure 4.13b with a serological pipet.

It should be stressed that with the serological pipet, every last trace of solution capable of being blown out must end up inside the receiving vessel. Some analysts find that this is more difficult and perhaps introduces more error than reading the meniscus twice with a Mohr pipet. For this reason, these analysts prefer to use a serological pipet as if it were a Mohr pipet. It is really a matter of personal preference. A double or single frosted ring circumscribing the top of a pipet (above the top graduation line) indicates the pipet is calibrated for blowout (Figure 4.14).

Disposable serological pipets are available. They are termed disposable because the calibration lines are not necessarily permanently affixed to the outside wall of the pipet. The calibration process is thus less expensive, resulting in a less expensive product that can be discarded after use.

Some pipets are calibrated "TC." Such pipets are used to transfer unusually viscous solutions such as syrups, blood, etc. With such solutions, the "wetness" remaining inside after delivery is a portion of the sample and would represent a significant nontransferred volume, which translates into a significant error by normal "TD" standards. With TC pipets, the calibration line is affixed at the factory so that every trace of solution contained within is transferred by flushing the solution out with a suitable solvent. Thus, the pipetted volume is contained within and then quantitatively flushed out. Such a procedure would actually be acceptable with any "TC" glassware, including the volumetric flask. Obviously, diluting the solution in the transfer process must not adversely affect the experiment.

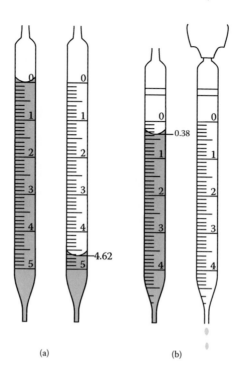

(a)   (b)

FIGURE 4.13 Delivery of 4.62 mL of solution with a Mohr pipet (a) and with a serological pipet (b). With the Mohr pipet, the meniscus is read twice (at 0.00 mL and again at 4.62 mL), but with a serological pipet, it can be read just once (at 0.38 mL) and the solution drained and blown out.

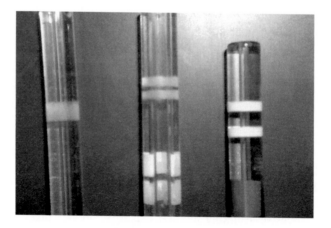

FIGURE 4.14 Tops of three pipets calibrated for blowout showing either a single ring (left) or a double ring (center and right) completely circumscribing the tops of the pipets.

4.8.3 BURET

The buret has some unique attributes and uses. One could call it a specialized graduated cylinder, having graduation lines that increase from top to bottom, with a usual precision of ±0.01 mL, and a stopcock at the bottom for dispensing a solution. There are some variations in the type of stopcock that warrant some discussion. The stopcock itself, as well as the barrel into which it fits, can be made of either glass or Teflon. Some burets have an all glass arrangement, some have a Teflon stopcock and a glass barrel, and some have the entire system made of Teflon. The three types are pictured in Figure 4.15.

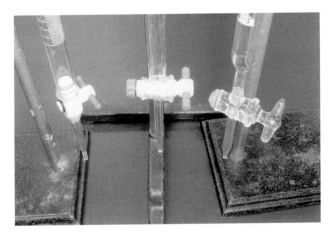

FIGURE 4.15 Three styles of buret stopcocks. (Left) All Teflon. (Center) Teflon stopcock and glass barrel. (Right) All glass.

When the all-glass system is used, the stopcock needs to be lubricated so it will turn with ease in the barrel. There are a number of greases on the market for this purpose. Of course, the grease must be inert to chemical attack by the solution to be dispensed. Also, the amount of grease used should be carefully limited so that excess grease does not pass through the stopcock and plug the tip of the buret. Any material stuck in a buret tip can usually be dislodged with a fine wire inserted from the bottom when the stopcock is open and the buret full of solution. The Teflon stopcocks are free of the lubrication problem. The only disadvantage is that the Teflon can become deformed, and this can cause leakage.

The correct way for a right-handed person to position the hands to turn the buret's stopcock during a titration is shown in Figure 4.16. The natural tendency with this positioning is to pull the stopcock in

FIGURE 4.16 Correct way for a right-handed person to position his or her hands when performing a titration.

as it is turned. This will prevent the stopcock from being pulled out, causing the titrant to bypass the stopcock. The other hand is free to swirl the flask as shown. This may appear a bit clumsy at first, but it is the most efficient way to perform a titration, eliminating the novice's usual one-handed technique.

4.8.4 CLEANING AND STORING PROCEDURES

The use of clean glassware is of utmost importance when doing a chemical analysis. In addition to the obvious need of keeping the solution free of contaminants, the walls of the vessels, particularly the transfer vessels (burets and pipets), must be cleaned so the solution will flow freely and not "bead up" on the wall as the transfer is performed. If the solution beads up, it is obvious that the pipet or buret is not delivering the volume of solution that one intends to deliver. It also means that there is a greasy film on the wall that could introduce contaminants. The analyst should examine, clean, and reexamine his/her glassware in advance so that the free flow of solution down the inside of the glassware is observed. For the volumetric flask, at least the neck must be cleaned in this manner so as to ensure a well-formed meniscus.

Both hand-washing and machine-washing/drying procedures are in use in industrial analytical laboratories. The variety of available soaps include alkaline phosphate-based, as well as phosphate-free. While phosphate-based soaps are quite satisfactory for cleaning purposes because phosphate helps to soften hard tap water, concerns for the environment as well as for phosphate contamination from soap residues are important considerations. In any case, thorough rinsing with distilled water, often at an elevated temperature, is important. Sometimes, to neutralize the effect of alkaline cleaners, acid rinses are important. However, acid residue must also be removed by thorough rinsing with distilled water. Programmable washing machines are available that can cycle the glassware through all of the above procedures.

Hand-washing typically involves the appropriate soap and a brush. For burets, a cylindrical brush with a long handle (buret brush) is used to scrub the inner wall. With flasks, a bottle or test tube brush is used to clean the neck. Also, there are special bent brushes available to contact and scrub the inside of the base of the flask.

Pipets pose a special problem. Brushes cannot be used because of the shape of some pipets and the narrowness of the openings. In this case, if soap is to be used, one must resort to soaking with a warm soapy water solution for a period of time proportional to the severity of the particular cleaning problem. Commercial soaking and washing units are available for this latter technique. Soap tablets are manufactured for such units and are easy to use.

In the past, chromic acid solutions have been used for cleaning. These solutions consist of concentrated sulfuric acid in which solid potassium dichromate has been dissolved. Because of safety concerns and concerns with chromium contamination in the environment, chromic acid has been essentially eliminated from use.

Once the glassware has been cleaned (by whatever method), one should also take steps to keep it clean. One technique is to rinse the items thoroughly with distilled water and then dry them either in the open air, or in a dryer. Following this, they are cooled and stored in a drawer. For shorter periods, it may be convenient to store them in a soaker under distilled water. This prevents their possible recontamination during dry storage. When attempting to use a soaker-stored pipet or buret, however, it must be remembered that the thin film of water is present on inner walls and this must be removed by rinsing with the solution to be transferred, being careful not to contaminate this solution in the process.

4.9 PIPETTERS, AUTOMATIC TITRATORS, AND OTHER DEVICES

4.9.1 PIPETTERS

Alternatives to the glass pipets themselves, called **pipetters,** are available and used rather extensively. Such devices are positive displacement devices, meaning that 100% of the liquid drawn in is forced out with a plunger rather than allowed to flow out by gravity. Volumes transferred using such devices usually do not exceed 10 mL and are most often used for very small volumes, such as from 0 to 500 μL. Those used for microliter volumes are appropriately called **micropipets** or **micropipetters**. The typical pipetter employs a bulb concealed within a plastic fabricated body, a spring-loaded push button at the top, and a nozzle at the bottom for accepting a plastic disposable tip. They may be fabricated for either single or variable volumes. In the latter case, a ratchet-like device with a digital volume scale is used to "dial in" the desired volume. Examples are shown in Figures 4.17 and 4.18. Figure 4.19 shows a micropipet that is electronic, meaning that the solution is drawn in, and also dispensed, by tapping a button on the top.

A term often used in reference to such devices is **repeater pipetters** because of the ease with which a given volume can be repeatedly transferred. In addition, some devices use multiple tips

FIGURE 4.17   One brand of commercial micropipet. The close-up view on the right shows that it is "dialed in" to deliver 500 μL, or 0.500 mL.

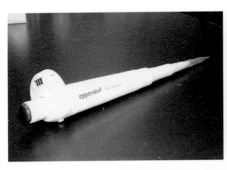
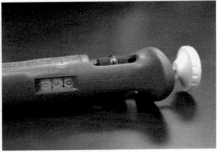

FIGURE 4.18   Another brand of commercial micropipet also set to deliver 500 μL.

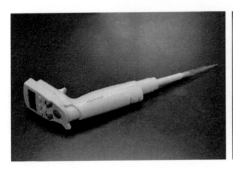

FIGURE 4.19   Battery-powered electronic micropipet also set to deliver 500 μL.

for transferring a given volume up to eight times in just one stroke. These are especially useful in conjunction with well plates that often have 96 small depressions (wells) into each of which a given volume must be transferred.

4.9.2 BOTTLE-TOP DISPENSERS

Bottle-top dispensers, often called **Repipets**™, are quite popular.* These are devices that fit on the top of reagent bottles threaded to receive screw-caps. The dispensers themselves have screw caps that screw onto the bottles. The caps, however, are fitted with a hand pump with a plunger that draws liquid from the bottle in the upstroke and then dispense a calibrated volume on the downstroke through a glass tip. Such devices are convenient and help prevent contamination of the reagent from the various pipets that the analyst might use for a transfer.

4.9.3 DIGITAL BURETS AND AUTOMATIC TITRATORS

There are also various devices that are commonly used for titrations in place of the glass burets previously described. A digital buret, for example, is an electronically controlled bottle-top dispenser that delivers 0.01-mL increments from a reagent bottle containing the titrant. There are also automatic titrators, such as that shown in Figure 4.20. Such titrators draw the titrant from a reagent bottle and store it in a built-in reservoir fitted with a plunger. The titration is performed by pressing a key on the keypad. One tap of the key delivers 0.01 mL to the titration flask.

FIGURE 4.20   Automatic titrator (or autotitrator). A pump draws the titrant from the reagent bottle on the left and fills the reservoir in the back of the unit. Pressing a key on the keypad in the foreground delivers titrant to the flask on the right as the solution is automatically stirred. The volume delivered is displayed on a digital readout.

---

* "Repipet" is a trademark of Barnstead-Thermolyne Company.

4.10   CALIBRATION OF GLASSWARE AND DEVICES

It is important to remember that whatever volume measurement device is used, the analyst depends on it to deliver the indicated volume, i.e., it must be properly calibrated. Analytical chemists and technicians often depend on Class A glassware to represent accurately calibrated glassware and hence do not usually feel the need for independent calibration. However, where non-Class A glassware and alternative transfer devices are used, there is legitimate concern over their proper calibration.

The calibration of glassware and devices involves the measurement of the weight of a delivered volume, since laboratory workers can be quite confident in the accuracy of weight measurements in the laboratory because of the common use of calibrated analytical balances. The usual procedure is to weigh an empty glass-weighing bottle, dispense the volume of water into the bottle, and then weigh it again. The weight of the water is then converted to volume using the known density of water at the temperature of the water. If the volume is not within the accepted precision of the device, the analyst concludes that the device is out of calibration and is discarded or returned to the manufacturer.

4.11   ANALYTICAL TECHNIQUE

Previously, we discussed the need to know when a weight measurement needs to be precise and when it does not need to be precise so that the appropriate balance is chosen for the measurement. We now repeat many of our previous comments but for volume measurements rather than weight.

The question of which piece of glassware to use for a given procedure, such as for solution preparation (pipet or graduated cylinder, beaker, or volumetric flask), remains. An analyst must be able to recognize when a high degree of precision is important and when it is not, so that precise volumetric glassware—pipets, burets, and volumetric flasks—is used when precision is important and regular glassware—graduated cylinders and beakers—is used when precision is not important.

First, volumetric glassware should **not** be used for any volume measurement when the overall objective is strictly qualitative or when quantitative results are to be reported to 2 significant figures or lower. Whether the volume measurements are for preparing solutions to be used in such a procedure, or for transferring an appropriate volume of solution or solvent for such an analysis, etc., they need not be accurate since the outcome will be either only qualitative or not necessarily accurate. Such volumes should, however, be measured with other marked glassware, such as graduated cylinders or marked beakers and flasks.

Second, if the results of a quantitative analysis are to be reported to 3 or more significant figures, then volume measurements that enter directly into the calculation of the results should be made with volumetric glassware so that the accuracy of the analysis is not diminished when the calculation is performed.

Third, if a volume measurement is only incidental to the overall result of an accurate quantitative analysis, then it need not be precise. This means that if the volume measurement to be performed has no bearing whatsoever on the quantity of the analyte tested, but is only needed to support the chemistry or other factor of the experiment, then it need not be precise.

Fourth, if a volume measurement does directly affect the numerical result of an accurate quantitative analysis (in a way other than entering directly into the calculation of the results), then volumetric glassware must be used.

These last two points require that the analyst carefully consider the purpose of the volume measurement and whether it will directly affect the quantity of analyte tested, the numerical value to be reported. It is often true that volume measurements taken during such a procedure need not be precise even though the analytical results are to be precisely reported.

For additional interesting information concerning volumetric glassware, see Application Note 4.1.

APPLICATION NOTE 4.1: GLASSWARE CALIBRATION PROCESS

Throughout this chapter, and especially here in Sections 4.10 and 4.11, we have discussed the need for analytical laboratory workers to have confidence that the volumes of solutions that are contained in volumetric flasks and transferred using volumetric pipets are accurate and highly precise volumes. We have discussed how we can come to have the confidence that the calibration marks found on these high-precision pieces of glassware are in exactly the right place (Section 4.10). We have also discussed what particular experimental situations call for this type of glassware (Section 4.11) and why such high precision is necessary in those situations. Two question remain: How do the glassware manufacturers determine exactly where on the neck of the flask and where on the top of the pipet the calibration mark needs to be imprinted, and then, what tools are used to actually place the mark right on that exact spot? After all, it does vary from flask to flask and from pipet to pipet.

To prove this latter point, compare two volumetric flasks of the same size by setting them together, side by side, on the laboratory bench. Are the calibration marks at exactly the same vertical location? Chances are they are not. So how is it determined that the calibration mark should be higher up on one flask and lower down on another? Check out the Web site below to get some answers.

Web URL: http://www.borosil.com/products/scientificindustrial/sc-borosil-advantage/calibration-process/

EXPERIMENTS

EXPERIMENT 8: HIGH-PRECISION GLASSWARE: A CALIBRATION EXPERIMENT

Various high-precision volume-measuring devices are tested to see how closely each comes to the volumes intended to be delivered or contained. The volume for each is 10 mL of water. The volumes delivered or contained will be weighed to see how closely in each case the weights match the theoretical weight of 10 mL of water at 20°C. The suggested devices, as indicated in Figure 4.21, are a 10-mL volumetric pipet, a 10-mL serological pipet, a 10-mL buret, a 50-mL buret, a digital buret, an autotitrator, a micropipet, and a 10-mL volumetric flask. Your instructor may choose any or all of the above and may also add others to the list. The weights will be determined on three different balances, a top-loader that has a precision of ±0.1 g, a top-loader that has a precision of ±0.01 g, and an analytical balance that has a precision of ±0.0001 g. With these three weights, you should be able to determine what kind of precision is needed in the weight measurement to conclude which balances do the best job of measuring 10 mL of water. The theoretical weight of 10 mL of water at 20°C is given in the table in Figure 4.21. Your instructor will demonstrate the use of each volume-measuring device with which you may not be familiar.

1. Prepare a table in the data section of your notebook similar to the table in Figure 4.21.
2. Obtain a clean and dry 50-mL beaker. Weigh the beaker on all three balances and record the weights in your table. Also obtain a clean and dry 10-mL volumetric flask and weigh it on all three balances, recording the weights in your table.
3. Select a station in your laboratory that has one of the volume-measuring devices that deliver the volume (all but the volumetric flask are intended to deliver a volume) and proceed to use the device to precisely deliver 10 mL of water to the beaker. Then, weigh the beaker containing the water on all three balances, recording each weight in your table. Subtract the weight of the empty beaker from the weight of the beaker containing the water to get the weight of the water. Record this weight of water in the table.

The "theoretical" weight of 10 mL of water at 20°C:

Balance A: <u>10.0 g</u>   Balance B: <u>9.98 g</u>   Balance C: <u>9.9820</u>

Weight of empty beaker: Balance A: _____   Balance B: _____   Balance C: _____

		Balance A	Balance B	Balance C
Item Used ↓				
10-mL vol. pipet	Weight of beaker + water			
	Weight of water			
10-mL buret	Weight of beaker + water			
	Weight of water			
50-mL buret	Weight of beaker + water			
	Weight of water			
Auto-titrator	Weight of beaker + water			
	Weight of water			
Digital buret	Weight of beaker + water			
	Weight of water			
Serological pipet	Weight of beaker + water			
	weight of water			
10-mL vol. flask	Weight of empty flask			
	Weight of flask + water			
	Weight of water			
Micropipetter	Weight of beaker + water			
	Weight of water			

FIGURE 4.21   Suggested data table for Experiment 8.

4. Repeat Step 3 for all devices intended to deliver 10 mL. If a micropipet is one of your choices, use it as many times as needed to deliver 10 mL. For example if a 2.5-mL micropipet is used, use it four times.

5. For the volumetric flask, fill it with water so that the meniscus rests on the calibration mark and then weigh on all three balances. Subtract the weight of the empty flask from the weight of the flask with the water to obtain the weight of the water alone. The 50-mL beaker is not used when testing the flask.

6. In the "Results" section of your notebook, write a short statement about each method of measuring 10 mL of water, noting for which method(s), if any, the analytical balance was needed to see any difference in the measured weight and the theoretical weight. In addition, note for which method(s) a difference between the theoretical weight and the measured weight could already be seen using the toploading balance that has a precision of 0.1 g. Include any conclusions you can draw from this information in terms of what device(s) are good method for measuring volume, which are mediocre, and which are not good.

EXPERIMENT 9: PREPARATION AND STANDARDIZATION OF HCl AND NaOH SOLUTIONS

Remember to use safety glasses.

1. For this experiment, you will need a minimum of 3–4 g of primary standard potassium hydrogen phthalate (KHP) for three titrations. Place at least 6 g of it in a weighing bottle and dry in a drying oven for 2 h. Your instructor may choose to dispense this to you.

2. Prepare CO_2-free water for a 0.10-M NaOH solution by boiling 1000 mL of distilled water in a covered beaker on a hotplate. While waiting for this water to boil, continue with Step 3.

3. Prepare 1 L of 0.10 M HCl by diluting the appropriate volume of concentrated HCl (12.0 M). Use a 1-L glass bottle and half-fill it with water before adding the concentrated acid. Use a graduated cylinder to measure the acid. Add more water to have about 1 L, shake well, and label.

4. Once the water from Step 2 has boiled remove from the heat and cool so that it is only warm to the touch. This can be accomplished by immersing the beaker in cold water (such as in a stoppered sink).

5. Using this freshly boiled water, prepare 1000 mL of a 0.10 M NaOH solution. Weigh out the appropriate number of grams of NaOH, and place it in a 1000-mL plastic bottle. Add the water to a level approximately equal to 1000 mL. Shake well to completely dissolve the solid. Allow to cool completely to room temperature before proceeding. Label the bottle.

6. Assemble the apparatus for a titration. The buret should be a 50-mL buret and should be washed thoroughly with a buret brush and soapy water. Clamp the buret to a ring stand with either a buret clamp or an ordinary ring stand clamp. The receiving flask should be a 250-mL Erlenmeyer flask. You should clean and prepare three such flasks. Place a piece of white paper (a page from your notebook will do) on the base of the ring stand. This will help you see the end point better.

7. Give both your acid and base solutions one final shake at this point to ensure their homogeneity. Rinse the buret with 5–10 mL of NaOH twice, then fill it to the top. Open the stopcock wide open to force trapped air bubbles from the stopcock and tip. Allow this excess solution to drain into a waste flask. Bring the bottom of the meniscus to the 0.00-mL line. Using a clean 25-mL pipet (volumetric) carefully place 25.00 mL of the acid solution into each of the three flasks. Add three drops of phenolphthalein indicator to each of the three flasks.

8. Titrate each acid sample with the base, one at a time. For the first sample, you can very easily make a good estimate of how much titrant should be required. Since both solutions are the same molar concentration and since the reaction is one-to-one in terms of moles (see Equation 4.2), the volume of the NaOH required should be approximately the same as the volume of the HCl pipetted, 25.00 mL. You should be able to open the stopcock and allow about 20 mL of the base to enter the flask without having to worry about overshooting the end point. From that point on however, you should proceed with caution so that the indicator changes color (from colorless to pink) upon the addition of the smallest amount of titrant possible; a fraction of a drop. At the point that you think the indicator will change color in this manner, rinse down the walls of the flask with water from a squeeze bottle. This will ensure that all of the titrant has had a chance to react with all of the substance titrated. This step can be performed more than once—addition of more water has no effect on the location of the end point. The fraction of a drop can then be added if necessary by (1) a very rapid rotation of the stopcock or (2) slowly opening the stopcock allowing the fraction of a drop to hang on the tip of the buret, and then washing it into the flask with water from a squeeze bottle. At the end point, a faint pink color will persist in the flask for at least 20 s. Read the buret to four significant figures, record the reading, and repeat with the next sample.

9. All three of your titrations should agree to within at least 0.05 mL in order to be acceptable. If they do not agree in this manner, more such titrations must be performed in order to have three good answers to rely on or until you have a precision satisfactory to your instructor.

10. After the 2-h drying period for the KHP has expired, remove from the drying oven and allow to cool to room temperature in your desiccator.

11. Prepare three solutions of the KHP, for titration. To do this, again clean three 250-mL Erlenmeyer flasks and give each a final thorough rinse with distilled water. Now, weigh into each flask, by difference, on the analytical balance, a sample of the KHP weighing between 0.7 and 0.9 g. Add approximately 50 mL of distilled water to each and swirl until completely dissolved.
12. Add three drops of phenolphthalein indicator to each flask and titrate each as before.
13. Calculate the exact normality of the NaOH solution from each set of the KHP data. Calculate the mean. Calculate the standard deviation and the parts per thousand RSD. If your three normalities do not agree to within 2.5 ppt RSD, repeat until you have three that do, or until you have a precision that is satisfactory to your instructor.
14. Calculate the exact normality of the HCl solution using the three volume readings from Step 8 and the average NaOH normality from Step 13. Compute the average of these three results.
15. Record in your notebook at least a sample calculation for each of these standardizations and all results.

QUESTIONS AND PROBLEMS

Introduction

1. How is titrimetric analysis different from gravimetric analysis?
2. Why is titrimetric analysis sometimes called volumetric analysis?
3. Compare Figure 3.1 with Figure 4.1 and tell how the analytical strategy for gravimetric analysis differs from the analytical strategy for titrimetric analysis.

Terminology

4. Define: standard solution, volumetric flask, volumetric pipet, standardization, buret, stopcock, titration, titrant, substance titrated, equivalence point, indicator, end point, automatic titrator.

Review of Solution Concentration

5. What is the molarity of the following?
   (a) 0.694 mol dissolved in 3.55 L of solution.
   (b) 2.19 mol of NaCl dissolved in 700.0 mL of solution.
   (c) 0.3882 g of KCl dissolved in 0.5000 L of solution.
   (d) 1.003 g of $CuSO_4 \cdot 5 H_2O$ dissolved in 250.0 mL of solution.
   (e) 30.00 mL of 6.0 M NaOH diluted to 100.0 mL of solution.
   (f) 0.100 L of 12.0 M HCl diluted to 500.0 mL of solution.
6. What is the equivalent weight of both reactants in each of the following?
   (a) $NaOH + HCl \rightarrow NaCl + H_2O$
   (b) $2 NaOH + H_2SO_4 \rightarrow Na_2SO_4 + 2 H_2O$
   (c) $2 HCl + Ba(OH)_2 \rightarrow BaCl_2 + 2 H_2O$
   (d) $3 NaOH + H_3PO_4 \rightarrow Na_3PO_4 + 3 H_2O$
   (e) $2 HCl + Mg(OH)_2 \rightarrow MgCl_2 + 2 H_2O$
   (f) $2 NaOH + H_3PO_4 \rightarrow Na_2HPO_4 + 2 H_2O$
   (g) $NaOH + Na_2HPO_4 \rightarrow Na_3PO_4 + H_2O$
   (h) $NaOH + H_3PO_4 \rightarrow NaH_2PO_4 + H_2O$
   (i) $Na_2CO_3 + 2 HCl \rightarrow 2 NaCl + H_2CO_3$

7. Calculate the normality of the following solutions.
   (a) 0.238 equivalents of an acid dissolved in 1.500 L of solution.
   (b) 1.29 mol of sulfuric acid dissolved in 0.5000 L of solution used for the following reaction:

$$H_2SO_4 + 2\ NaOH \rightarrow Na_2SO_4 + 2\ H_2O$$

   (c) 0.904 mol of H_3PO_4 dissolved in 250.0 mL of solution used for the following:

$$H_3PO_4 + 3\ KOH \rightarrow K_3PO_4 + 3\ H_2O$$

   (d) 0.827 mol of $Al(OH)_3$ dissolved in 0.2500 L of solution used for the following reaction:

$$3\ HCl + Al(OH)_3 \rightarrow AlCl_3 + 3\ H_2O$$

   (e) 1.38 g of KOH dissolved in 500.0 mL of solution used for the chemical reaction in part (c) above.
   (f) 2.18 g of NaH_2PO_4 dissolved in 1.500 L used for the following reaction:

$$NaH_2PO_4 + Al(OH)_3 \rightarrow Al(NaHPO_4)_3 + H_2O$$

   (g) 0.728 g of KH_2PO_4 dissolved in 250.0 mL of solution used for the following reaction:

$$KH_2PO_4 + Ba(OH)_2 \rightarrow KBaPO_4 + 2\ H_2O$$

8. An H_3PO_4 solution is to be used to titrate an NaOH solution as in the equation in the following reaction. If the normality of the H_3PO_4 solution is 0.2411 N, what is its molarity?

$$H_3PO_4 + KOH \rightarrow KH_2PO_4 + H_2O$$

Review of Solution Preparation

9. Tell how you would prepare each of the following.
   (a) 500.0 mL of a 0.10-M solution of KOH from pure solid KOH?
   (b) 250.0 mL of a 0.15-M solution of NaCl from pure solid NaCl?
   (c) 100.0 mL of a 2.0-M solution of glucose from pure solid glucose ($C_6H_{12}O_6$)?
   (d) 500.0 mL of a 0.10-M solution of HCl from concentrated HCl that is 12.0 M?
   (e) 100.0 mL of a 0.25-M solution of NaOH from a solution of NaOH that is 2.0 M?
   (f) 2.0 L of a 0.50-M solution of sulfuric acid from concentrated sulfuric acid, which is 18.0 M?
10. Tell how you would prepare each of the following.
   (a) 500.0 mL of 0.20 N KH_2PO_4 using pure, solid KH_2PO_4 and used for the following reaction:

$$KH_2PO_4 + 2\ NaOH \rightarrow KNa_2PO_4 + 2\ H_2O$$

   (b) 500.0 mL of 0.11 N H_2SO_4 from concentrated H_2SO_4 (18.0 M) used for the following reaction:

$$H_2SO_4 + Ca(OH)_2 \rightarrow CaSO_4 + 2\ H_2O$$

(c) 750.0 mL of 0.11 N $Ba(OH)_2$ from pure solid $Ba(OH)_2$ given the following reaction:

$$2\ Na_2HPO_4 + Ba(OH)_2 \rightarrow Ba(Na_2PO_4)_2 + 2\ H_2O$$

(d) 200.0 mL of a 0.15-N solution of the base using the pure, solid chemical and to be used in the following reaction:

$$2\ HBr + Na_2CO_3 \rightarrow 2\ NaBr + H_2O + CO_2$$

(e) 700.0 mL of a 0.25-N solution of the acid using the pure solid chemical and to be used in the following reaction:

$$2\ NaHCO_3 + Mg(OH)_2 \rightarrow Mg(NaCO_3)_2 + 2\ H_2O$$

(f) 700.0 mL of a 0.30-N solution of $Ba(OH)_2$ from a 15.0-N solution of $Ba(OH)_2$ used for the following reaction:

$$2\ H_3PO_4 + Ba(OH)_2 \rightarrow Ba(H_2PO_4)_2 + 2\ H_2O$$

(g) 300.0 mL of 0.15 N solution of H_3PO_4 from concentrated H_3PO_4 (15 M) used for the following reaction:

$$H_3PO_4 + Al(OH)_3 \rightarrow AlPO_4 + 3\ H_2O$$

11. How would you prepare 250.0 mL of a 0.35-M solution of NaOH using
    (a) a bottle of pure, solid NaOH?
    (b) a solution of 6.0 M NaOH?
12. How many milliliters of a KOH solution, prepared by dissolving 60.0 g of KOH in 100.0 mL of solution, are needed to prepare 450.0 mL of a 0.70-M solution?
13. How many milliliters of a KCl solution, prepared by dissolving 45 g of KCl in 500.0 mL of solution, are needed to prepare 750.0 mL of a 0.15-M solution?
14. A solution of KNO_3, 500.0 mL with 0.35 M concentration, is needed. Tell how you would prepare this solution:
    (a) from pure, solid KNO_3.
    (b) from a solution of KNO_3 that is 4.5 M.
15. To prepare a certain solution, it is determined that acetic acid must be present at 0.17 M and that sodium acetate must be present at 0.29 M, both in the same solution. If the sodium acetate to be used to prepare this solution is a pure solid chemical and the acetic acid to be used is a concentrated solution (17 M), how would you prepare 500.0 mL of this buffer solution?
16. How many milliliters of a NaH_2PO_4 solution prepared by dissolving 0.384 g in 500.0 mL are needed to prepare 1.000 L of a 0.00200-N solution given the following reaction?

$$NaH_2PO_4 + Ca(OH)_2 \rightarrow CaNaPO_4 + 2\ H_2O$$

Stoichiometry of Titration Reactions

17. Explain why Equation 4.16 is a correct equation.
18. Explain how Equation 4.19 can be derived from Equation 4.16.
19. Explain why Equation 4.20 is a correct equation.
20. Explain how Equation 4.23 follows from Equation 4.20.

Standardization

21. Explain how Equation 4.26 follows logically from Equation 4.19.
22. Explain how Equation 4.27 follows logically from Equation 4.23.
23. Suppose 0.7114 g of KHP was used to standardize a $Mg(OH)_2$ solution as in the following reaction:

$$Mg(OH)_2 + 2\ KHC_8H_4O_4 \rightarrow Mg(KC_8H_4O_4)_2 + 2\ H_2O$$

If 31.18 mL of the $Mg(OH)_2$ were needed, what is the molarity of the $Mg(OH)_2$?
24. A NaOH solution was standardized against a H_3PO_4 solution as in the following reaction:

$$H_3PO_4 + 3\ NaOH \rightarrow Na_3PO_4 + 3\ H_2O$$

If 25.00 mL of 0.1427 M H_3PO_4 required 40.07 mL of the NaOH, what is the molarity of the NaOH?
25. A solution of KOH is standardized with primary standard KHP ($KHC_8H_4O_4$). If 0.5480 g of the KHP exactly reacted with 25.41 mL of the KOH solution, what is the molarity of the KOH?

$$KOH + KHP \rightarrow K_2P + H_2O$$

26. What is the normality of a solution of HCl, 35.12 mL of which were required to titrate 0.4188 g of primary standard Na_2CO_3?

$$2\ HCl + Na_2CO_3 \rightarrow 2\ NaCl + CO_2 + H_2O$$

27. What is the normality of a solution of sulfuric acid that was used to titrate a 0.1022-N solution of KOH as in the following reaction if 25.00 mL of the base was exactly neutralized by 29.04 mL of the acid?

$$H_2SO_4 + 2\ KOH \rightarrow K_2SO_4 + 2\ H_2O$$

28. Primary standard tris(hydroxymethyl)aminomethane, also known as THAM or TRIS (formula weight = 121.14 g/mol), is used to standardize a hydrochloric acid solution. If 0.4922 g of THAM are used and 23.45 mL of the HCl are needed, what is the normality of the HCl?

$$(HOCH_2)_3CNH_2 + HCl \rightarrow (HOCH_2)_3CNH_3^+ + Cl^-$$

29. Suppose a sulfuric acid solution, rather than the hydrochloric acid solution as in problem 28, is standardized with primary standard THAM. Does the calculation change in any way? Explain.
30. What is a primary standard?
31. What are three requirements of a primary standard?
32. What is a SRM? What is a CRM? What is NIST an acronym for?
33. Define titer.
34. What is the titer (expressed in mg/mL) of a solution of disodium dihydrogen EDTA with respect to calcium carbonate if 17.29 mL of it were needed to titrate 0.0384 g of calcium carbonate.

Percent Analyte Calculations

35. Explain how Equations 4.19 and 4.31 combine to produce Equation 4.35.
36. Explain how Equations 4.25 and 4.31 combine to produce Equation 4.38.
37. What is the percent of K_2HPO_4 in a sample when 46.79 mL of 0.04223 M $Ca(OH)_2$ exactly neutralizes 0.9073 g of the sample according to the following equation?

$$2 \ K_2HPO_4 + Ca(OH)_2 \rightarrow Ca(K_2PO_4)_2 + 2 \ H_2O$$

38. What is the percent of $Al(OH)_3$ in a sample when 0.3792 g of the sample is exactly neutralized by 23.45 mL of 0.1320 M H_3PO_4 according to the following equation?

$$3 \ H_3PO_4 + Al(OH)_3 \rightarrow Al(H_2PO_4)_3 + 3 \ H_2O$$

39. What is the percent of NaH_2PO_4 in a sample if 24.18 mL of 0.1032 N NaOH was used to titrate 0.3902 g of the sample according to the following?

$$NaH_2PO_4 + 2 \ NaOH \rightarrow Na_3PO_4 + 2 \ H_2O$$

40. A 0.1057-N HCl solution was used to titrate a sample containing $Ba(OH)_2$. If 35.78 mL of HCl were required to exactly react with 0.8772 g of the sample, what is the percentage of $Ba(OH)_2$ in the sample?

$$2 \ HCl + Ba(OH)_2 \rightarrow BaCl_2 + 2 \ H_2O$$

Volumetric Glassware

41. Tell which statements are true and which are false.
    (a) "TC" means "to control."
    (b) The volumetric pipet has graduation lines on it, much like a buret.
    (c) The volumetric flask is never used for delivering an accurate volume of solution to another vessel.
    (d) Volumetric pipets are not calibrated for "blowout."
    (e) Two frosted rings found near the top of a pipet mean that the pipet is a Mohr pipet.
    (f) The volumetric flask has the letters "TD" imprinted on it.
    (g) The volumetric pipet is a type of "measuring" pipet.
    (h) The serological pipet is a type of volumetric pipet.
42. Completion
    (a) The kind of pipet that has graduations on it much like a buret is called the _____ pipet.
    (b) To say that a given pipet is not calibrated for blowout means that _____
    _____.
    (c) Burets and pipets that are not dry should be rinsed first with the solution to be used because _____.
43. Which is more accurate, a Mohr pipet or a volumetric pipet? Why?
44. Should a 100-mL volumetric flask be used to measure out 100 mL of solution to be added to another vessel? Why or why not?
45. What does a frosted ring near the top of a pipet indicate?
46. How does a volumetric flask differ from an Erlenmeyer flask?
47. Explain the reasons for rinsing a pipet as directed in Step 3 of Box 4.8.

48. With a Mohr pipet, the meniscus must be read twice, but with a serological pipet, the meniscus may be read only once, if desired. Explain this.

49. A student is observed using a 50-mL volumetric flask to "accurately" transfer 50 mL of a solution from one container to another. What would you tell the student (a) to explain his/her error and (b) to help him do the "accurate" transfer correctly?

50. Why must a pipet that has been stored in distilled water be thoroughly rinsed with the solution to be transferred before use?

51. A technician is directed to prepare accurately to 4 significant figures 100 mL of a solution of Na_2CO_3 that is around 0.25 M. The laboratory supervisor provides a solution of Na_2CO_3 that is 4.021 M and also some pure, solid Na_2CO_3 and tells the technician to proceed by whichever method is easier. Give specific details as to how you would prepare the solution by two different methods, one by dilution and one by weighing the pure chemical, including how many grams or milliliters to measure (show calculation), what type of glassware (include the size and kind of pipets and flasks, if applicable) is used, and how they are used.

52. For one of the methods in problem 51, you needed a pipet. Is the pipet you would choose calibrated for blowout? How can you tell by looking at the markings on the pipet? What is the name of the pipet you would choose? Give specific details as to how the volume is delivered using this pipet.

53. The concentration of solutions can be known accurately to four significant figures either directly through their preparation or by standardization.
    (a) If the solute is a pure solid, give specific instructions that would ensure such accuracy directly through its preparation.
    (b) What is meant by "standardization"?

54. Compare a volumetric pipet with a serological measuring pipet in terms of
    (a) the number of graduation lines on the pipet.
    (b) whether it is calibrated for "blowout."
    (c) which one to select if you need to deliver 3.72 mL.
    (d) whether it is calibrated "TC" or "TD."

55. Some pipets are calibrated "TC." Why would one ever want a pipet calibrated "TC" rather than "TD"?

Pipetters, Automatic Titrators, and Other Devices

56. Briefly describe what is meant by each of the following: pipetter, micropipet, micropipetter, repipet, digital buret, and automatic pipet.

Calibration of Glassware and Devices

57. Do Class A volumetric pipets need to be checked for proper calibration? Explain.
58. What does checking the calibration of flasks and pipets generally involve.

Analytical Technique

59. Consider an experiment in which a pure solid chemical is weighed on an analytical balance into an Erlenmeyer flask. The nature of the experiment is such that this substance must be dissolved before proceeding. The experiment calls for you to dissolve it in 75 mL of water. In the subsequent calculation of the results, which are to be reported to 4 significant figures, the weight measurement is needed but the 75 mL measurement is not. Should the 75 mL of water be measured with a pipet, or is a graduated cylinder good enough? Explain.

5 Applications of Titrimetric Analysis

5.1 INTRODUCTION

The standardization and percent analyte examples in Chapter 4 (Examples 4.10–4.13, 4.15, 4.16, and Experiment 9) involved acid–base reactions. In this chapter, we discuss the chemistry of these and others.

For a titrimetric analysis to be successful, the equivalence point must be easily and accurately detected; the reaction involved must be fast; and the reaction must be quantitative. If an equivalence point cannot be detected (i.e., if there is no acceptable indicator or other detection method), then the correct volume of titrant cannot be determined. If the reaction involved is not fast, then the end point cannot be detected immediately upon adding the last fraction of a drop of titrant and there would be some doubt if the end point has been reached. If the reaction is not quantitative, meaning that if every trace of reactant in the titration flask is not consumed by the titrant at the end point, then again the correct volume of titrant cannot be determined. This latter point means that equilibrium reactions that do not go essentially to completion immediately are not acceptable reactions for this type of analysis. Thus, not all reactions are acceptable reactions.

In this chapter, we investigate individual types of reactions that meet all the requirements. We will also discuss "back titrations" and "indirect" titrations in which some of the limitations that we may encounter are solved. Our discussions in Chapter 4 involved acid–base reactions. Acid–base reactions will be discussed here, but we will see that there are others that are applicable reactions.

5.2 ACID–BASE TITRATIONS AND TITRATION CURVES

Various acid–base titration reactions are discussed in this section, including a number of scenarios of base in the buret and acid in the reaction flask and vice versa and also various monoprotic and polyprotic acids titrated with a strong base and various weak monobasic and polybasic bases titrated with strong acids. A **monoprotic acid** is an acid that has only one hydrogen ion (or proton) to donate per formula. Examples are hydrochloric acid, HCl, a strong acid, and acetic acid, $HC_2H_3O_2$, a weak acid. A **polyprotic acid** is an acid that has two or more hydrogen ions to donate per formula. Examples include sulfuric acid, H_2SO_4, a **diprotic acid**, and phosphoric acid, H_3PO_4, a **triprotic** acid.

A **monobasic base** is one that will accept just one hydrogen ion per formula. Examples include sodium hydroxide, NaOH, a strong base, ammonium hydroxide, NH_4OH, a weak base, and sodium bicarbonate, $NaHCO_3$, also a weak base. A **polybasic base** is one that will accept two or more hydrogen ions per formula. Examples include sodium carbonate, Na_2CO_3, a **dibasic base**, and sodium phosphate, Na_3PO_4, a **tribasic base**.

5.2.1 TITRATION OF HYDROCHLORIC ACID

A graphic picture of what happens during an acid–base titration is easily produced in the laboratory. Consider again what is happening as a titration proceeds. Consider, specifically, NaOH as the titrant and HCl as the substance titrated. In the titration flask, the following reaction occurs when titrant is added:

$$H^+ + OH^- \rightarrow H_2O \qquad (5.1)$$

As H^+ ions are consumed in the reaction flask by the OH^- added from the buret, the pH of the solution in the flask will change, since pH = $-\log [H^+]$. In fact, the pH should increase as the titration proceeds, since the number of H^+ ions decreases due to the reaction with OH^-. The lower the $[H^+]$, the higher is the pH. This increase in pH can be monitored with the use of a pH meter. Thus, if we were to measure the pH in this manner after each addition of NaOH and graph the pH vs. volume (mL) of NaOH added we would have a graphical display of the experiment. Figure 5.1a shows the results of such an experiment for the case in which 0.10 N HCl is titrated with 0.10 N NaOH. The graph is called a **titration curve**. The sharp increase in the pH at the center of the graph occurs at the equivalence point of the titration or the point at which all the acid in the flask has been neutralized by the added base. The point at which the slope of a titration curve is a maximum (a sharp change in pH such as this, whether an increase or decrease) is called an **inflection point**.

An initial response to this curve might be: "Why the strange shape? Why the steady, although slight, increase in the beginning? Why is there an inflection point? Why the steady increase at the end?" An acid–base titration curve is not without theoretical foundation. The entire curve can also

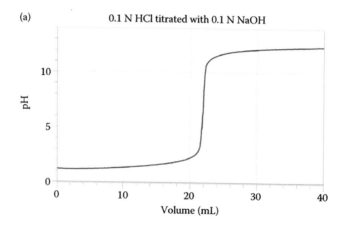

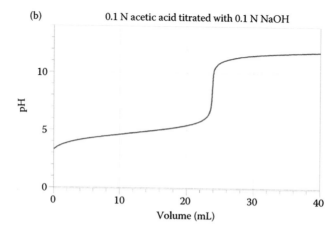

FIGURE 5.1 Acid–base titration curves. (a) 0.10 N HCl (strong acid) titrated with 0.10 N NaOH (strong base). (b) 0.10 N acetic acid (weak acid) titrated with 0.10 N NaOH.

be recorded independent of the pH meter experiment by calculating the [H⁺] and the pH after each addition and plotting the results. Such calculations and plotting are beyond our scope here but can be done. The calculations would indicate a steady and slight increase in pH over a wide range of volume of NaOH (in milliliters) added in the beginning. This would be followed by the inflection point, which in turn is followed by a steady but slight increase again over a broad range—exactly what Figure 5.1a shows.

We should emphasize that the concentrations of the acid and base are important to consider. These concentrations in the discussions thus far have been 0.10 N, 0.10 N NaOH in the buret and 0.10 N HCl in the reaction flask. The lower the concentration of the acid, the fewer H^+ ions are present, the higher the initial pH and the higher the pH level of the initial steady increase. The lower the concentration of the base, the lower the level of the pH after the inflection point.

5.2.2 Titration of Weak Monoprotic Acids

It is interesting to compare the strong acid titrated with a strong base case in the last section with the titration of a *weak* acid, such as acetic acid, with a strong base, such as NaOH, i.e., a weak acid titrated with a strong base. The difference between this and the strong acid–strong base case just discussed is that, for the same 0.10 N concentrations, the pH starts and continues to the equivalence point at a higher pH level in a manner similar to the lower concentration of HCl (see Figure 5.1b). This should not be unexpected, since a solution of a weak acid is being measured, and once again there are fewer H^+ ions in the solution. Still weaker acids start at even higher pH values.

The weak acid curves can also be calculated. This involves the use of the equilibrium constant expression for a weak monoprotic acid ionization.

$$HA \leftrightarrows H^+ + A^- \tag{5.2}$$

$$K_a = \frac{[H^+][A^-]}{[HA]} \tag{5.3}$$

In these equations, HA symbolizes a weak acid and A^- symbolizes the anion of the weak acid. The calculations are again beyond our scope. However, we can correlate the value of the equilibrium constant for a weak acid ionization, K_a, with the position of the titration curve. The weaker the acid, the smaller the K_a and the higher the level of the initial steady increase. Figure 5.2 shows

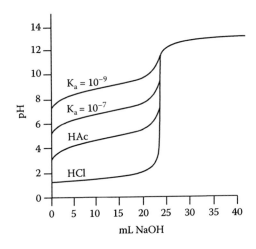

FIGURE 5.2 Family of acid–base titration curves for a 0.10-N strong acid (HCl) and three weak acids as indicated (0.10 N each) titrated with a 0.10-N NaOH (strong base). "HAc" is a representation of acetic acid.

a family of curves representing several acids at a concentration of 0.10 N titrated with a strong base. The curves for HCl and acetic acid (represented as "HAc") are shown as well as two curves for two acids even weaker than acetic acid (the K_a's are indicated).

5.2.3 TITRATION OF MONOBASIC STRONG AND WEAK BASES

Now consider an alternative in which the acid is in the buret and the base is in the reaction flask. If the titration of 0.10 N NaOH with a 0.10-N HCl (a strong base titrated with strong acid) is considered, we have a curve that starts at a high pH value (a solution of a base has a high pH) and ends at a low pH–just the opposite to that observed when titrating an acid with a base (see Figure 5.3a). Likewise, the curves for 0.10 N weak bases, titrated with a strong acid, such as ammonium hydroxide titrated with HCl, start out at a lower pH compared with the strong base, just as those for 0.10 N weak acids started out at a higher pH compared with the strong acid (see Figure 5.3b). A family of curves for the titration of bases with acids is given in Figure 5.4. These curves have a theoretical foundation as well and can be calculated.

5.2.4 EQUIVALENCE POINT DETECTION

Before continuing with other examples, it is important to consider how the equivalence point in an acid–base titration is found and what relationship this has with titration curves. As we have said, the

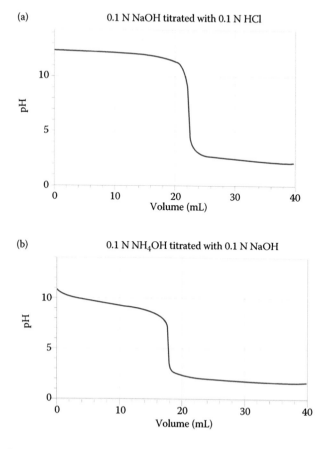

FIGURE 5.3 Titration curves for the titration of (a) 0.10 N NaOH titrated with 0.10 N HCl and (b) 0.10 N NH₄OH titrated with 0.10 N HCl.

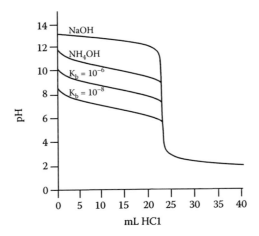

FIGURE 5.4 Family of titration curves for various bases titrated with a strong acid.

inflection point at the center of these curves occurs at the equivalence point, the point at which all of the substance titrated has been exactly consumed by the titrant. The exact position for this in the strong acid titrated with strong base case, and also the strong base titrated with strong acid case, is at pH 7, exactly in the middle of the inflection point. For the weak acid with strong base case, the equivalence point is again in the middle of the inflection point, but this occurs at a pH higher than pH 7 (refer to Figure 5.2). In fact, the weaker the acid, the higher the pH value that corresponds to the equivalence point. The opposite is observed in the case of the weak base with strong acid cases (see Figure 5.4). The equivalence point occurs at progressively lower pH values, the weaker the base. The exact pH at these equivalent points can be calculated as indicated previously.

The problem the analyst has is to choose indicators that change color close enough to an equivalence point so that the accuracy of the experiment is not diminished, which really means at any point during the inflection point (refer to Chapter 4, Section 4.2 for the definitions of equivalence point and end point). It almost seems like an impossible task, since there must be an indicator for each possible acid or base to be titrated. Fortunately, there are a large number of indicators available, and there is at least one available for all acids and bases with the exception of only the extremely weak acids and bases. Figure 5.5a–d addresses the use of two such indicators, phenolphthalein and bromocresol green.

Phenolphthalein changes color in the pH range 8–10, whereas bromocresol green changes color in the pH range 4–5.5. Notice that in Figure 5.5a, both of these have their color change range at the inflection point. This means that both will change their color near the equivalence point. Thus, both are suitable indicators for 0.10 N HCl (or any strong acid) titrated with 0.010 N NaOH.

Contrast this with the weak acid titrated with a strong base case in Figure 5.5b, 0.10 N acetic acid titrated with 0.10 N NaOH. Because the first part of the curve is at a higher pH level compared with the strong acid titrated with strong base case, the bromocresol green color change range now appears too low. The color change range is not at the inflection point. As the acid becomes progressively weaker, the pH range available for the indicator to change color becomes progressively narrower, and a smaller number of indicators are useful. The color change must take place at the inflection point.

A similar observation is made when titrating bases with acids. Once again, both indicators' color change ranges are at the inflection point when both the acid and base are strong, as noted in Figure 5.5c. Thus, either would work as the indicator. However, for the weak base example, the phenolphthalein color change range is not at the inflection point, as you can clearly see in Figure 5.5d, but bromocresol green's color change range *is* at the inflection point. The conclusion to draw is that bromocresol green will work, whereas phenolphthalein will not.

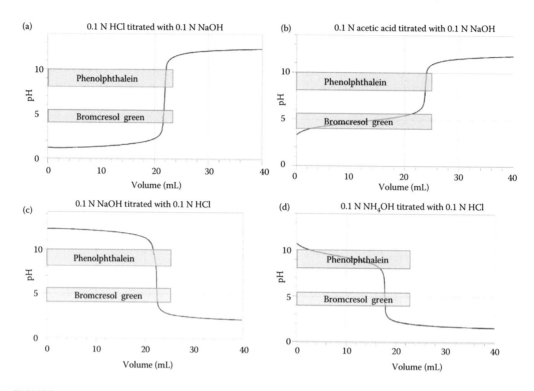

FIGURE 5.5 Representations of why one indicator works for a given titration, whereas another may not. (a) Strong acid titrated with a strong base. Both phenolphthalein and bromocresol will work. (b) Weak acid titrated with a strong base. Phenolphthalein will work, but bromocresol green will not. (c) Strong base titrated with a strong acid. Again, both phenolphthalein and bromocresol green will work, although the color change is the opposite of what it is in (a). (d) Weak base titrated with a strong acid. Bromocresol green will work, but phenolphthalein will not.

Equivalence points do not have to be found with indicators necessarily. The appearance of an inflection point on the titration curve would be evidence enough in many cases. This means that we could monitor the pH as the titrant is added, as we do when we measure a titration curve, and the end point would occur when the pH increases or decreases sharply at the inflection point. Some precision in the volume of titrant measurement (the critical measurement in any titration) may be lost, however, unless rather than monitoring the pH we would monitor the first or second derivative of the pH. The first derivative of the pH behavior is a measure of the rate of change of the pH signal as the titrant is added. Of course, as can be seen in any titration curve, the greatest of change occurs in the middle of the inflection point. Thus, the first derivative signal would be a maximum at that point (see Figure 5.6a). Yet better precision can be achieved with the second derivative (the rate of change of the rate of change, so to speak), i.e., by measuring and displaying the rate of change of the first derivative (see Figure 5.6b). The peak of the first derivative (the equivalence point of the titration) would be where the second derivative crosses the x-axis.

5.2.5 TITRATION OF POLYPROTIC ACIDS: SULFURIC ACID AND PHOSPHORIC ACID

Titrations curves for polyprotic acids have an inflection point for each hydrogen in the formula if the dissociation constant (the K_a) for each hydrogen is very different from the others and if any dissociation constant is not too small. The titration curves of the polyprotic acids H_2SO_4 and H_3PO_4 are shown in Figures 5.7 and 5.8. Sulfuric acid has essentially one inflection point (like hydrochloric

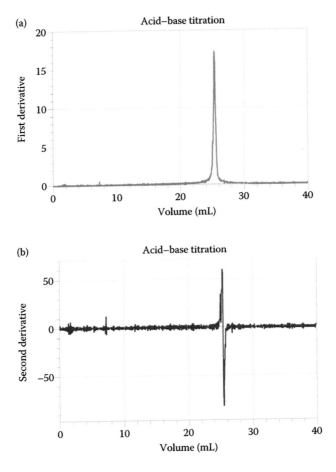

FIGURE 5.6 (a) First derivative of the titration curve of a strong acid titrated with a strong base. (b) Second derivative of the titration curve of a strong acid titrated with a strong base.

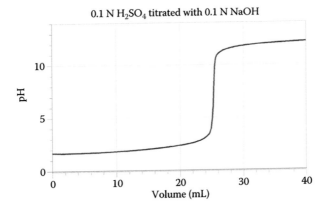

FIGURE 5.7 Titration curve of 0.10 N H_2SO_4 titrated with 0.10 N NaOH.

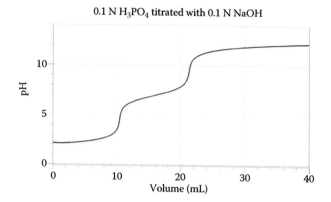

FIGURE 5.8   Titration curve of 0.10 N H_3PO_4 titrated with 0.10 N NaOH.

acid—compare with Figure 5.1a), whereas phosphoric acid has two apparent inflection points. Both hydrogens on the sulfuric acid molecule are "strongly acidic" (the dissociation of each in water solution is very nearly complete) and are neutralized simultaneously, therefore just one inflection point, so that it gives the same appearance as a strong monoprotic acid. The dissociation constants for the three hydrogens on the phosphoric acid molecule are very different from each other, and therefore, we would expect that they are neutralized one at a time. However, the third dissociation constant is very small, hence just two inflection points.

Thus, for sulfuric acid, there is essentially one reaction along the way to the lone inflection point:

$$H_2SO_4 + 2\ OH^- \rightarrow SO_4^{2-} + 2\ H_2O \tag{5.4}$$

For phosphoric acid, there are three reactions along the way. The three reactions are

$$H_3PO_4 + OH^- \rightarrow H_2PO_4^- + H_2O \tag{5.5}$$

$$H_2PO_4^- + OH^- \rightarrow HPO_4^{2-} + H_2O \tag{5.6}$$

$$HPO_4^{2-} + OH^- \rightarrow PO_4^{3-} + H_2O \tag{5.7}$$

The dissociation constants for the three hydrogens are

$$H_3PO_4 \leftrightarrows H_2PO_4^- + H^+ \quad K_{a1} = 7.11 \times 10^{-3} \tag{5.8}$$

$$H_2PO_4^- \leftrightarrows HPO_4^{2-} + H^+ \quad K_{a2} = 6.32 \times 10^{-8} \tag{5.9}$$

$$HPO_4^{2-} \leftrightarrows PO_4^{3-} + H^+ \quad K_{a3} = 7.1 \times 10^{-13} \tag{5.10}$$

Compare this information with the curves in Figure 5.2 and the fact that as the K_a gets smaller, the initial range of steadily but gradually increasing pH is at a higher pH range (and the inflection point range is narrowed. The result of this is two inflection points, one for each of the hydrogen as they react with the base one at a time.

5.2.6 TITRATION OF POTASSIUM BIPHTHALATE

For standardizing a base solution, primary standard grade potassium biphthalate is a popular choice. Also called potassium hydrogen phthalate, potassium acid phthalate, or simply KHP, it is the salt representing partially neutralized phthalic acid and is a monoprotic weak acid. The true formula is $KHC_8H_4O_4$. Figure 5.9 shows the chemical structure of phthalic acid and KHP. The reaction with a base is

$$KHC_8H_4O_4 + OH^- \rightarrow K_2C_8H_4O_4 + H_2O \tag{5.11}$$

KHP is a white crystalline substance of high formula weight (204.23), stable at oven-drying temperatures, and available in a very pure form. It is a weak acid with a titration curve similar to acetic acid (see Figure 5.10; compare with acetic acid, Figure 5.1b). KHP was used in Experiment 9 (Chapter 4) to standardize a 0.1-N solution NaOH solution. In Experiment 11 in this chapter, KHP is the analyte in an experiment in which the sample is impure KHP. In Experiment 9, the equivalence point is detected with the use of phenolphthalein, whereas in Experiment 11, the equivalence point is detected by actually measuring the pH with a pH meter. It is only a slightly weaker acid than acetic acid and so the titration curve for a 0.10-N solution is similar to that of acetic acid shown in Figure 5.1b.

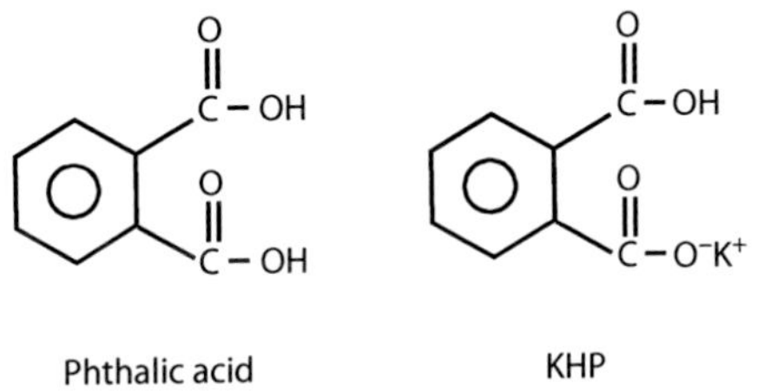

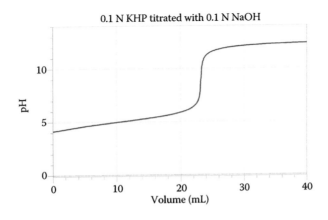

Phthalic acid   KHP

FIGURE 5.9   Molecular structures of phthalic acid and potassium hydrogen phthalate (KHP).

0.1 N KHP titrated with 0.1 N NaOH

FIGURE 5.10   Titration curve for the titration of KHP with NaOH.

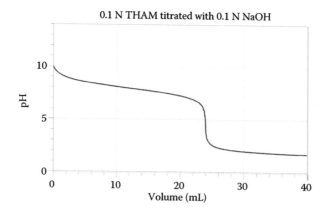

FIGURE 5.11   Titration curve for the titration of THAM with 0.10 N NaOH.

5.2.7 TITRATION OF TRIS-(HYDROXYMETHYL)AMINO METHANE

For standardizing acid solutions, primary standard tris-(hydroxymethyl)amino methane, "THAM" (also sometimes referred to as "TRIS"), can be used. Its formula is

$$(HOCH_2)_3CNH_2$$

It has a high-formula weight (121.14), is stable at moderate drying oven temperatures, and is available in pure form. The reaction with an acid involves the acceptance of one hydrogen by the amine group, $-NH_2$.

$$(HOCH_2)_3CNH_2 + H^+ \rightarrow (HOCH_2)_3CNH_3^+ \tag{5.12}$$

It is a weak base, and the end point occurs at a pH between 4.5 and 5. It is a slightly stronger base than ammonium hydroxide, and so its titration curve would appear similar to that of ammonium hydroxide in Figure 5.3b (see Figure 5.11).

5.2.8 TITRATION OF SODIUM CARBONATE

Primary standard sodium carbonate may also be used to standardize acid solutions. Sodium carbonate also possesses all the qualities of a good primary standard, like KHP and THAM. When titrating sodium carbonate, carbonic acid, H_2CO_3, is one of the products and must be decomposed with heat to push the equilibria below to completion to the right.

$$Na_2CO_3 + 2\ HCl \leftrightarrows 2\ NaCl + H_2CO_3 \tag{5.13}$$

$$H_2CO_3 \leftrightarrows CO_2 + H_2O \tag{5.14}$$

Heat also will eliminate CO_2 from the solution, which aids further in this "completion push."

The $H_2CO_3 \leftrightarrows CO_2$ equilibrium (Equation 5.14) can be a problem in all base solutions because CO_2 from the air dissolves in the solution, forming carbonic acid due to the reverse reaction represented by Equation 5.14, resulting in the formation of carbonate since carbonic acid is a weak acid.

$$H_2CO_3 \leftrightarrows 2\ H^+ + CO_3^{2-} \tag{5.15}$$

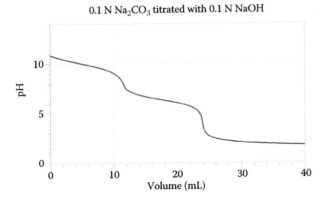

FIGURE 5.12   Titration curve for Na_2CO_3 titrated with 0.10 N NaOH.

The carbonic acid would then also react with whatever acid may be used in titrimetric procedures, thus creating an error. The water used to prepare such solutions is boiled in advance (or freshly distilled) so that all CO_2 is eliminated such that there is no extra carbonate to be neutralized. Such solutions may have to be restandardized, however, due to CO_2 from the air redissolving in the solution over time.

Figure 5.12 represents the titration curve of sodium carbonate titrated with a strong acid. Notice that there are two inflection points. This is because sodium carbonate is a dibasic base—there are two hydrogen ions are being accepted by the carbonate. On the way to the first inflection point, hydrogen ions are being accepted by the carbonate to form bicarbonate as in the following:

$$CO_3^{2-} + H^+ \rightarrow HCO_3^- \tag{5.16}$$

This reaction is complete at the first inflection point. On the way to the second inflection point, the bicarbonate from the first reaction reacts with more hydrogen ions from the buret to form carbonic acid as in the following:

$$HCO_3^- + H^+ \rightarrow H_2CO_3 \tag{5.17}$$

The titration of Na_2CO_3 (see Experiment 10) usually uses the second inflection point for the end point. At that pH range, either bromocresol green or similar indicator red is usually used (refer back to Figure 5.5d). The end point, however, is not sharp (the drop in pH is not sharp) because of a buffering effect due to the presence of a large concentration of H_2CO_3, the product of the titration to the second inflection point (Equation 5.17). If the solution is heated (boiled) close to the end point, however, the $H_2CO_3 \leftrightarrows CO_2$ equilibrium (Equation 5.14) is pushed to the right (recall previous discussion), consuming the H_2CO_3 and thus greatly diminishing this buffering effect. This end point then becomes sharp and very satisfactory at a pH of approximately 4.5. Boiling at the end point is therefore what is done in Experiment 10 and also in acid standardization experiments in which sodium carbonate is the primary standard.

5.3   EXAMPLES OF ACID/BASE DETERMINATIONS

In the real world of analytical chemistry, there are many examples of the need to determine the acid or base content of samples. Let us take a close look at some of these.

5.3.1 ALKALINITY OF WATER OR WASTEWATER

One important application of acid/base titrations is the determination of the alkalinity of various kinds of samples. It is an especially important measurement for the proper treatment of municipal water and wastewater. **Alkalinity** of a water sample is defined as its acid-neutralizing capacity. It is determined by titrating the water sample with standard acid until a particular pH is achieved. The alkalinity value depends on the pH used for the end point. The **total alkalinity** of a water sample is determined by titrating the sample usually to a pH of 4.5. The hydroxides, carbonates, bicarbonates, and other bases in the water are all neutralized when a pH of 4.5 is reached during the titration, and thus this pH is considered to be the equivalence point. Note that it is the same pH as the second inflection point in the titration of carbonate discussed in Section 5.2.8.

Alkalinity is usually expressed as the millimoles of H^+ required to titrate 1 L of water. Millimoles is calculated by multiplying the molarity of the acid (millimoles per milliliter) by buret reading in milliliters as shown in the solution to Example 5.1 below.

Example 5.1

What is the alkalinity of a water sample if 100.0 mL of the water required 5.29 mL of 0.1028 N HCl solution to reach a pH of 4.5?

Solution 5.1

$$\text{alkalinity} = \frac{\text{mmol of } H^+}{\text{L of water}} = \frac{0.1028\dfrac{\text{mmol}}{\text{mL}} \times 5.29 \text{ mL}}{0.1000 \text{ L}} = 5.44 \text{ mmol/L}$$

5.3.2 BACK TITRATION APPLICATIONS

Sometimes a special kind of titrimetric procedure known as a **back titration** is required. In this procedure, the analyte is consumed using excess titrant (the end point is intentionally overshot), and the end point is then determined by titrating the excess with a second titrant. Thus, two titrants are used, and the exact concentration of each is needed for the calculation.

Figure 5.13 shows a graphic representation of a back titration using the titration curve of a strong acid titrated with a strong base to illustrate. The top arrow (point to the right) represents the amount of titrant added initially. It is apparent that this amount is beyond what was needed to reach the equivalence point, which as we discussed previously, coincides with the inflection point, the sharp

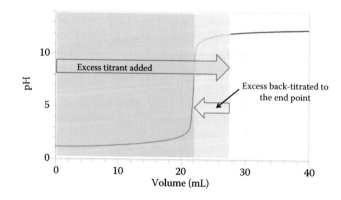

FIGURE 5.13   Illustration of a back titration.

increase in pH. To determine precisely how much was added in excess, this excess is titrated with a second titrant, as the left pointing arrow illustrates. We say we "come back" to the equivalence point after having gone beyond it. The difference between the total equivalents of the first titrant and the total equivalents of the second titrant is the number of equivalents that actually reacted with the analyte, represented by the darker shaded block in Figure 5.13. It is this number of equivalents that is needed for the calculation. The calculation therefore uses the following equation derived from Equation PA.2 in Box 4.7 in Chapter 4. This equation is given as Equation PA.3 in Box 4.7.

$$\% \text{ analyte} = \frac{(L_T \times N_T - L_{BT} \times N_{BT}) \times EW_{analyte}}{\text{sample weight}} \times 100 \tag{5.18}$$

in which "BT" symbolizes the "back titrant."

It may seem strange that we would ever want to perform an experiment of this kind. First of all, it would be used in the event of a slow reaction taking place in the reaction flask. Perhaps the sample is not dissolved completely, and the addition of the titrant causes dissolution to take place over a period of minutes or hours. Adding an excess of the titrant and back titrating it later would seem an appropriate course of action in that case. An example would be the determination of the calcium carbonate in an antacid tablet. The tablet and the analyte, often calcium carbonate, do not dissolve in water, but it does dissolve in a standard solution of HCl according to the following neutralization reaction:

$$2 \text{ HCl} + CaCO_3 \rightarrow CaCl_2 + CO_2 + H_2O$$

Example 5.2

What is the percentage of $CaCO_3$ in an antacid given that a tablet that weighed 1.3198 g reacted with 50.00 mL of 0.4486 N HCl that subsequently required 3.72 mL of 0.1277 N NaOH for back titration? Also report the milligrams of $CaCO_3$ in the tablet.

Solution 5.2

Equation 5.18 is used as follows.

$$\% \text{ CaCo}_3 = \frac{(L_{HCl} \times N_{HCl} - L_{NaOH} \times N_{NaOH}) \times EW_{CaCO3}}{\text{sample weight}} \times 100$$

$$\% \text{ CaCo}_3 = \frac{(0.05000 \times 0.4486 - 0.00372 - 0.1277) \times 100.086/2}{1.3198} \times 100$$

$$= 83.25\%$$

The grams of $CaCO_3$ is the numerator above, or 1.0987 g.

It is also possible that the analyte must be calculated from a gaseous product of another reaction. In this case, one would want the gas to react with the titrant as soon as it is formed, since it could escape into the air because of a high vapor pressure. Thus, an excess of the titrant would be present in a solution through which the gas is bubbled. After the gas-forming reaction has stopped, the excess titrant in this "bubble flask" could be titrated with the back titrant and the results calculated. This latter experiment is one form of the classical **Kjeldahl titration** for determining protein (or nitrogen) in grains and products derived from grains. In this case, the proteins are broken down in a dissolving/digestion step, and the nitrogen in the protein molecules converted to the ammonium ion, NH_4^+. Addition of NaOH to the solution converts the NH_4^+ to NH_3 (ammonia), a base, and the gas that would be bubbled through the solution containing a standard solution, the titrant.

Example 5.3

In a Kjeldahl analysis, a flour sample weighing 0.9857 g was digested in concentrated H_2SO_4 for 45 min. A concentrated solution of NaOH was added such that all of the nitrogen was converted to NH_3. Following this, the NH_3 was distilled into a flask containing 50.00 mL of 0.1011 N H_2SO_4. The excess required 5.12 mL of 0.1266 N NaOH for titration. What is the percentage of nitrogen in the sample?

Solution 5.3

$$\% \ N = \frac{(L_{H_2SO_4} \times N_{H_2SO_4} - L_{NaOH} \times N_{NaOH}) \times 14.00}{\text{sample weight}} \times 100$$

$$\% \ N = \frac{(0.05000 \times 0.1011 - 0.00512 \times 0.1266) \times 14.00}{0.9857} \times 100$$

$$= 6.259\% \ N$$

5.3.3 Indirect Titration Applications

There exists an alternate procedure for the classic Kjeldahl analysis, and it is an "indirect" procedure. An **indirect titration** is one in which the product of a reaction involving the analyte is titrated and not the analyte itself (direct titration) or the excess of a titrant (back titration) (see Figure 5.14).

The indirect Kjeldahl method uses boric acid (H_3BO_3) in the bubble flask and the ammonia reacts with the boric acid, producing a partially neutralized salt of boric acid ($H_2BO_3^-$):

$$NH_3 + H_3BO_3 \rightarrow H_2BO_3^- + NH_4^+ \tag{5.19}$$

which can then be titrated with a standardized acid:

$$H_2BO_3^- + H^+ \rightarrow H_3BO_3 \tag{5.20}$$

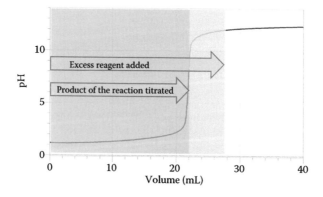

FIGURE 5.14 Illustration of an indirect titration.

The amount of standardized acid needed is proportional to the amount of ammonia that bubbled through. It is an indirect method because the ammonia is determined but is not titrated. It is determined indirectly by titration of the $H_2BO_3^-$. In a direct titration, the analyte would be reacted directly with the titrant, as per the discussion in Chapter 4, Section 4.6. The concentration of the boric acid in the receiving vessel does not enter into the calculation and need not be known. Equation PA.2 in Box 4.7 in Chapter 4 is used for the calculation.

Example 5.4

In a Kjeldahl analysis, a grain sample weighing 1.1033 g was digested in concentrated H_2SO_4 for 40 min. A concentrated solution of NaOH was added such that all of the nitrogen was converted to NH_3. The NH_3 was then distilled into a flask containing a solution of boric acid. Following this, the solution in the receiving flask was titrated with 0.1011 N HCl, requiring 24.61 mL. What is the percentage of nitrogen in the sample?

Solution 5.4

$$\% \, N = \frac{L_{HCl} \times N_{HCl} \times 14.00}{\text{sample weight}} \times 100$$

$$= \frac{0.02461 \times 0.1011 \times 14.00}{1.1033} \times 100$$

$$= 3.157\%$$

5.4   OTHER ACID/BASE APPLICATIONS

Box 5.1 lists other possible acid/base titration applications (see also Application Note 5.1).

5.5   BUFFER SOLUTION APPLICATIONS

Titration curves help us to understand a very important concept in many areas of chemistry and biochemistry. This concept is that of buffering and buffer solutions. A **buffer solution** is a solution that resists changes in pH even when a strong acid or base is added or when the solution is diluted with water.

BOX 5.1   OTHER EXAMPLES OF ACID/BASE TITRIMETRIC ANALYSES

Phosphoric acid in soda pop
Acetic acid in vinegar
Ammonia (or ammonium hydroxide) in household cleaners
Acidity of various beverages, alcoholic and nonalcoholic
Assays of organic carboxylic acids used in pharmaceutical preparations: acetylsalicylic acid (aspirin), ascorbic acid (Vitamin C), etc.
Assays of less well-known acids used in pharmaceutical preparations: benzoic acid, salicylic acid, citric acid, etc.

APPLICATION NOTE 5.1: ACIDITY OF CHEESE BY ACID/BASE TITRATION

The acidity of cheese may be determined by titration with standardized sodium hydroxide to the phenolphthalein end point. The sodium hydroxide solution is around 0.1 N and 10–12 g of the finely divided cheese are used. The cheese sample is shaken in an Erlenmeyer flask with 80 mL of warm distilled water. This solution is then quantitatively transferred to a 100-mL volumetric flask, shaken, and filtered. For the titration, 25-mL aliquots of this filtered solution are used. The reaction is as follows:

$$HL + NaOH \rightarrow NaL + H_2O$$

in which "L" is the lactate ion and HL is lactic acid.

The acidity is reported as the percent of lactic acid. Equation PA.1 in Box 4.7 is used. Also, since one fourth of the sample solution is titrated (25 mL out of 100), the calculated percent must be multiplied by 4.

(*Reference:* James, C., *Analytical Chemistry of Foods*, Chapman & Hall (1995), Boca Raton, FL, USA, pp. 168–9.)

5.5.1 CONJUGATE ACIDS AND BASES

The typical composition of a buffer solution includes a weak acid or base of a particular concentration and its conjugate base or acid of a particular concentration. A **conjugate base** is the product of an acid neutralization that is a base because it can gain back a hydrogen that it lost during the neutralization. Conversely, a **conjugate acid** is the product of a base neutralization that is an acid because it can lose the hydrogen that it gained during the neutralization. A **conjugate acid–base pair** would consist of the acid and its conjugate base or the base and its conjugate acid. An important example of a conjugate base is the acetate ion formed during the neutralization of acetic acid ($HC_2H_3O_2$) with hydroxide.

$$HC_2H_3O_2 + OH^- \rightarrow C_2H_3O_2^- + H_2O \tag{5.21}$$

Acetate is a conjugate base because it can gain a hydrogen and become acetic acid again (the reverse of Equation 5.21). Acetic acid and acetate ion (such as from sodium acetate) would constitute a conjugate acid–base pair.

An important example of a conjugate acid is the ammonium ion formed by the neutralization of ammonia with an acid.

$$H^+ + NH_3 \rightarrow NH_4^+ \tag{5.22}$$

The ammonium ion is a conjugate acid because it can lose a hydrogen and become ammonia again (the reverse of Equation 5.22) and thus ammonia (or ammonium hydroxide) and ammonium ion (such as from ammonium chloride) together constitute another conjugate acid–base pair.

In the process of a weak acid or weak base neutralization titration, a mixture of a conjugate acid–base pair exists in the reaction flask in the time period of the experiment leading up to the inflection point. For example, during the titration of acetic acid with sodium hydroxide, a mixture of acetic acid and acetate ion exists in the reaction flask prior to the inflection point. In that portion of the titration curve, the pH of the solution does not change appreciably even upon the addition of more sodium hydroxide. Thus, this solution is a buffer solution as we have defined it at the beginning of this section.

Buffer solutions can become very effective when the concentrations of the conjugate acid–base pair are higher. When the concentrations are higher, the portion of the curve leading up to the inflection point is extended and thus the **buffer capacity**, or the ability to resist neutralization, is extended. Such a mixture would require even more strong acid or base before it is rendered neutral. The region of a titration curve leading up to the inflection point is often called the **buffer region**.

5.5.2 HENDERSON–HASSELBALCH EQUATION

Buffer solutions can be prepared quite simply by appropriate combination of conjugate acid–base pairs, as in the above examples. Although commercially prepared buffer solutions are available, these are most often utilized solely for pH meter calibration and not for adjusting or maintaining a chemical reaction system at a given pH. It is not surprising, therefore, that the analyst often needs to prepare his/her own solutions for this purpose. It then becomes a question of what proportions of the conjugate acid and base should be mixed to give the desired pH.

The answer is in the expression for the ionization constant, K_a or K_b, where the ratio of the conjugate acid and base concentrations are found. In the case of a weak monoprotic acid, HA, we have the following:

$$HA \leftrightarrows H^+ + A^- \tag{5.23}$$

$$K_a = \frac{[H^+][A^-]}{[HA]} \tag{5.24}$$

where [HA] approximates the weak acid concentration and [A$^-$] is the conjugate base concentration.

Knowing the value of K_a or K_b for a given weak acid or base and knowing the desired pH value, one can calculate the ratio of conjugate base (or acid) concentration to acid (or base) concentration that will produce the given pH. Rearranging Equation 5.24, for example, would give the following:

$$[H^+] = K_a \times \frac{[HA]}{[A^-]} \tag{5.25}$$

Taking the negative logarithm of both sides would give the following:

$$pH = pK_a - \log\frac{[HA]}{[A^-]} \tag{5.26}$$

or

$$pH = pK_a + \log\frac{[A^-]}{[HA]} \tag{5.27}$$

where pK_a is defined as the negative logarithm of the K_a.

It may be more convenient in the case of a weak base to use Equation 5.27 and think of the conjugate acid to be the weak acid and to use its pK_a. In that case, we could write the following:

$$pH = pK_a + \log\frac{[B]}{[BH^+]} \tag{5.28}$$

In this equation, [BH⁺] is the concentration of the conjugate acid of the base and the pK_a is the pK_a for this acid.

Thus, for a weak acid with a given K_a (or pK_a) and a given ratio of conjugate base concentration to acid concentration, the pH may be calculated. Or, given the desired pH and the K_a (pK_a), the ratio of salt concentration to acid concentration can be calculated and the buffer subsequently prepared. Equations 5.25 through 5.28 are each a form of the **Henderson–Hasselbalch equation** for dealing with buffer solutions.

Example 5.5

What is the pH of an acetic acid/sodium acetate solution if the acid concentration is 0.10 N and the acetate concentration is 0.20 N? (The K_a of acetic acid is 1.8×10^{-5}.)

Solution 5.5

Using Equation 5.27, we have the following:

$$pH = -\log(1.8 \times 10^{-5}) + \log \frac{[0.20]}{[0.10]}$$

$$pH = 5.04$$

Example 5.6

What concentration of THAM hydrochloride is required to have a buffer solution of pH 7.98 if the THAM concentration is 0.20 N? (The K_a for THAM hydrochloride is 8.4×10^{-9}.)

Solution 5.6

Using Equation 5.28, we have

$$7.98 = -\log(8.41 \times 10^{-5}) + \log \frac{[0.20]}{[\text{THAM hyd.}]}$$

$$7.98 = 8.075 + \log \frac{[0.20]}{[\text{THAM hyd.}]}$$

$$-0.095 = \log \frac{[0.20]}{[\text{THAM hyd.}]}$$

$$0.8035 = \frac{0.20}{[\text{THAM hyd.}]}$$

$$[\text{THAM hyd.}] = \frac{0.20}{0.80} = 0.25$$

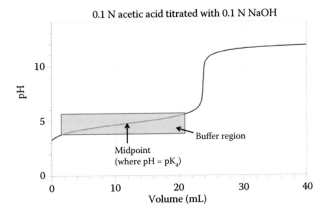

FIGURE 5.15 Drawing of the titration curve for a weak acid titrated with a strong base pointing out the buffer region and midpoint.

It should be stressed that since the K_a enters into the calculation, how weak the acid is dictates what is a workable pH range for that acid. This range can be seen graphically on the titration curve for the weak acid or base—it is the pH range covered by the shallow increase (in the case of a weak acid) or decrease (in the case of a weak base) leading up to the inflection point (see Figure 5.15). The pH value at the midpoint of the pH range (the so-called **midpoint of a titration**) equals the pK_a value as can be seen in the following derivation starting with Equation 5.29.

$$pH = pK_a + \log \frac{[A^-]}{[HA]} \tag{5.29}$$

At the midpoint of the titration, the weak acid concentration equals the conjugate base concentration (half of the acid has been converted to the conjugate base).

$$[A^-] = [HA] \tag{5.30}$$

Thus, ratio of $[A^-]$ to $[HA]$ is equal to 1.

$$\frac{[A^-]}{[HA]} = 1 \tag{5.31}$$

As a result, we have the following:

$$pH = pK_a + \log (1) \tag{5.32}$$

Because the logarithm of 1 is zero, we have the following:

$$pH = pK_a \tag{5.33}$$

See Figure 5.15.

Table 5.1 gives commonly used examples of conjugate acid–base pair combinations and for what pH range each is useful. This range corresponds to the pH range defined by buffer region in the titration curve for each and the middle of the range corresponds to the midpoint of each titration.

TABLE 5.1

Commonly Used Examples of Conjugate Acid–Base Combinations and Corresponding Useful pH Ranges

Combination	pH Range
Chloroacetate/chloracetatic acid ($K_a = 1.36 \times 10^{-3}$)	1.8–3.8
Acetate/acetic acid ($K_a = 1.8 \times 10^{-5}$)	3.7–5.7
Monohydrogen phosphate/dihydrogen phosphate ($K_a = 6.23 \times 10^{-8}$)	6.1–8.1
THAM/THAM hydrochloride ($K_a = 8.4 \times 10^{-9}$)	7.5–9.5
Ammonium hydroxide/ammonium ion ($K_a = 5.7 \times 10^{-10}$)	8.3–10.3

Example 5.7

Suppose a buffer solution of pH 2.0 is needed. Suggest a conjugate acid/base pair for this solution and calculate the ratio of the concentration of conjugate base to acid that is needed to prepare it.

Solution 5.7

From Box 5.1, we can see that a particular chloroacetic acid/chloroacetate combination would give a pH of 2.00. Rearranging Equation 5.27, we have

$$\log\frac{[A^-]}{[HA]} = pH - pK_a = 200 - 2.87 = -0.87$$

$$\frac{[A^-]}{[HA]} = 0.13$$

It should also be stressed that the pH value of an actual buffer solution prepared by mixing quantities of the weak acid or base and its conjugate base or acid based on the calculated ratio will likely be different from what was calculated. The reason for this is the use of approximations in the calculations. For example, the molar concentration expressions found in Equations 5.22 through 5.28, e.g., $[H^+]$, are approximations. Also, to be thermodynamically correctly, the **activity** of the chemical should be used rather than the concentration. Activity is directly proportional to concentration, the **activity coefficient** being the proportionality constant.

$$a = \gamma C \tag{5.34}$$

In this equation, a is the activity, γ (Greek letter gamma) is the activity coefficient, and C is the molar concentration. However, for most applications, especially at small concentrations, the activity coefficient is assumed to be 1, and the activity therefore is equal to the concentration. In addition, the values for the un-ionized acid and base concentrations in the denominators in Equations 5.24 and 5.27 are actually approximations. For example, C_{HA}, the total concentration of acid, is often substituted into Equation 5.27 rather than $[HA]$, the un-ionized acid concentration. This was the case in our Examples 5.5 and 5.6. A better method of accurate buffer solution preparation would be to prepare a solution (the concentration is not important) of the conjugate base (or acid) and then add a solution of a strong acid (or base) until the pH, as measured by a pH meter, is the desired pH. At that point, the required conjugate acid–base pair combination required is in the container.

For example, to prepare a pH 9 buffer solution, one would prepare a solution of ammonium chloride (refer to Table 5.2) and then add a solution of sodium hydroxide while stirring and monitoring

TABLE 5.2

Recipes for Some of the More Popular Buffer Solutions

pH = 4.0 phthalate buffer	Dissolve 10.12 grams of potassium hydrogen phthalate (KHP) in 1 L of solution.
pH = 6.9 phosphate buffer	Dissolve 3.39 grams of potassium dihydrogen phosphate and 3.53 grams of dried sodium monohydrogen phosphate in 1 L of solution.
pH = 10.0 ammonia buffer	Dissolve 70.0 grams of ammonium chloride and 570 mL of concentrated ammonium hydroxide (ammonia) in 1 L of solution.

the pH with a pH meter. The preparation is complete when the pH reaches 9. The required conjugate acid–base pair would be NH_3/NH_4^+. Recipes for standard buffer solutions can be useful, however. Table 5.2 gives specific directions for preparing some popular buffer solutions.

Many applications for buffer solutions are found in the analytical laboratory. It is frequently required to have solutions that do not change pH during the course of an experiment. An example is cited in Section 5.6.3 and used in Experiment 12.

5.6 COMPLEX ION FORMATION REACTIONS

5.6.1 INTRODUCTION

Acid–base (neutralization) reactions are only one type of many that are applicable to titrimetric analysis. There are reactions that involve the formation of a precipitate. There are reactions that involve the transfer of electrons. There are reactions, among still others, that involve the formation of a complex ion. This latter type typically involves transition metals and is often used for the qualitative and quantitative colorimetric analysis (Chapters 7 and 8) of transition metal ions, since the complex ion that forms can be analyzed according to the depth of a color that it imparts to a solution. In this section, however, we are concerned with a titrimetric analysis method in which a complex ion-forming reaction is used.

5.6.2 COMPLEX ION TERMINOLOGY

A complex ion is a polyatomic charged aggregate consisting of a positively charged metal ion combined with either a neutral molecule or negative ion. The neutral molecule or negative ion in this aggregate is called a **ligand**. The ligand can consist of a monatomic, negative ion such as F^-, Cl^-, etc., or a polyatomic molecule or ion, such as H_2O, CN^-, CNS^-, NH_3, CN^-, etc. Some simple examples of the complex ion-forming reaction are presented in Table 5.3. Notice that complex ions can be either positively or negatively charged.

Ligands can be classified according to the number of bonding sites that are available for forming a coordinate covalent bond to the metal ion. A **coordinate covalent bond** is one in which the shared electrons are contributed to the bond by only one of the two atoms involved. In the case of

TABLE 5.3

Examples of Reactions between Metal Ions and Ligands to Form Complex Ions

Metal Ion		Ligand		Complex Ion
Cu^{2+}	+	$4NH_3$	$\leftrightarrows$	$Cu(NH_3)_4^{2+}$
Fe^{3+}	+	CNS^-	$\leftrightarrows$	$Fe(CNS)^{2+}$
Co^{2+}	+	$4Cl^-$	$\leftrightarrows$	$CoCl_4^{2-}$

complex ion-forming reactions, both electrons are always donated by the ligand. If only one such pair of electrons is available per molecule or ion, we say that the ligand is **monodentate**. If two are available per molecule or ion, it is described as **bidentate** etc. Thus, we have the following terminology:

No. of Sites	Descriptive Term
1	Monodentate
2	Bidentate
3	Tridentate
6	Hexadentate

Examples of monodentate ligands are all of those given in Table 5.2 (Cl^-, CN^-, NH_3, etc.). Nitrogen-containing ligands are especially evident in reactions of this kind. This is true because of the pair of electrons occupying the nonbonding orbital found on the nitrogen in many nitrogen-containing compounds. This pair of electrons is present in the Lewis electron dot structure for the nitrogen atom shown in Figure 5.16. It is easy to see that nitrogen would form three covalent bonds, since it has three unpaired electrons, thus often leaving the non-bonded pair of electrons available for coordinate covalent bonding. An example is ammonia, NH_3, as shown in Figure 5.16.

A good example of a bidentate ligand is the 1,10-phenanthroline molecule, which, since it forms a stable complex ion with Fe^{2+} ions that has a deep orange color, is used in the colorimetric analysis of iron(II) ions (see Experiment 15 in Chapter 7). This ligand is shown in Figure 5.17a. The two bonding sites are pointed out with the arrows. Another bidentate ligand is ethylene-diamine shown in Figure 5.17b. These ligands, while bidentate, will form a complex ion with only one metal ion. It is also common for more than one such ligand to combine with a single metal ion.

Complex ions that involve ligands with two or more bonding sites (bidentate, tridentate, etc.) are also called **chelates** and the ligands **chelating agents**. Thus, the ligands shown in Figure 5.17 are examples of chelating agents and the complex ions formed are examples of chelates. Another

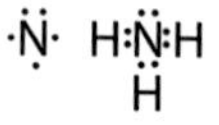

$$\cdot \ddot{N} \cdot \quad H \colon \ddot{N} \colon H$$
$$\underset{H}{\overset{}{|}}$$

FIGURE 5.16 Electron dot structure of a nitrogen atom on the left and an ammonia molecule on the right. The pair of electrons above the nitrogen is the nonbonding pair available for coordinate covalent bonding.

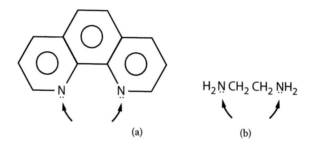

(a)

(b)

FIGURE 5.17 Two bidentate ligands: (a) 1,10 phenanthroline and (b) ethylenediamine. The arrows point out the bonding sites.

term associated with complex ion chemistry is **masking**. Masking refers to the use of ligands and complex ion formation reactions for the purpose of avoiding interferences. When the complex ion formation equilibrium lies far to the right such that the equilibrium constant (more often called **formation constant** in the case of complex ion formation) is very large, the complex ion formed is very stable. This has the effect of "tying up" or **masking** the metal ion such that the interference does not occur. The ligand used in this application is called the **masking agent**. A good example is in the water hardness analysis to be studied in the next section. Dissolved iron ions can interfere with this analysis. This interference can be removed if the iron ions are reacted with the cyanide ligand (CN^-). The formation constant of the $Fe(CN)_6^{3-}$ complex ion is large, meaning that the iron ions are effectively removed from the solution, or "masked," cyanide being the "masking agent." The reaction involved and the formation constant, K_f, are as follows:

$$Fe^{3+} + 6CN^- \leftrightarrows Fe(CN)_6^{3-} \tag{5.35}$$

$$K_f = \frac{[Fe(CN)_6^{3-}]}{[Fe^{2+}][CN^-]^6} = 10^{31} \tag{5.36}$$

A very important ligand (or chelating agent) for titrimetric analysis is the ethylenediaminetetraacetate (EDTA) ligand. It is especially useful in reacting with calcium and magnesium ions in hard water such that water hardness can be determined. The next section is devoted to this subject.

5.6.3 EDTA and Water Hardness

EDTA is a hexadentate ligand. Its structure is shown in Figure 5.18. Again, the bonding sites are pointed out by the arrows. In addition to the four charged sites at the carboxyl group oxygen, each of the two nitrogen has an unshared pair of electrons making six electron pairs available to form coordinate covalent bonds. In forming a complex ion with calcium ions, for example, all six bonding sites bond to calcium forming a large aggregate consisting of a single EDTA ligand wrapped around the calcium ion as shown in Figure 5.19.

Thus, a one-to-one reaction is involved, and in fact, all reactions of EDTA with metal ions (and most metals do indeed react with EDTA) are one to one. We will thus not be concerned with a formal scheme for determining equivalent weights as we were with acid–base reactions.

The usual source of EDTA for use in metal analysis is the disodium, dihydrogen salt of ethylenediaminetetraacetic acid. This is the partially neutralized salt of ethylenediaminetetraacetic acid and is shown in Figure 5.20.

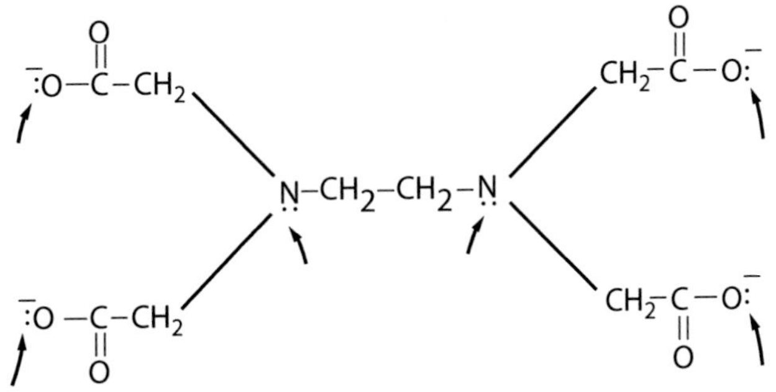

FIGURE 5.18   Structure of the EDTA ligand. The bonding sites are pointed out with arrows.

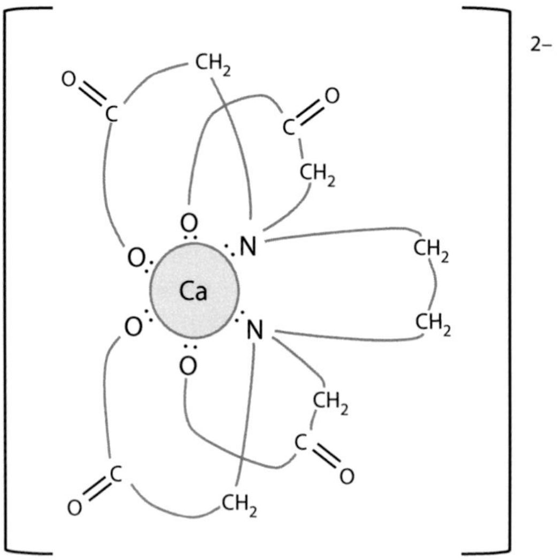

FIGURE 5.19 Representation of the calcium–EDTA complex ion. Each of the six bonding sites on the EDTA is represented by a pair of dots (electrons). The EDTA ligand wraps itself around the calcium ion such that all six pairs of electrons are being shared by this single calcium ion. The four oxygens and the two nitrogens that are bonded to the calcium are printed in larger letters for clarity.

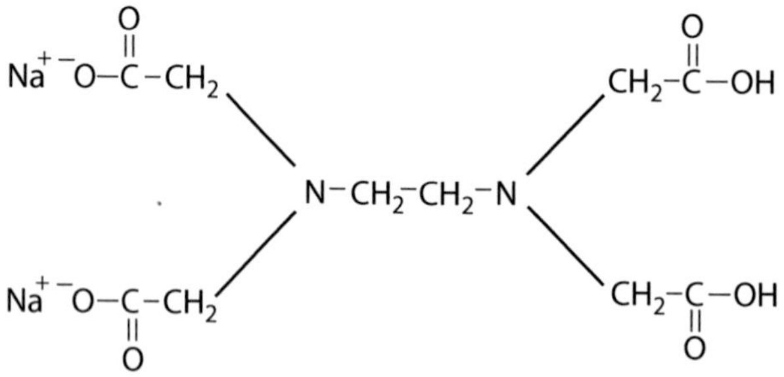

FIGURE 5.20 Disodium dihydrogen salt of ethylenediaminetetracetic acid.

Ethylenediaminetetraacetic acid is frequently symbolized H_4Y, and the disodium salt, Na_2H_2Y. The hydrogens in these formulas are the acidic hydrogens associated with the carboxyl groups, as in any weak organic carboxylic acid, and they dissociate from the EDTA ion in a series of equilibrium steps:

$$\text{Step 1 } H_4Y \rightleftharpoons H^+ + H_3Y^- \tag{5.37}$$

$$\text{Step 2 } H_3Y^- \rightleftharpoons H^+ + H_2Y^{2-} \tag{5.38}$$

$$\text{Step 3 } H_2Y^{2-} \rightleftharpoons H^+ + HY^{3-} \tag{5.39}$$

$$\text{Step 4 } HY^{3-} \rightleftharpoons H^+ + Y^{4-} \tag{5.40}$$

The exact position of this equilibrium is established by controlling the pH of the solution. With extremely basic pH values, Y^{4-} will predominate, whereas in extremely acidic pH values, H_4Y will predominate. The intermediate species will predominate at intermediate pH values (see Figure 5.21).

When Na_2H_2Y is dissolved in water and mixed with a solution of a metal ion, such as calcium, at a basic pH, the metal ion will react with the predominant EDTA species such that the following equilibrium is established:

$$Ca^{2+} + Y^{4-} \leftrightarrows CaY^{2-} \tag{5.41}$$

A problem exists with this procedure, however, in that at basic pH values, many metal ions precipitate as the hydroxide, e.g., $Mg(OH)_2$, and thus would be lost to the analysis. This occurs with the magnesium in the water hardness procedure alluded to earlier. Luckily, a happy medium exists. At pH 10, the reaction of the metal ion with the predominant HY^{3-} and Y^{4-} species (Figure 5.21) is shifted sufficiently to the right for the quantitative requirement to be fulfilled, while at the same time, the solution is not basic enough for the magnesium ions to precipitate appreciably. Thus, all solutions in the reaction flask in the water hardness determination are buffered at pH 10, meaning that a conjugate acid/base pair is added to each solution so that the pH 10. In this case, the buffer is the ammonia/ammonium ion system mentioned previously (see Tables 5.1 and 5.2).

EDTA titrations are routinely used to determine water hardness in a laboratory. Raw well water samples can have a significant quantity of dissolved minerals that contribute to a variety of problems associated with the use of such water. These minerals consist chiefly of calcium and magnesium carbonates, sulfates, etc. The problems that arise are mostly a result of heating or boiling the water over a period of time such that the water is evaporated and the calcium and magnesium salts become concentrated and precipitate in the form of a "scale" on the walls of the container, hence the term "hardness." This kind of problem is thus evident in boilers, domestic and commercial water heaters, humidifiers, tea kettles, and the like.

Consumers and industrial companies can install water softeners to eliminate the problem. Water softeners work on an ion-exchange principle (see Chapter 10) and remove the metal ions that cause the hardness properties, replacing them with sodium ions.

A second problem with hard water is that these metals react with soap molecules and form a "scum" to which bathtub rings and others are attributed. Hard water is therefore not the best

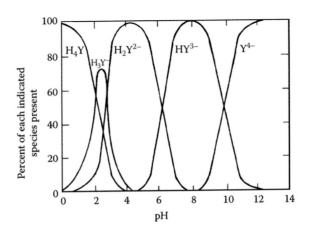

FIGURE 5.21   Predominance of EDTA species as a function of pH.

water to use for efficient soapy water cleaning processes, since the metal–soap precipitation reaction competes with the cleaning action. Water softeners assist with solving this problem too.

Water from different sources can have very different hardness values. Water samples can vary from simply being "extremely soft" to being "extremely hard." While this description of hardness is anything but quantitative, a quantitative description based on an EDTA titration can be given.

The EDTA determination of water hardness results from the reaction of the EDTA ligand with the metal ions involved, calcium and magnesium. An interesting question is: "How are the results reported?" Hardness is not usually reported precisely as so much calcium, plus so much magnesium, etc. There is no distinction made between the metals involved. All species reacting with the EDTA are considered one species and the results reported as an amount of one species, calcium carbonate ($CaCO_3$). That is, in the calculation, when a formula weight is used to convert moles to grams, the formula weight of calcium carbonate is used, and thus a quantity of $CaCO_3$ equivalent to the sum total of all contributors to the hardness is what is reported. Hardness is most often expressed as either ppm $CaCO_3$ or grains per gallon $CaCO_3$.

The indicator that is most often used is called **Eriochrome Black T** (EBT). EBT is actually a ligand that also reacts with the metal ions, like EDTA. In the free uncombined form, it imparts a sky-blue color to the solution, but if it is a part of a complex ion with either calcium or magnesium ions, it is a wine red color. Thus, before adding any EDTA from a buret, the hard water sample containing the pH 10 ammonia buffer and several drops of EBT indicator will be wine red. As the EDTA solution is added, the EDTA ligand reacts with the free metal ions and then actually reacts with the metal–EBT complex ion, complexing the metal and resulting in the free EBT ligand, which, as mentioned earlier, gives a sky blue color to the solution. The color change then is the total conversion of the wine red color to the sky blue color, with every trace of red disappearing at the end point.

It is known that this color change is quite sharp when magnesium ions are present. In cases in which magnesium ions are not present in the water samples, the end point will not be sharp. Because of this, a small amount of magnesium chloride is added to the EDTA as it is prepared and thus a sharp end point is assured.

5.6.4   EXPRESSING CONCENTRATION USING PARTS PER MILLION

As indicated in the last section, the amount of hardness in a water sample is often reported as parts per million (ppm) $CaCO_3$. **Parts per million** is a unit of concentration similar to molarity, percent, etc. A solution that has a solute content of 1 ppm has 1 part of solute dissolved for every million (10^6) parts of solution. A "part" can be a weight measurement, a volume measurement, or both. It if is a weight measurement, it might mean, for example, 1 mg/10^6 mg or 1 mg/kg. A conversion of units here indicates that it might also mean 1 μg/10^6 μg or 1 μg/g. If it is a volume measurement, it might mean 1 mL/10^6 mL, 1 mL/1000 L or 1 μL/L. If the intent is to represent a concentration in a water solution, and this is frequently the case, we can recognize that 1 g of water (and dilute water solutions) is 1 mL of the water, and that 1 L of water weighs 1 kg. Thus, 1 ppm might also mean 1 mg/L.

Most commonly, if the sample to be analyzed is a solid material (for example soil), ppm is expressed as mg/kg (or μg/g), since the weight (as opposed to volume) of the material is most easily measured. However, if the volume is most easily measured (such as with a liquid sample like water or water solutions), the ppm is expressed as mg/L (or μg/mL). Thus, if a certain solution is described as being 10 ppm iron, this means 10 mg of iron dissolved per liter of solution. If, however, an analysis of soil for potassium is reported as 250 ppm K, this means 250 mg of potassium per kg of soil (see Table 5.4). Some of the solution preparation procedures that a technician may encounter involving the ppm unit are discussed below (see Box 5.2).

TABLE 5.4

The ppm Unit for Solid Samples and Liquid Samples and Solutions

For Solid Samples	For Liquid Samples and Solutions
1 ppm = 1 mg/kg	1 ppm = 1 mg/L
= 1 μg/g	= 1 μg/mL

BOX 5.2   PREPARATION OF PPM SOLUTIONS

If the substance the ppm is expressed as is the same substance as that is to be weighed,

$$\text{ppm}_{\text{Desired}} \times L_{\text{Desired}} = \text{mg to be weighed}$$

If the substance the ppm is expressed as is not the same substance as that to be weighed,

$$\text{ppm}_{\text{Desired}} \times L_{\text{Desired}} \times \text{gravimetric factor} = \text{mg to be weighed}$$

If a more concentrated solution is to be diluted,

$$V_B = \frac{C_A \times V_A}{C_B}$$

Summary of the possible calculations to prepare ppm solutions.

5.6.4.1   Solution Preparation

If you wanted to prepare 1 L of a 10-ppm solution of zinc, Zn, you would weigh out 10 mg of zinc metal, place it in a 1-L flask, dissolve it, and dilute to the 1-L mark with distilled water. If you need to prepare a volume other than 1 L, the ppm concentration is multiplied by the volume in liters:

$$\text{ppm}_{\text{Desired}} \times L_{\text{Desired}} = \text{mg to be weighed} \tag{5.42}$$

$$\frac{\text{mg}}{L} \times L = \text{mg} \tag{5.43}$$

Example 5.8

Tell how you would prepare 500.0 mL of a 25.0-ppm copper solution from pure copper metal.

Solution 5.8

$$\frac{25.0 \text{ mg}}{L} \times 0.5000 \text{ L} = 12.5 \text{ mg}$$

Weigh 12.5 mg of copper into a 500-mL flask, dissolve (in dilute nitric acid), and dilute to the mark with water.

Another example of the preparation of ppm solutions is by dilution. It is a very common practice to purchase solutions of metals that are fairly concentrated (1000 ppm) and then dilute them to obtain the desired concentration. This is done to save solution preparation time in the laboratory. As per our Chapter 4 discussion (see Equation 4.14 and accompanying discussion), the concentration is multiplied by the volume both before and after dilution. Using the ppm unit, we have

$$ppm_B \times V_B = ppm_A \times V_A \tag{5.44}$$

where A and B refer to "before dilution" and "after dilution" as in Chapter 4.

Example 5.9

How would you prepare 100.0 mL of a 20.0-ppm iron solution from a 1000.0-ppm solution?

Solution 5.9

Using Equation 5.44, we have

$$1000.0 \text{ ppm} \times V_B = 20.0 \text{ ppm} \times 100.0 \text{ mL}$$

$$V_B = \frac{20.0 \times 100.0}{1000.0} = 2.00 \text{ mL}$$

Dilute 2.00 mL of 1000.0 ppm to 100.0 mL

There is a third type of solution preparation problem that could be encountered with the ppm unit. This is the case in which the solution of a metal is to be prepared by weighing a metal salt, rather than the pure metal, when only the ppm of the metal in the solution is given. In this case, the weight of the metal needs to be converted to the weight of the metal salt via a gravimetric factor (see Chapter 3, Section 3.7.3) so that the weight of the metal salt is known.

$$ppm_D \times L_D \times \text{gravimetric factor} = \text{mg to be weighed} \tag{5.45}$$

Example 5.10

How would you prepare 250.0 mL of a 50.0-ppm solution of nickel from pure solid nickel chloride hexahydrate, $NiCl_2 \cdot 6H_2O$, FW = 237.71 g/mol?

Solution 5.10

Using Equation 5.45, we have

$$\frac{50.0 \text{ mg}}{L} \times 0.2500 \text{ L} \times \frac{NiCl_2 \cdot 6H_2O}{Ni} = \text{mg to be weighed}$$

$$\frac{50.0 \text{ mg}}{L} \times 0.2500 \text{ L} \times \frac{237.71}{58.69} = 50.6 \text{ mg to be weighed}$$

Weigh 50.6 mg of $NiCl_2 \cdot 6H_2O$, place in a 250-mL volumetric flask, dissolve with water, and dilute to the mark.

Box 5.2 summarizes the calculations for solution preparation using the ppm unit.

5.6.5 WATER HARDNESS CALCULATIONS

As mentioned earlier, the hardness of a water sample is often reported as ppm of calcium carbonate. Using our definition of ppm for liquid solutions (Table 5.4), we have the following:

$$ppm_{CaCO_3} = \frac{mg\ of\ CaCO_3}{liters\ of\ water} \tag{5.46}$$

At the end point of the titration, the moles of $CaCO_3$ reacted are the same as the moles of EDTA added (it is a one-to-one reaction; see Equation 5.41).

$$moles_{CaCO_3} = moles_{EDTA} \tag{5.47}$$

The moles of EDTA, and therefore the moles of $CaCO_3$, are computed from the titration data.

$$moles_{CaCO_3} = moles_{EDTA} = L_{EDTA} \times M_{EDTA} \tag{5.48}$$

The grams of $CaCO_3$ can be computed by multiplying by the formula weight of $CaCO_3$ (100.03 g/mol) and converted to milligrams by multiplying by 1000 mg/g.

$$L_{EDTA} \times M_{EDTA} \times FA_{CaCO_3} \times 1000 = mg_{CaCO_3} \tag{5.49}$$

The ppm is then calculated by dividing by the liters of water used in the titration. Combining Equations 5.46 and 5.49, we have the following:

$$ppm_{CaCO_3} = \frac{L_{EDTA} \times M_{EDTA} \times FW_{CaCO_3} \times 1000}{L_{water\ used}} \tag{5.50}$$

Example 5.11

What is the hardness in a water sample, 100.0 mL of which required 27.95 mL of a 0.01266-N EDTA solution for titration?

Solution 5.11

Substituting into Equation 5.52, we have the following:

$$ppm_{CaCO_3} = \frac{0.02795\ L \times 0.01266\ M \times 100.086 \times 1000}{0.1000\ L} = 354.2\ ppm\ CaCO_3$$

Multiplying by the formula weight of $CaCO_3$ in the numerator has the effect of reporting all metals that reacted with the EDTA as $CaCO_3$.

The molarity of the EDTA solution, M_{EDTA}, in Equation 5.6, can be known directly through its preparation with the use of an analytical balance and a volumetric flask. That is, one can purchase pure disodium dihydrogen EDTA and use it as a primary standard. In that case, the solution is prepared and the concentration calculated according to the discussion in Chapter 4 (see Sections 4.3, especially Example 4.2, and Section 4.4.1).

Example 5.12

The following EDTA solution is needed in a water hardness laboratory: 500.0 mL of a 0.01000-N solution of disodium dihydrogen EDTA dihydrate (FW = 372.23 g/mol). How many grams would be needed if you were to prepare it to be exactly 0.01000 N? How would you prepare this solution?

Solution 5.12

The grams of solute required is calculated according to Equation 4.11 in Chapter 4, in which L_D is the liters of solution desired, M_D is the molarity desired, and FW_{SOL} is the formula weight of the solute.

$$grams\ to\ weigh = L_D \times M_D \times FW_{SOL}$$

$$= 0.5000\ L \times 0.01000\ mol/L \times 372.23\ g/mol$$

$$= 1.861\ g$$

The solution is prepared by weighing 1.861 g of dried disodium dihydrogen EDTA dihydrate on an analytical balance, transferring to a clean 500-mL volumetric flask, adding water to dissolve, diluting to the mark with water and making homogeneous by shaking.

Example 5.13

Suppose you followed the procedure in Example 5.12, but you did not weigh exactly 1.861 g, but came close with 1.9202 g. What is the molarity of the EDTA solution prepared with this weight of solute?

Solution 5.13

The exact molarity can be calculated using Equation 4.1 in Chapter 4 as follows.

$$molarity = \frac{moles\ of\ solute}{liters\ of\ solution} = \frac{grams\ of\ solute/FW\ of\ solute}{liters\ of\ solution} = \frac{1.9202/372.23}{0.5000} = 0.01032$$

Note: Standardization with a primary standard may still be important as a cross-check (see discussion and examples below).

The molarity of the EDTA solution may also become known via standardization. In this case, the solution was prepared with an ordinary balance and a vessel other than a volumetric flask such that its molarity is not known to more than two significant figures. Just as with acids and bases (see Chapter 4, Section 4.5), the standardization can take place by either weighing an appropriate primary standard accurately into the reaction flask or by accurately pipetting a standard solution of a metal into the flask. The calculations are similar to what was discussed in Chapter 4, and we can derive equations similar to Equations 4.17 and 4.18 but using molarity instead of normality. Starting with Equation 5.48, we have the following at the end point of a titration using calcium carbonate.

$$moles_{EDTA} = moles_{CaCO_3} \qquad (5.51)$$

In the case of standardization with a solution of calcium carbonate, we have the following:

$$V_{EDTA} \times M_{EDTA} = V_{CaCO_3} \times M_{CaCO_3} \qquad (5.52)$$

In the case of standardization with a weighed quantity of calcium carbonate, we have the following:

$$L_{EDTA} \times M_{EDTA} = \frac{grams\ of\ CaCO_3}{FW\ of\ CaCO_3} \qquad (5.53)$$

It is also possible to prepare a standard solution of calcium carbonate accurately using an analytical balance and a volumetric flask, as was suggested previously for EDTA, and pipetting an **aliquot** (a portion of a larger volume) of this solution into the reaction flask in preparation for the standardization of the EDTA solution. In that case, the concentration of the calcium carbonate solution is first calculated from the weight/volume preparation data and then the molarity of the EDTA solution is determined using Equation 5.52.

Example 5.14

A $CaCO_3$ solution is prepared by weighing 0.5047 g of $CaCO_3$ into a 500-mL volumetric flask, dissolving with HCl, and diluting to the mark. If a 25.00-mL aliquot of this solution required 28.12 mL of an EDTA solution, what is the normality of the EDTA?

Solution 5.14

The molarity of the $CaCO_3$ solution is first calculated from the weight and volume data given.

$$M_{CaCO_3} = \frac{grams\ of\ CaCO_3/FW\ of\ CaCO_3}{L_{solution}}$$

$$= \frac{0.5047/100.086}{0.5000}$$

$$= 0.01009\ M$$

Next, using Equation 5.52, we have

$$28.12 \times M_{EDTA} = 25.00 \times 0.01008$$

$$M_{EDTA} = \frac{25.00 \times 0.01008}{28.12}$$

$$= 0.008962\ M$$

5.6.6 OTHER USES OF EDTA TITRATIONS

EDTA titrations may also be used for determining the calcium in pharmaceutical preparations, such as calcium carbonate and calcium citrate tablets. In this case, the powdered solid material is dissolved in hydrochloric acid, diluted with distilled water, and, after adding the pH = 10 buffer and the EBT (or other) indicator, titrated with EDTA. Experiment 12, Part D, is an example of such an analysis (see also Application Note 5.2 for an additional application of EDTA titrations).

**APPLICATION NOTE 5.2: THE DETERMINATION OF
MAGNESIUM IN MILK OF MAGNESIA**

The determination of water hardness is not the only use of EDTA as a titrant. Many pharmaceutical preparations that are calcium or magnesium compounds may be assayed by dissolving them and titrating with EDTA. Examples include calcium citrate (Citracal™) tablets, magnesium sulfate (Epsom Salt™), and milk of magnesia, which is a suspension of magnesium hydroxide.

In the case of milk of magnesia, the magnesium hydroxide must first be dissolved. Since it is a base, it may be dissolved using an acid. The official procedure uses 3 N HCl for this. The excess HCl is then neutralized to a pH of 7 using 1 N NaOH. At that point, the pH 10 ammonia buffer and the eriochrome black T indicator are added. The solution is then titrated with EDTA to the same end point as in the water hardness titration.

(*Reference: The Official Compendia of Standards*, U.S. Pharmacopeia and National Formulary (2000), Rockville, MD, USA, p. 1000.)

5.7　OXIDATION–REDUCTION REACTIONS

Oxidation–reduction reactions represent yet another type of reaction that titrimetric analysis can utilize. In other words, a solution of an oxidizing agent can be in the buret and a solution of a reducing agent can be in the reaction flask (and vice versa). In this section, we review the fundamentals of oxidization–reduction chemistry and discuss the titrimetric analysis applications.

5.7.1　Review of Basic Concepts and Terminology

Many chemical species have a tendency to either give up or take on electrons. This tendency is based on the premise that a greater stability, or lower energy state, is achieved as a result of this electron donation or acceptance. A hypothetical example is the reaction between a sodium atom and a chlorine atom

$$\text{sodium atom (Na)} + \text{chlorine atom (Cl)} \rightarrow \text{sodium chloride formula unit (NaCl)} \quad (5.54)$$

A sodium atom with only one electron in its outermost energy level would achieve a lower energy state if this electron were released to the chlorine atom, which would also achieve a lower energy state as a result. Both atoms would become ions (Na^+ and Cl^-) and each would have a stable, filled outermost energy level identical with those of the noble gases, neon in the case of sodium and argon in the case of chlorine. Thus, the "electron transfer" does take place and sodium chloride, NaCl, is formed.

Those reactions that this sodium–chlorine case typifies are called **oxidation–reduction reactions**. The term **oxidation** refers to the loss of electrons, whereas the term **reduction** refers to the gain of electrons. A number of oxidation–reduction reactions (also called **redox reactions**) are useful in titrimetric analysis, and many are encountered in other analysis methods.

It may appear strange that the term reduction is associated with a gaining process. Actually, the term reduction was coined as a result of what happens to the oxidation number of the element when the electron transfer takes place. The **oxidation number** of an element is a number representing the state of the element with respect to the number of electrons the element has given up, taken on, or contributed to a covalent bond. For example, pure sodium metal has neither given up, taken on nor shared electrons, and thus its oxidation number is zero. In sodium chloride, however, the sodium has given up an electron and becomes a +1 charge, and thus, its oxidation number is +1. The chlorine in NaCl has taken on an electron and becomes a −1 charge, and thus its oxidation number is −1. Chlorine also had an oxidation number of 0 prior to the reaction. Thus, although it is true that the chlorine gained an electron, it is also true that its oxidation number became lower (from 0 to −1), hence the term "reduction."

As we shall see, it is useful to be able to determine the oxidation number of a given element in a compound. Since most elements can exist in a variety of oxidation states, it is necessary to adopt a set of rules or guidelines for this determination. These are listed in Table 5.5.

A discussion of several examples should clarify the general scheme. First, rules 1 through 9 cover many situations in which, with few exceptions, the oxidation number is a set number. For example, in $BaCl_2$, barium is +2 (rule 3) and Cl is −1 (rule 5). In K_2O, potassium is +1 (rule 2) and O is −2 (rule 8). In CaH_2, calcium is +2 (rule 3) and H is −1 (rule 7).

Next, rule 10 is used for determining all oxidation numbers of all elements that rules 1 through 8 do not cover (this must be all but one element in a formula), and then assigning the remaining elements oxidation numbers, knowing that the total must add up to either zero or the ionic charge (Table 5.5).

Example 5.15

What is the oxidation number of sulfur in H_2SO_4?

Solution 5.15

Oxygen is −2 and $4 \times (−2) = −8$
Hydrogen is +1 and $2 \times (+1) = +2$
Sum $= −8 + 2 = −6$

Therefore, sulfur must be +6 in order for the total to be 0.

Example 5.16

What is the oxidation number of manganese in $KMnO_4$?

TABLE 5.5
Rules for Assigning Oxidations Numbers

Element	Oxidation Number
1. Any uncombined element	0
2. Combined alkali metal (group I)	+1
3. Combined alkaline earth metal (group II)	+2
4. Combined aluminum	+3
5. Halogen (group VIIA) in binary compound with a metal or hydrogen	−1
6. Hydrogen combined with nonmetal	+1
7. Hydrogen combined with metal (hydrides)	−1
8. Combined oxygen (except peroxides)	−2
9. Oxygen in peroxides	−1
10. All others—oxidation numbers are determined from the fact that the sum of all oxidation numbers within the molecule or polyatomic ion must add up to either zero (in the case of a neutral molecule) or the charge on the ion (in the case of a polyatomic ion)	

Solution 5.16

Potassium is +1 and $1 \times (+1) = +1$
Oxygen is –2 and $4 \times (-2) = -8$
Sum $= -8 + 1 = -7$

Manganese must be +7 in $KMnO_4$ in order for the total to be 0.

Example 5.17

What is the oxidation number of nitrogen in the nitrate ion NO_3^{-1}?

Solution 5.17

Oxygen is –2 and $3 \times (-2) = -6$
Sum $= -6 + 5 = -1$

Nitrogen must be +5 in order for the net charge to be –1.

The usefulness of determining the oxidation number in analytical chemistry is twofold. First, it will help determine if there was a change in oxidation number of a given element in a reaction. This always signals the occurrence of an oxidation–reduction reaction. Thus, it helps tell us whether a reaction is a redox reaction or some other reaction. Second, it will lead to the determination of the number of electrons involved, which will aid in balancing the equation. These latter points will be discussed in the later sections.

Example 5.18

Which of the following is a redox reaction?

(a) $Cl_2 + 2NaBr \rightarrow Br_2 + 2NaCl$
(b) $BaCl_2 + K_2SO_4 \rightarrow BaSO_4 + 2KCl$

Solution 5.18

The oxidation numbers of Cl and Br in (a) have changed. Cl has changed from 0 to –1, while Br has changed from –1 to 0. (In diatomic molecules, the elements are considered to be in the uncombined state; thus, rule 1 applies.) There is no change in any oxidation number of any element in (b). Thus, (a) is redox.

The terms oxidation and reduction, as should be obvious from the discussion to this point, can be defined in two different ways—according to the gain or loss of electrons or according to the increase or decrease in oxidation number:

Oxidation – the loss of electrons or the increase in oxidation number
Reduction – the gain of electrons or the decrease in oxidation number

To say that a substance has been oxidized means that the substance has lost electrons. To say that a substance has been reduced means that the substance has gained electrons. To say that substance A has oxidized substance B means that substance A has caused B to lose electrons. Similarly, to say that substance A has reduced substance B means that substance A has caused B to gain electrons. When a substance (A) causes another substance (B) to be oxidized, substance A is called the **oxidizing agent**. When substance A causes substance B to be reduced, substance A is a **reducing agent**.

To say that substance A "causes" oxidation or reduction means that substance A either removes electrons from or donates electrons to substance B. Thus, oxidation is always accompanied by reduction and vice versa. Also, every redox reaction has an oxidizing agent (which is the substance reduced) and a reducing agent (which is the substance oxidized).

Example 5.19

Tell what is oxidized and what is reduced and tell what the oxidizing agent and the reducing agent are in the following

$$Zn + 2HCl \rightarrow ZnCl_2 + H_2$$

Solution 5.19

$$\text{Oxidation: } Zn \rightarrow Zn^{2+}$$

(The oxidation number of Zn has increased from O to +2.)

$$\text{Reduction: } 2H^+ \rightarrow H_2$$

(Oxidation number of H has decreased from +1 to 0.)

Zn has been oxidized. Therefore, it is the reducing agent. H^+ has been reduced. Therefore, it is the oxidizing agent.

5.7.2 THE ION-ELECTRON METHOD FOR BALANCING EQUATIONS

We have seen how analytical calculations in titrimetric analysis involve stoichiometry (Chapter 4, Sections 4.5 and 4.6). We know that a balanced chemical equation is needed for basic stoichiometry. With redox reactions, balancing equations by "inspection" can be quite challenging, if not impossible. Thus, several special schemes have been derived for balancing redox equations. The **ion-electron method** for balancing redox equations takes into account the electrons that are transferred, since these must also be balanced. That is, the electrons given up must be equal to the electrons taken on. A review of the ion-electron method of balancing equations will therefore present a simple means of balancing redox equations.

The method makes use of only those species, dissolved or otherwise, which actually take part in the reaction. The so-called **spectator ions**, or ions that are present but play no role in the chemistry, are not included in the balancing procedure. Solubility rules are involved here, since spectator ions result only when an ionic compound dissolves and ionizes. Also, the scheme is slightly different for acid and base conditions. Our purpose, however, is to discuss the basic procedure, and thus, spectator ions will be absent from all examples from the start, and acidic conditions will be the only conditions considered. The step-wise procedure we will follow is

Step 1: Look at the equation to be balanced and determine what is oxidized and what is reduced. This involves checking the oxidation numbers and discovering which have changed.

Step 2: Write a **half-reaction** for both the oxidation process and the reduction process and label as "oxidation" and "reduction." These half-reactions show only the species being oxidized (or the species being reduced) on the left side with only the product of the oxidation (or reduction) on the right side.

Step 3: If oxygen appears in any formula on either side in either equation, it is balanced by writing H_2O on the opposite side. This is possible since the reaction mixture is a water solution. The hydrogen in the water is then balanced on the other side by writing H^+, since we are dealing with acid solutions. Now balance both half-reactions for all elements by inspection.

Step 4: Balance the charges on both sides of the equation by adding the appropriate number of electrons (e⁻) to whichever side is deficient in negative charges. The charge balancing is accomplished as if the electron is like any chemical species—place the appropriate multiplying coefficient in front of the "e⁻."

Step 5: Multiply through both equations by appropriate coefficients, so that the number of electrons involved in both half-reactions is the same. This has the effect of making the total charge loss equal to the total charge gain and thus eliminates electrons from the final balanced equation as you will see in Step 6.

Step 6: Add the two equations together. The number of electrons, being the same on both sides, cancels out and thus does not appear in the final result. One can also cancel out some H^+ and H_2O, if they appear on both sides at this point.

Step 7: Make a final check to see that the equation is balanced.

Example 5.20

Balance the following equation by the ion-electron method

$$MnO_4^- + Fe^{2+} \rightarrow Fe^{3+} + Mn^{2+}$$

Solution 5.20

Step 1: Fe^{2+} oxidized from +2 → +3 (loss of electrons)

MnO_4^- reduced from +7 → +2 (gain of electrons)

Step 2: Oxidation: $Fe^{2+} \rightarrow Fe^{3+}$

Reduction: $MnO_4^- \rightarrow Mn^{2+}$

Step 3: Oxidation: $Fe^{2+} \rightarrow Fe^{3+}$

Reduction: $8H^+ + MnO_4^- \rightarrow Mn^{2+} + 4H_2O$

Step 4: Oxidation: $Fe^{2+} \rightarrow Fe^{3+} + 1e^-$

Reduction: $5e^- + 8H^+ + MnO_4^- \rightarrow Mn^{2+} + 4H_2O$

Step 5: Oxidation: $5(Fe^{2+} \rightarrow Fe^{+3} + 1e^-)$

Reduction: $1(5e^- + 8H^+ + MnO_4^- \rightarrow Mn^{2+} + 4H_2O)$

Step 6: Oxidation: $5(Fe^{2+} \rightarrow Fe^{3+} + 1e^-)$

Reduction: $1(5e^- + 8H^+ + MnO_4^- \rightarrow Mn^{2+} + 4H_2O)$

$5Fe^{2+} + 8H^+ + MnO_4^- \rightarrow 5Fe^{3+} + Mn^{2+} + 4H_2O$

Step 7: 5 Fe on each side

8 H on each side

1 Mn on each side

4 O on each side

The equation is balanced.

5.7.3 ANALYTICAL CALCULATIONS

Following are some examples of analytical calculations involving redox titrations.

Example 5.21

What is the molarity of a solution of $K_2Cr_2O_7$ if 0.6729 g of ferrous ammonium sulfate hexahydrate (sometimes referred to as FAS, FW = 392.14 g/mol) are exactly consumed by 24.92 mL of the solution as in the following equation?

$$Fe^{2+} + Cr_2O_7^{2-} \rightarrow Fe^{3+} + Cr^{3+}$$

Solution 5.21

Balancing the equation using the ion-electron method results in the following (check it):

$$6\ Fe^{2+} + 14\ H^+ + Cr_2O_7^{2-} \rightarrow 6\ Fe^{3+} + 2\ Cr^{3+} + 7\ H_2O$$

This is an example of standardization with a primary standard. Calculating the molarity involves Equation 4.32 from Chapter 4.

$$L_T \times M_T \times \text{mole ratio (PS/T)} = \frac{\text{grams}_{PS}}{FW_{PS}}$$

$$L_{K_2Cr_2O_7} \times M_{K_2Cr_2O_7} \times \text{mole ratio (Fe}^{2+}/Cr_2O_7^{2-}) = \frac{\text{grams}_{FAS}}{FW_{FAS}}$$

$$0.12492\ L \times M_{K_2Cr_2O_7} \times 6/1 = \frac{0.6729}{392.14}$$

$$M_{K_2Cr_2O_7} = 0.01148\ N$$

Example 5.22

What is the percent of hydrogen peroxide (H_2O_2) in a sample if 23.01 mL of 0.04739 N $KMnO_4$ was needed to titrate 10.0232 g of a hydrogen peroxide solution according to the following?

$$H_2O_2 + MnO_4^- \rightarrow O_2 + Mn^{2+}$$

Solution 5.22

Balancing the equation using the ion-electron method results in the following (check it):

$$5\ H_2O_2 + 6\ H^+\ 2\ MnO_4^- \rightarrow 5\ O_2 + 2\ Mn^{2+} + 8\ H_2O$$

Calculating the percent hydrogen peroxide utilizes Equation 4.37 from Chapter 4.

$$\% \ H_2O_2 = \frac{L_{KMnO_4} \times M_{KMnO_4} \times \text{mole ratio (H}_2O_2 \div KMnO_4) \times FW_{H_2O_2}}{\text{weight of sample}}$$

$$\% \ H_2O_2 = \frac{0.02301 \times 0.04739 \times 5 \div 2 \times 34.01}{10.0232}$$

$$= 0.9250\%$$

5.7.4 APPLICATIONS

5.7.4.1 Potassium Permanganate

An oxidizing agent that has a significant application in redox titrimetry is potassium permanganate, $KMnO_4$. As discovered earlier, the manganese in $KMnO_4$ has a +7 oxidation number. It is as if the manganese has contributed all seven of its outermost electrons ($4s^2$, $3d^5$) to a bonding situation, an observation that implies significant instability. Manganese atoms are in lower energy states if they are found with +4 or +2 oxidation numbers. Thus, like sodium and chlorine in the zero state, manganese in the +7 state is very unstable and will take electrons, given the opportunity, and be reduced. Hence, it is a strong (relatively speaking) oxidizing agent. Some notable examples of its oxidizing powers include Fe^{2+} to Fe^{3+}, $H_2C_2O_4$ to CO_2, As (III) to As (V), and H_2O_2 to O_2. Organic compounds could also be included in this list, and in fact, potassium permanganate solutions, which are deep purple, are used to test qualitatively for alkenes, the reaction being the reduction of MnO_4^- to MnO_2. The purple color disappears in this test, and the brown precipitate, MnO_2, forms and indicates a positive test.

Working with $KMnO_4$ presents some special problems because of its significant oxidizing properties. The keys to successful redox titrimetry using $KMnO_4$ are (1) to prepare solutions well in advance so that any oxidizable impurities (usually organic in nature) in the distilled water used are completely oxidized and (2) to protect the standardized solution from additional oxidizable materials (such as lint, fingerprints, rubber, etc., so that its concentration remains constant until the solution is no longer needed). One additional problem is that once some MnO_2 has formed through the oxidation of such organic substances as those listed, it can catalyze further decomposition and thereby cause further changes in the $KMnO_4$ concentration. It is obvious that if the solutions are not carefully protected, the concentration of the standardized solution cannot be trusted. Even ordinary light can catalyze the reaction, and for this reason, solutions must be protected from light. The MnO_2 present from the initial reactions (prior to standardization) is filtered out so that the catalytic reactions are minimized.

There are three major points to be made concerning the actual titrations using $KMnO_4$. First, since the solutions are unstable when first prepared due to the presence of oxidizable materials in the distilled water, it cannot be used as a primary standard. Thus, $KMnO_4$ solutions must always be standardized prior to use. Second, all titrations are carried out in acid solution. Thus, Mn^{2+} is the product, rather than MnO_2, and no brown precipitate forms in the reaction flask. Third, the fact that the color of the $KMnO_4$ solution is a deep purple means that the solution may serve as its own indicator. At the point when the substance titrated is exactly consumed, the $KMnO_4$ will no longer react, and since it is a highly colored chemical species, the slightest amount of it in the reaction flask is easily seen. Thus, the end point is the first detectable pink color due to unreacted permanganate.

$KMnO_4$ solutions are standardized using primary standard-grade reducing agents. Typical reducing agents include sodium oxalate $Na_2C_2O_4$, and ferrous ammonium sulfate hexahydrate, $Fe(NH_4)_2(SO_4)_2 \cdot 6H_2O$, also known as "Mohr's salt" (see Application Note 5.3).

5.7.4.2 Iodometry: An Indirect Method

Another important reactant in redox titrimetry is potassium iodide KI. KI is a reducing agent ($2I^- \rightarrow I_2 + 2e^-$) that is useful in analyzing for oxidizing agents. The interesting aspect of the iodide–iodine chemistry is that it is most often used as an indirect method (recall the indirect Kjeldahl titration involving boric acid discussed previously). This means that the oxidizing agent analyte is not measured directly by a titration with KI, but is measured indirectly by the titration of the iodine that forms in the reaction. The KI is actually added in excess, since it need not be measured at all. The experiment is called iodometry. Figure 5.22 shows the sequence of events. Thus, the percent of the oxidizing agent ("O" in Figure 5.22) is calculated indirectly from the amount of titrant since the titrant actually reacts with I_2 and not "O." This titrant is normally sodium thiosulfate ($Na_2S_2O_3$).

The sodium thiosulfate solution must be standardized. Several primary standard oxidizing agents are useful for this. Probably the most common one is potassium dichromate, $K_2Cr_2O_7$. Primary

APPLICATION NOTE 5.3: HYDROGEN PEROXIDE TOPICAL SOLUTION

In hydrogen peroxide, H_2O_2, the oxidation number of the oxygen is −1. In oxides, it is −2 and in oxygen gas, O_2, it is 0. An oxidation number of −1 for oxygen is quite unstable. The assay for the 3% hydrogen peroxide topical solution product sold in pharmacies involves a titration with potassium permanganate. In this titration, the oxidation number of the oxygen in the H_2O_2 changes from −1 to 0, meaning that it is the substance oxidized (the reducing agent) in the titration reaction and oxygen gas forms as a product of the reaction. The reaction takes place in acid solution and the permanganate is reduced from MnO_4^- to Mn^{2+}. The topical solution is pipetted (2.0 mL) into the reaction flask, diluted to 20 mL with distilled water, and 20 mL of 2 N H_2SO_4 are added prior to beginning the titration. The permanganate concentration is 0.1 N.

(*Reference: The Official Compendia of Standards*, U.S. Pharmacopeia and National Formulary (2000), The location is Rockville, MD, USA, p. 836.)

standard potassium bromate, $KBrO_3$, or potassium iodate, KIO_3, can also be used. Even primary standard iodine, I_2, can be used (but because solid iodine releases corrosive fumes, it should not be weighed on an analytical balance). Usually in the standardization procedures, KI is again added to the substance to be titrated ($Cr_2O_7^{2-}$ and others) and the liberated iodine titrated with thiosulfate. If I_2 is the primary standard, it is titrated directly. The end point is usually detected with the use of a starch solution as the indicator. Starch, in the presence of iodine, is a deep blue color. It is not added, however, until near the end point after the color of the solution changes from mahogany to straw yellow. Upon adding starch, the color changes to the deep blue. The addition of the thiosulfate is then continued until one drop changes the solution color from blue to colorless. Some important precautions concerning the starch, however, are to be considered. The starch solution should be fresh, should not be added until the end point is near, cannot be used in strong acid solutions, and cannot be used with solution temperatures above about 40°C.

An important application of iodometry can be found in many wastewater treatment plant laboratories. Chlorine, Cl_2, is used in a final treatment process prior to allowing the wastewater effluent to flow into a nearby river. Of course, the chlorine in both the free and combined forms can be just as harmful environmentally as many components in the raw wastewater. Thus, an important measurement for the laboratory to make is the amount of residual chlorine remaining unreacted in the effluent. Such chlorine, which is an oxidizing agent, can be determined by iodometry. It is the "O" in Figure 5.22.

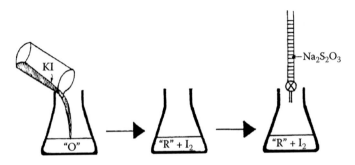

FIGURE 5.22 In iodometry, a solution of KI is added to a solution of the analyte, represented here by "O." The products are "R" and iodine, I_2. The iodine formed is then titrated with standardized sodium thiosulfate. It is an indirect titration, since the I_2 is titrated, but the analyte is "O." Refer back to Figure 5.14 for an illustration of an indirect acid–base titration.

5.7.4.3   Prereduction and Preoxidation

Perhaps the most important application of redox chemicals in the modern laboratory is in oxidation or reduction reactions that are required as part of a preparation scheme. Such "preoxidation" or "prereduction" is also frequently required for certain instrumental procedures for which a specific oxidation state is required to measure whatever property is measured by the instrument. An example in this textbook can be found in Experiment (the hydroxylamine hydrochloride keeps the iron in the +2 state). Also in wastewater treatment plants, it is important to measure dissolved oxygen ("DO"). In this procedure, $Mn(OH)_2$ reacts with the oxygen in basic solution to form $Mn(OH)_3$. When acidified and in the presence of KI, iodine is liberated and titrated. This method is called the "Winkler method."

5.8   OTHER EXAMPLES

Precipitation reactions are used for some determinations. These involve principally reactions using the highly insoluble nature of silver compounds. Two example reactions are the Volhard method for silver ($Ag^+ + CNS^- \rightarrow AgCNS$) and the Mohr method for chloride (also called the chloride method for silver ($Ag^+ + Cl^- \rightarrow AgCl$). End points in these can be detected in any one of several ways. A color change resulting from the first excess of the titrant may be utilized. In this case, the color change is due to the titrant reacting with another added component, the product being colored. With the precipitate also present in the solution, this color may be difficult to see. In these cases, a blank correction is needed.

End points may also be detected using so-called ion-selective electrodes. Such titrations are called potentiometric titrations and will be discussed in Chapter 14.

EXPERIMENTS

EXPERIMENT 10: TITRIMETRIC ANALYSIS OF A COMMERCIAL SODA ASH UNKNOWN FOR SODIUM CARBONATE

Note: Safety glasses are required.

1. Obtain a soda ash sample from your instructor and dry, as before, for 2 h. Allow to cool and store in your desiccator.
2. Prepare the flasks and weigh, by difference, as before, three samples of the soda ash, each weighing between 0.4 and 0.5 g. Dissolve in 75 mL of water.
3. Perform this on each of your samples one at a time, starting with the one of least weight. Add three drops of bromocresol green indicator and titrate with your standard HCl solution from Experiment 9. The color will change slowly from blue to a light green. When the light green color is apparent, stop the titration place a watch glass on the flask, and bring to boil on a hot plate (see explanation in Section 5.2.8). Boil for 2 min. The color of the solution will turn back to blue. Cool to room temperature (you can use a cold water bath) and resume the titration (do not refill the buret!). The color change now should be sharp from blue to greenish yellow. Record the buret reading next to the corresponding weight.
4. Calculate the percentage of Na_2CO_3 in the sample, and record at least a sample calculation in your notebook along with all results and the average. If the calculated percentages do not fall within 10-ppt deviation from the average, you should titrate additional samples until you have three that agree in this manner or until you have a precision that is satisfactory to your instructor.

EXPERIMENT 11: TITRIMETRIC ANALYSIS OF A COMMERCIAL KHP UNKNOWN FOR KHP

Note: This experiment calls for you to use a pH meter and a combination pH electrode (Chapter 14) to detect the end point of a titration. Your instructor may choose to have you use an indicator instead. Safety glasses are required.

1. Obtain an unknown KHP sample from your instructor and dry, as before, for 2 h. Allow to cool and store in your desiccator.
2. Prepare three 250-mL beakers and weigh, by difference, three samples of the unknown KHP, each weighing between 1.0 and 1.3 g, into the beakers. Dissolve in 75 mL of distilled water. Place watch glasses on the beakers.
3. Prepare and standardize a pH meter with a combination probe for pH measurement. Your instructor may provide special instructions for the pH meter you are using. You will use an automatic stirrer with magnetic stirring bar to stir the solution in the beaker while you are titrating. Mount the electrode in a ringstand clamp on a ringstand so that it is just immersed in the solution in one of the beakers. The beaker should be positioned on the stirrer with the stirring bar in the center of the beaker and the pH probe off to one side so as to not contact the stirring bar. The stirring speed should be slow enough so as not to splash the solution but fast enough to thoroughly mix the added titrant quickly.
4. The titrant is the standardized 0.10 N NaOH from Experiment 9. The procedure is to monitor the pH as the titrant is added. A sharp increase in pH will signal the end point. You will want to determine the midpoint of the sharp increase as closely as possible, since that will be the end point. The titrant is added very slowly when the pH begins to rise, since only a very small volume will be required at that point to reach the end point. You can add the titrant rapidly at first, but slow to a fraction of a drop when the pH begins to rise. The pH at the end point will be in the range of 8–10.
5. Record the buret readings at the end points for all three beakers. Calculate the percent KHP in the sample for all three titrations and include at least a sample calculation in your notebook along with all results and the average. If the calculated percentages do not fall within 10-ppt deviation from the average, you should titrate additional samples until you have three that do agree in this manner or until you have a precision satisfactory to your instructor.

EXPERIMENT 12: EDTA TITRATIONS

Note: All Erlenmeyer and volumetric flasks used in this experiment must be rinsed thoroughly with distilled water prior to use. Ordinary tap water contains hardness minerals that will contaminate. The pH 10 ammonia buffer required can be prepared by dissolving 35 g of NH_4Cl and 285 mL of concentrated ammonium hydroxide in water and diluting to 500 mL. The EBT indicator should be fresh and prepared by dissolving 200 mg in a mixture of 15 mL of triethanolamine and 5 mL of ethyl alcohol. Remember to wear safety glasses.

PART A: PRELIMINARY PREPARATIONS

1. Weigh 4.0 g of disodium dihydrogen EDTA dihydrate and 0.10 g of magnesium chloride hexahydrate into a 1-L glass bottle. Add one pellet of NaOH and fill to approximately 1 L with water. Shake well. These ingredients will require some time to dissolve, so it is recommended that this solution be prepared one laboratory session ahead of its intended use.
2. Dry a quantity (at least 0.5 g) of primary standard $CaCO_3$ for one hour. Your instructor may choose to dispense this to you. While drying continue with either Part B or Part C below. If Part D is to be performed, also obtain your powdered, solid unknown and dry in the same way.

PART B: TITRATION OF A WATER SAMPLE

3. Obtain a water sample. It will be contained in a 1-L volumetric flask. Dilute to the mark with distilled water. Shake well.
4. Pipet 100.0-mL aliquots of the water sample (an **aliquot** is a portion of a solution) into each of three clean 500-mL Erlenmeyer flasks. Add 5.0 mL (graduated cylinder) of the pH 10 buffer, and 3 drops of EBT indicator to each flask.

5. Give your EDTA solution one final shake to ensure its homogeneity. Then, clean a 50-mL buret, rinse it several times with your EDTA solution (the titrant), and fill, as usual, with your titrant. Alternatively, your instructor may suggest using an auto-titrator. Be sure to eliminate the air bubbles from the stopcock and tip. Titrate the solutions in your flask, one at a time, until the last trace of red disappears in each with a fraction of a drop. This final color change will be from a violet color to a deep sky blue, but should be a sharp change. All three titrations should agree to within 0.05 mL of each other. If they do not, repeat until you have three that do. Record all readings in your notebook.

PART C: STANDARDIZATION OF EDTA

6. With this step, we begin the EDTA standardization process. Weigh accurately 0.3 to 0.4 g of the dried and cooled $CaCO_3$ by difference into a clean, dry, short-stemmed funnel in the mouth of a 500-mL volumetric flask. Tap the funnel gently to force the $CaCO_3$ into the flask. Wash any remaining $CaCO_3$ into the flask with a squeeze bottle. Add a small amount of concentrated HCl (less than 2.0 mL total) to the flask through the funnel such that the funnel is rinsed thoroughly in the process. Rinse the funnel into the flask one more time with distilled water and remove the funnel. Rinse the neck of the flask with distilled water. Swirl the flask until all the $CaCO_3$ is dissolved and effervescence has ceased. Dilute to the mark with distilled water and shake well. If the solution is warm at this point, cool in a cold water bath and then add more distilled water so as to bring the meniscus back to the mark. Shake well again.
7. Pipet a 25.00-mL aliquot of this standard calcium solution into each of three 250-mL Erlenmeyer flasks, add the buffer solution, and titrate each to the same end point as before, but use only 2 drops of EBT. All buret readings should agree to within 0.05 mL. If they do not, repeat until you have three that do.
8. Calculate the molarity of your EDTA for each of the three titrations and calculate the average. Record in your notebook. Also calculate the ppm $CaCO_3$ in the sample and compute the average. Record, as usual, in your notebook.

PART D: ANALYSIS OF A SOLID, POWDERED UNKNOWN FOR CALCIUM OXIDE

9. Prepare a solution of the unknown you dried in Part A in the same way you prepared a solution of the pure calcium carbonate in Step 6, but use a 250-mL volumetric flask instead of a 500-mL flask, and use about half as much HCl.
10. Titrate three 25-mL aliquots of the solution from Step 10 as you did the pure calcium carbonate solution in Step 7. Again, all buret readings should agree to within 0.05 mL. If they do not agree in this way, repeat until you have three that do.
11. Calculate the percent of calcium oxide (CaO) in your unknown. Remember that each titrated sample contains 1/10 of what was weighed into the 250-mL volumetric flask.

PART E: DESIGN AN EXPERIMENT: EFFECT OF pH ON THE DETERMINATION OF WATER HARDNESS

A significant point was made in this chapter concerning the need for the pH of the titrated water sample to be around pH 10 and what would happen if the pH were not at pH 10. Design an experiment in which the hardness titration is done at various pH values across the entire pH range with the goal of proving that pH 10 is indeed the optimum. Write a step-by-step procedure as in other experiments in this book for preparing the needed solutions and performing a series of titrations. After your instructor approves your procedure, perform the experiment in the laboratory.

QUESTIONS AND PROBLEMS
Introduction

1. What are three attributes of a successful titrimetric analysis?

Acid–Base Titrations and Titration Curves

2. Define *monoprotic acid, polyprotic acid, monobasic base, polybasic base, titration curve,* and *inflection point.*

3. Compare the titration curves for 0.10 N hydrochloric acid and 0.10 N acetic acid each titrated with 0.10 N sodium hydroxide. What parts of the titrations curves are the same and what parts are different? Why? Compare the inflections points for the two curves and tell what impact the differences have on indicator selection.

4. Repeat question 3, but compare the titration curves of 0.10 N sodium hydroxide with 0.10 N ammonium hydroxide titrated with 0.10 N hydrochloric acid.

5. Repeat question 3, but compare the titration curves of 0.10 N sulfuric acid and 0.10 N phosphoric acid titrated with 0.10 N sodium hydroxide.

6. Phenolphthalein indicator changes color in the pH range 8–10, Methyl orange changes color in the pH range 3–4.5. Roughly sketch two titration curves as follows:
   (a) One that represents a titration in which phenolphthalein would be useful, but methyl orange would not.
   (b) One that represents a titration in which a methyl orange would be useful, but phenolphthalein would not.

7. Roughly sketch the following three titration curves:
   (a) A weak acid titrated with a strong base
   (b) A strong base titrated with a strong acid
   (c) A weak base titrated with a strong acid

8. Bromocresol green was used as the indicator for the Na_2CO_3 titration in Experiment 10. Would phenolphthalein also work? Explain.

9. Would bromocresol green be an appropriate indicator for an acetic acid titration? Explain.

10. How do the first and second derivatives of a titration curve help us to determine the equivalence point of a titration?

11. Look at Figures 5.5 and 5.8 and tell what indicator, phenolphthalein or bromocresol, you would recommend for the titration of phosphoric acid at the second inflection point. Explain.

12. What is the "P" in KHP? Draw the structure of KHP. Tell why it is useful as a primary standard chemical.

13. What is the name of the chemical often referred to as THAM or TRIS? What is its structure? It does not contain OH^- ions, yet it is a base. Explain. Tell why it is useful as a primary standard chemical.

14. Why does the titration curve of sodium carbonate have two inflection points? Why does this titration require that the solution be boiled as you approach the second equivalence point? Why can bromocresol green be used as the indicator and not phenolphthalein?

Examples of Acid/Base Determinations

15. Define *total alkalinity.*

16. What is the alkalinity of a water sample if 50.00 mL of the water sample required 3.09 mL of 0.09928 N HCl to reach a pH of 4.5?

17. Describe in your own words what a back titration is.

18. Why is a back titration useful in the case of the analysis of an antacid tablet containing calcium carbonate as the active ingredient?

19. What is the percent of $CaCO_3$ in an antacid given that a tablet that weighed 1.2918 g reacted with 50.00 mL of 0.4501 N HCl that subsequently required 3.56 mL of 0.1196 N NaOH for back titration? Also report the milligrams of $CaCO_3$ in the tablet.

$$2\ HCl + CaCO_3 \rightarrow CaCl_2 + CO_2 + H_2O$$

20. Concerning the Kjeldahl method for nitrogen,
    (a) What is NaOH used for?
    (b) What might boric acid be used for?
21. In the calculation of the percent analyte when using a back titration, the following appears in the numerator:

$$(L_T \times N_T - L_{BT} \times N_{BT})$$

    Why is it necessary to do this subtraction?
22. A grain sample was analyzed for nitrogen content by the Kjeldahl method. If 1.2880 g of the grain were used and 50.00 mL of 0.1009 N HCl were used in the receiving flask, what is the percentage of nitrogen in the sample when 5.49 mL of 0.1096 N NaOH were required for back titration?
23. A flour sample was analyzed for nitrogen content by the Kjeldahl method. If 0.9819 grams of the flour were used and 35.10 mL of 0.1009 N HCl were used to titrate the boric acid solution in the receiving flask, what is the percent of nitrogen in the sample?

Other Acid/Base Applications

24. Name some chemicals or consumer products in which acid and bases are present that one could analyze using acid/base titrations.

Buffer Solution Applications

25. Define *buffer solution, conjugate acid, conjugate base, conjugate acid–base pair, buffer capacity*, and *buffer region*.
26. Under what circumstances can the acetate ion be thought of as a conjugate base? Under what circumstances can the ammonium ion be thought of as a conjugate acid?
27. What is the Henderson–Hasselbalch equation? Tell how it is useful in the preparation of buffer solutions.
28. What is the pH of a solution of chloroacetic acid (0.25 N) and sodium chloroacetate (0.20 N)? The K_a of chloroacetic acid is 1.36×10^{-3}.
29. What is the pH of a solution that is 0.30 N in iodoacetic acid and 0.40 N in sodium iodoacetate? The K_a of iodoacetic acid is 7.5×10^{-4}.
30. What is the pH of a solution of THAM and THAM hydrochloride if the THAM concentration is 0.15 N and the THAM hydrochloride concentration is 0.40 N? The K_a of THAM hydrochloride is 8.41×10^{-9}.
31. What ratio of the concentrations of acetic acid to sodium acetate are needed to prepare a buffer solution of pH 4.00? The K_a of acetic acid is 1.76×10^{-5}.
32. A buffer solution of pH 3.00 is needed. From Table 5.1, select a weak acid/conjugate base combination that would give that pH and calculate the ratio of acid to conjugate base concentrations that would give that pH.
33. To prepare a certain buffer solution, it is determined that acetic acid must be present at 0.17 N and that sodium acetate must be present at 0.29 N, both in the same solution. If the sodium acetate ($NaC_2H_3O_2$) is a pure solid chemical and the acetic acid to be measured out is a concentrated solution (17 N), how would you prepare 500 mL of this buffer solution?
34. Tell how you would prepare 500 mL of the buffer solution in question 30. Both THAM and THAM hydrochloride are pure solid chemicals. The formula weight of THAM is 121.14 g/mol. The formula weight of THAM hydrochloride is 157.60 g/mol.

35. In Experiment 12, a recipe for a pH 10 buffer solution is given. This recipe calls for dissolving 35 g of NH_4Cl and 285 mL of concentrated ammonium hydroxide (15 N) in water in the same container and diluting with water to 500 mL. Calculate the pH of the buffer using the data given and confirm that it really is pH 10. The K_a for the ammonium ion is 5.70×10^{-10}.

Complex Ion Formation Reactions

36. Define *monodentate, bidentate, hexadentate, ligand, complex ion, chelate, chelating agent, masking, masking agent, formation constant, coordinate covalent bond, water hardness,* and *aliquot.*
37. Define *ligand* and *complex ion.* Give an example of each.
38. Given the following reaction, tell which of the three species is the ligand and which is the complex ion:

$$Co^{2+} + 4\ Cl^- \leftrightarrows CoCl_4^{2-}$$

39. Give one example each (either structure of name) of a monodentate ligand and a hexadentate ligand. Explain what is meant by *monodentate.*
40. Is the ligand ethylenediamine, $H_2NCH_2CH_2NH_2$, monodentate, bidentate, tridentate, or what? Explain your answer.
41. Consider the reaction shown in Figure 5.23.
    (a) Which chemical species is a ligand?
    (b) Which chemical species is a complex ion?
    (c) Is the ligand monodentate, bidentate, or what? Explain your answer.
    (d) Is the complex ion a chelate? Explain.
42. Concerning the EDTA ligand:
    (a) How many bonding sites on this molecule bond to a metal ion when a complex ion is formed?
    (b) How many EDTA molecules will bond to a single metal ion?
    (c) What is the word describing the property pointed out in (a)?
43. Explain why water samples titrated with EDTA need to be buffered at pH 10 and not at pH 12 or pH 8.
44. In the water hardness titration:
    (a) What chemical species is the wine red color at the beginning of the titration due to?
    (b) What chemical species is the sky blue color at the end point due to?
    (c) What does it mean to say that cyanide is a "masking agent"?
45. Explain why the pH 10 ammonia buffer is required in EDTA titrations for water hardness.
46. How would you prepare each of the following?
    (a) 250.0 mL of a 25-ppm solution of magnesium from pure magnesium metal?
    (b) Using pure silver metal, 750.0 mL of a solution that is 30.0 ppm silver?

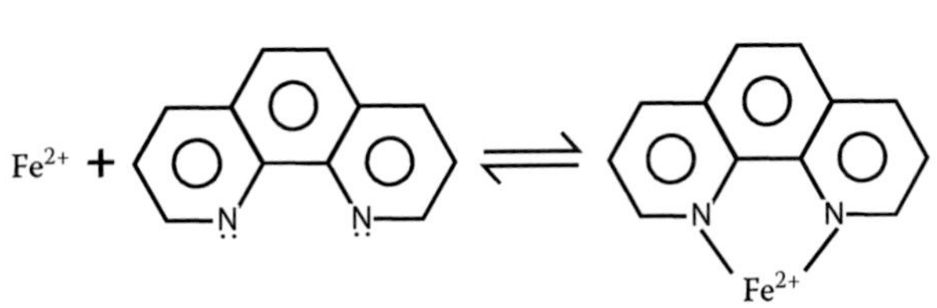

FIGURE 5.23   Reaction for question 41.

(c) 600.0 mL of a solution that is 40.0 ppm aluminum, using pure Al metal?

(d) 500.0 mL of a 15-ppm Mg solution using pure magnesium for the solute.

(e) Using pure iron metal for the solute, 250.0 mL of a 30.0-ppm solution of iron?

(f) 100.0 mL of a solution that is 125 ppm copper using pure copper metal?

47. How many milliliters of a 1000.0-ppm solution of the metal are needed to prepare each of the following?

(a) 100.0 mL of a 125-ppm in solution of copper?

(b) 600.0 mL of a 15-ppm solution of zinc?

(c) 250.0 mL of a 25 ppm solution of sodium?

48. How would you prepare 500.0 mL of a 50.0-ppm Na solution

(a) from pure, solid NaCl?

(b) from a solution that is 1000.0 ppm Na?

49. How would you prepare 250.0 mL of a 30.0-ppm solution of iron

(a) Using pure iron wire as the solute?

(b) Using solid $Fe(NO_3)_3 \cdot 9H_2O$ (FW = 404.02) as the solute?

(c) If you needed to dilute a 1000.0-ppm solution of iron?

50. Tell how you would prepare 100.0 mL of a 50.0-ppm solution of copper

(a) using pure solid copper sulfate pentahydrate, $CuSO_4 \cdot 5H_2O$

(b) using pure copper metal

(c) by diluting a 1000.0-ppm copper solution?

51. A technician wishes to prepare 500 mL of a 25.0-ppm solution of barium.

(a) How many milliliters of 1000.0 ppm barium would be required if he/she were to prepare this by dilution?

(b) If he/she would be able to prepare this from pure barium metal, how many grams would be required?

(c) If he/she were to prepare this from pure, solid $BaCl_2 \cdot 2H_2O$, how many grams would be required?

52. How many milligrams of the solute is needed to prepare the following volumes of solution?

(a) Solute is KBr and 600.0 mL of a 40.0-ppm bromide solution are needed.

(b) Solute is K_2HPO_4 450.0 mL of a 10.0-ppm phosphorus solution are needed.

(c) Solute is KNO_3 100.0 mL of a solution that is 50.0 ppm nitrogen are needed?

53. Tell how you would prepare 500.0 mL of a 0.0250-N solution of the solid disodium dihydrogen EDTA dihydrate for use as a standard solution without having to be standardized.

54. How many grams of disodium dihydrogen EDTA dihydrate are required to prepare 1000 mL of a 0.010-N EDTA solution?

55. What is the molarity of an EDTA solution given the following standardization data?

(a) If 10.0 mg of the Mg required 40.08 mL of the EDTA.

(b) If 0.0236 g of solid $CaCO_3$ were dissolved and exactly consumed by 12.01 mL of an EDTA solution.

(c) If 30.67 mL of it reacts exactly with 45.33 mg of calcium metal?

(d) If 34.29 mL of it is required to react with 0.1879 g of $MgCl_2$.

(e) If a 100.0-mL aliquot of a zinc solution required 34.62 mL of it. (The zinc solution was prepared by dissolving 0.0877 g of zinc in 500.0 mL of solution.)

(f) If a solution of primary standard $CaCO_3$ was prepared by dissolving 0.5622 g of $CaCO_3$ in 1000 mL of solution a 25.00-mL aliquot of it required 21.88 mL of the EDTA.

(g) If 25.00 mL of a solution prepared by dissolving 0.4534 g of $CaCO_3$ in 500.0 mL of solution, reacts with 34.43 mL of the EDTA solution.

(h) If a solution has 0.4970 g of $CaCO_3$ dissolved in 500.0 mL and 25.00 mL of it reacts exactly with 29.55 mL of the EDTA solution.

(i) If 25.00 mL of a $CaCO_3$ solution reacts with 30.13 mL of the EDTA solution and there are 0.5652 g of $CaCO_3$ per 500.0 mL of the solution.

56. What is the hardness of the water sample in ppm $CaCO_3$ in each of the following situations?
    (a) If a 100.0-mL aliquot of the water required 27.62 mL of 0.01462 N EDTA for titration?
    (b) If 25.00 mL of the water sample required 11.68 mL of 0.01147 N EDTA.
    (c) If 12.42 mL of a 0.01093-N EDTA solution were needed to titrate 50.00 mL of the water sample.
    (d) If, in the experiment for determining water hardness, 75.00 mL of the water sample required 13.03 mL of an EDTA solution that is 0.009242 N.
    (e) If the EDTA solution used for the titrant was 0.01011 N, a 150.0-mL sample of water and 16.34 mL of the titrant were needed.
    (f) If 14.20 mL of an EDTA solution, prepared by dissolving 4.1198 g of $Na_2H_2EDTA \cdot 2H_2O$ in 500.0 mL of solution, were needed to titrate 100.0 mL of a water sample.
    (g) When 100.0 mL of the water required 13.73 mL of an EDTA solution prepared by dissolving 3.8401 g of $Na_2H_2EDTA \cdot 2H_2O$ in 500.0 mL of solution?

Oxidation–Reduction Reactions

57. Define *oxidation, reduction, oxidation number, oxidizing agent,* and *reducing agent.*
58. What is the oxidation number of the following:
    (a) P in H_3PO_4
    (b) Cl in $NaClO_2$
    (c) Cr in CrO_4^{2-}
    (d) Br in $KBrO_3$
    (e) I in IO_4^-
    (f) N in N_2O
    (g) S in H_2SO_3
    (h) S in H_2SO_4
    (i) N in NO_2^-
    (j) P in PO_3^{3-}
59. What is the oxidation number of bromine (Br) in each of the following?
    (a) HBrO
    (b) NaBr
    (c) BrO_3^-
    (d) Br_2
    (e) $Mg(BrO_2)_2$
    (f) BrO_3^-
60. What is the oxidation number of chromium (Cr) in each of the following?
    (a) $CrBr_3$
    (b) Cr
    (c) CrO_3
    (d) CrO_4^{2-}
    (e) $K_2Cr_2O_7$
61. What is the oxidation number of iodine (I) in each of the following?
    (a) HIO_4
    (b) CaI_2
    (c) I_2
    (d) IO_2^-
    (e) $Mg(IO)_2$
62. What is the oxidation number of sulfur (S) in each of the following?
    (a) SO_2
    (b) H_2S
    (c) S

(d) SO_4^{2-}

(e) K_2SO_3

(f) K_2S

(g) H_2SO_4

(h) SO_3^{2-}

(i) SO_2

(j) SF_6

63. What is the oxidation number of phosphorus (P) in each of the following:

(a) P_2O_5

(b) Na_3PO_3

(c) H_3PO_4

(d) PCl_3

(e) HPO_3^{2-}

(f) PO_4^{3-}

(g) P^{3-}

(h) $Mg_2P_2O_7$

(i) P

(j) NaH_2PO_4

64. In the following redox reactions, tell what has been oxidized and what has been reduced and explain your answers:

(a) $3CuO + 2NH_3 \rightarrow 3Cu + N_2 + 3H_2O$

(b) $Cl_2 + 2KBr \rightarrow Br_2 + 2KCl$

65. In each of the following reactions, tell what is the oxidizing agent and what is the reducing agent and explain your answers.

(a) $Mg + 2HBr \rightarrow MgBr_2 + H_2$

(b) $4Fe + 3O_2 \rightarrow 2Fe_2O_3$

66. Which of the following unbalanced equations represent redox reactions? Explain your answers.

(a) $H_2SO_4 + NaOH \rightarrow Na_2SO_4 + H_2O$

(b) $H_2S + HNO_3 \rightarrow S + NO + H_2O$

(c) $Na + H_2O \rightarrow NaOH + H_2$

(d) $H_2SO_4 + Ba(OH)_2 \rightarrow BaSO_4 + H_2O$

(e) $K_2CrO_4 + Pb(NO_3)_2 \rightarrow 2KNO_3 + PbCrO_4$

(f) $K + Br_2 \rightarrow 2KBr$

(g) $2KClO_3 \rightarrow 2KCl + 3O_2$

(h) $KOH + HCl \rightarrow KCl + H_2O$

(i) $BaCl_2 + Na_3PO_4 \rightarrow Ba_3(PO_4)_2 + NaCl$

(j) $Mg + HCl \rightarrow MgCl_2 + H_2$

67. Consider the following two unbalanced equations.

(a) $KOH + HCl \rightarrow KCl + H_2O$

(b) $Cu + HNO_3 \rightarrow Cu(NO_3)_2 + NO + H_2O$

Which represents a redox reaction, (a) or (b)? In the redox reaction, what has been oxidized and what has been reduced? In the redox reaction, what is the oxidizing agent and what is the reducing agent?

68. One of the following unbalanced equations represents a redox reaction and one represents a reaction that is not a redox reaction. Select the one that is a redox reaction and answer the questions that follow:

(1) $Pb(NO_3)_2 + K_2CrO_4 \rightarrow PbCrO_4 + 2KNO_3$

(2) $Zn + HCl \rightarrow ZnCl_2 + H_2$

(a) Which one is redox, (1) or (2)?

(b) What is the oxidizing agent?

    (c) What has been oxidized?

    (d) Did the reducing agent lose or gain electrons?

69. Balance the following equations by the ion-electron method:

    (a) $Cl^- + NO_3^- \rightarrow ClO_3^- + N^{3-}$

    (b) $Cl^- + NO_3^- \rightarrow ClO_2^- + N_2O$

    (c) $ClO^- + NO_3^- \rightarrow ClO_3^- + NO_2^-$

    (d) $ClO^- + NO_3^- \rightarrow ClO_2^- + NO$

    (e) $ClO_3^- + SO_4^{2-} \rightarrow ClO_4^- + S^{2-}$

    (f) $BrO_3^- + SO_4^{2-} \rightarrow BrO_4^- + SO_3^{2-}$

    (g) $IO_3^- + SO_3 \rightarrow IO_4^- + S^{2-}$

    (h) $Cl^- + SO_4^{2-} \rightarrow ClO^- + SO_2$

    (i) $Cl^- + SO_4^{2-} \rightarrow ClO_4^- + S$

    (j) $Br^- + SO_3 \rightarrow Br_2 + S$

    (k) $I^- + NO_3^- \rightarrow IO_2^- + N_2$

    (l) $P + IO_4^- \rightarrow PO_4^{3-} + I^-$

    (m) $SO_2 + BrO_3^- \rightarrow SO_4^{2-} + Br^-$

    (n) $Fe + P_2O_5 \rightarrow Fe^{3+} + P$

    (o) $Cr + PO_4^{3-} \rightarrow Cr^{3+} + PO_3^{3-}$

    (p) $Ni + PO_4^{3-} \rightarrow Ni^{2+} + P$

    (q) $MnO_4^- + H_2C_2O_4 \rightarrow Mn^{2+} + CO_2$

    (r) $I^- + Cr_2O_7^{2-} \rightarrow I_2 + Cr^{3+}$

    (s) $Cl_2 + NO_2^- \rightarrow Cl^- + NO_3^-$

    (t) $S^{2-} + NO_3^- \rightarrow S + NO$

    (u) $SO_3^{2-} + NO_3^- \rightarrow SO_4^{2-} + NO_2^-$

70. (a) Balance the following equation

$$S_2O_3^{2-} + Cr_2O_7^{2-} \rightarrow S_4O_6^{2-} + Cr^{3+}$$

    (b) If 0.5334 g of $K_2Cr_2O_7$ were titrated with 24.31 mL of the $Na_2S_2O_3$ solution, what is the molarity of the $Na_2S_2O_3$?

71. Consider the reaction of Fe^{2+} with $K_2Cr_2O_7$ according to the following:

$$Fe^{2+} + Cr_2O_7^{2-} \rightarrow Fe^{3+} + Cr^{3+}$$

What is the exact molarity of a solution of $K_2Cr_2O_7$ if 1.7976 g of Mohr's salt [an Fe^{2+} compound, $Fe(NH_4)_2(SO_4)_2 \cdot 6H_2O$] were exactly reacted with 22.22 mL of the solution?

72. Consider the standardization of a solution of KIO_4 with Mohr's salt (see #70) according to the following:

$$Fe^{2+} + IO_4^- \rightarrow Fe^{3+} + I^-$$

What is the exact molarity of the solution if 1.8976 g of Mohr's salt were exactly reacted with 24.22 mL of the solution?

73. Consider the standardization of a solution of $K_2Cr_2O_7$ with iron metal according to the following:

$$Fe + Cr_2O_7^{2-} \rightarrow Fe^{3+} + Cr^{3+}$$

What is the exact molarity of the solution if 0.1276 g of iron metal were exactly reacted with 48.56 mL of the solution?

74. What is the percent SO_3 in a sample if 45.69 mL of a 0.2011-N solution of KIO_3 are needed to consume 0.9308 g of sample according to equation (g) of Question 68?

75. What is the percent of K_2SO_4 in a sample if 35.01 mL of 0.09123 N $KBrO_3$ solution are needed to consume 0.7910 g of sample according to equation (f) of Question 68?

76. What is the percent of Fe in a sample titrated with $K_2Cr_2O_7$ according to the following equation if 2.6426 g of the sample required 40.12 mL of 0.1096 N $K_2Cr_2O_7$?

$$Fe^{2+} + Cr_2O_7^{2-} \rightarrow Fe^{3+} + Cr^{3+}$$

77. What is the percent of Sn in a sample of ore if 4.2099 g of the ore were dissolved and titrated with 36.12 mL of 0.1653 N $KMnO_4$?

$$MnO_4^- + Sn^{2+} \rightarrow Sn^{4+} + Mn^{2+}$$

78. What does it mean to say that potassium permanganate is its own indicator?

79. (a) Why is potassium permanganate, $KMnO_4$, termed an oxidizing agent?
    (b) Why must standardized potassium permanganate solutions be carefully protected from oxidizable substances if you expect it to remain standardized?

80. Explain the difference between an indirect titration and a back titration.

81. One redox method we have discussed is called iodometry. What is iodometry and why is it called an "indirect" method?

82. Briefly explain the use of the following substances in iodometry.
    (a) KI
    (b) $Na_2S_2O_3$
    (c) $K_2Cr_2O_7$

Other Examples

83. How might an end point be detected when the titration reaction involves the formation of a precipitate?

Report

84. For this exercise, your instructor will select a "real-world" titrimetric analysis method, such an application note from an instrument vendor or from a methods book, a journal article, or Web site, and will give it to you as a handout. Write a report giving the details of the method according to the following scheme. On your paper, write (or type) "A," "B," etc., to clearly present your response to each item.
    (a) Method title: Give a more descriptive title than what is given on the handout.
    (b) Type of material examined: Is it water, soil, food, a pharmaceutical preparation, or what?
    (c) Analyte: Give both the name and the formula of the analyte.
    (d) Sampling procedure: This refers to how to obtain a sample of the material. If there is no specific information given as to obtaining a sample, make something up. If you want to be correct in what you write, go to a library, or other search to discover a reasonable response. Be brief.
    (e) Sample preparation procedure: This refers to the steps taken to prepare the sample for the analysis. It does not refer to the preparation of standards or any associated solutions.
    (f) What is the titrant and how is it standardized (including what primary standards are used)?

(g) What other solutions are needed and how are they prepared?

(h) What glassware is needed and for what? Was an autotitrator used?

(i) What end-point detection method is used for both the standardization and analysis steps?

(j) What reactions (write balanced equations) are involved in both the standardization and analysis steps?

(k) Is it a direct titration, an indirect titration, or a back titration (both the standardization and analysis steps)?

(l) Any special procedures?

(m) Data handling and reporting: Once you have the data for the standardization and analysis, then what? Be sure to present any post-run calculations that might be required.

(n) References: Did you look at any other references sources to help answer any of the above? If so, write them here.

6 Introduction to Instrumental Analysis

6.1 REVIEW OF THE ANALYTICAL PROCESS

In our introduction to analytical chemistry in Chapter 1, we presented a flowchart (Figure 1.1) outlining the general process of analytical chemistry. As shown in this figure, the analytical process consists of five parts: (1) sampling, (2) sample preparation, (3) the analytical method, (4) data handling, and (5) the calculation and reporting of results. This is a brief summary of what is involved: (1) the sample is obtained from the bulk system under consideration; (2) this sample is appropriately prepared for the analysis; (3) the chosen method is executed; (4) the data obtained from the method are analyzed; and (5) the results are calculated and reported.

As stated previously, all analytical schemes begin with proper sampling and sample preparation procedures. These procedures were studied in detail in Chapter 2. However, we mention them in this brief review again to emphasize their importance. In terms of sampling, we indicated that it is important for the sample to represent the entire bulk system for which an analytical result is to be reported. We also indicated that it is important to maintain the integrity of the sample at all times, so that the sample is not inappropriately altered prior to the analysis, and to document the chain of custody of all samples so that any doubts about sample integrity can be answered. The sampling procedures are central to the success of a chemical analysis and the importance of proper sampling procedures should never be discounted. An analytical result is only as good as the sample that was used.

In terms of sample preparation, we indicated that in nearly every case, a pre-analysis procedure(s) must be performed to get the sample into a form that can be utilized by the chosen analytical method. Common sample preparation schemes can involve any number of physical or chemical processes, such as drying, dissolving, extracting, or chemical alteration.

Because the basics of sampling and sample preparation were covered in detail in the earlier chapters, they will only appear in the coming chapters if there is a need to discuss the state of a sample for the particular method under consideration, or if there is some other special relationship between a sampling/sample preparation procedure and the method. This fact should not imply that the importance of these topics is diminished to any degree. *Sampling and sample preparation procedures are very important to the success of all analytical work.*

The analysis methods we have covered to this point (Chapters 3, 4, and 5) are the classical wet chemical analysis methods of gravimetric analysis and titrimetric analysis. The data handling procedures for these were also discussed. With these methods now behind us, we begin a thorough discussion of the methods, data handling, and reporting of results generally relating to instrumental analysis. The general analytical process flowchart for quantitative instrumental analysis is presented in Figure 6.1. Note especially the "carry out the analysis method" box in this figure. In quantitative instrumental analyses, we pay special attention to the calibration of the instrument. Most often, this involves preparing a series of reference standard solutions to be measured by the instrument, and this is followed by the testing of samples.

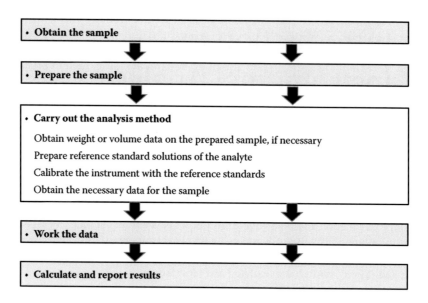

FIGURE 6.1   Steps characteristic of instrumental analysis methodology.

6.2   INSTRUMENTAL ANALYSIS METHODS

Quantitative instrumental analysis mostly involves sophisticated electronic instrumentation that generates an electronic signal(s) that is related to some property of the analyte and proportional to the analyte's concentration in a solution. In other words, as depicted in Figure 6.2, the standards and samples are provided to the instrument, the instrument measures the property, the electronic signal is generated, and the desired readout is displayed. A simple example with which you may be familiar is the pH meter. In this case, the pH electrode is immersed into the solution, an electrical signal proportional to the pH of the solution is generated, electronic circuitry converts the signal to a pH value, and the pH is then displayed.

Most instrumental analysis methods can be classified in one of three general categories: spectroscopy, which uses instruments generally known as a "spectrometers," chromatography, which uses instruments generally known as "chromatographs" and electroanalytical chemistry. These are the three categories that are emphasized in the remainder of this book. Spectroscopic methods involve the use of light and measure either the amount of light absorbed (absorbance) or the amount of light emitted by solutions of the analyte under certain conditions. Chromatographic methods involve more complex samples in which the analyte is separated from interfering substances using specific instrument components and electronically detected, with the electrical signal generated by any one of a number of detection devices. Electroanalytical methods involve the measurement of a voltage or current resulting from electrodes immersed into the solution. The example of a pH meter mentioned above is an example of an electroanalytical instrument.

FIGURE 6.2   General principle of analysis with electronic instrumentation.

Let us briefly return to the concept of calibration. Any device or instrument that provides the analyst with a measurement must be calibrated. In general terms, **calibration** is a procedure by which any instrument or measuring device is tested with a standard in order to determine its response for an analyte in a sample for which the true response is either already known or needs to be established. In cases in which the true response is already known, such as the use of a known weight when calibrating a balance (Chapter 3) or the pH of a buffer solution when calibrating a pH meter, the device may either be removed from service if the response is not the correct response (e.g., the balance), or the electronics of the instrument may be tweaked to display the correct response (e.g., the pH meter). In other cases, the true response may need to be established, and this is done by measuring the responses of the standards (the known quantities). If a single standard is used, the calibration often results in a calibration constant to be used in subsequent calculations. If a series of standards is used, a **calibration curve** (or **standard curve**) is usually plotted. This is a plot of the instrument response vs. the concentration, or other known quantity, of the standards. The response of the sample is then applied to the standard curve, and the concentration, or other desired quantity, is determined. Details of calibration and of the calibration curve are presented in Section 6.4.

6.3 BASICS OF INSTRUMENTAL MEASUREMENT

Now, let us dwell on Figure 6.2 for a moment. The standards and sample solutions are introduced to the instrument in a variety of ways. In the case of a pH meter, and other electroanalytical instruments, the tips of one or two "probes" are immersed in the solution. In the case of an automatic digital Abbe refractometer (Chapter 7), a small quantity of the solution is placed on a prism at the bottom of a sample well inside the instrument. In an ordinary spectrophotometer (Chapters 7 and 8), the solution is held in a round (like a test tube) or square container called a "cuvette," which fits in a holder inside the instrument. In an atomic absorption spectrophotometer (Chapter 9), or in instruments utilizing an "**auto-sampler**," the solution is sucked or aspirated into the instrument from an external container. See Application Note 6.1. In instrumental chromatography (Chapters 11 and 12), the solution is usually injected into the instrument with the use of a small-volume syringe, although auto-samplers may be used there as well. Once inside, or otherwise in contact with the instrument,

APPLICATION NOTE 6.1: THE AUTO-SAMPLER

A common accessory seen on many laboratory instruments in busy analytical laboratories is the so-called auto-sampler. Although the design can vary, the typical auto-sampler is a circular carousel, as shown on the right, having round slots to hold vials that contain the solutions to be analyzed—standards, samples, and controls. In operation, the carousel slowly rotates and a robotic sampling tube dips into the vials one at a time and draws the solutions into the instrument to which it is attached. The automatic action allows the operator to perform other tasks while the measurements are being taken.

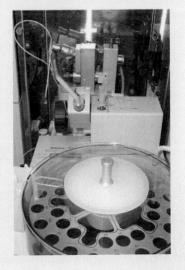

the instrument is designed to act on the solution. We now address the processes that occur inside the instrument in order to produce the electrical signal that is seen at the readout.

6.3.1 SENSORS, SIGNAL PROCESSORS, READOUTS, AND POWER SUPPLIES

Refer to Figure 6.3 for the following discussion. In general, an instrument consists of four components: a **sensor** that converts a property of the solution into a weak electrical signal, a **signal processor** that amplifies or scales the signal and converts it to a useable form, a **readout device** that displays the signal for the analyst to see, and a **power supply** to provide the power to run these three components. The information flow within the instrument occurs with the movement of electrons, or electrical current.

A sensor is a kind of translator. It receives specific information about the system under investigation and transmits this information in the form of a weak electrical signal. Sensors are specific to a given property of the system under investigation. Some sensors are sensitive to temperature, others are sensitive to light, still others to pH, to pressure, etc.

The signal processor is also measurement specific. In some cases, the weak signal generated by the sensor may need to simply be amplified. In other cases, a mathematical treatment of this signal, such as logarithmic conversion, may be needed. Modern instruments that utilize infrared light (Chapter 8) require that a complex mathematical operation be performed known as the Fourier transformation. So we "process" the signal generated by the sensor in order to display exactly what the operator desires as a readout. In the modern laboratory, the processing is often conducted with computer software.

The readout device is a translator like the sensor. It translates the electrical signal produced by the signal processor to something the analyst can understand. This can be a number on a digital display, the position of a needle on a meter, a computer monitor display, etc. The readout device is not specific to the measurement. It can take the signal from any signal processor and display it.

The power supply is also not measurement specific. Almost all sensors, signal processors, and readouts operate from similar voltages. The power supply converts the AC line power (or battery power) to voltage levels needed to operate the other functional elements of the instrument.

6.3.2 CALIBRATION OF AN ANALYTICAL INSTRUMENT

We mentioned in Section 6.2 that the response of an instrument used for chemical analysis is proportional to the concentration of the analyte in a solution. This proportionality can be expressed as follows:

$$R = KC \tag{6.1}$$

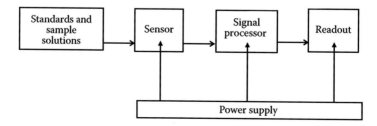

FIGURE 6.3 Four instrument components for translating a property of a solution to a readout for the analyst to observe.

In this equation, R is the instrument readout, C is the concentration of the analyte, and K is the proportionality constant. The most common example of this is **Beer's law**, which will be introduced in Chapter 7. If the object of the experiment is the concentration, C, of the analyte in the sample solution, then the R and the K for this solution must be known so that C can be calculated.

$$C = \frac{R}{K} \tag{6.2}$$

We also indicated in Section 6.2 that the calibration of an instrument for quantitative analysis can utilize a single standard, resulting in a **calibration constant**, or a series of standards, resulting in a **standard curve**. If a single standard is used, the value of K is the calibration constant. It is found by determining the instrument response for a standard solution of the analyte and then calculating K.

$$R_S = KC_S \tag{6.3}$$

$$K = \frac{R_s}{C_s} \tag{6.4}$$

The concentration of the unknown sample solution is then calculated from its instrument response and the K determined from Equation 6.4.

$$C_U = \frac{R_U}{K} \tag{6.5}$$

In these equations, R_S is the readout for the standard solution, C_S is the concentration of the analyte in the standard solution, R_U is the readout for the unknown sample solution, C_U is the concentration of the analyte in the unknown solution, and the proportionality constant, K, is the calibration constant.

There are two limitations with the above process: (1) the analyst is "putting all the eggs in one basket" by comparing the sample to just one standard (this is not very statistically sound), and (2) the calibration constant, K, must truly be constant at the two concentration levels, C_S and C_U (possible, but not guaranteed). Because of these limitations, the concept of the standard curve is used most of the time.

The concept of a series of standards refers to an experiment in which a series of standard solutions is prepared covering a concentration range within which the unknown concentration is expected to fall. For example, if an unknown concentration is thought to be around 4 ppm, then a series of standards bracketing this value, such as 1, 3, 5, and 7 ppm, are prepared. The readout for each of these is then measured. The standard curve is a plot of the readout vs. concentration. The unknown concentration is determined from this plot.

The standard curve procedure is free from the limitations involved in comparing an unknown to a single standard because (1) if the unknown concentration is compared with more than one standard, the results are more statistically sound, and (2) if the curve is a straight line, the value of K is constant through the concentration range used. If the curve is a straight line free of bends or curves, the value of K is the slope of the line, and is constant (see Figure 6.4). If a portion of the standard curve is linear, this linear portion may be used to determine the unknown concentration.

We should indicate at this point that, while the limitations of the single standard are eliminated, there is still considerable opportunity for experimental error in this procedure (error in solution preparation and instrument readout) and this error may make it appear that the data points are "scattered" and not linear. This problem is addressed in the next two subsections.

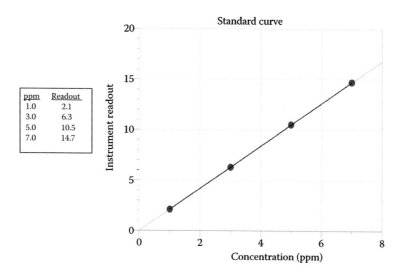

ppm	Readout
1.0	2.1
3.0	6.3
5.0	10.5
7.0	14.7

FIGURE 6.4 Sample data and standard curve showing the desired linearity. The value of K is constant at all concentrations at 2.1.

This "series of standard solutions" method is commonplace in an instrumental analysis laboratory for the vast majority of all quantitative instrumental procedures. Examples of this abound for many spectrophotometric, chromatographic, and other techniques.

6.3.3 Mathematics of Linear Relationships

The mathematical equation representing a straight-line relationship between variables x and y is the following equation:

$$y = mx + b \tag{6.6}$$

If the variables x and y bear a linear relationship to one another, then m and b are constants defining the slope (m) of the line (how "steep" it is) and the y-intercept (b), the value of y when x is zero. The slope of a line can be determined very simply if two or more points, (x_1, y_1) and (x_2, y_2), that make up the line are known.

$$m = \frac{\Delta y}{\Delta x} = \frac{(y_2 - y_1)}{(x_2 - x_1)} \tag{6.7}$$

The y-intercept can be determined by observing what the y-value is on the graph where the line crosses the y axis, or if the slope and one point are known, may be calculated using Equation 6.6.

The important observation for instrumental analysis is that the K in Equation 6.1 is the same as the m in Equation 6.6. Thus, in a graph of R vs. C, the slope of the line is the "K" value, the proportionality constant.

It would seem that the y-intercept would always be zero, since if the concentration is zero, the instrument readout would logically be zero, especially since a "blank," is often used. The blank, a solution prepared so as to have all sample components in it except the analyte (See Section 6.6), is a solution of zero concentration. Such a solution is often used to "zero" the readout, meaning that the instrument is manually set to read zero when the blank is being read by the instrument. This represents a calibration step in which the result is known and the instrument readout can be "tweaked."

However, "real data" sometimes does not fit a straight line graph all the way to the y-intercept and thus the y-intercept is **not** always zero and the (0,0) point is usually not included in the standard curve.

Another problem with "real" data is that due to random indeterminate errors (Chapter 1), the analyst cannot expect the measured points to fit a straight line graph exactly. Thus, it is often true that we draw the *best* straight line that can be drawn through a set of data points, and the unknown is determined from *this* line. A "linear regression," or "least squares," procedure is then done to obtain the correct position of the line and therefore the correct slope, etc.

6.3.4 METHOD OF LEAST SQUARES

The linearity of instrument readout vs. concentration data must be determined. As stated in Section 6.3.2, random indeterminate errors during the solution preparation and during the measurement of R will likely cause the resulting points to appear "scattered" to some extent. If the data appear to be linear despite the scatter, a method must be adopted that will fit a straight line to the data as well as possible. It may happen that some (or even all) the points may not fall exactly on the line because of these random errors, but a straight line is still drawn, since the random errors are indeed random and cannot be compensated for directly. Thus, the best straight line possible is drawn through the points (see Figure 6.5). If only a portion of the curve appears to be linear, then this portion is plotted and the best straight line is drawn through these points.

In the modern laboratory, standard curves are generated with computer programs utilizing the method of least squares, or linear regression, which determines the placement of the straight line mathematically. By this method, the best straight line fit is obtained when the sum of the squares of the individual y-axis value deviations (deviations between the plotted y values and the values on the proposed line) are at a minimum. This "proposed line" is actually calculated from the given data, a slope and y-intercept (m and b, respectively, in Equation 6.6) are then obtained, and the deviations $(y_{point} - y_{line})$ for each given x are calculated. Finding the values of the slope and the y-intercept that minimize the sum of the squares of the deviations involves calculations with which we will not concern ourselves. Computers and programmable calculators, however, handle this routinely in the modern laboratory (see Experiment 14 for a procedure using Excel spreadsheet), and the results are

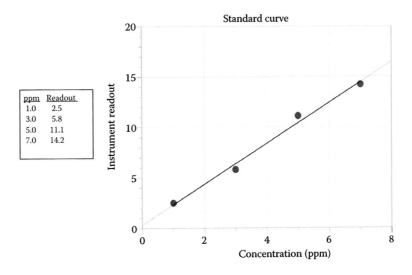

ppm	Readout
1.0	2.5
3.0	5.8
5.0	11.1
7.0	14.2

FIGURE 6.5 Example of a standard curve in which, due to random errors, the data points do not lie on the line.

very important to a given analysis since the line that is determined is statistically the most correct line that can be drawn with the data obtained. The concentrations of unknown samples are also readily obtainable on the calculator or computer since the equation of the straight line, including the slope and y-intercept, become known as a result of the least squares procedure. Other statistically important parameters are readily obtainable as well, including the "correlation coefficient."

Example 6.1

The equation of the straight line determined from a least square fit procedure for an experiment in which the instrument readout, R, was plotted on the y-axis and concentration in ppm was plotted on the x-axis is

$$R = 1.37\ C + 0.029 \tag{6.8}$$

What is the analyte concentration in the unknown if the instrument readout for the unknown is 0.481?

Solution 6.1

Solving Equation 6.8 for C, we get the following.

$$C = \frac{(R - 0.029)}{1.37} = 0.330$$

6.3.5 THE CORRELATION COEFFICIENT

The correlation coefficient is one measure of how well the straight line fits the analyst's data—how well a change in one variable "correlates" with a change in another. A correlation coefficient of exactly "1" indicates perfectly linear data. This, however, rarely occurs in practice. It occurs if all instrument readings increase by exactly the same factor from one concentration to the next, as in the data in Figure 6.4. Due to the indeterminate errors that are present in all analyses, such data are quite rare. Data that approaches such linearity will show a correlation coefficient less than 1, but very nearly equal to 1. Numbers such as 0.9997 or 0.9996 are considered excellent and attainable correlation coefficients for many instrumental techniques. Good pipetting and weighing technique when preparing standards and well-maintained and calibrated instruments can minimize random errors and can produce excellent correlation coefficients and therefore accurate results. The analyst usually strives for **at least** two 9s, or possibly three, in their correlation coefficients, depending on the particular instrumental method used. Again, these coefficients can be determined on programmable calculators and laboratory computers using the appropriate software. A step-by-step procedure using Microsoft Excel spreadsheet software is presented in Experiment 14. The usual symbol for the correlation coefficient is r. The square of the correlation coefficient, R^2, may also be reported.

6.4 PREPARATION OF STANDARDS

The series of standard solutions of the analyte can be prepared in several ways. Probably, the most common is through several **dilutions of a stock standard solution**. The stock standard may either be prepared by the analyst or purchased from a chemical vendor. It may be prepared by weighing a quantity of the primary standard analyte chemical (using an analytical balance) into a volumetric flask and diluting to the mark (as previously discussed in Chapters 4 and 5). It may also be a secondary standard prepared by diluting a primary stock that was purchased from a chemical vendor or prepared in the laboratory. The series of standards is then prepared by diluting various volumes of this stock (as determined by dilution calculations discussed in Chapter 4) to volume in a series of volumetric flasks. These dilutions require diligent solution transfer technique, such as with one of the various pipetting devices discussed in the previous chapters, to minimize the indeterminate errors.

Another is a so-called **serial dilution** procedure. In this procedure, the second solution is prepared by diluting the first, the third by diluting the second, the fourth by diluting the third, and so forth. Once again, volumetric flasks and suitable solution transfer devices are required.

Example 6.2

How would you prepare 50.00 mL each of a series of standards that have analyte ppm concentrations of 1.0, 3.0, 5.0, and 7.0 from a stock standard that is 100.0 ppm?

Solution 6.2

We use the dilution equation (Equation 4.1) for each solution to be prepared. The dilution equation is the following.

$$C_B \times V_B = C_A \times V_A$$

In this equation, C_B is the concentration before dilution, V_B is the volume to be diluted, C_A is the concentration after dilution, and V_A is the final solution volume.

The 1.0-ppm solution: 100.0 ppm $\times V_B$ = 1.0 ppm $\times$ 50.00 mL; V_B = 0.50 mL

The 3.0-ppm solution: 100.0 ppm $\times V_B$ = 3.0 ppm $\times$ 50.00 mL; V_B = 1.5 mL

The 5.0-ppm solution: 100.0 ppm $\times V_B$ = 5.0 ppm $\times$ 50.00 mL; V_B = 2.5 mL

The 7.0-ppm solution: 100.0 ppm $\times V_B$ = 7.0 ppm $\times$ 50.00 mL; V_B = 3.5 mL

The solutions are prepared by pipetting the calculated volumes of the 100.0-ppm solutions into separate 50.00 mL volumetric flasks. Each is then diluted to the mark and made homogeneous by thorough mixing.

6.5 BLANKS AND CONTROLS

In addition to the series of standard solutions needed for an instrumental analysis, there are often other solutions needed for the procedure. We have already briefly mentioned the need for and use of a "blank" (Section 6.3.3). As stated previously, the **blank** is a solution that contains all the substances present in the standards and the unknown (if possible) except for the analyte. The readout for such a solution should be zero, and as we have indicated, the readout is often manually adjusted to read zero when this blank is being measured. Thus, the blank is useful as a sort of precalibration step for the instrument.

6.5.1 REAGENT BLANKS

A blank such as the one described above is appropriately called a **reagent blank**. While an analyst may be tempted to use distilled water or other pure sample solvent as a blank, this may not be desirable because other chemicals (reagents) may have been added to enhance the readout for the analyte in the standards and unknowns and these chemicals may independently affect the readout in some small way. The blank must take into account the effects of all the reagents (and any contaminants in the reagents) used in the analysis. Thus, if the reagents or their contaminants contribute to an instrument's readout when the analyte is being measured, such effects would be cancelled out as a result of the zeroing step. The value of a reagent blank is thus obvious.

6.5.2 SAMPLE BLANKS

While a reagent blank is frequently prepared and used as described above, a sample blank is sometimes appropriate instead. A **sample blank** takes into account any chemical changes that may take place as the sample is taken and/or prepared. For example, the analysis of an air sample for a component in the particulates in the air involves drawing the air sample through a filter in order to capture the particulates so that they can be dissolved. This dissolution step involves not just the particulates but also the filter itself. Thus, the dissolved filter changes the chemistry of the sample and may itself contribute to the instrument readout, and this would not be taken into account by a simple reagent blank. The answer to this problem is to take a clean filter and carry it through the dissolution procedure alongside the sample and to use the resulting solution as the blank. Such a blank is called a sample blank.

Of course, anytime the dissolving of a sample includes a heating step, or any other step that may produce chemical changes in the solution, the blank should undergo the same steps. The reagent blank then becomes a sample blank, and the resulting solution would represent a matrix as close to the composition of the sample solution as possible, thus enhancing accuracy when reading the samples.

6.5.3 CONTROLS

A **control** is a standard solution of the analyte prepared independently, often by other laboratory personnel, for the purpose of cross-checking the analyst's work. If the concentration found for such a solution agrees with the concentration it is known to have (within acceptable limits based on statistics), then this increases the confidence a laboratory has in the answers found for the real samples. If, however, the answer found differs significantly from the concentration it is known to have, then this signals a problem that would not have otherwise been detected. The analyst then knows to scrutinize his or her work for the purpose of discovering an error.

The results of the analysis of a control are often plotted on a **control chart** in order to visualize the history of the analysis in the laboratory so that a date and time can be identified as to when the problem was first detected. Thus, the problem can be traced to a bad reagent, instrument, or other component of the procedure if such a component was first put into use the day the problem was first detected. Your instructor may want you to use controls and a control chart in various experiments in this text.

6.6 POST-RUN CALCULATIONS IN INSTRUMENTAL ANALYSIS

The concentration obtained from the standard curve is rarely the final answer in a real-world instrumental analysis. In most procedures, the sample has undergone some form of preanalysis treatment prior to the actual measurement. In some cases, the sample must be diluted prior to the measurement. In other cases, a chemical must be added prior to the measurement, possibly changing the analyte's concentration. In still other cases, the sample is a solid and must be dissolved or extracted prior to the measurement.

The instrument measurement is the measurement of the solution tested, and the concentration found is the concentration in that solution. What the concentration is in the original, untreated solution or sample must then be calculated based on what the pretreatment involved. Often, this is merely a "dilution factor." It may also be a calculation of the grams of the constituent from the ppm or molar concentration of the solution or the calculation of the ppm in a solid material based on the weight of the solid taken and the volume of extraction solution used and whether or not the extract was diluted to the mark of a volumetric flask or if a portion of it was diluted. We address each of these possibilities below. Remember that ppm for a solute in dilute water solutions is mg/L (or μg/mL), whereas for an analyte in solid samples is mg/kg (or μg/g). Review Chapter 5 for more information about the ppm unit.

6.6.1 CALCULATION OF PPM ANALYTE IN A SOLUTION GIVEN MASS AND VOLUME DATA

The ppm concentration of an analyte in a solution is calculated by dividing the mass of the analyte in a given volume of solution by that volume. The mass must be in milligrams and the volume in liters in order to have the unit mg/L (ppm) in the answer. It may be necessary to convert the mass to milligrams and the volume to liters.

Example 6.3

What is the ppm concentration of lead in a water sample if there are 0.00890 g of lead dissolved in 500.0 mL of sample?

Solution 6.3

$$\text{ppm Pb} = \frac{\text{mg Pb}}{\text{L of sample}} = \frac{0.00890 \text{ g} \times 1000 \text{ mg/g}}{500.0 \text{ mL} \div 1000 \text{ mL/L}} = 17.8 \text{ ppm Pb}$$

6.6.2 CALCULATION OF PPM ANALYTE IN A SOLID SAMPLE GIVEN MASS DATA

The ppm concentration of an analyte in a solid sample is calculated by dividing the mass of the analyte in a given mass of solid sample by that mass of solid sample. The mass of the analyte must be in milligrams and the mass of the sample in kilograms in order to have the unit mg/kg (ppm) in the answer. It may be necessary to convert the mass to milligrams and the volume to liters.

Example 6.4

What is the ppm concentration of mercury in fish tissue if, after testing with an instrument, it was found that there are 0.000475 g of mercury in 2.250 g of the tissue?

Solution 6.4

$$\text{ppm Hg} = \frac{\text{mg Hg}}{\text{kg sample}} = \frac{0.000475 \text{ g} \times 1000 \text{ mg/g}}{2.250 \text{ g} \div 1000 \text{ g/kg}} = 211 \text{ ppm Hg}$$

6.6.3 CALCULATION OF THE MASS OF ANALYTE FOUND IN AN EXTRACT

The milligrams of analyte found in a solution used to extract an analyte from a solid or liquid sample (i.e., the "extract") is calculated by multiplying the ppm concentration by the liters of the extract used. It is the same calculation regardless of whether the sample is a liquid or a solid.

Example 6.5

How many milligrams of iron are dissolved 100.0 mL of an extract if, after testing with an instrument, the ppm concentration of iron in the extract was found to be 4.58 ppm?

Solution 6.5

$$\text{milligrams} = \frac{mg}{L} \times L = 4.58 \, \frac{mg}{L} \times (100.0 \text{ mL} \div 1000 \text{ mL/L}) = 0.458 \text{ mg}$$

6.6.4   CALCULATION OF PPM ANALYTE IN A LIQUID OR SOLID THAT WAS EXTRACTED

The calculation of the ppm concentration of an analyte in a liquid or solid for which sample and extraction data are given is a combination of the calculations in Sections 6.6.1 through 6.6.3. In other words, the milligrams of analyte found in the extract is calculated, and this number of milligrams is then divided by the liters or kilograms of sample used. We will assume here that the analytes are 100% extracted.

Example 6.6

A 0.5693-g sample of insulation was analyzed for formaldehyde residue by extracting with 25.00 mL of extracting solution. After extraction, the sample was filtered and the filtrate analyzed by an instrument without further dilution. The concentration of formaldehyde in this extract was determined to be 4.20 ppm. What is the concentration of formaldehyde in the insulation?

Solution 6.6

Step 1: Calculation of the milligrams of formaldehyde in the extract.

$$\text{milligrams of formaldehyde in extract} = \frac{mg}{L} \times L = ppm \times liters$$
$$= 4.20 \text{ ppm} \times 0.02500 \text{ L}$$
$$= 0.105 \text{ mg}$$

Step 2: Calculation of the ppm concentration in the sample.

$$\text{ppm in sample} = \frac{mg \text{ in extract}}{kilograms \text{ of sample}} = \frac{0.105 \text{ mg}}{0.0005693 \text{ kg}} = 184 \text{ ppm}$$

6.6.5   CALCULATION WHEN A DILUTION IS INVOLVED

There are two possible scenarios involving a dilution in these types of analyses: (1) when the entire extract is transferred to a volumetric flask and diluted to the mark and (2) when only a portion of the extract is transferred to a volumetric flask and diluted to the mark.

In the first scenario, the calculation is the same as in Section 6.6.4, except the final dilution volume is the volume used in the calculation, since the entire amount extracted is in this volume. In the second scenario, a dilution factor must be applied.

Example 6.7

A soil sample weighing 2.00 g was tested for phosphorus content by extracting the phosphorus with 20.00 mL of a solution containing HCl and NH_4F and adjusted to pH = 2.6. The resulting muddy mixture was filtered and the filtrate (extract) captured in a beaker. The soil was then rinsed with more extracting solution and the rinsings were added to the extract. Ultimately, the extract/filtrate was diluted to 100.0 mL. If this diluted solution was tested with an instrument and found to have a phosphorus concentration of 1.24 ppm, what is the concentration of phosphorus in the soil?

Solution 6.7

Step 1: Calculation of the milligrams of phosphorus in the diluted extract.

$$\text{milligrams of phosphorus} = \frac{mg}{L} \times L = ppm \times liters$$
$$= 1.24\ ppm \times 0.1000\ L$$
$$= 0.124\ mg$$

Step 2: Calculation of the ppm concentration in the sample.

$$ppm\ in\ sample = \frac{mg\ in\ extract}{kilograms\ of\ sample} = \frac{0.124\ mg}{0.00200\ kg} = 62.0\ ppm$$

Example 6.8

A water sample was tested for iron content, but was diluted prior to obtaining the instrument reading. This dilution involved taking 10.00 mL of the sample and diluting it to 100.00 mL. If the instrument reading gave a concentration of 0.891 ppm for this diluted sample, what is the concentration in the undiluted sample?

Solution 6.8

Step 1: Calculation of the milligrams of iron in the diluted sample.

$$\text{milligrams of phosphorus} = \frac{mg}{L} \times L = ppm \times liters$$
$$= 0.891\ ppm \times 0.1000\ L$$
$$= 0.0.0891\ mg$$

Step 2: Calculation of the ppm concentration in the sample. The volume of the original sample was 10.00 mL or 0.01000 L.

$$ppm\ in\ sample = \frac{mg\ in\ solution}{liters\ of\ sample} = \frac{0.0891\ mg}{0.01000\ L} = 8.91\ ppm$$

This same answer (8.91 ppm) is found if you consider that the sample was diluted by a factor of 10 (from 10.00 to 100.0 mL). Thus, the dilution factor is 10 and 0.891 × 10 = 8.91 ppm.

Example 6.9

A 5.000-g soil sample was analyzed for potassium content by extracting the potassium using 10.00 mL aqueous ammonium acetate solution. Following the extraction, the soil was filtered and rinsed. The filtrate with rinsings was diluted to exactly 50.00 mL. Then, 1.00 mL of this solution was diluted to 25.00 mL, and this dilution was tested with an instrument. The concentration in this 25.00 mL was found to be 3.18 ppm. What is the concentration of the potassium in the soil in ppm?

Solution 6.9

Step 1: Calculation of the milligrams of potassium in the diluted extract.

$$\text{milligrams of phosphorus} = \frac{\text{mg}}{\text{L}} \times \text{L} = \text{ppm} \times \text{liters}$$
$$= 3.18 \text{ ppm} \times 0.0.02500 \text{ L}$$
$$= 0.0795 \text{ mg}$$

This is the amount of potassium in the 25.00-mL solution. It is also the amount of potassium in the 1.00-mL solution, since no additional potassium was added when diluted to 25.00 mL. However, the amount in the 1.00 mL is only one fiftieth of what was in the soil since the 1.00-mL solution was taken from the 50.00-mL solution, and the 50.00-mL solution contained all the potassium in the sample. Thus, we must multiply the 0.0795 mg by 50.00 in order to have the weight (milligrams) that was in the soil sample.

$$\text{mg of potassium in soil} = 0.0795 \times 50.00 = 3.975 \text{ mg}$$

Step 2: Calculation of the ppm concentration in the sample.

$$\text{ppm in sample} = \frac{\text{mg in extract}}{\text{kilograms of sample}} = \frac{3.975 \text{ mg}}{0.00500 \text{ kg}} = 795 \text{ ppm}$$

6.7 LABORATORY DATA ACQUISITION AND INFORMATION MANAGEMENT

In this chapter, we have advocated the use of a computer for plotting the standard curve and performing the least squares fit procedure. Indeed, computers play a central role in the analytical laboratory for acquiring and manipulating data generated by instruments and for information management.

6.7.1 DATA ACQUISITION

It is very commonplace today for the analyst to use a computer for acquiring data. The term *data acquisition* refers to the fact that the data generated by an instrument can be fed directly into and acquired by a computer via an electronic connection between the two. Often, the computer monitor then serves as the readout device in real time. For example, a continuously changing readout can be monitored as a function of time on the screen, with the signal tracing across the screen as this signal is being generated by the instrument's signal processor. Such data are thereby stored in short term in the computer's memory and, in the long term, on magnetic storage media, such as disks. Hard copies can be generated through connection to a printer.

The actual mode of connection between instrument and computer varies depending on the type of signal generated and on the design of the instrument. The connection can be made via a serial, parallel, or USB port. The electronic circuitry required is either built into the instrument's internal readout electronics or into an external "box" used for conditioning the instrument's output signal. In some cases, the instrument will not operate without the computer connection and is switched on and off as the computer is switched on and off. In other cases, the entire computer is built into the instrument and is not designed for any other application.

When in actual use, a software program is run on the computer, which establishes the sampling time interval and other parameters at the discretion of the operator, and begins the data acquisition at the touch of a key on the computer keyboard or a click of the mouse.

6.7.2 Laboratory Information Management

Modern laboratories are complex multifaceted units with vast amounts of information passing to and from instruments and computers and to and from analysts and clients daily. The development of high-speed, high-performance computers has provided laboratory personnel with the means to handle the situation with relative ease. Software written for the purpose has meant that ordinary personal computers can handle the chores. The hardware and software system required has come to be known as the **Laboratory Information Management System**, or **LIMS**.

LIMS systems vary in sophistication. However, all systems perform basic information management as it relates to the sample labeling and tracking, laboratory tests, personnel assignments, analytical results, report writing, and client communication. More sophisticated versions include verifications, validations, and approvals. They produce reports on work in progress. They maintain a backlog. They incorporate data for productivity and resources. They also provide the means for conducting an audit, computing costs, and maintaining an archive.

The LIMS computer is located on the site, and several terminals may be provided for entry of data from notebooks and instrument readouts and for the retrieval of information. Bar coding for sample tracking and access codes for laboratory personnel are parts of the system. Instruments may be interfaced directly with the LIMS computer to allow direct data entry without help from the analyst. The LIMS may also incorporate statistical methods and procedures, including statistical control and control chart maintenance.

EXPERIMENTS

EXPERIMENT 13: PLOTTING A STANDARD CURVE USING EXCEL SPREADSHEET SOFTWARE (VERSION 2010)

Note: Use hypothetical data to practice plotting a standard curve with the Excel spreadsheet software procedure below. Later experiments will refer you back to this procedure to plot standard curves for real experiments. Launch Excel (or click "file" then "new" if already launched) to begin.

1. Type in the data for the standard curve in the A and B spreadsheet columns. Use the "A" column for the concentrations and the "B" column for the corresponding instrument read-out values. For the unknown, type in the instrument readout value in the "B" column cell below the standards, but leave the corresponding concentration cell blank. When finished, the A and B columns should appear as in Table 6.1 in which there are four standards with concentrations of 1, 2, 3, and 4 ppm and one unknown.
2. Highlight the standards data in the A and B columns.
3. Select the "Insert" tab, click "Scatter," and then select the upper left icon. The graph should now be on the screen.

TABLE 6.1

**Example of the A–B Columns
in the Spreadsheet in Step 1**

	A	B
1	1	3
2	2	5.9
3	3	9.1
4	4	12
5		7.2

4. Select the "Layout" tab and click "Axis Titles." Enter the titles for both the horizontal and vertical axes (for example, ppm or % for the horizontal axis and refractive index or absorbance for the y-axis). For the vertical title, use "Rotated Title." Also add a "Chart Title," such as "Experiment 20" or other descriptive title, above the graph.
5. While still on the "Layout" tab, Click "Trendline." Select "More Trendline Options." Then select "Linear," "Display Equation on Chart," and "Display r-squared on Chart." Click "Close." Move (click and drag) the box containing the equation and the r^2 value to the upper right corner of the graph so that you can clearly see the slope (m), the y-intercept (b), and the r^2 value.
6. Calculate the concentration of the unknown as follows: (a) Highlight the "A" cell for the unknown and then click on the white space to the right of "fx." The formula for calculating the concentration is $x = (y - b)/m$. In the space to the right of the "fx," type "=" followed by the rest of the equation using the values that y, b, and m are now known to have. Press "Enter." The concentration of the unknown should now be in the "A" cell for the unknown.
7. Move (click and drag) the graph to a location directly below the data. Print.

QUESTIONS AND PROBLEMS
Review of the Analytical Process

1. What are the five parts of the general analytical process?

Instrumental Analysis Methods

2. Distinguish between "wet" methods of analysis and "instrumental" methods of analysis. What do the analytical processes for the wet chemical methods and instrumental methods have in common?
3. Why are sampling and sample preparation activities important no matter what analytical method is chosen?
4. What is different about the analytical process for instrumental analysis compared with wet chemical analysis?
5. Explain the general principle of analysis with electronic instrumentation and cite an example.
6. What are three major instrumental analysis classifications? What are the general names of the instruments?
7. Distinguish between spectroscopic methods and chromatographic methods.
8. Define "calibration."
9. Briefly explain the differences between calibrating an analytical balance and calibrating a pH meter and what happens if the measured standard does not give the correct result.
10. What is a "standard curve?"
11. Briefly explain how the plotting of a standard curve is a calibration process.

Basics of Instrumental Analysis

12. Describe the various methods by which a sample is introduced into an instrument.
13. What is a "sensor?"
14. What are the four instrument components for translating a property of a solution to an instrument readout?
15. What are the advantages of calibrating an instrument with a single standard solution compared with a series of standard solutions? Which is preferred and why?
16. The proportionality constant between an instrument readout and concentration is 54.2. Assuming a linear relationship between the readout and concentration, what is the numerical value of the concentration of a solution when the instrument readout is 0.922?
17. What is the numerical value of the concentration in a solution that gave an instrument readout of 53.9 when the proportionality constant is 104.8?
18. An instrument reading for a standard solution whose concentration is 8.0 ppm is 0.651. This same instrument gave a reading of 0.597 for an unknown. What is the concentration of the unknown?
19. Calculate the concentration for the unknown given the following data:

R	C (ppm)
72.0	0.693
68.1	C_u

20. Using spreadsheet software, plot the following data and give the concentration of the unknown solution:

R	C (ppm)
8.2	2.00
17.0	4.00
24.9	6.00
31.9	8.00
40.5	10.00
26.7	C_u

21. Why must the instrument readout for an unknown fall within the range of the series of readouts for a series of standard solutions in order to be accurate?
22. Calculate the slope of the straight line defined by the following two points: (0.20, 0.439) and (0.50, 0.993).
23. Given the following data, calculate the proportionality constant, K, between A and C. (Assume that the y-intercept is zero.)

A	C
0.419	3.00
0.837	6.00

24. Given the data and results from problem 23, what is the concentration in a solution that gave an A value of 0.677?
25. When plotting the results of the measurement of a series of standard solutions, why do we draw "the best straight line" possible through the points rather than just connect the points?

26. What is the method of least squares, and why is it useful in instrumental analysis?
27. What is it about the calculations involved in the method of least squares that gives this method its name?
28. What is meant by "linear regression" analysis?
29. Give four parameters that are readily obtainable as a result of the method of least squares treatment of a set of data.
30. What is meant by perfectly linear data? What value of what parameter would indicate that a data set is perfectly linear?
31. What are some realistic values of a correlation coefficient that would indicate to a laboratory worker that the error associated with his or her data is probably minimal?

Preparation of Standards

32. What is meant by "serial dilution?"
33. The "series of standard solutions" method works satisfactorily most of the time. When does it **not** work well?
34. A stock standard solution of copper (1000.0 ppm) is available. Tell how you would prepare 50.00 mL each of five standard solutions, each required to be 1000.0 ppm, and how you would proceed with the preparation, including the kind of pipette needed. The concentrations of the standards should be 1.00, 2.00, 3.00, 4.00, and 5.00 ppm.
35. How many milliliters of a 100.0-ppm stock are needed to prepare 25.00 mL each of four standards with concentrations 2, 4, 6, and 8 ppm?
36. What is a "blank?" What is the difference between a "reagent blank" and a "sample blank?"
37. What is a "control" sample and how is it useful as an accuracy check?

Post-run Calculations in Instrumental Analysis

38. Explain why the concentration of an unknown obtained from a graph of a series of standard solutions may not be the final answer in an instrumental analysis.
39. Using the information given in each row of the table below, calculate the item marked with a question mark (?).

	Nature of Sample	Weight of Analyte in the Sample	Volume or Weight of Sample	Concentration of Analyte in the Sample (ppm)
(a)	Water	3.6 mg	2.45 L	?
(b)	Water	6.2 mg	500.0 mL	?
(c)	Water	0.0381 g	100.0 mL	?
(d)	Soil	43.6 mg	0.0290 kg	?
(e)	Soil	0.238 mg	1.50 g	?
(f)	Soil	0.00284 g	0.850 g	?

40. Using the information given in each row of the table below, calculate the item marked with a question mark (?).

	Nature of Sample	Concentration of Analyte Found in Extract (ppm)	Volume of Extraction Solution Used	Weight of Analyte Extracted (mg)	Weight or Volume of Sample Used	Concentration of Sample (ppm)
(a)	xxxxx	289	0.0250 L	?	xxxxx	xxxxx
(b)	xxxxx	62.0	10.0 mL	?	xxxxx	xxxxx

(c)	Water	9.63	0.0500 L	?	0.250 L	?
(d)	Water	16.3	25.00 mL	?	0.5000 L	?
(e)	Water	27.2	50.00 mL	?	200.0 mL	?
(f)	Water	70.6	0.1000 L	?	500.0 mL	?
(g)	Soil	5.60	0.02500 L	?	0.00250 kg	?
(h)	Soil	11.6	20.00 mL	?	1.50 g	?
(i)	Soil	22.9	0.0100 L	?	0.00300 kg	?
(j)	Soil	6.40	5.00 mL	?	1.00 g	?
(k)	Water	13.6	10.00 mL	xxxxx	100.0 mL	?
(l)	Soil	43.0	15.00 mL	xxxxx	5.00 g	?

41. An unknown solution of riboflavin, contained in a 25-mL volumetric flask, was determined to have a concentration of 0.525 ppm. How many milligrams of riboflavin are in the flask?

42. After analysis for iron content, a water sample was found to have a concentration of 4.62 ppm iron. How many mg of iron are contained in 100.0 mL of the water?

43. A 2.000-g sample of a soil is found to yield 3.73 mg of phosphorus after extraction. What is the concentration of phosphorus in the soil in ppm?

44. After a 5.000-g sample of concrete was dissolved, the resulting solution was found to contain 0.229 g of manganese. What is the concentration of manganese in the concrete in ppm?

45. A certain water sample was diluted from 1 to 50 mL with distilled water. After an analysis for zinc was performed, this diluted sample was found to contain 10.7 ppm zinc. What is the zinc concentration in the undiluted sample?

46. A rag that a farmer was using was analyzed for pesticide residue. The rag weighed 49.22 g and yielded 25.00 mL of an extract solution that was determined to have a pesticide concentration of 102.5 ppm. How many grams of pesticide were in the rag, and what is the concentration of pesticide in the rag in ppm?

47. The soil around an old gasoline tank buried in the ground is analyzed for benzene. If 10.00 g of soil shows the concentration of benzene in 100.0 mL of soil extract to be 75.0 ppm, what is the concentration of benzene in the soil in ppm?

48. Using the information given in each row of the table below, calculate the item marked with a question mark (?).

	Nature of Sample	Weight or Volume of Sample Used	Volume of Extraction Solvent Used (mL)	Extract Diluted from ___ to ___ (mL)	Concentration of Analyte in Diluted Extract (ppm)	Concentration of Analyte in Original Sample (ppm)
(a)	Water	200.0 mL	20.00	20.00 to 100.0	5.48	?
(b)	Water	100.0 mL	5.00	1.00 to 25.00	8.77	?
(c)	Water	1.000 L	10.00	1.00 to 50.00	3.84	?
(d)	Soil	5.00 g	25.00	5.00 to 25.00	17.4	?
(e)	Soil	1.50 g	50.00	10.00 to 100.0	0.568	?
(f)	Soil	0.00200 kg	100.0	20.00 to 250.0	0.409	?

49. Suppose 4.272 g of a soil sample undergo an extraction with 50 mL of extracting solvent to remove the potassium. This 50 mL was then diluted to 250 mL and tested with an instrument. The concentration of potassium in this diluted extract was found to be 35.7 ppm. What is the potassium concentration, in ppm, in the untreated soil sample?

50. A city's water supply is found to be contaminated with carbon tetrachloride. The chemical analysis procedure involves the extraction of the carbon tetrachloride from the water with hexane. A 4.00-L sample of water is extracted with 10 mL of hexane. If this hexane

solution is diluted from 1 to 25 mL and the concentration of carbon tetrachloride in the diluted solution found to be 10.4 ppm, what is the concentration in the original water?

Laboratory Data Acquisition and Information Management

51. What is meant by data acquisition by computer?
52. Why has computerized data storage developed into an important function of computers in the modern laboratory?
53. Name one way how a spreadsheet computer software can be useful in the modern laboratory.
54. What is LIMS an acronym for? Explain how a LIMS is useful in the modern laboratory.

7 Introduction to Spectrochemical Methods

7.1 INTRODUCTION

Many instrumental methods of analysis involve the refraction, absorption, or emission of light. These can be referred to as **spectrochemical methods**. The science that deals with the refraction of light is called **refractometry** and the instrument used is called the **refractometer**. The science that deals with the absorption and emission of light is called **spectroscopy** or **spectrometry**. The broad term for the instruments used is **spectrometer**, whereas a slightly more specific term (when a light sensor known as a phototube is used) is **spectrophotometer**. In spectrochemical analysis procedures, the degree to which light is refracted or absorbed, or the intensity of light that is emitted, is related to the amount of an analyte present in the sample tested. Thus, the degree of light refraction, absorption, and the intensity of light emission are the critical measurements. The electrical signal readout referred to in Figure 6.2 is an electrical signal that is related to the degree of light refraction, absorption or the intensity of light emission. The instrument readings mentioned in Figure 6.3 are the readings generated by the instrument as a result of this refraction, absorption, or emission (see Figure 7.1).

In this chapter, we expand the above brief summary so that all aspects of refractometry and spectroscopy as analytical methods can be clearly understood and practiced. This will mean providing a full discussion of the nature and parameters of light, including energy, wavelength, frequency, and wavenumber. We will address exactly what is meant by light refraction, absorption, and emission. The discussion includes the spectral differences between atoms (**atomic spectroscopy**) and molecules (**molecular spectroscopy**). The different effects caused by ultraviolet light, visible light, and infrared light are covered. Included also are the instrument designs and what exactly gives rise to the electrical signals that are generated. The fine points of instrument and experiment design are discussed so that experimental results can be optimized. Finally, the discussion includes the instrument readings and how they are related to the amount of analyte present.

7.2 CHARACTERIZING LIGHT

The modern characterization of light is that it has a dual nature. This means that some qualities of light are best explained if we describe light as consisting of moving particles, often called **photons** or **quanta** (referred to as the **particle theory** of light). Other qualities are best explained if we describe light as consisting of moving electromagnetic disturbances called as **electromagnetic waves** (the **wave theory** of light). Such a dual nature is similar to the modern description of electrons, a description that you likely encountered in your previous studies of chemistry. Electrons may be described as particles in order to explain some aspects of their behavior and as entities of energy and not particles in order to explain other aspects.

The wave theory of light states that light travels in a fashion similar to a series of repeating waves of water, as in a wave pool at an amusement park. However, water waves are mechanical waves and require matter, such as water, to exist. Light waves are **electromagnetic waves**, meaning that they are moving wave disturbances that have an electrical component and a magnetic component and do not require matter to exist. They therefore can (and do) travel through a vacuum, such as outer space, where no matter is present (see Figure 7.2).

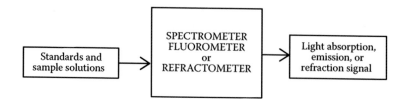

FIGURE 7.1 Principle of analysis when the instrument is a refractometer, a spectrometer, or a fluorometer (compare with Figure 6.2).

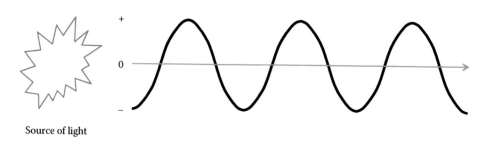

Source of light

FIGURE 7.2 Illustration of electromagnetic waves moving away from a source of light. The "+" and "–" indicate that the waves have a positive and negative amplitude. Amplitude is related to the intensity, or brightness, of the light.

7.2.1 WAVELENGTH, SPEED, FREQUENCY, ENERGY, AND WAVENUMBER

The water waves in a wave pool at an amusement park can be long waves or short waves, depending on the kind of motion at their origin. The same is true of electromagnetic waves. The length of an electromagnetic wave is called its **wavelength** and its symbol is the Greek lowercase of lambda, "λ." In a set of repeating waves, wavelength is the physical distance from a point on one wave, such as the crest of the wave, to the same point (the crest) on the next wave (see Figure 7.3). In science, it is measured in metric system units. These can be meters, centimeters, nanometers, etc. Wavelengths of electromagnetic waves vary from as short as atomic diameters to as long as several miles.

The speed with which electromagnetic waves move is called the **speed of light** and is given the symbol "c." The speed of light in a vacuum is approximately 3.00×10^{10} cm/s. This extraordinarily

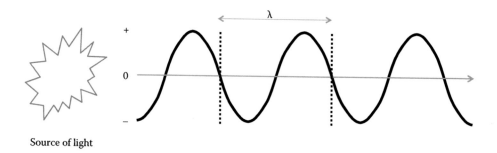

Source of light

FIGURE 7.3 Wavelength, λ, is the distance covered by one complete $\pm$ amplitude cycle. Alternatively, one can say that it is the distance from a point on one wave to the same point on the next wave, such as from crest to crest.

fast speed accounts for the seemingly instantaneous manner with which light fills a room when the light switch is switched on. This is one huge difference between mechanical water waves, which travel almost infinitely slower, and electromagnetic waves. It is important to know that all electromagnetic waves travel at the same speed in a vacuum regardless of their wavelength.

The number of the moving electromagnetic wavelengths (defined above) that pass a fixed point in one second is called the **frequency** of the light (see Figure 7.4 for an illustration of frequency). Its symbol is the lower case of the Greek letter nu, "ν." It is expressed in waves (most often called cycles) per second, or Hertz (Hz). In mathematical formulas, the unit is reciprocal seconds, s^{-1}.

Mathematically, wavelength, speed, and frequency are related by the formula

$$c = \lambda \nu \tag{7.1}$$

Multiplying the wavelength expressed in centimeters by the frequency in reciprocal seconds gives the units of speed, cm/s.

$$cm \times s^{-1} = cm/s \tag{7.2}$$

With speed of light in a vacuum being the same for all light (approximately equal to 3.00×10^{10} m/s as stated previously), it is clear that wavelength and frequency are inversely proportional. This means that as one increases the other decreases such that when one is multiplied by the other, the same number, the speed of light, always results. It is also clear that a given wavelength is always associated with a particular frequency, and given one, the other can be calculated by rearranging Equation 7.1.

$$\nu = \frac{c}{\lambda} \tag{7.3}$$

$$\lambda = \frac{c}{\nu} \tag{7.4}$$

See Figure 7.5 for an illustration of how the number of waves that pass a fixed point in one second is greater when the wavelength is shorter.

Light is a form of **energy** and each wavelength or frequency has a certain amount of energy associated with it. This energy is considered to be the energy associated with a single photon of the light. Thus, the particle theory and the wave theory are linked via the energy. The relationship between energy and frequency is as follows:

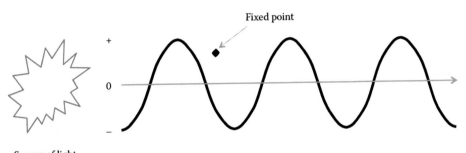

FIGURE 7.4 Illustration of the definition of frequency. With the waves moving left to right as shown, frequency is the number of wavelengths (cycles) that pass a fixed point in one second.

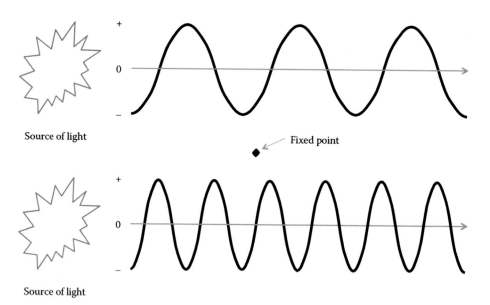

FIGURE 7.5 More wavelengths pass the fixed point in one second when the wavelength is shorter. Assume both waves are moving left to right at the speed of light.

$$E = h\nu \qquad (7.5)$$

In this equation, E is energy, ν is the frequency, and h is a proportionality constant known as "Planck's constant," after the famous physicist Max Planck. The value of "h" depends on what unit of energy is used. If the energy is to be expressed in joules (a unit of energy in the metric system), Planck's constant is 6.63×10^{-34} J s.

Combining Equations 7.1 and 7.3, the relationship between energy and wavelength is as follows:

$$E = \frac{hc}{\lambda} \qquad (7.6)$$

If the wavelength is expressed in centimeters, waves of light are sometimes characterized by the reciprocal of this wavelength. This parameter is known as the wavenumber and is given the symbol $\bar{\nu}$ (Greek letter nu with a bar).

$$\bar{\nu} = \frac{1}{\lambda \ (cm)} \qquad (7.7)$$

It has the units cm^{-1} and is used especially in conjunction with infrared light, as we shall see in Chapter 8. The relationships between wavenumber and frequency and between wavenumber and energy are as follows:

$$\nu = c\bar{\nu} \qquad (7.8)$$

The frequency, energy, and wavenumber are directly proportional, as can be seen in Equations 7.3, 7.8, and 7.9. This means that as frequency increases, energy and wavenumber also increase. If we are dealing with a very high energy wave, then we are dealing with a wave of high frequency and high wavenumber. Wavelength, on the other hand, is inversely proportional to frequency, energy, and wavenumber, as can be seen from Equations 7.3, 7.6, and 7.7. If we are dealing with a very short wavelength, then we are dealing with a high frequency, high energy, and high wavenumber.

7.3   THE ELECTROMAGNETIC SPECTRUM

Wavelengths can vary from distances as little as fractions of atomic diameters to distances as long as several miles. This suggests the existence of an extremely broad spectrum of wavelengths. This **electromagnetic spectrum** of light is so broad that we break it down into "regions." Most of us are familiar with at least a few of these regions. The region of wavelengths that we see with our eyes is called the **visible region**. It is located somewhat in the middle of the electromagnetic spectrum and has wavelengths that vary from approximately 350 nm to approximately 750 nm—a very "narrow" region compared with the entire spectrum and to the other regions. Others include the ultraviolet region, the infrared region, the X-ray region, and the radio wave region. These regions are considerably broader, as you can see from the representations in Figure 7.6. In this figure, wavelength increases left to right and energy, frequency, and wavenumber decrease left to right. The approximate borders of the various regions in nanometers are shown. In the ultraviolet, visible and infrared regions, which are the regions that are emphasized in this chapter, the nanometer (nm) and the micrometer (μm) are the most commonly used units of wavelength (see Figure 7.7).

The wavelength pictured in Figure 7.2 is several centimeters long. Most wavelengths that we deal with, however, are either much shorter than that, or much longer. Radio waves, for example, are on the long end of the electromagnetic spectrum and are on the order of kilometers long. These are very low energy waves (remember: long wavelength = low energy) that do no harm as far as our health and safety are concerned. That is a good thing, because the atmosphere in which we live is full of these wavelengths transporting the radio sound from every studio on earth ultimately to our personal radios. Microwaves have wavelengths on the order of a centimeter and are also of low energy, but they can be dangerous because their absorption causes the generation of much internal heat. Microwave ovens are used in our kitchens and may be used in an analytical laboratory for heating samples.

The infrared region is another portion of the spectrum in which the wavelengths are extremely short by comparison. Wavelengths in this range are so short that we cannot represent them on paper or measure them with common measuring tools. Although the wavelengths are shorter and have

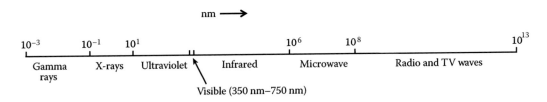

FIGURE 7.6   Illustration of the electromagnetic spectrum. The exponential numbers across the top are approximate wavelength values, which increase left to right. Energy, frequency, and wavenumber increase right to left.

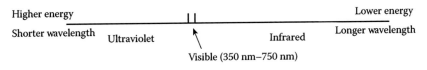

FIGURE 7.7   Illustration of the ultraviolet/visible/infrared regions of the electromagnetic spectrum.

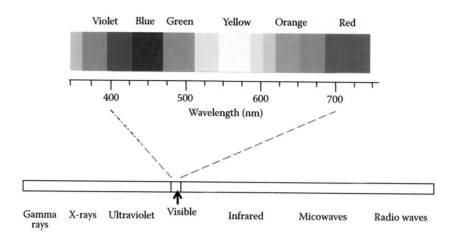

FIGURE 7.8   Visible region of the electromagnetic spectrum.

higher energy than radio waves or microwaves, they also cause us no harm. Indeed, the remote controls we use to control our stereos and televisions utilize infrared light.

When we consider the ultraviolet region, the wavelengths are shorter still, meaning more energy. They are known to cause harm, such as sunburn and skin cancer. X-rays are extremely short wavelengths of extremely high energy, penetrating our skin and tissue and causing harm, hence the reason for the lead aprons used by doctors and dentists when our bodies are "X-rayed." Gamma rays have wavelengths on the order of atomic diameters and cause extreme damage to the human body due to their extremely high energy.

The wavelengths within the visible region of the electromagnetic spectrum are associated with the colors we see. Consider a rainbow. When the light from the sun moves through the earth's atmosphere after a rainstorm, a rainbow may appear in the sky. The reason for this is that the different wavelengths of visible light present in the white light travel through the atmospheric water at different speeds. The result is the visible region of the electromagnetic spectrum—the violet, blue, green, yellow, orange, red sequence—displayed for all to see (see Figure 7.8).

7.4   REFRACTOMETRY

The technique known as refractometry is based on the fact that the speed of light is slightly slower in air, glass, or a liquid solution, compared with a vacuum, which is, as was indicated in the previous section, 3.00×10^{10} cm/s. The effect of this change of speed is seen, as shown in Figure 7.9, as a "disconnect" or a "bending" of an object when viewed through water, for example, compared with when viewed through air.

A parameter known as the **refractive index** is sometimes measured in order to characterize an analyte solution. This refractive index, symbolized by either "n" or "η" (Greek letter "eta"), is very simply defined as the ratio of the speed of light in a vacuum divided by the speed of light in the material medium as shown here.

$$\eta = c/v \tag{7.9}$$

In this equation, v is the speed of light in the material medium. Since the speed of light in a material medium is slower than in a vacuum, the refractive index for any material medium is a number greater than one. The refractive index of air is 1.0003, which means that the speed of light through air is only slightly slower than that through a vacuum. The refractive index of water (at 20°C) is 1.33. The refractive index of a sugar solution with a concentration of 30% (at 20°C) is 1.38. Refraction

FIGURE 7.9 Photographs illustrating refraction. On the left, a spoon in an empty glass. On the right, a spoon in the same glass filled with water. Notice the "disconnect" on the right.

index also depends on temperature, which is why we specified 20°C for the values of the refractive index of water and the sugar solution.

Refractive index also depends on wavelength, and that is why a beam of visible light traveling through a glass prism or other dispersing element, is seen to separate into the component colors. However, this also means that when we measure refractive index, we must use a standard wavelength for the measurement. The standard wavelength commonly used is referred to as the sodium D line, which is 589.3 nm—a wavelength in the yellow region of the visible spectrum and the wavelength emitted by the so-called sodium lamp. More information about wavelength "lines" is given in Section 7.5.3 and in Chapter 9.

The instrument used to measure the refractive index is called a **refractometer**. You might wonder how a refractometer functions, since the refractive index is defined as a ratio of two speeds of light. Does the instrument actually measure the speed of light? The answer is "no." The refractometer makes use of the "bending" of the light, that is observed when a beam of light moves from one medium to another, as mentioned at the beginning of this section. The slower the light moves through the medium in question, the greater angle of bending (see Figure 7.10). Therefore, the

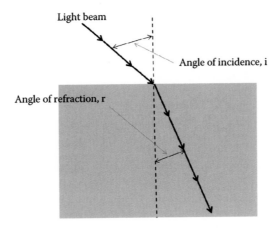

FIGURE 7.10 Definitions of the angle of incidence and the angle of refraction.

refractive index is related to the angle of bending as well as the speed. For a beam of light moving from one medium to another, we can write the following expression as an alternate definition of refractive index.

$$\eta = \sin i / \sin r \qquad (7.10)$$

In this equation, "i" is the angle of incidence and "r" is the angle of refraction as defined in Figure 7.10. Therefore, using the sodium D line, a refractometer determines the refractive index based on these angles.

There are several designs of refractometers. There are inexpensive handheld instruments. There are benchtop instruments in which the refractive index is read on a scale of numbers seen when looking through an eyepiece. There are also benchtop instruments in which a small volume of the liquid is placed on a sapphire prism inside the instrument and the refractive index read on a digital display (no eyepiece) (see Figure 7.11). This latter design is interesting. Rather than measuring the angles by passing the light through the sample, the digital refractometer reflects the light off an interface between the sample and a sapphire prism. The reflection angle when the sample is on the prism is different from the reflection angle when there is no sample on the prism and the angle depends on the refractive index of the sample (see Figure 7.12). A special sensor consisting of an array of diodes (a diode is a small device that converts light intensity into an electrical signal) is used, and when the angle of reflection changes, a different diode receives the light and generates the signal as shown in the figure. This results in a measurement of the angle of reflection and then, by calculation, the refractive index of the sample.

Refractive index is useful for both qualitative analysis and quantitative analysis. The refractive index of a pure liquid at a given temperature is unique to that liquid. Hence, it can be used to help identify a liquid. The refractive index of an analyte solution depends on the concentration of the analyte. For sugar-in-water solution, it is a linear proportion. Thus, refractive index is routinely used to determine the sugar concentration in water solutions (see Experiment 15 at the end of this chapter for such an experiment). Refractometers are also used as detectors for liquid chromatography instruments. These will be discussed in Chapter 12.

FIGURE 7.11   Photograph of a refractometer in use. The design pictured is such that the sample is placed on a sapphire prism and the refractive index read as a digital readout. Also very common is a refractometer in which the refractive index is read through an eyepiece.

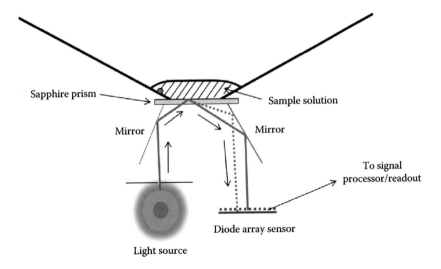

FIGURE 7.12 Illustration of the optical path for the refractometer pictured in Figure 7.11. The small dot on the left side of the sample well under the surface of the sample solution is a thermistor.

7.5 ABSORPTION AND EMISSION OF LIGHT

As stated previously, spectrochemical methods of analysis also involve the absorption and emission of light.

7.5.1 BRIEF SUMMARY

The visible region of the spectrum provides a good starting point to understanding the process of light absorption. Visual evidence of light absorption abounds in our world because of our ability to see the colors that are the visible wavelengths. Objects display a color because some visible light wavelengths are absorbed. Why does a red sheet of paper appear red? All the wavelengths of visible light from the sun and the light bulbs on in the room are absorbed except for the red wavelengths, and these are reflected to our eye (see Figure 7.13). Why does a solution of potassium permanganate appear to be a deep purple color? The wavelengths of visible light that are incident on the solution from the light in the room are all absorbed except for those in the violet and red ends of the visible region. The result is an intense purple color. Thus, those wavelengths of light that are not absorbed

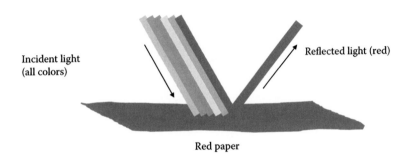

FIGURE 7.13 Piece of red paper appears to be red because red light is not absorbed by the paper, whereas the other wavelengths are absorbed.

by a sample of matter are the wavelengths that are reflected, or transmitted, to our eye, and give the sample its color.

What is the nature of the interaction between light and matter that causes certain wavelengths of light to be absorbed? The answer lies in the structure of the atoms and molecules of which matter is composed. First, consider atoms. The modern theory of the atom states that electrons exist in energy levels around the nucleus. The energy associated with ultraviolet and visible light is comparable to the energy differences between the outermost occupied electron energy level of an atom and the unoccupied levels immediately above it (see Figure 7.14). Electrons can be moved from the lower energy level to the higher one if conditions are right. For example, the outermost electron of a sodium atom (which has electron configuration $1s^22s^22p^63s^1$) can be moved from the 3s level to the vacant 3p level if conditions are right. These conditions consist of (1) the addition of a specific amount of energy to the electron (in the case of sodium, this is the energy difference between the 3s and the 3p levels) and (2) a vacancy for the electron with this greater energy in a certain higher energy level (in the case of sodium, the 3p level is vacant). In other words, if an electron absorbs the energy required for it to be promoted (or elevated) to a higher vacant energy level then it is promoted (or elevated) to that level. If this energy is supplied in the form of light, then the electron will gain that specific amount of energy and move to the higher level. The light then no longer exists—it has been absorbed (see Figure 7.15).

Atoms in which no electrons are in the higher vacant level are said to be in the **ground state**. This state is designated in energy level diagrams as E_0. Atoms in which there is an electron in the

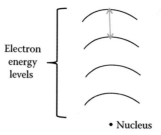

FIGURE 7.14   Difference between the outermost occupied electron energy level of an atom and the next highest (vacant) level (this difference designated here by the double arrow) is on the order of ultraviolet or visible light.

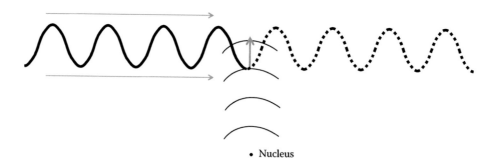

FIGURE 7.15   If the energy difference between the outermost energy level of an atom and the empty level immediately above matches the energy of a beam of light incident upon it, then the atom will take that energy and the light will no longer exist—it is absorbed.

higher level are said to be in an **excited state**. Excited states are designated in energy level diagrams as E_1, E_2, E_3, etc. An **energy level diagram** consists of short horizontal lines representing the levels or states with each line labeled as E_0, E_1, etc. Often, an energy level diagram shows the movement of electrons between levels with longer vertical arrows. The movement of an electron between electron energy levels is called an **electronic energy transition** (see Figure 7.16). It is important to keep in mind that the light coming in must be exactly the same energy as the energy difference between the two electronic levels; otherwise the atom will not absorb it (see Figure 7.17). If it is not absorbed, then, if it is it in the visible region, we see it. It becomes part of the light that is reflected or transmitted and therefore detected by our eyes. The absorption of light by atoms consists of the absorption of only a few very specific wavelengths because the energy difference between two levels is very specific. It is like climbing a ladder. If there is no rung to stand on, then we cannot stand at the level. If there is no energy level for the electron to go to when supplied with light energy, then that light cannot be absorbed.

We cited sodium previously. Gaseous sodium atoms absorb in the visible region of the spectrum. The "few very specific" wavelengths for sodium are 589.0 and 589.6 nm. Both of these represent transitions from the 3s level to the 3p level. The 3p level is actually split narrowly into two levels due to the effect of two possible spin states for the electron in this level, hence the observation of

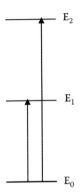

FIGURE 7.16 Energy level diagram as described in the text.

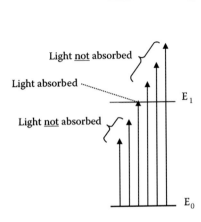

FIGURE 7.17 Energy level diagram of an atom showing the fact that some wavelengths possess too much or too little energy to be absorbed, while another possesses the exact energy required and is therefore absorbed.

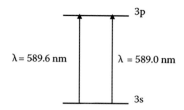

FIGURE 7.18 Electronic energy transitions for gaseous sodium atoms. There are two slightly different transitions between the two levels because of the effect of two spin states that differ slightly in energy.

two transitions that are nearly equal in energy. The energy level diagram for gaseous sodium is illustrated in Figure 7.18. To repeat, the energy associated with electronic energy transitions in atoms is equivalent to the energy of visible and ultraviolet light and thus atoms absorb light in these regions.

What about molecules and complex ions? The absorption of light by molecules and complex ions results in the promotion of electrons to higher energy states in the same way as atoms. However, it is more complicated because molecules and complex ions have energy states that atoms do not.

7.5.2 ATOMS VS. MOLECULES AND COMPLEX IONS

Molecules and complex ions exist in vibrational and rotational energy states as well as electronic states. A vibrational energy state represents a particular state of covalent bond vibration that a molecule can have. This concept will be discussed in Chapter 8. A rotational energy state represents a particular state of rotation of a molecule. The vibrational energy states exist in each electronic state and rotational energy states exist in each vibrational state. Energy level diagrams for molecules show additional horizontal lines, or levels, within ("superimposed" on) each electronic level to represent these vibrational states. The vibrational states can be labeled V_0, V_1, V_2, etc. (see Figure 7.19). The rotational states are superimposed on the vibrational states in the same way. However, unless we are dealing with the rotational states directly, they are most often not depicted on energy level diagrams in order to keep the diagrams relatively simple. The rotational levels are not considered in our discussion here.

For molecules and complex ions, an electronic transition can refer to a transition from any vibrational level in one electronic level to any vibrational level in another electronic level. As with atoms,

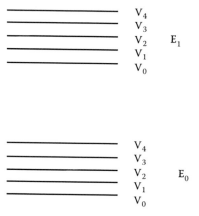

FIGURE 7.19 Energy level diagram for a molecule or complex ion showing the vibrational energy levels superimposed on the electronic levels.

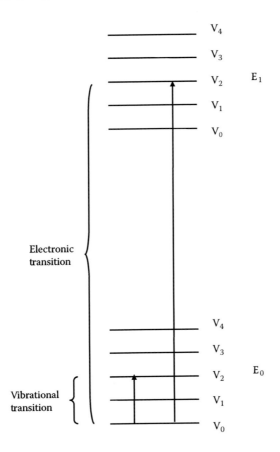

FIGURE 7.20 Energy level diagram for a molecule or complex ion expanded to show an electronic energy transition and a vibrational energy transition. The electronic transition involves visible or ultraviolet light, whereas the vibrational transition involves infrared light.

the amount of energy required for such a transition is found in either the visible or the ultraviolet regions of the electromagnetic spectrum and thus involves either visible or ultraviolet light. A vibrational transition refers to a transition from the lowest vibrational level within a certain electronic level to another vibrational level in the same electronic level. Since such a transition does not involve another electronic level, it requires much less energy and involves the infrared region of the electromagnetic spectrum (see Figure 7.20).

7.5.3 ABSORPTION SPECTRA

Many instruments that are used to measure the absorption of light by atoms, molecules, and complex ions are constructed to measure the ultraviolet and visible regions of the electromagnetic spectrum. Others are constructed to measure the infrared region of the spectrum. All of these instruments have a light source from which a beam of light is formed. The sample being measured is contained such that the light beam is passed through it and the absorption of the wavelengths present in the light beam is measured by a light sensor and signal processor (see Figure 7.21 for a simple drawing of the instrument for the case of ultraviolet and visible light absorption). There are sufficient differences in the mechanics of measuring molecules by ultraviolet and visible wavelengths as opposed to infrared wavelengths to warrant separate discussions of each. Also, there are sufficient differences in the mechanics of measuring atoms as opposed to molecules and complex ions in the ultraviolet

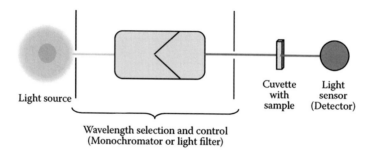

Light source

Cuvette
with
sample

Light
sensor
(Detector)

Wavelength selection and control
(Monochromator or light filter)

FIGURE 7.21 Instruments manufactured to measure ultraviolet or visible light absorption by a test sample have a light source from which a beam of light is formed. The intensities of the wavelengths of light present in the light beam are measured at the light sensor and converted to a readout so that the amount of light absorbed by the sample at selected wavelengths becomes known.

and visible regions so as to also warrant separate discussions of these. For example, the atoms must be in the gas phase and there are a number of different methods used for converting metal ions in solution to atoms in the gas phase. If molecules are being measured in the ultraviolet and visible regions, the technique is referred to as **ultraviolet/visible (or UV/VIS) molecular spectrophotometry**. This is developed fully in Chapter 8. If molecules are being measured in the infrared region, the technique is referred to as **infrared (IR) spectrometry**. This is also developed fully in Chapter 8. If atoms are being measured, the technique is referred to as **atomic spectroscopy**, and this, including a number of subtechniques, is developed fully in Chapter 9. In this section, we follow up the fundamentals discussed in Section 7.5.2 to describe one particular property important for both molecules and atoms, that of the absorption spectrum.

An **absorption spectrum** is a plot of the amount of light absorbed by a sample vs. the wavelength of the light. The amount of light absorbed is called the **absorbance**. It is symbolized as "A" and will be clearly defined in Section 7.6. It is obtained by using a spectrometer to scan a particular wavelength region and to observe the amount of light absorbed by the sample along the way. Consider a solution of copper sulfate. This is an example of a solution of a complex ion, $Cu(H_2O)_4^{2+}$ (see Chapter 5, Section 5.3). It displays a blue color, which means that the blue portion of the visible region is not absorbed but transmitted to our eyes while the red portion is absorbed. The absorption spectrum of this solution in the visible region is shown in Figure 7.22. This spectrum clearly shows

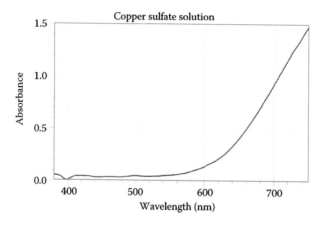

FIGURE 7.22 Absorption spectrum, visible region, of a copper sulfate solution.

that wavelengths in the blue and violet regions (350–500 nm) are not absorbed while wavelengths in the red region (650–750 nm) are absorbed.

Notice also that this absorption spectrum is a **continuous spectrum**, meaning that the spectrum is a smooth curve, left to right. It does not display any breaks or sharp peaks of absorption at particular wavelengths but rather shows that a smooth "band" of wavelengths in a given region, such as the red region, is absorbed.

Compare this with the absorption spectrum of gaseous copper atoms (Figure 7.23). First, the wavelength region shown (320 to 330 nm) is in the ultraviolet region. Gaseous copper atoms do not absorb in the visible region. Second, sharp "lines" of absorption are observed, one at 324.8 nm and one at 327.4 nm. It is a **line spectrum** meaning that individual absorption *lines* are observed rather than a continuous, unbroken "band" observed for the copper sulfate solution.

These observations are explained via energy level diagrams. With atoms, only specific energies (wavelengths), represented by the precisely defined energy differences between two atomic electronic levels can get absorbed. Thus, Figure 7.16 translates into Figure 7.23—two absorption lines corresponding to two electronic transitions. With molecules or complex ions, an entire range of wavelengths can get absorbed, due to the presence of the vibrational and rotational levels superimposed on the electronic levels. The transitions possible between electronic levels E_0 and E_1, for example, each with five vibrational levels, are shown in Figure 7.24. Figure 7.24 translates into

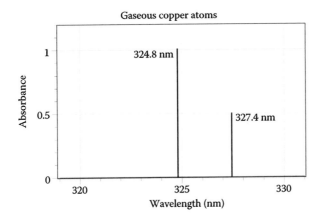

FIGURE 7.23   Absorption spectrum, in a narrow portion of the ultraviolet region, of gaseous copper atoms.

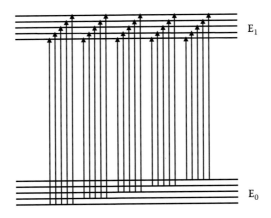

FIGURE 7.24   Energy transitions possible between two electronic states, each with five vibrational levels superimposed.

Figure 7.22—many wavelengths absorbed each corresponding to a particular transition. When the higher electronic levels are considered, and when the rotational levels superimposed on each vibrational level in each electronic level are considered, virtually all wavelengths may be absorbed and the result is the continuous absorption pattern of Figure 7.22.

The absorption vs. wavelength information for molecules and complex ions may also be displayed as a **transmission spectrum** rather than an absorption spectrum by plotting the amount of light transmitted by a sample rather than absorbed. The parameter that is plotted on the y-axis for this is the **transmittance**, or **percent transmittance** rather than the absorbance. Transmittance is symbolized as "T." Transmittance will be specifically defined in Section 7.6. For molecules and complex ions, the transmission spectrum is similar to the corresponding absorption spectrum, but upside down, since a high absorbance corresponds to a low transmittance and vice versa. For example, the transmission spectrum of the copper sulfate solution is shown in Figure 7.25 (compare this with Figure 7.22).

A molecule or complex ion of one compound is different from a molecule or complex ion of another compound in terms of the energy differences between the ground state and the various excited states. Therefore the absorption pattern differs from compound to compound and from complex ion to complex ion. This results in a unique absorption spectrum for each compound or ion, and thus, the absorption spectrum is a "molecular fingerprint." Similarly, the atom of one element is different from the atom of every other element and the result is a unique line spectrum for that element, and thus, the line spectrum for a given element is also a fingerprint for that element. This means that absorption and transmission spectra are useful for the identification and detection of impurities or other sample components.

Previously, we mentioned that it makes sense that copper sulfate solutions have a blue color by noticing what regions of the visible spectra are absorbed and which are not. Figure 7.26 shows the molecular absorption spectra in the visible region for yellow, blue, green, and red food coloring. Examine the spectra in Figure 7.26. See how well you can explain why each has the color it has. Notice that the spectrum for green food coloring includes a pattern similar to the spectrum of blue food coloring. What is it that you might be able to conclude from that?

Many compounds absorb in the ultraviolet and infrared regions of the spectrum and not in the visible region. Such compounds therefore do not display a color. However, the absorption spectra of such compounds can be measured. We have ultraviolet and infrared light sources that are used and we have light sensors that are capable of measuring absorbance in these other regions of the electromagnetic spectrum. More information on ultraviolet and infrared techniques are presented in Chapter 8, and there are several experiments at the end of Chapter 8 that are designed to provide experience with these spectra.

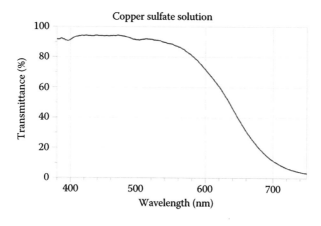

FIGURE 7.25   Transmission spectrum, visible region, of a copper sulfate solution.

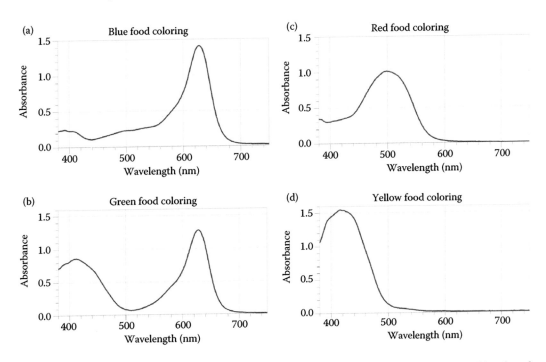

FIGURE 7.26 Absorption spectra of solutions of various food colorings: (a) blue, (b) green, (c) red, and (d) yellow.

7.5.4 LIGHT EMISSION

Under certain conditions, atomic, molecular, and ionic analytes present in laboratory samples will emit light, and this light can be useful for qualitative and quantitative analysis. For example, most processes used to obtain free ground state atoms in the gas phase from liquid phase solutions of their ions (Chapter 9) result in a small percentage of the atoms being elevated to the excited state even if no light beam is used. Whether a light beam is used or not, excited atoms return to the ground state because the ground state is the lowest energy state. The energy loss that occurs when the atoms return to the ground state may be dissipated as heat, but it may also involve the emission of light because the difference in energy between the ground state and the excited state is equivalent to light energy. Energy level diagrams can be used to depict such a process and an **emission spectrum**, the plot of emission intensity vs. wavelength, may be plotted (see Figure 7.27). The process is the reverse

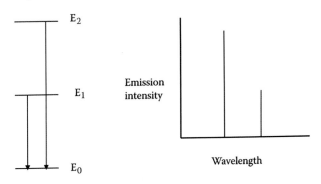

FIGURE 7.27 (Left) Energy level diagram with arrows pointing downward indicating the loss of energy and the return to the ground state. (Right) Line emission spectrum (emission intensity vs. wavelength) showing two emission lines that may result from the process on the left.

of the light absorption process by atoms and so downward-pointing arrows are used to indicate the return to the ground state and the wavelengths emitted are the same wavelengths as those that are absorbed. Thus, the lines in an emission spectrum of a metal are often at the same wavelengths as the lines in the absorption spectrum for the same metal. More details are given in Chapter 9.

Similarly, molecules and complex ions may also emit light under certain conditions, a phenomenon known as **fluorescence**. Specifically, it is when the absorption of a wavelength of light in the UV region is followed by the emission of light of a longer wavelength, such as in the visible region. As with atomic emission, it involves the loss in energy from an excited state to a lower state, and this loss corresponds to the energy of light. More details of this phenomenon are presented in Chapter 8.

7.6 ABSORBANCE, TRANSMITTANCE, AND BEER'S LAW

Finally, let us return to the absorption phenomenon again and precisely define absorbance and transmittance, two terms previously used in this chapter. The intensity of light striking the light sensor (more often called a "detector") in Figure 7.21 when a "blank" solution (no analyte species) is held in the path of the light is given the symbol "I_o." The blank, as discussed in Section 6.6, is a solution that contains all chemical species that are present in the standards and samples to be measured (at equal concentration levels) except for the analyte species. Such a solution does not display any absorption due to the analyte, and thus, I_o represents the maximum intensity that can strike the detector at any time. When the blank is replaced with a solution of the analyte, a less intense light beam will be detected because it is absorbed by the analyte. The intensity of the light for this solution is given the symbol "I" (see Figure 7.28). The fraction of light transmitted is thus I/I_o. This fraction is defined as the "transmittance," "T."

$$T = \frac{I}{I_o} \tag{7.11}$$

The "percent transmittance" is similarly defined:

$$\%T = T \times 100 \tag{7.12}$$

Transmittance and percent transmittance are two parameters that can be displayed on the readout.

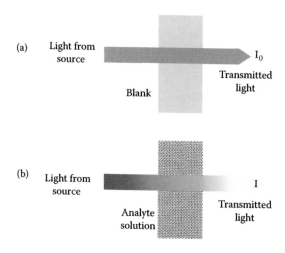

FIGURE 7.28 Illustration of the definitions of I and I_o. The intensity of the light that has passed through a blank solution is designated I_o. The intensity of the light that has passed through the analyte solution is less than I_0 and is designated I.

An important fact to recognize concerning transmittance is that it is not linear with concentration. In other words, a plot of T or %T vs. concentration would not be an acceptable standard curve because it is not a straight line. Rather, it is a logarithmic relationship. Because it is logarithmic, the logarithm of the transmittance can be expected to be linear with concentration. Thus, we define the parameter of "absorbance" as being the negative logarithm of the transmittance and give it the symbol "A":

$$A = -\log T \qquad (7.13)$$

Absorbance is a parameter then that increases linearly with concentration and is the parameter that is important for quantitative analysis. If the analyst measures transmittance, he/she must convert it to absorbance via Equation 7.13 before plotting the standard curve. Instrument readouts can display either transmittance or absorbance and either absorption spectra or transmission spectra can be displayed. The appropriate signal processor is often activated with a press of a key on the instrument keypad or with a click of a mouse.

Example 7.1

What is the absorbance of a sample if the transmittance is 0.347?

Solution 7.1

$$A = -\log T$$

$$A = -\log (0.347)$$

$$A = 0.460$$

Example 7.2

What is the absorbance of a sample that has a %T of 49.6%?

Solution 7.2

$$\%T = T \times 100$$

$$T = \frac{\%T}{100} = \frac{49.6}{100} = 0.496$$

$$A = -\log T$$

$$A = -\log (0.496) = 0.305$$

The conversion of A to T, or %T, the reverse of the above, may also be important.

Example 7.3

What is the %T if A = 0.774?

Solution 7.3

$$A = -\log T$$

$$0.774 = -\log T$$

$$T = 0.168$$

$$\%T = T \times 100$$

$$\%T = 16.8\%$$

The equation of the straight line A vs. C standard curve is known as the **Beer–Lambert law**, or simply as **Beer's law**. A statement of Beer's law is

$$A = abc \tag{7.14}$$

where A is the absorbance, a is known as the **absorptivity**, or **extinction coefficient**, b is the **path-length**, and c the concentration. Absorbance was defined previously in this section. Pathlength is the distance the light travels through the measured solution. It is the inside diameter of the sample container placed in the light path. Pathlength is measured in units of length, usually centimeters or millimeters. Absorptivity is the inherent ability of a chemical species to absorb light and is constant at a given wavelength, pathlength, and concentration. It is a characteristic of the absorbing species itself. Concentration can be expressed in any concentration unit with which you are familiar. Usually, however, it is expressed in molarity, ppm, or grams per 100 mL. The value and units of absorptivity depends on the units of these other parameters, since absorbance is a dimensionless quantity. When the concentration is in molarity and the pathlength is in centimeters, the units of absorptivity must be liters mol^{-1} cm^{-1}. Under these specific conditions, the absorptivity is called the "molar absorptivity," or the "molar extinction coefficient," and is given a special symbol, the Greek letter epsilon, ε. Beer's law is therefore sometimes given as:

$$A = \varepsilon bc \tag{7.15}$$

Example 7.4

The measured absorbance of a solution with a pathlength of 1.00 cm is 0.544. If the concentration is 1.40×10^{-3} M, what is the molar absorptivity for this analyte?

Solution 7.4

$$A = \varepsilon bc \quad (\text{b in cm, c in molarity})$$

$$\varepsilon = \frac{A}{bc}$$

$$= \frac{0.544}{100 \, cm \times 1.40 \times 10^{-3} \, mol \, L^{-1}}$$

$$= 389 \, L \, mol^{-1} \, cm^{-1}$$

The nature of the sample container varies according to the specific method. In UV/VIS molecular spectrophotometry, it is typically a small test tube or a square tube with an inside diameter (pathlength) of 1 cm. Such a container is called a **cuvette**. Cuvettes with different pathlengths, such as 1 mm, are available. More information on cuvette selection and handling is given in Chapter 8. In IR molecular spectrophotometry, the container is called the **IR liquid sampling cell**. In this cell, the liquid sample is contained in a space between two salt plates created with a thin spacer between the plates and the pathlength is the thickness of the spacer. More information is given in Chapter 8. In atomic absorption spectroscopy, the pathlength is defined by the device used to convert metal ions in solution to atoms in the gas phase. The traditional device used is a burner with a **flame**. In this case, the sample is contained in the flame and the width of the flame where the light beam passes is the pathlength. More information is given in Chapter 9 (see Figure 7.29).

Defining the molar absorptivity parameter presents analytical chemists with a standardized method of comparing one spectrochemical method with another. The larger the molar absorptivity, the more sensitive the method. (It is not unusual for molar absorptivity values to be as large as 10,000 L mol^{-1} cm^{-1} and higher.) In addition, it was stated above that the absorptivity is constant at a given wavelength, implying that it changes as the wavelength changes. The greatest analytical **sensitivity** (the ability to detect and measure small concentrations accurately) occurs at the wavelength at which the absorptivity is a maximum. This is the same wavelength at which the absorbance is a maximum in the molecular absorption spectrum for that species. This wavelength is often referred to as the **wavelength of maximum absorbance** (refer to Figure 7.30 for an example). As you can see, for blue food coloring, the wavelength of maximum absorbance is approximately 628 nm.

Most quantitative analyses by Beer's law involve preparing a series of standard solutions, measuring the absorbance of each in identical containers (or the same container), and plotting the measured absorbance vs. concentration, thus creating a standard curve as defined in Chapter 6 (Section 6.4.2). The absorbance of an unknown solution is then measured and its concentration determined from the standards data. The standard curve in this case is sometimes called a **Beer's law plot** (see Figure 7.31; see Application Note 7.1 for a real-world example of a quantitative analysis by visible spectrophotometry).

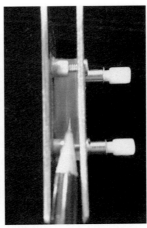

FIGURE 7.29 Examples of sample containers. (Left) "Cuvettes" used in UV/VIS spectrophotometry (pathlength is the inner diameter—approximately 1 cm). (Center) Liquid sampling cell used in IR spectrometry (pathlength is thickness of spacer to the left of the pencil tip—approximately 0.1 mm). (Right) Atomic absorption flame (pathlength is the width of the flame—approximately 4 in.).

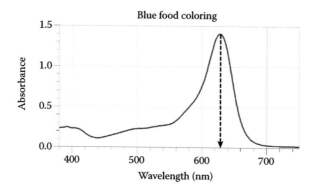

FIGURE 7.30   Wavelength of maximum absorption. In this case, it is approximately 628 nm.

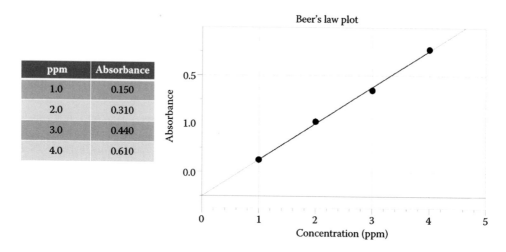

FIGURE 7.31   Example of data and standard curve for a quantitative spectrophotometric analysis. The standard curve in this case is also known as a "Beer's law plot."

APPLICATION NOTE 7.1: ANALYSIS OF GROUND AND SURFACE WATER FOR BROMIDE USING VISIBLE SPECTROPHOTOMETRY

Bromide in ground and surface water can be determined by visible spectrophotometry. This is accomplished by causing the bromide in the water to undergo a chemical reaction in which the benzene rings in the structure of the acid–base indicator phenol red are "bromated." To be brominated means that the bromide, after first being oxidized to bromine, attaches to the benzene ring. This brominated product gives the solution a color with a maximum absorbance at 590 nm. In the procedure, the water sample is mixed with a solution that contains both the oxidizing agent and the phenol red and the color-forming reaction occurs over a 20-min time frame. The standard curve is created by making dilutions of a stock potassium bromide solution so as to have concentrations of 0.20, 0.40, 0.60, 0.80, and 1.00 ppm bromide.

(*Reference: Standard Methods for the Examination of Water and Wastewater*, Method 4500-Br⁻, 19th Edition, APPH, Washington, DC, AWWA, Denver, CO [2000].)

7.7   EFFECT OF CONCENTRATION ON SPECTRA

From the discussion in Section 7.6, it is clear that the absorbance at the wavelength of maximum absorbance increases as the concentration increases and decreases as the concentration decreases. This is true not just at the wavelength of maximum absorbance but at *all* wavelengths. This means that an entire absorption spectrum is affected by concentration changes. It is important to note, however, that the *pattern* of the absorption—the characteristic of the spectrum that makes it useful for *qualitative* analysis—does *not* change. The complete absorption spectra for a given absorbing species for a series of different concentration levels shows that the absorption pattern is the same for all concentrations but the level of the absorption is different (see Figure 7.32).

EXPERIMENTS

EXPERIMENT 14: PERCENTAGE OF SUGAR IN SOFT DRINKS BY REFRACTOMETRY

1. Prepare solutions of sucrose in water that are 4%, 8%, 12%, and 16% sucrose (all w/v percent) by diluting the available 20%. Use 25-mL volumetric flasks.
2. Using a refractometer, read and record the refractive indexes of these solutions, the soft drink unknowns, and a control sample if one is provided.
3. Plot the standard curve using the spreadsheet procedure used in Experiment 13 and obtain the correlation coefficient and the concentrations of the unknowns and control.

EXPERIMENT 15: COLORIMETRIC ANALYSIS OF PREPARED AND/OR REAL WATER SAMPLES FOR IRON

Remember to wear safety glasses.

1. Prepare 100 mL of a 100-ppm Fe stock solution from a 1000-ppm stock solution that is available commercially. Use a clean 100-mL volumetric flask and a clean volumetric pipette. Shake well.
2. Prepare calibration standards of 1.0, 2.0, 3.0, and 4.0 ppm iron from the 100-ppm stock in 50-mL volumetric flasks, but do not dilute to the mark until Step 6. Also, prepare a flask for the blank (no iron). If real water samples are being analyzed and you expect the iron content to be low, prepare two additional standards that are 0.1 and 0.5 ppm.
3. Prepare the real water samples by pipetting 25.00 mL of each into 50-mL volumetric flasks. As with the standards, do not dilute to the mark until Step 6. If an instructor-prepared unknown is provided, do not dilute it to the mark until as directed in Step 6 for all flasks. Your instructor may also provide you with a control sample. Prepare the control as if it were a real water sample by pipetting 25.00 mL of it into a 50-mL flask.

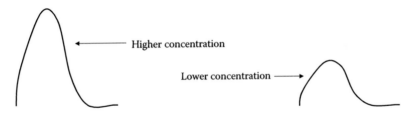

FIGURE 7.32   Effect of concentration on an absorption spectrum. It is the same pattern of absorption at all concentrations (the "fingerprint"), but the level of absorption is different.

4. To each of the flasks (standards, samples, blank, and control), add
   (a) 0.5 mL of 10% hydroxylamine hydrochloride
   (b) 2.5 mL of 0.1% o-phenanthroline solution
   (c) 4.0 mL of 10% sodium acetate

   These reagents are required for proper color development. Hydroxylamine hydrochloride is a reducing agent, which is required to keep the iron in the +2 state. The o-phenanthroline is a ligand that reacts with the Fe^{2+} to form the orange-colored complex ion. This ion is the "absorbing species." In addition, since the reaction is pH dependent, sodium acetate is needed for buffering at the optimum pH.

5. Dilute each to the mark with distilled water and shake well.

6. Using one of your standards and the Spectronic 20, or other single-beam visible spectrophotometer, obtain an absorption (or transmittance) spectrum of the Fe/o-phenanthroline complex ion (instructor will demonstrate use of instrument). Determine the wavelength of maximum absorbance from the spectrum and use this wavelength to obtain absorbance readings of all the solutions. (Use the blank for 100% T setting.)

7. Plot the standard curve using the spreadsheet procedure used in Experiment 13 and obtain the correlation coefficient and the concentrations of the unknowns and control.

8. If real water samples and a control were analyzed, remember that these samples were diluted by a factor of 2. Multiply the concentrations found in Step 7 by 2 to get the final answers in that case.

9. Maintain the logbook for the instrument used in this experiment. Record the date, your name(s), the experiment name or number, the R^2 value, and the results for the control sample. Also, if directed by your instructor, plot the control sample results on a control chart for this experiment posted in the laboratory.

EXPERIMENT 16: DESIGN AN EXPERIMENT: A STUDY OF THE EFFECT OF pH ON THE ANALYSIS OF WATER SAMPLES FOR IRON

In Experiment 15, the pH was controlled by adding 4 mL of 10% sodium acetate to each of the solutions. Presumably, this created the optimum pH for the absorbance measurements. Design an experiment in which the pH is varied broadly, while the other variables are identical to those in Experiment 15. The end goal is a plot of absorbance vs. pH to see if indeed the pH of the solutions as prepared in Experiment 15 is the optimum. Write a step-by-step procedure as in other experiments in this book for preparing the needed solutions and using the spectrophotometer to measure the absorbances of each. After your instructor approves your procedure, perform the experiment in the laboratory.

EXPERIMENT 17: DESIGN AN EXPERIMENT: DETERMINING THE CONCENTRATION AT WHICH A BEER'S LAW PLOT BECOMES NONLINEAR

You should have noticed that the Beer's law plot constructed in Experiment 15 is linear between concentrations of 1 ppm Fe and 4 ppm Fe. At some point beyond 4 ppm, there will be a deviation from Beer's law. Design an experiment that will precisely determine the concentration at which this occurs. Write a step-by-step procedure as in other experiments in this book for preparing the needed solutions and using the spectrophotometer to measure the absorbances of each. After your instructor approves your procedure, perform the experiment in the laboratory.

EXPERIMENT 18: THE DETERMINATION OF PHOSPHORUS IN ENVIRONMENTAL WATER

Note: In this experiment, the phosphorus in water samples is determined by visible spectrometry following a reaction of the phosphates in the sample with potassium antimonyl tartrate, ammonium molybdate, and ascorbic acid. All glassware should be washed in phosphate-free detergent prior to use to avoid phosphate contamination. Remember to wear safety glasses.

PREPARATION OF REAGENT SOLUTIONS

1. Potassium antimonyl tartrate solution. In a 250-mL Erlenmeyer flask, dissolve 0.45 g of potassium antimonyl tartrate in about 50-mL of distilled water. Swirl gently until dissolved, and then dilute to 100 mL and swirl to make homogeneous. Prepare just prior to the run.
2. Ammonium molybdate solution. Dissolve 6 g of ammonium molybdate in distilled water in a 250-mL Erlenmeyer flask and dilute to 100 mL with distilled water and swirl to make homogeneous. Prepare just prior to the run.
3. Ascorbic acid solution. Weigh 2.7 g of ascorbic acid into a 250-mL Erlenmeyer flask, dilute to 100 mL with distilled water and swirl to make homogeneous. Prepare just prior to the run.
4. Dilute sulfuric acid. Prepare 100 mL of 2.52 M H_2SO_4 from concentrated sulfuric acid.
5. Combined reagent. To prepare 100 mL of the color-producing reagent, mix, in the following order, the following portions, swirling gently after each addition: 50 mL of the dilute sulfuric acid, 5 mL of the potassium antimonyl tartrate solution, 15 mL of the ammonium molybdate solution, and 30 mL of the ascorbic acid solution. Use a 125-mL Erlenmeyer flask. If this solution turns blue after several minutes, phosphate contamination is indicated and all solutions used will need to be reprepared. This combined reagent is stable for a few hours.

PREPARATION OF STANDARD SOLUTIONS

6. Your instructor has prepared a 50-ppm phosphorus stock solution by dissolving 0.228 g of dried KH_2PO_4 in 1000 mL of solution.
7. Prepare 4 calibration standards that are 0.1, 0.5, 1.0, and 1.5 ppm P from the stock standard prepared in Step 6. Use 100-mL volumetric flasks.

ABSORBANCE MEASUREMENTS AND RESULTS

8. For each of calibration standard, unknown, control sample and distilled water for the blank, pipette 50.00 mL into a clean, dry 125-mL Erlenmeyer flask. Add 8.0 mL of the combined reagent and mix thoroughly. After at least 10 min, but no more than 30 min, measure the absorbance of each at 880 nm.
9. Create the standard curve, using the procedure practiced in Experiment 13, by plotting absorbance vs. concentration and obtain the concentration of the unknowns and control.
10. Maintain the logbook for the instrument used in this experiment. Record the date, your name(s), the experiment name or number, the R^2 value, and the results for the control sample. Also plot the control sample results on a control chart for this experiment posted in the laboratory.

QUESTIONS AND PROBLEMS

Introduction

1. Define *spectrochemical methods*, *spectroscopy*, *spectrometry*, *spectrometer*, and *spectrophotometer*.
2. Define: refractometry and refractometer. I will also add the answer to Appendix 4.

Characterizing Light

3. What is meant by the "dual nature" of light?
4. Light waves are "electromagnetic" and not "mechanical." What does this have to do with the fact that light can travel through outer space?
5. Define *wavelength*, *frequency*, *speed of light*, *energy of light*, and *wavenumber*.

6. Which has a greater frequency, light of wavelength 627 Å or light of wavelength 462 nm?
7. Which has the longer wavelength, light with a frequency of 7.84×10^{13} s^{-1} or light with a frequency of 5.13×10^{13} s^{-1}?
8. Which has the greater energy, light of wavelength 591 nm or light with a wavelength of 238 nm?
9. Which has the greater wavenumber, light with an energy of 7.34×10^{-13} J or light with an energy of 5.23×10^{-14} J?
10. Which has the lower energy, light with a wavenumber of 1.7×10^{3} cm^{-1} or light with a wavenumber of 1.91×10^{4} cm^{-1}?
11. A certain light, A, has a greater frequency than a second light, B.
    (a) Which light has the greater energy, A or B?
    (b) Which light has the shorter wavelength, A or B?
    (c) Which light has the higher wavenumber, A or B?
12. If the wavelength used in an instrument is changed from 460 to 560 nm,
    (a) Has the energy been increased or decreased?
    (b) Has the frequency been increased or decreased?
    (c) Has the wavenumber been increased or decreased?

The Electromagnetic Spectrum

13. Compare IR light with UV light in terms of wavelength, frequency, and wavenumber.
14. List the following in order of increasing energy, frequency, and wavelength: X-rays, infrared light, visible light, radio waves, ultraviolet light.
15. What are the upper and lower wavelength limits of visible light?
16. Compare infrared and ultraviolet radiation,
    (a) In terms of energy?
    (b) In terms of the type of disturbance they cause within molecules?

Refractometry

17. Give two mathematical definitions of refractive index.
18. Does a refractometer measure the speed of light? Explain.
19. Explain how the measurement of the angle of incidence and the angle of refraction is sufficient to calculate the refractive index of a material.
20. Is refractive index useful for qualitative analysis, quantitative analysis, or both? Explain.
21. What is a diode? What is the design of the diode array sensor depicted in Figure 7.11, and what is its function in a refractometer?

Absorption and Emission of Light

22. Why does a yellow sweatshirt appear yellow and not some other color?
23. What is an energy level diagram? What is meant by "ground state?" What is meant by "excited state?"
24. Explain briefly the phenomenon of light absorption in terms of the energy associated with light and in terms of electrons and the energy levels in atoms and molecules.
25. What is meant by each of the following: electronic transition, vibrational transition, rotational transition? Which of these transitions requires the most energy? Which requires the least energy?
26. Describe the basic instrument for measuring light absorption.
27. Distinguish among UV/VIS spectrophotometry, IR spectrometry, and atomic spectroscopy.
28. What is an "absorption spectrum?" What is the difference between a molecular absorption spectrum and an atomic absorption spectrum, and why does this difference exist?

29. What is a line spectrum? Why is an atomic absorption spectrum a "line" spectrum?
30. Why is it that with atoms, only certain specific wavelengths get absorbed (resulting in a line spectra), whereas with molecules, broad bands of wavelengths get absorbed (resulting in continuous spectra)?
31. Why is a molecular absorption spectrum called a molecular "fingerprint?"
32. What is a "transmission spectrum?" What is an "emission spectrum?"
33. Differentiate between an energy level diagram used to depict atomic absorption from one used to depict atomic emission.

Absorbance, Transmittance, and Beer's Law

34. What is the mathematical definition of transmittance, T? Define the parameters that are found in this mathematical definition.
35. What absorbance corresponds to a transmittance of
    (a) 0.821?
    (b) 0.492?
    (c) 0.244?
36. What absorbance corresponds to a percent transmittance of
    (a) 46.7% T?
    (b) 28.9% T?
    (c) 68.2% T?
37. What transmittance corresponds to an absorbance of
    (a) 0.622?
    (b) 0.333?
    (c) 0.502?
38. What is the percent transmittance, given that the absorbance is
    (a) 0.391?
    (b) 0.883?
    (c) 0.490?
39. What is Beer's law? With one word each, tell what each of the parameters is.
40. What is the absorbance given that the absorptivity is 2.30×10^4 L mol^{-1} cm^{-1}, the pathlength is 1.00 cm, and the concentration is 0.0000453 M?
41. A sample in a 1-cm cuvette gives an absorbance reading of 0.558. If the absorptivity for this sample is 15,000 L mol^{-1} cm^{-1}, what is the molar concentration?
42. If the transmittance for a sample having all the same characteristics as in question 49 is measured as 72.6%, what is the concentration?
43. The transmittance of a solution, measured at 590 nm in a 1.5-cm cuvette was 76.2%.
    (a) What is the corresponding absorbance?
    (b) If the concentration is 0.0802 M, what is the absorptivity of this species at this wavelength?
    (c) If the absorptivity is 10,000 L mol^{-1} cm^{-1}, what is the concentration?
44. Calculate the transmittance of a solution in a 1.00-cm cuvette given that the absorbance is 0.398. What additional information, if any, would you need to calculate the molar absorptivity of this analyte?
45. What is the molar absorptivity given that the absorbance is 0.619, the pathlength is 1.0 cm, and the concentration is 4.23×10^{-6} M?
46. Calculate the concentration of an analyte in a solution given that the measured absorbance is 0.592, the pathlength is 1.00 cm, and the absorptivity is 3.22×10^4 L mol^{-1} cm^{-1}.
47. What is the concentration of an analyte given that the %T is 70.3%, the pathlength is 1.0 cm, and the molar absorptivity is 8382 L mol^{-1} cm^{-1}?
48. What is the pathlength in cm when the molar absorptivity for a given absorbing species is 1.32×10^3 L mol^{-1} cm^{-1}, the concentration is 9.23×10^{-4} M, and the absorbance is 0.493?

49. What is the pathlength in cm when the transmittance is 0.692, the molar absorptivity is 7.39×10^4 L mol^{-1} cm^{-1}, and the concentration is 9.23×10^{-5} M?

50. What is the transmittance when the molar absorptivity for a given absorbing species is 2.81×10^2 L mol^{-1} cm^{-1}, the pathlength is 1.00 cm, and the concentration is 1.87×10^{-4} M?

51. What is the molar absorptivity when the percent transmittance is 56.2%, the pathlength is 2.00 cm, and the concentration is 7.48×10^{-5} M?

52. In each of the following, enough data are given to calculate the indicated parameter(s). Show your work.
    (a) Calculate "A" given that "T" is 0.551.
    (b) Calculate the molar absorptivity given that "A" is 0.294, "b" is 1.00 cm, and "c" is 0.0000351 M.
    (c) Calculate the transmittance given that the absorbance is 0.297.
    (d) Calculate T given that %T is 42.8%.
    (e) Calculate the percent transmittance, given that the absorptivity is 12562 L mol^{-1} cm^{-1}, the pathlength is 1.00 cm, and the concentration is 3.55×10^{-6} M.

53. In each of the following, enough data are given to calculate the indicated parameter(s).
    (a) Calculate "A" if "T" is 0.651.
    (b) Calculate "a" if "A" is 0.234, "b" is 1.00 cm, and "c" is 0.0000391 M.
    (c) Calculate T if A is 0.197.
    (d) Calculate T if %T is 62.8%.
    (e) Calculate %T if a is 13562 L mol^{-1} cm^{-1}, b is 1.00 cm, and c is 3.55×10^{-6} M.

54. A standard 5-ppm iron sample gave a transmittance reading of 52.8%. What is the concentration of an unknown iron sample if its transmittance is 61.7%?

55. A series of five standard copper solutions is prepared, and the absorbances measured as indicated below. Plot the data and determine the concentration of the unknown.

A	C (ppm)
0.104	1
0.198	2
0.310	3
0.402	4
0.500	5
0.334	Unknown

56. Match the items in the left-hand column to items in the right-hand column by writing the appropriate letter in the blank. Each is used only once.

    ___T   (a) The intensity of light after having passed through a solution of an absorbing species.

    ___A   (b) A = abc

    ___I   (c) molar absorptivity

    ___I$_o$ (d) pathlength

    ___a   (e) I/I$_o$

    ___b   (f) –log T

    ___Beer's law (g) The intensity of light after having passed through a blank solution.

    ___% T   (h) absorptivity

    ___ε (i) T × 100

57. Compare the common sample containers for UV/VIS work, IR work, and atomic absorption work.

58. State the importance of the wavelength of maximum absorbance.

Effect of Concentration on Spectra

59. We have described UV/VIS spectra as fingerprints, meaning that a given spectrum is a unique pattern—unique to a given compound. Does changing the concentration of a compound change this pattern? Explain.

Report

60. For this exercise, your instructor will select a "real-world" visible spectrophotometric analysis method, such as an application note from an instrument vendor or from a methods book, a journal article, or Web site, and will give it to you as a handout. Write a report giving the details of the method according to the following scheme. On your paper, write (or type) "a," "b," etc., to clearly present your response to each item.
    (a) Method Title: Give a more descriptive title than what is given on the handout.
    (b) Type of Material Examined: Is it water, soil, food, a pharmaceutical preparation, or what?
    (c) Analyte: Give both the name and the formula of the analyte.
    (d) Sampling Procedure: This refers to how to obtain a sample of the material. If there is no specific information given as to obtaining a sample, make something up. If you want to be correct in what you write, do a library or other search to discover a reasonable response. Be brief.
    (e) Sample Preparation Procedure: This refers to the steps taken to prepare the sample for the analysis. It does not refer to the preparation of standards or any associated solutions.
    (f) Standard Curve: Is a standard curve suggested? If so, what is plotted on the y-axis and what is plotted (give units) on the x-axis?
    (g) What Color-Development Reagent Is Used (if any)? This refers to what is added to the samples and standards in order to develop the required color.
    (h) Reactions Involved to Obtain the Color: If a chemical equation is given, write it down here. If no chemical equation is given, state in words what the reaction is. If this information is not given, write "not given."
    (i) What Wavelength Is Used?
    (j) How Are the Data Gathered? The "data" are the absorbance measurements. In other words, you have got your blank, your standards, and your samples ready to go. Now what? Be brief.
    (k) Concentration Levels for Standards. This refers to what the concentration range is for the standard curve. Be sure to include the units. If no standard curve is used, state this.
    (l) How Are the Standards Prepared? Are you diluting a stock standard? If so, how is the stock prepared and what is its concentration? If not, how are the standards prepared?
    (m) Potential Problems: Is there anything mentioned in the method document that might present a problem? Be brief.
    (n) Data Handling and Reporting: Once you have the data for the standard curve, then what? Be sure to present any post-run calculations that might be required.
    (o) References: Did you look at any other references sources to help answer any of the above? If so, write them here.

8 UV-Vis and IR Molecular Spectrometry

8.1 REVIEW

Spectrochemical methods of analysis were introduced in Chapter 7. UV-Vis molecular spectrochemical methods utilize light in the ultraviolet and visible regions of the electromagnetic spectrum to analyze laboratory samples for molecular compounds and complex ions. Qualitative analysis (identification of unknowns and detection of impurities in knowns) is accomplished by comparing absorption or transmission spectra (molecular fingerprints) with known spectra. Quantitative analysis is accomplished with the use of Beer's law (Section 7.6). The fundamental instrument design is illustrated in Figure 7.21, and the sample containers are the "cuvettes" pictured in the left photograph in Figure 7.29.

UV-Vis spectra result from electronic transitions occurring in the analyte molecules and complex ions. Since vibrational levels are superimposed on the electronic levels, many electronic transitions are possible (refer to Figure 7.24 and accompanying discussion), resulting in continuous spectra rather than the line spectra that typify spectra of atoms.

8.2 UV-VIS INSTRUMENTATION

The fundamental instrument design illustrated in Figure 7.21 is meant only as an illustration of the basic concept of the instrument design. As we will see in the sections to follow, the path from light source, through the wavelength selector and sample to the detector can involve significant other optical components in addition to the instrument components illustrated in Figure 7.21. We now proceed to describe the instruments in more detail.

8.2.1 SOURCES

First we discuss the possibilities for the source of the light used. Special light sources are used in order to provide an optimum quality light beam for the region of the spectrum utilized. An ideal light source is one that emits an intense continuous spectrum of light across an entire region of the spectrum, such as the visible region, while also exhibiting a long life. A light source used frequently for visible light absorption studies is the tungsten filament source. If an instrument is meant strictly for visible light studies, then this lamp is the only one present in the instrument. Such an instrument can be referred to as a **colorimeter**.

A light source used frequently for ultraviolet absorption studies is the deuterium lamp. If an instrument is meant strictly for ultraviolet work, then the deuterium lamp is the only light source present and the instrument is called a **UV spectrophotometer**. It is possible that both a tungsten filament lamp and a deuterium lamp are present and are individually selectable. Also, instead of having two independently selectable sources, a light source that can be used for both ultraviolet and visible studies, the xenon arc lamp, may be present. In these latter two cases, the instrument is called a **UV-Vis spectrophotometer**.

8.2.1.1 Tungsten Filament Lamp

For the visible region, many instruments utilize a light bulb with a tungsten filament. Such a source is very bright and emits light over the entire visible region and into the near infrared region. The

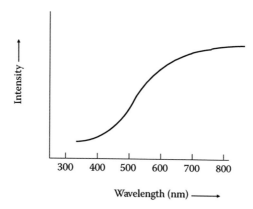

FIGURE 8.1   Graph showing approximately how the intensity of light from a tungsten filament source varies with wavelength.

intensity of the light varies dramatically across this wavelength range, however (Figure 8.1). This creates a bit of a problem for the analyst, and we will discuss this later.

8.2.1.2   Deuterium Lamp

"Deuterium" is the name given to the isotope of hydrogen that has one neutron in the atomic nucleus (as opposed to zero neutrons for "hydrogen" and two neutrons for "tritium," the other hydrogen isotopes). The lamp contains deuterium at a low pressure. Electricity applied to electrodes in the lamp results in a continuous UV emission due to the presence of the deuterium. Its wavelength output ranges from 185 to about 375 nm, satisfactory for most UV analyses. Here again the intensity varies with wavelength.

8.2.1.3   Xenon Arc Lamp

This lamp contains xenon at a fairly high pressure and the light is formed via a discharge across a pair of electrodes. The lamp is often called a xenon "arc" lamp. A continuous ultraviolet and visible emission is emitted due to the presence of the xenon. In some instruments, the electronic circuitry creates regular pulses of light that are very intense and therefore more useful. This also results in a longer life for the lamp. Again, the intensity varies with wavelength.

8.2.2   Wavelength Selection

In order to plot the absorption spectrum of a compound or complex ion, we must be able to carefully control the wavelengths from the broad spectrum of wavelengths emitted by the source so that we can measure the absorbance at each wavelength. Additionally, in order to perform quantitative analysis by Beer's law, we need to be able to carefully select the wavelength of maximum absorption also from this broad spectrum of wavelengths in order to plot the proper absorbance at each concentration. These facts dictate that we must be able to "filter out" the unwanted wavelengths and allow only the wavelength of interest to pass.

As a point of clarification, however, we must recognize that, in reality, there is no such thing as a "single wavelength." In the visible region, the spectrum of colors is continuous, meaning that there is no sharp delineation between green light and blue light, for example, or where one wavelength ends and the adjacent wavelength begins. The electromagnetic spectrum is a continuous wavelength "band." Thus, wavelength selection in a spectrochemical instrument actually consists of the selection of a narrow wavelength band from the larger band. The width of the band that is allowed to pass is called the **bandwidth**. The narrowness of the band that is allowed to pass varies from one design to another and is called the **resolution**. Thus, a high resolution (narrow bandwidth) is the ideal, although many applications do not require the best available resolution and money is often saved by purchasing instruments with a low resolution (wide bandwidth).

8.2.2.1 Absorption Filters

The most inexpensive way to isolate a wavelength band is with the use of absorption filters. For visible light, such a filter would consist of colored glass, the color of the glass indicating what region of the visible spectrum is passed. Thus, if the wavelength called for in a given method is in the red region of the visible spectrum, a red colored glass filter is chosen (see Figure 8.2). A photograph of an absorption filter is shown in Figure 8.3 (see also Application Note 8.1).

8.2.2.2 Monochromators

The word "monochromator" is derived from the Latin language, "mono," meaning "one," and "chromo," meaning "color." It is a device more sophisticated than an absorption filter that isolates the narrow band of wavelengths from visible and ultraviolet sources (refer to Figure 8.4a as you read the following description).

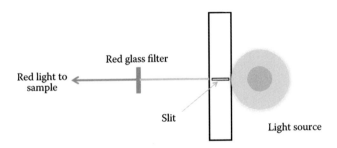

FIGURE 8.2 Illustration of how a red glass filter isolates the red region of the visible spectrum.

FIGURE 8.3 Photograph of a colored glass filter about to be inserted into a spectrophotometer. The color in this case is yellow.

APPLICATION NOTE 8.1: FLOW INJECTION ANALYSIS (FIA)

If you have performed the iron and/or the phosphate analyses in Chapter 7, you have undoubtedly noticed that sometimes, in order to cause an analyte to display a color so that it can be measured by visible spectrometry, several reagents need to be added to the analyte solution. An instrument has been developed that automates this process. It is called flow injection analysis and is abbreviated FIA. This instrument uses small diameter tubing to carry and mix reagents and samples in a continuous

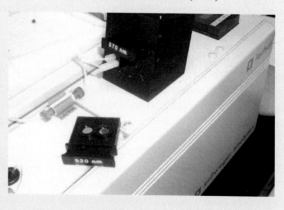

flow system. Following this mixing, the test samples are channeled to a "flow cell" where a fiber optic light source and light sensor are used to measure the absorbance and concentration of the analytes as the solution flows through. The flow cell and fiber optic and sensor electronics are housed in a detector module with inlet and outlet pathways for the flowing solution. In addition, glass absorption filters are used to isolate the region of the spectrum required and are interchangeable, meaning a different filter is used for each different analyte determined. Note the filters shown in the photograph on the right. Also notice the value of the wavelength inscribed on the edge of each filter.

A monochromator is made up of three parts: an entrance slit, a dispersing element, and an exit slit. In addition to these, there is often a network of mirrors situated for the purpose of aligning or collimating a beam of light before and after it contacts the dispersing element. A **slit** is a small circular or rectangular hole cut in an otherwise opaque plate, such as a black metal plate. The size of the opening is often variable—a "variable slit width." The entrance slit is where light enters the monochromator from the source. Its purpose is to create a unidirectional beam of light of appropriate intensity from the multidirectional light emanating from the source. Its slit width is usually variable so that the intensity of the beam can be adjusted—the wider the opening, the more intense the beam.

After passing through the entrance slit, the beam encounters a **dispersing element**. The dispersing element disperses the light into its component wavelengths. For visible light, for example, this would mean that a beam of white light is dispersed into a spray of rainbow colors, the violet/blue wavelengths on one end to the red wavelengths on the other, with the green and yellow in between, like a rainbow. The narrow band of the spectrum is then selected by the **exit slit** (see Figure 8.4a). As the dispersing element is rotated, the spray of colors moves across the exit slit such that a different narrow wavelength range emerges from the exit slit at each position of rotation (see Figure 8.4b and compare with Figure 8.4a). The exit slit width can be variable too, but making it wider would result in a wider wavelength band (a wider **bandwidth**) passing through, which can be undesirable (see Figure 8.5 for illustrations of a narrow band being selected from a wide band). In any case, the light emerging from this exit slit is therefore "monochromatic" and is passed on to the sample compartment to pass through the sample. The concept is the same for UV light. The rotation of the dispersing element is accomplished by either manually turning a knob on the face of the instrument or internally by programmed scanning controls. The position of the knob is coordinated with the wavelength emerging from the exit slit such that this wavelength is read from a scale of wavelengths

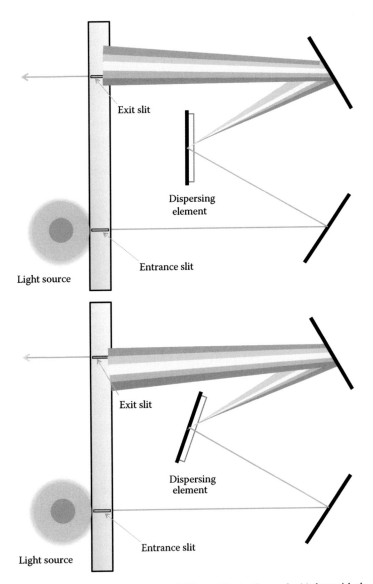

FIGURE 8.4 (a) Illustration of a monochromator. (b) Same illustration as in (a), but with the dispersing element rotated such that the wavelength emerging from the exit slit is different than in (a).

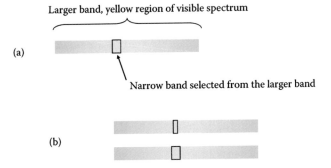

FIGURE 8.5 (a) Illustration of a narrow wavelength band being selected from a wider band. (b) Illustration of how the selected band can be narrow or wide depending on the width of the exit slit.

on the face of the instrument, on a readout meter, or on a computer screen. For manual control, the operator thus simply dials in the desired wavelength. In more sophisticated instruments, the wavelength information is set with software on a computer screen. The bandwidth for slit/dispersing element/slit monochromators varies depending on the quality of the dispersing element and the wavelength range in question. Some instruments improve (narrow) the bandwidth by using filters ahead of the dispersing element.

The dispersing element is either a diffraction grating or a prism. A **prism** is a three-dimensional triangularly shaped glass or quartz block. When the light beam strikes one of the three faces of the prism, the light emerging through another face is dispersed.

A diffraction grating is used more often than a prism. The dispersing element pictured in Figure 8.4a and 8.4b is a diffraction grating. A **diffraction grating** is like a highly polished mirror that has a large number of precisely parallel lines or grooves scribed onto its surface. Light striking this surface is reflected, diffracted, and dispersed into the component wavelengths as indicated in Figure 8.4a and 8.4b.

8.2.3 SAMPLE COMPARTMENT

Following the wavelength selection by the monochromator, the beam passes on to the sample compartment where the sample solution, held in the cuvette, is positioned in its path. The sample compartment is an enclosure with a "lid" that can be opened and closed in order to insert and remove the cuvette. When the lid is closed, the compartment should be relatively free of **stray light**, although this is not a requirement if a xenon arc lamp is used as the source. This is because of the high intensity of the xenon arc lamp. The cuvette is held snugly in a spring-loaded holder.

Some spectrophotometers are "single-beam" instruments and some are "double-beam" instruments. In a double-beam instrument, the light beam emerging from the monochromator is "split" into two beams at some point between the monochromator and the detector. The double beam design provides certain advantages that we will discuss shortly.

8.2.3.1 Single-Beam Spectrophotometer

In a single-beam spectrophotometer, the monochromatic light beam created by the monochromator passes directly through the sample solution held in the cuvette, then proceeds to the detector (see Figure 8.6). This is the most inexpensive design and is especially useful for routine absorbance

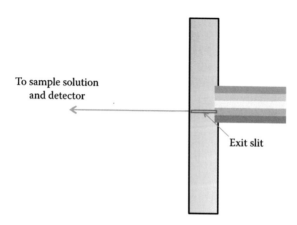

FIGURE 8.6   In a single-beam spectrophotometer, the light emerging from the exit slit passes directly into the sample compartment as a single beam.

measurements for which the wavelength of maximum absorbance (the wavelength to be used) is known in advance without having to scan a particular wavelength range to determine it.

In Chapter 6, we briefly discussed the need for a blank in instrumental analysis (all sample components present except the analyte), saying that it is needed for a precalibration step before calibrating the instrument with standards. For UV-Vis spectrophotometry, this precalibration step would consist of placing the blank in the cuvette in the path of the light. This would produce the most intense light beam possible at the detector and thus a readout of 100%T. If the readout is not 100%T, the electronics is "tweaked" so as to produce this readout, much like a pH meter's electronics is tweaked to produce a readout expected for a given buffer solution. The precalibration may also include a "dark current" control. In this case, with no cuvette in the cuvette holder, a shutter blocks all light from the detector and thus produces a 0%T reading. Having performed these precalibration steps at the wavelength to be used, the instrument is ready for making the measurements on the standards and samples.

If the wavelength of maximum absorbance is not known in advance, then it must be determined by scanning the wavelength range involved. With a single-beam instrument, this is a manual, tedious process because the precalibration step must be repeated each time the wavelength is changed. The reason for this is that the intensity of the light from the source (and therefore the intensity of the light at the detector after having passed through the blank) is not the same at all wavelengths. It is also true that the detector response is not constant at all wavelengths (refer again to Figure 8.1 and accompanying discussion concerning source variations). That being the case, the intensity of light that has passed through the blank will produce a %T readout different from 100% when the instrument is switched to the new wavelength, and so the electronics must again be tweaked to produce the 100%T reading assumed for the blank.

The tedious process (set wavelength, read blank, tweak to read 100%, read sample, change wavelength, read blank, tweak to read 100%, read sample, change wavelength, etc.), can be eliminated by one of two ways. One is to use a single-beam instrument with rapid scanning capability. Rapid scanning occurs by programming the instrument to steadily rotate the grating with a motor while simultaneously measuring the absorption data. In this case, the absorbance-vs.-wavelength information for the blank is quickly acquired when the blank is in the light path and the results stored and used later to adjust the results for the sample via computer for a similar rapid scan with the sample in the cuvette. While single-beam instruments with the rapid scanning capability exist, the double-beam designs to be discussed below are more popular and offer an additional advantage.

For precise work, a second problem exists with a single-beam design, both with manual blank/sample switching and with single-beam rapid scanning. This is that the precalibration with the blank occurs at least 10 s, and perhaps as long as several minutes or more, before the sample is read. This may be a problem because there can be minor intensity fluctuations or drift in the light source or detector (even if the wavelength is not changed) that can undo the precalibration before reading a sample. The double-beam designs discussed below offer an advantage for this problem as well.

8.2.3.2 Beam Splitting and Chopping

"Double-beam" means that the light beam produced by the monochromator is split into two beams. This is usually done in one of two ways. One is to use a slotted mirror, called a **beamsplitter**, set at an angle in the path of the light (see the drawing and photograph in Figure 8.7a and 8.7b). In this way, half the light intensity passes through the slots and half is reflected to another path (a second beam), creating a "double beam in space." Another is to utilize a rotating partial mirror (often called a light **chopper**). A chopper creates a "strobe" effect, or a "double beam in time" as the chopper rapidly rotates. The two beams rapidly alternate. One beam follows a path through the open area of the chopper at one moment in time and the other reflects to another path

(a)

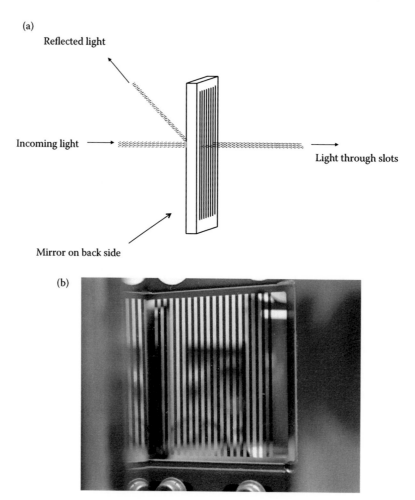

FIGURE 8.7 (a) Drawing of a beamsplitter. Half of the light beam passes through the slots while the other half reflects. (b) Photograph of a beamsplitter as seen inside a spectrophotometer.

at another moment in time (see Figure 8.8a and 8.8b). Photographs of a chopper are shown in Figure 8.9a and 8.9b.

8.2.3.3 Double-Beam Designs

In one double-beam design, the light emerging from the monochromator is chopped into two beams that take parallel but separate paths through the sample compartment. In the sample compartment are two cuvette holders, one to hold the blank and one to simultaneously hold the sample being measured. One of the two beams passes through the blank and is called the "reference beam." The other passes through the sample and is called the "sample beam." The beams are recombined with a second chopper prior to reaching the detector (see Figure 8.10). The detector is programmed to the chopping frequency and can differentiate between the two beams. This design allows the instrument's electronics, or computer software, to self-adjust for the blank reading at each wavelength a split second before taking the sample reading. Thus, both disadvantages of the single-beam systems (the tedium and the time factor) are done away with while also rapidly scanning the wavelength

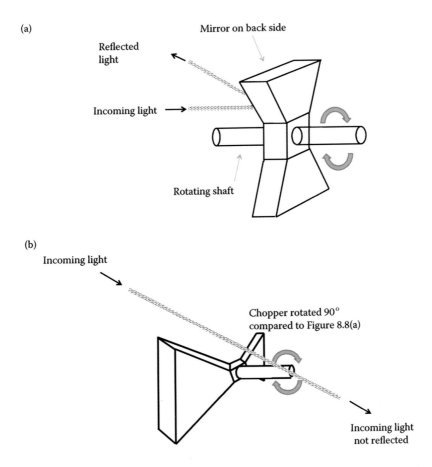

FIGURE 8.8   Drawings of a rotating chopper. (a) At one moment in time, a mirror is in the path of the light causing the beam to reflect. (b) At the next moment in time, an open area of the chopper is in the path of the light enabling the light beam to pass through. The chopper rapidly rotates around the shaft and a strobe effect is created.

range of interest (by motor-driven dispersing element rotation) to obtain the molecular absorption spectrum.

Another double-beam design is one in which the beam emerging from the monochromator is split with a beamsplitter (Figure 8.11). However, the second beam is directed immediately to a second detector rather than through the blank. This design requires only one cuvette holder in the sample compartment. Such a design adjusts for the different intensities at different wavelengths in a manner identical to the single-beam scanning instrument, i.e., by scanning the blank at a separate time via motor driven dispersing element rotation, and then adjusting the sample spectrum with the blank spectrum. Thus, there is no advantage for the problem of different intensities at different wavelengths. However, monitoring the intensity of the second beam in the manner described does assist with the problem of source drift and fluctuations, or with changes that may occur because the operator changed the scan speed, because the instrument can immediately adjust.

8.2.3.4   Diode Array Design

A **diode array** is a series of several hundred photodiode light sensors (mentioned briefly in Chapter 7, Section 7.4, and also in Section 8.2.4.2) arranged in a linear array. Single-beam

(a)

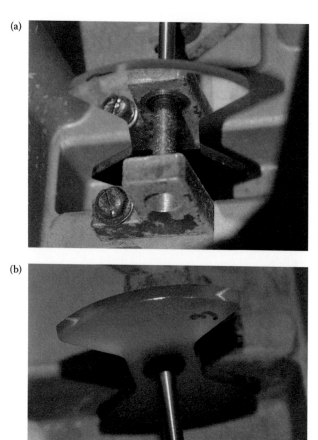

(b)

FIGURE 8.9   (a) Photograph of the mirrored side of a chopper as seen inside a spectrophotometer. (b) Photograph of the back side of a chopper as seen inside a spectrophotometer.

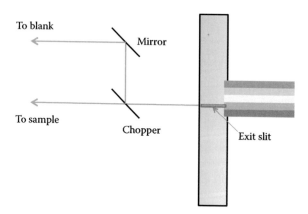

FIGURE 8.10   Illustration of the double-beam design in which both beams pass into the sample compartment with one passing through the sample while the other passes through the blank. In this design, the second beam is created using a chopper.

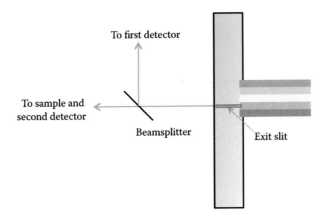

FIGURE 8.11 Illustration of the double-beam design in which one beam passes into the sample compartment while a second beam passes directly to a detector. In this design, the two beams are created using a beamsplitter.

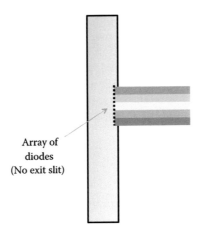

FIGURE 8.12 Illustration of the diode array design. There is no exit slit. Rather, there is an array of diodes lined up where the exit slit would normally be such that all wavelengths are measured simultaneously. The light beam passes through the sample (or blank) before being dispersed.

spectrophotometers have been invented that utilize a diode array as the detector. In this case, the cuvette is positioned between the source and the dispersing element. Then, following the dispersion of the light, there is no exit slit. The spray of wavelengths created by the diffraction grating fall instead across the diode array and the entire spectrum of light is measured at once (see Figure 8.12). The diode array spectrophotometer is a powerful instrument whenever a very rapid measurement of the absorption spectrum is needed, such as in the liquid chromatography instruments discussed in Chapter 12 (see Application Note 8.2). The blank is read at a separate time like other single-beam instruments.

8.2.3.5 Summary
The characteristics of the various UV-Vis instruments described in this section are summarized in Box 8.1.

APPLICATION NOTE 8.2: ANALYSIS OF HEADACHE MEDICINES

Some headache medicines have more than one ingredient. One is the pain reliever (analgesic), such as aspirin, but there can also be a fever reducer (antipyretic), such as acetaminophen, and also a stimulant, such as caffeine. The quality assurance of such products would include an analysis for all these ingredients. The three mentioned here, aspirin, acetaminophen, and caffeine, are all organic compounds that absorb light in the ultraviolet region of the spectrum, but all of them absorb in the UV range 225 to 300 nm and their spectra all overlap. This overlap means that no matter what wavelength is chosen for one, there is going to be an interference from the others. It would seem, using the information presented in this chapter, that an accurate measurement of each ingredient is therefore impossible with UV spectroscopy.

The modern-day solution to this problem is to use a separation procedure ahead of the UV absorbance measurement. The separation method of choice is liquid chromatography, which will be discussed in full in Chapter 12. Briefly, the separation takes place as a solution of the mixture flows through a tube, the chromatography column. The flowing solution then enters a flow-through cuvette where the absorbance measurements are made. A single wavelength may be used, or a diode array instrument may be used, in which case, since the diode array can record the absorption spectrum in a matter of seconds, the entire spectrum can be measured while each of the analytes is in the flow-through cuvette. The data then include everything that is needed for a quantitative analysis free of the interference problem.

(*Reference:* Pedjie, N., *Analysis of Drug Substances in Headache Medicines with the PerkinElmer Flexar FX-15 System Equipped with a PDA Detector*, PerkinElmer, Shelton, CT [2011].)

8.2.4 Detectors

Photomultiplier tubes or photodiodes (light sensors) are used as detectors in UV-Vis spectrophotometers, while thermocouples (heat sensors) are used as detectors for IR spectrometry. This is the reason UV-Vis instruments are called spectrophotometers while IR instruments are called spectrometers.

8.2.4.1 Photomultiplier Tube

A photomultiplier tube is a light sensor combined with a signal amplifier. The light emerging from the sample compartment strikes the photosensitive surface and the resulting electrical signal is amplified.

The photomultiplier tube consists of a "**photocathode**," an anode, and a series of "**dynodes**" for multiplying the signal, hence its name. A dynode is an electrode that, when struck with electrons, emits other secondary electrons. The dynodes are situated between the photocathode and the anode. A high voltage is applied between the photocathode (an electrode that emits electrons when light strikes it) and the anode. When the light beam from the sample compartment strikes the photocathode, electrons are emitted and accelerated, because of the high voltage, to the first dynode where more electrons are emitted. These electrons pass on to the second dynode, where even more electrons are emitted, etc. When the electrons finally reach the anode, the signal has been sufficiently multiplied as to be treated as any ordinary electrical signal able to be amplified by a conventional amplifier. This amplified signal is then sent on to the readout in one form or another. Figure 8.13 illustrates this process.

BOX 8.1   SUMMARY OF THE VARIOUS UV-VIS INSTRUMENTS DESCRIBED IN THIS TEXT

Single-beam, nonscanning
  Slow, tedious process to obtain absorption spectrum
  Manual replacement of sample with blank after each wavelength increment
  Blank checked and instrument zeroed at least 10 s before sample is read
  Source drift/fluctuation within this time frame not compensated
Single-beam, scanning
  Blank scanned at separate time; information in computer memory
  Sample scanned, automatic compensation for blank
  Time between blank scan and sample scan can be considerable
  Source drift/fluctuation within this time frame not compensated
Double-beam, scanning, second beam directly to detector
  Blank scanned at separate time; information in computer memory
  Sample scanned, automatic compensation for blank
  Time between blank scan and sample scan can be considerable, but …
  Source drift/fluctuation immediately compensated
Double-beam, scanning, second beam through blank, then to detector
  No scanning of the blank
  Sample scanned, automatic compensation for blank
  Virtually no time between blank and sample readings
  Source drift/fluctuation immediately compensated
Diode array
  No scanning of either sample or blank
  Blank read at separate time
  Absorbance at all wavelengths read simultaneously via an array of diodes
  Approximately 10 s between blank and sample readings
  Very fast, so very useful for a detector for liquid chromatography instruments

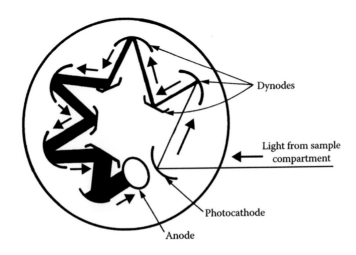

FIGURE 8.13   Photomultiplier tube.

8.2.4.2 Photodiodes

Photodiodes make use of the unique properties of semiconductors, such as silicon. Silicon can be "doped" with impurities to either make it "electron-rich" (an **n-type semiconductor**) or "electron poor" (a **p-type semiconductor**). When an n-type semiconductor is in contact with a p-type semiconductor, electronic changes occur at the boundary, or "junction." A photodiode is a p–n junction constructed with the top "p" layer so thin that it is transparent to light. Light shining though the "p" layer creates additional free electrons in the "n" layer that can diffuse to the "p" layer, thus creating an electrical current that depends on the intensity of the light. This small current is easily amplified and measured.

8.3 CUVETTE SELECTION AND HANDLING

Cuvettes used for UV-Vis spectrophotometry must be transparent to all wavelengths of light for which it is used. If visible light is used, this means that the material must be completely clear and colorless, which means that inexpensive materials, such as colorless plastic and ordinary colorless glass, are perfectly suitable. A case of 500 plastic, 1-cm square cuvettes may cost as little as $50. However, ordinary colorless glass and plastic are not transparent to light in the ultraviolet region. For the ultraviolet spectrophotometry, the cuvettes must be made of quartz, which is more expensive. A matched set of two cuvettes to be used in a double-beam spectrophotometer may cost as much as $400.

If two or more different cuvettes are used in an analysis, one should be sure that they are "matched." Matched cuvettes are identical with respect to pathlength and reflective and refractive properties in the area where the light beam passes. If the pathlengths were different, or if the wall of one cuvette reflects more or less light than another cuvette, then the absorbance measurement could be different for that reason, and not because the solution concentration is different. Thus, there would be an error.

FIGURE 8.14 Photograph of a sample compartment with a cuvette partially inserted. The vertical line on the cuvette is lined up with the raised plastic on the cuvette holder to ensure that the cuvette is inserted at the same rotational position as previous insertions.

A general guideline with respect to matching cuvettes is that a solution that transmits 50% of the incident light should not have a %T reading differing by more than 1% in any cuvette.

Also, for precise work, a cuvette should be placed in the instrument in exactly the same rotational position each time, since the pathlength and reflective/refractive properties can change by rotating the cuvette. Some instruments are constructed to ensure the rotational position is the same each time the cuvette is inserted. One style of instrument, for example, has a raised plastic "line" that is meant to coincide with a line imprinted on the cuvette (see Figure 8.14).

Obviously, cuvettes that have scratches, or cleaning procedures that may cause scratching, should be avoided. Cotton swabs can be used for scrubbing rather than metal-handled brushes. Any liquid or fingerprints adhering to the outside wall must be removed with a soft lint-free cloth or towel.

In addition to scratches, fingerprints, water spots, etc., the analyst should be aware that any foreign particulate matter suspended in the sample or standards would also be a problem. Particulates in the path of the light would reflect and scatter the light, lessening the intensity of the transmitted light, and would result in an error in the reading. Extracts of solids, such as soils, are especially susceptible to this difficulty. The problem is solved by making sure the samples are well filtered. Some spectrometric experiments, however, are designed to actually *measure* such particles. The analysis of samples for suspended particulate matter by spectrophotometry is an analysis for the **turbidity** of those samples.

8.4 INTERFERENCES, DEVIATIONS, MAINTENANCE, AND TROUBLESHOOTING

8.4.1 INTERFERENCES

Interferences are quite common in qualitative and quantitative analysis by UV-Vis spectrophotometry. An interference is a contaminating substance that gives an absorbance signal at the same wavelength or wavelength range selected for the analyte. For qualitative analysis, this would show up as an incorrect absorption spectrum, thus possibly leading to erroneous conclusions if the contaminant was not known to be present. For quantitative analysis, this would result in a higher absorbance than one would measure otherwise. Absorbances are additive. This means that the total absorbance measured at a particular wavelength is the sum of absorbances of all absorbing species present. Thus, if an interference is present, the correct absorbance can be determined by subtracting the absorbance of the interference at the wavelength used, if it is known. The modern solution to these problems is to utilize separation procedures such as liquid–liquid extraction or liquid chromatography to separate the interfering substance from the analyte prior to the spectrophotometric measurement. Liquid–liquid extraction was discussed in Chapter 2 as a sample preparation technique. Liquid chromatography will be discussed in Chapters 10 and 12.

8.4.2 DEVIATIONS

Deviations from Beer's law are in evidence when the Beer's law plot is not linear. This is probably most often observed at the higher concentrations of the analyte, as indicated in Figure 8.15. Such deviations can be either chemical or instrumental.

Instrumental deviations occur because it is not possible for an instrument to be accurate at extremely high or extremely low transmittance values—values that are approaching either 0%T or 100%T. The normal working range is between 15% and 80%, corresponding to absorbance values between 0.10 and 0.82. It is recommended that standards be prepared so as to measure in this range and that unknown samples be diluted if necessary.

Deviations due to chemical interferences occur when a high or low concentration of the analyte causes chemical equilibrium shifts in the solution that directly or indirectly affect its absorbance. It

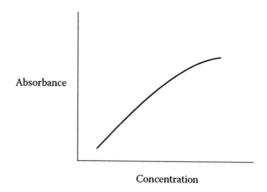

Concentration

FIGURE 8.15 Deviations from Beer's law often manifest themselves by a nonlinear portion of the Beer's law plot at the higher concentrations.

may be necessary in these instances to work in a narrower concentration range than expected. This means that unknown samples may also need to be further diluted as in the instrumental deviation case.

8.4.3 MAINTENANCE

The basic considerations for spectrophotometer maintenance are safety and cleanliness. Maintaining a safe and clean work environment ensures a longer life for the instruments and a lower likelihood for problems relating to contamination, broken or misaligned parts, or injury. Solutions spilled on sensitive electronic circuits can render them inoperative. Any spills should be cleaned up immediately according to local safety protocols.

Instrument operators may also want to conduct periodic wavelength calibrations. It is important to know that the wavelength displayed or "dialed in" really is the wavelength passing through the sample compartment. One way to check for proper wavelength calibration is to prepare a solution of an analyte for which the wavelength of maximum absorbance is well known. If the maximum absorbance does not occur at this wavelength, it may be that the wavelength control is out of calibration. In that case, the operator should contact the manufacturer's service organization.

8.4.4 TROUBLESHOOTING

Troubleshooting of spectrophotometers and procedures involving spectrophotometers can involve checking electrical components, checking the mechanics of a procedure, and checking for contamination. The failure of electrical components can involve something as simple as checking whether the instrument, or its component, is plugged in and switched on, whether a fuse has blown, whether a lamp switch is on, or whether the light source has burned out, etc. But it may also involve much more complicated problems, such as the failure of an entire electronic module or circuit board due to a faulty component, such as an amplifier. In the latter case, the operator may need to contact the manufacturer's service organization.

Problems with the mechanics of a procedure can involve an improperly diluted sample (perhaps manifested by an absorbance reading that is greater than specified or expected), an obstruction in the sample or reference beam, an improperly aligned source or mirror, the incomplete programming of a scan, or improper or inappropriate software entry. In these cases, the operator will need to carefully examine his/her technique or procedure, or instrumental parameters, such as the optical path, perhaps with the help of the instrument troubleshooting guide, to solve the problem.

Contamination, as indicated previously, can manifest itself in the absorption spectrum. If a sample is contaminated with a chemical that has its own absorption spectrum in the range studied, the absorption pattern (spectrum) of the analyte will not match the expected pattern. If a contaminant is suspected, it may originate from any of the chemicals used in the procedure, including the solvent, the analyte used to prepare standards, or any other chemical added to the solutions tested. In that case, repeating the solution preparation using chemicals from fresh, unopened containers may help solve the problem.

8.5 FLUOROMETRY

Fluorometry is an analytical technique that utilizes the ability of some substances to exhibit luminescence. **Luminescence** is a phenomenon in which a substance appears to glow when a light shines on it. In other words, light of a wavelength different from that of the irradiating light is released or "emitted" following the absorption of this light. The casual observer notices luminescence when the irradiating light is ultraviolet light and the emitted light is visible light, such as what can be seen, for example, when a "black" light shines on a poster or other material with fluorescent dyes in it. The phenomenon is explained based on light absorption theory and what can happen to a chemical species in order to revert back to the ground state once the absorption—the elevation to an excited state—has taken place. Luminescence can occur with molecules, complex ions, and atoms. The present discussion will focus on molecules and complex ions. Atomic fluorescence will be mentioned in Chapter 9.

All atoms, molecules, and complex ions seek to exist in their lowest possible energy state at all times. When a molecule is raised to an excited electronic energy state through the absorption of light, it is no longer in its lowest possible energy state and will subsequently lose the energy it gained to return to the lowest (ground) state. Most often, the energy is lost through mechanical means, such as through molecular collisions. However, there can be a direct jump back to the ground state with only intermediate stops at some lower vibrational states. Luminescence caused by such a direct jump is called **fluorescence**. Luminescence resulting from a jump back to the ground state after routing through other electronic states is called **phosphorescence**. Phosphorescence is often referred to as "delayed" fluorescence. In either case, the energy the molecule gained as a result of the absorption process is lost in the form of light. Since it is light of less energy due to the accompanying small energy losses in the form of vibrational loss, the wavelength of emission is longer than the wavelength of absorption and the material appears to glow. A simplified energy level diagram is shown in Figure 8.16. We will be referring to fluorescence and phosphorescence collectively as fluorescence.

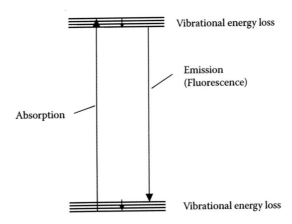

FIGURE 8.16 Energy level diagram depicting the phenomenon of fluorescence in a molecule or complex ion.

Figure 8.16 is a *simplified* energy level diagram because it shows only one wavelength of light being absorbed and only one wavelength being emitted. In reality, *many* wavelengths can be absorbed (giving rise to the molecular absorption spectrum as discussed in Chapter 7) and *many* wavelengths of light can be emitted (giving rise to a so-called **fluorescence spectrum**) due to the presence of many vibrational levels in the each of the electronic levels.

In the laboratory, solutions of analytes that fluoresce are tested by measuring the intensity of the light emitted. The instrument for measuring fluorescence intensity is called a **fluorometer**. Inexpensive instruments used for routine work utilize absorption filters similar to what was described previously for absorption spectrophotometers (see Figures 8.2 and 8.3) and are called **filter fluorometers**. Two such filters are needed—one to isolate the wavelength from the source to be absorbed, the wavelength of maximum absorbance, and the other to isolate the wavelength of maximum fluorescence to be detected. In addition, the instrument components are arranged in a **right-angle configuration** so that the fluorescence intensity measurement is free of interference from transmitted light from the source (see Figure 8.17). Research-grade instruments called **spectrophotofluorometers** utilize the slit/dispersing element/slit monochromators and are considerably more expensive. Such instruments are used for more precise work and can measure the absorption and fluorescence spectra.

When the concentration of a fluorescing analyte is small so as to result in a small absorbance value, the intensity of the resulting fluorescence is proportional to concentration and is therefore measured for quantitative analysis. Thus, the usual procedure for quantitative analysis consists of the measurement of a series of standard solutions of the fluorescing analyte or other species proportional to the analyte. A graph of fluorescence intensity vs. concentration is expected to be linear in the concentration range studied.

The types of compounds that can be analyzed by fluorometry are rather limited. Benzene ring systems, such as the vitamins riboflavin (Figure 8.18) and thiamine, are especially highly fluorescent compounds and are analyzed in foods and pharmaceutical preparations by fluorometry. Metals can be analyzed by fluorometry if they are able to form complex ions by reaction with a ligand having a benzene ring system.

Fluorometry and absorption spectrophotometry are competing techniques in the sense that both are techniques that analyze for molecular species and complex ions. Each offers its own advantages and disadvantages, however. As stated above, the number of chemical species that exhibit fluorescence is very limited. However, for those species that do fluoresce, the fluorescence is generally very

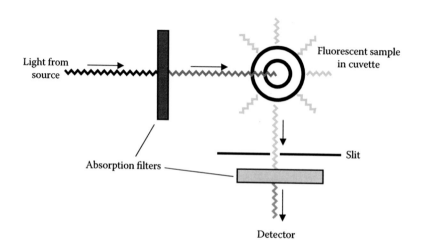

FIGURE 8.17   Illustration of a filter fluorometer.

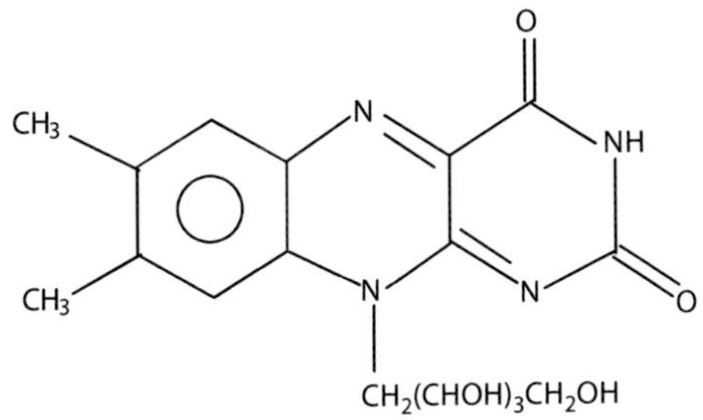

FIGURE 8.18   Structure of riboflavin.

intense. Thus, we can say that while absorption spectrophotometry is much more universally applicable, fluorometry suffers less from interferences and is usually much more sensitive. Therefore, when an analyte does exhibit the somewhat rare quality of fluorescence, fluorometry is likely to be chosen for the analysis. The analysis of foods and pharmaceutical preparations for vitamin content is an example, since vitamins such as riboflavin exhibit fluorescence and fluorometry would be relatively free of interference and would be very sensitive.

8.6   INTRODUCTION TO IR SPECTROMETRY

IR spectrometry differs from UV/Vis spectrophotometry in the following ways:

1. Absorption of IR light results in vibrational energy transitions rather than electronic transitions. The concept of vibrational energy transitions was introduced in Chapter 7 (see Figure 7.20).
2. While liquid solutions are often analyzed in IR work, pure liquids and undissolved solids, including polymer films, are also often analyzed, as are gases.
3. In IR work, the containers that hold liquids or liquid solutions in the path of the light are not called cuvettes. They are called liquid sampling cells.
4. The liquid sampling cells have extremely short pathlengths (often fractions of a millimeter), defined by the thickness of a thin polymer film "spacer" (see the center photograph in Figure 7.29).
5. Liquid sampling cells utilize large polished inorganic salt crystals as "windows" (cell walls) for the IR light. Water, in which these windows are soluble, must be scrupulously avoided, meaning that analyte solvents must be water-free organic liquids that do not dissolve the inorganic salt crystals. Glass and plastic have significant disadvantages and are not usually used.
6. IR spectra are usually transmission spectra rather than absorption spectra and wavenumber, rather than wavelength, is plotted on the x-axis.
7. IR spectra are characterized by rather sharp absorption bands and each such band is characteristic of a particular covalent bond in the sample molecule. An example is shown in Figure 8.19. Thus, while IR spectra are "molecular fingerprints" as are UV/Vis spectra, they have a greater worth to a qualitative analysis scheme due to the specificity of the absorption bands.
8. Modern IR spectrometers do not use light dispersion to acquire spectra. Rather, they utilize a device called an "interferometer" between the source and the sample. This design requires a signal processing circuit that performs a mathematical operation called a "Fourier transformation" to obtain the spectra.

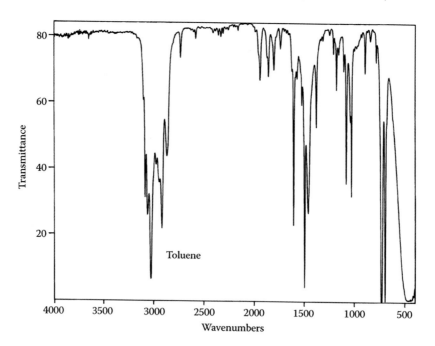

FIGURE 8.19 IR spectrum of toluene. Note that the x-axis is wavenumber. Also note the sharpness of the absorption bands.

8.7 IR INSTRUMENTATION

Let us begin with the instrumentation. Dispersive IR instruments, similar to the double-beam instruments described for UV/Vis spectrophotometry, have been used in the past but have become all but obsolete. While some laboratories may still use these instruments, we will not discuss them here.

The methodology that involves instruments that utilize the interferometer and Fourier transformation mentioned in Section 8.6, has come to be known as **Fourier transform infrared spectrometry**, or **FTIR**.

In the FTIR instrument, the undispersed beam of IR light from the source first enters an interferometer. An **interferometer** is a device that creates a pattern of light resulting from the combined constructive and destructive interference of all component wavelengths. This combined interference is caused by first splitting the beam and then directing the two beams through paths of two different lengths, one of which is fixed and one of which is variable. The fixed length path utilizes a mirror that is fixed in a position. The variable length path utilizes a mirror that is movable. When the two beams are recombined at the same beamsplitter (mirrored on both sides), they may be in phase (when the two paths are of the same length) or out of phase (when the two paths are not of the same length) (see Figure 8.20).

When the two beams are in phase, all wavelengths combine constructively, resulting in a maximum intensity burst of radiant power. However, even a slight change in the variable length path due to a movement of the movable mirror results in a dramatic decrease in the intensity. A complete traversing of the movable mirror through its path gives a pattern to the light at the detector directly attributable to the combined constructive and destructive interference of all the wavelengths in the beam as a function of mirror position. This pattern is called an **interferogram**. The pattern is the same each time, provided there is nothing in the path of the light to absorb any given wavelength. When an absorbing sample is placed in the path of the light in the sample compartment, some wavelengths are absorbed sharply and this creates a different interferogram at the detector. Comparing the two interferograms using the Fourier transformation signal processing results in the absorption spectrum of the sample.

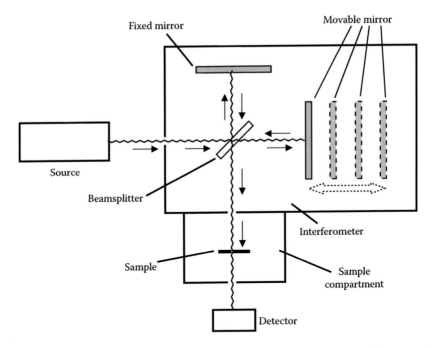

FIGURE 8.20 Illustration of an FTIR instrument showing the light source, the interferometer, the sample compartment, and the detector.

The advantages of the FTIR over the dispersive technique are (1) it is faster, making it possible to be incorporated into chromatography schemes as we will see briefly in Chapters 12 and 13, and (2) the energy reaching the detector is much greater thus increasing the sensitivity.

8.8 SAMPLING

Sampling, in the context of infrared spectrometry, refers to the method by which a sample is held in the path of the IR light. Since liquids (including pure liquids and liquid solutions), solids, and gases can be analyzed by infrared spectrometry, we have **liquid sampling**, **solid sampling**, and **gas sampling**. In any case, it is important to point out that glass and plastic are undesirable materials for the cells. The reason is that glass and plastic are molecular (covalent) materials and would absorb IR light and interfere with reading the sample. Inorganic salts, such as NaCl and KBr, are ionic materials and do not absorb IR light because ionic bonds cannot undergo vibrational energy transitions. Covalent bonds can vibrate, whereas ionic bonds cannot.

8.8.1 LIQUID SAMPLING

The liquid sampling cell previously mentioned (Items 3–5 in Section 8.6 and Figure 7.29) is the primary means for liquid sampling. In other words, for pure liquids (often referred to as **neat liquids**) and liquid solutions, sandwiching a thin layer of liquid between two large NaCl or KBr crystals (the windows) is the classic procedure for mounting the sample in the path of the light. The windows are positioned such that the IR light beam from the interferometer passes directly through the assembly from one window, through the sample, and then through the other window to the detector as shown in Figure 8.21.

Typical dimensions for such windows are about 2 cm wide × 3 cm long × 0.5 cm thick. Positioning or holding the windows (often called salt "plates") in place is done using either a "sealed cell," a "demountable cell," or a combination "sealed demountable" cell. **Sealed cells** are intended to be permanent fixtures for the windows and should not be disassembled during their useful lifetime.

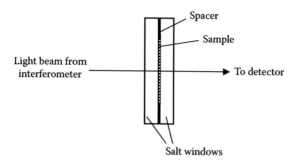

FIGURE 8.21   Illustration of a liquid sampling cell placed in the path of the light.

They have a fixed pathlength and are very useful for quantitative analysis, since the pathlength is reproduced (same cell, same pathlength) from one standard to another and also for the unknown. The fixed pathlength is defined by the thin polymer film spacer mentioned in item 4 in Section 8.6 and also illustrated in Figure 8.21. The spacer leaves a space in the center to be filled by a very small volume of sample as shown in Figure 8.21. Sealed cells have inlet and outlet ports on a metal "frame" for introducing the sample via a Luer-lock syringe through one salt plate to the space between the salt plates. The window on the side of the ports has holes drilled in it to facilitate moving the sample from the syringe, through the inlet port to the space and then out the outlet port (see Figure 8.22).

   Demountable cells can be disassembled and may or may not use a spacer. Without a spacer, the liquid sample is simply applied to one salt plate and the second plate is positioned over the first so as to smear the liquid out between the plates. Such a method cannot be useful for quantitative analysis because the pathlength is undefined and not reproducible. Also, the two salt plates may be positioned in the path of the light without a frame if there is a way to hold them there. If a spacer is used, it is positioned over one of the plates. The sample is applied to this plate in the cut-out space in the spacer reserved for it and the second plate is then positioned over the first with both spacer and sample in between. There are no inlet/outlet ports since the sample is not introduced that way. However, the plates, with sample, are placed in a "frame," or holder, similar in appearance to the sealed cell but without the ports (see Figure 8.23). Such a cell is also undesirable for quantitative analysis because of the difficulty in obtaining an identical pathlength each time the cell is reassembled.

   A **sealed demountable cell** is a combination of the sealed cell and the demountable cell, meaning that it is like a sealed cell but is meant to be disassembled if desired. Disassembly would be for the purpose of changing the spacer and not for introducing the sample, since the sample is introduced through the inlet port with the syringe. A sealed demountable is useful for both qualitative and quantitative

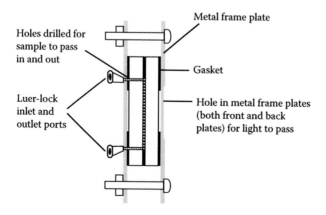

FIGURE 8.22   Illustration of the sealed cell assembly.

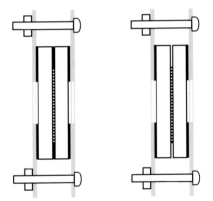

FIGURE 8.23 (Left) A demountable cell with spacer. (Right) A demountable cell without spacer. Note no drilled holes for the sample and no inlet or outlet ports. Both metal frame plates (front and back) have holes for light to pass.

analysis since the pathlength is defined and is reproducible as long as the cell is not disassembled between standards and unknowns. A sealed demountable looks identical to a sealed cell (Figure 8.22).

All three cells utilize neoprene gaskets to cushion the fragile salt crystals from the metal frame. In the case of the sealed and sealed demountable cells, the top neoprene gasket and window have holes drilled in them to coincide with the inlet and outlet ports to facilitate filling the space, created by the spacer, with the liquid sample using the syringe. The path of the liquid sample is shown in the figures. Photographs of a sealed cell, or an assembled sealed demountable cell, are shown in Figure 8.24.

Filling the cells with sample and eliminating the sample when finished can be troublesome. The sample inlet and outlet ports are each tapered to receive a syringe with a Luer (tapered) tip that locks in place. The useful procedure for filling is to raise the outlet end by resting it on a pencil or similar object (to eliminate air bubbles) and then to use a pressure/vacuum system with the use of two syringes, one in the outlet port and one in the inlet port, as shown in Figure 8.25. While pushing on the plunger of the syringe containing the liquid sample in the inlet port and pulling up on the plunger of the empty syringe in the outlet port, the cell can be filled without excessive pressure on the inlet

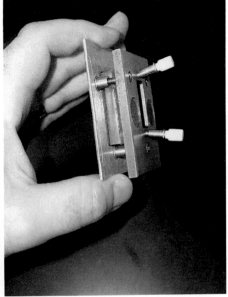

FIGURE 8.24 Photographs of a sealed (or sealed demountable) cell.

side. This reduces the possibility of damaging the cell (by causing a leak in the seal between the salt plates and spacer, for example) due to the excessive pressure that may needed, especially when working with unusually viscous samples and short pathlengths. If the sample can be loaded without excess pressure, then the outlet syringe can be eliminated. Tapered Teflon plugs are used to stopper the ports immediately after filling. The cell may be emptied and readied for the next sample using two empty syringes and the same push–pull method. When refilling, an excess of liquid sample may be used to rinse the cell and eliminate the residue from the previous one. Alternatively, the cell may be rinsed with a dry volatile solvent and the solvent evaporated before introducing the next sample.

The analyst must be careful to protect the salt crystals from water during use and storage. Sodium chloride and potassium bromide are, of course, highly water soluble and the crystals may be severely damaged with even the slightest contact with water. All samples introduced into the cell must be dry. This is important for another reason. Water contamination will be seen on the measured spectrum (as an alcohol; see Section 8.10) and cause erroneous conclusions. If the windows are damaged with traces of water, they will become "fogged" and will appear to become nontransparent. The windows may be repolished if this happens. Depending on the extent of the damage, various degrees of abrasive materials may be used, but the final polishing step must utilize a polishing pad and a very fine abrasive. Polishing kits are available for this. Finger cots should be used to protect the windows from finger moisture.

Liquids can be sampled as either the "neat" liquid (pure) or mixed with a solvent (solution). Neat liquids are tested when the purpose of the experiment is either identification or the determination of purity.

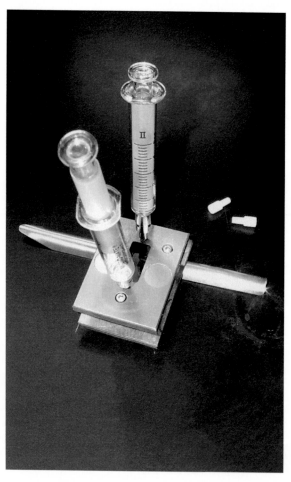

FIGURE 8.25   Filling a liquid sampling cell.

Identification is possible because the spectrum is a fingerprint when no solvent or contaminant is present. Impurities are found when extraneous absorption bands, or distortions in analyte absorption bands, appear.

When a solution is tested, both analyte and solvent absorption bands will be present in the spectrum and identification, if that is the purpose of the experiment, is hindered. Some solvents have rather simple IR spectra and are thus considered more desirable as solvents for qualitative analysis. Examples are carbon tetrachloride (CCl_4; only C–Cl bonds), choloroform ($CHCl_3$), and methylene chloride (CH_2Cl_2). The infrared spectra of carbon tetrachloride and methylene chloride are shown in Figure 8.26. There is a problem with toxicity with these solvents, however. For quantitative analysis, such absorption band

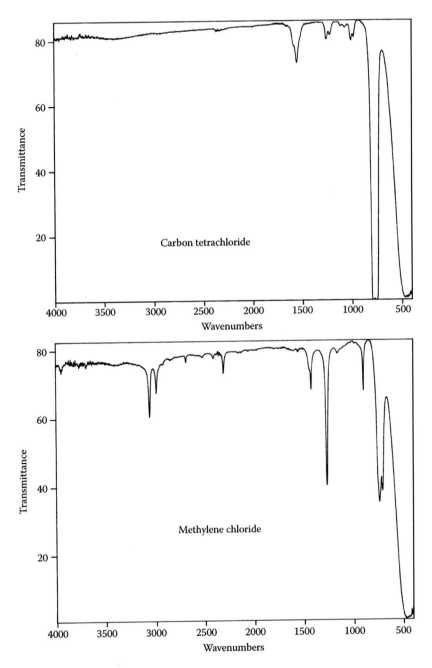

FIGURE 8.26 Infrared spectra of carbon tetrachloride (top) and methylene chloride (bottom).

interference is less of a problem because one only needs to have a single absorption band of the analyte isolated from the other bands. This one band can be the source of the data for the standard curve since the peak absorption increases with increasing concentration (see Section 8.11 and Experiment 25).

8.9 SOLID SAMPLING

Infrared spectrometry is one of the few analytical techniques that can measure both dissolved and undissolved solids. Thus, there are some unique and interesting solid sampling methods.

8.9.1 SOLUTION PREPARED AND PLACED IN A LIQUID SAMPLING CELL

The most straightforward method for analyzing a solid material by infrared spectrometry is to dissolve it in a suitable solvent and then to measure this solution using a liquid sampling cell such as one the several described in Section 8.8. Thus, it becomes a liquid sampling problem and the experimental details of this have already been discussed (Section 8.8). It is the only method of solid sampling suitable for quantitative analysis because it is the only one that has a defined and reproduced pathlength.

8.9.2 THIN FILM FORMED BY SOLVENT EVAPORATION

A simple method for solids is to prepare a solution of it, place several drops of the solution on surface of a single salt plate, and then allow, or force, the solvent to evaporate, leaving a thin film of the solid on the plate. If this plate is fixed in the path of the light, the spectrum of the solid can be measured by passing the light through it as described for the liquid sampling devices (see Figure 8.27). The so-called disposable IR cards are also available for this. Such cards have polyethylene windows on which the sample solution may be applied and the solvent evaporated. Unless they are subtracted out, IR spectra from such cards would show the polyethylene absorption bands.

The solvent evaporation method is especially useful for polymers that can be "cast" onto the surface of the salt plate. Casting involves dissolving the polymer in an appropriate solvent at an elevated temperature and evaporating the solvent on the salt plate by heating the salt plate on a hot plate or under a heat lamp.

8.9.3 KBR PELLET

The KBr pellet technique is based on the fact that *dry*, finely powdered potassium bromide has the property of being able to be squeezed under very high pressure into a transparent disc—transparent to both infrared light and visible light. It is important for the KBr to be dry both in order to obtain a good pellet and to avoid absorption bands due to water in the spectrum.

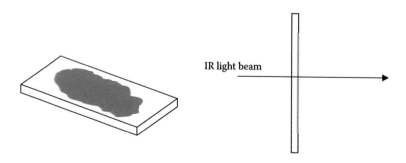

IR light beam

FIGURE 8.27   A solid may be prepared by dissolving it in an organic solvent, placing it on a salt crystal (left) and allowing the solvent to evaporate such that the a "film" of the solid is "cast" on the surface of the salt plate, which can then be placed in the path of the light (right).

A small amount of the dry solid analyte (0.1% to 2.0%) is added to the KBr prior to pressing, then a wafer (pellet) can be formed, from which the spectrum of the solid can be obtained. Such a wafer is simply placed in the path of the light in the instrument such that the IR beam passes from the interferometer, through the wafer to the detector.

Pressing the KBr/sample mixture to make the transparent wafer is done in a special die made for this purpose. The die is placed in a laboratory press also made for this purpose. One rather simple such die and press consists of a threaded body and two bolts with polished faces. One bolt is turned completely into the body of the die. A small amount of the powdered KBr/sample mixture, enough to cover the face of the bolt inside, is added and the other bolt turned down onto the sample, squeezing it into the pellet or wafer (see Figure 8.28). The two bolts are then carefully removed. If the pellet is well formed, it will remain in the die and appear transparent, or nearly so. The die is then placed in the instrument so that the light beam passes directly through the center of the die and through the pellet.

In addition, compression methods that utilize hydraulics and levers may be used. These sometimes have pressure gauges that allow the analyst to apply a certain optimal force in order to maximize the chances of making a quality pellet. To make a quality pellet, in addition to using the optimum pressure, it is important for the KBr and the sample to be dry, finely powdered, and well mixed. An agate mortar and pestle is recommended for the grinding and mixing of the KBr and sample prior to compression (see Figures 8.29 and 8.30).

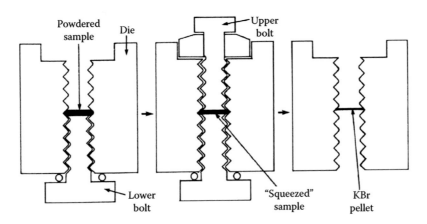

FIGURE 8.28 Drawing of a cross section of one type of pellet die showing the preparation of a KBr pellet.

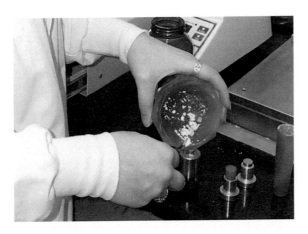

FIGURE 8.29 Photograph showing a powdered KBr/analyte sample, prepared by grinding with a mortar and pestle, being transferred to a KBr pellet die.

FIGURE 8.30 Photograph showing the pellet die from (a) positioned in a laboratory press to squeeze the KBr/sample mixture into a transparent wafer known as the KBr pellet.

8.9.4 Nujol Mull

The Nujol (mineral oil) mull is also often used for solids. In this method, a small amount of the finely divided solid analyte (1- to 2-μm particles) is mixed together with an amount of mineral oil to form a mixture with a toothpaste-like consistency. This mixture is then placed (lightly squeezed) between two NaCl or KBr windows, which are similar to those used in the demountable cell discussed previously for liquids. If the particles of solid are not already the required size when received, they must be finely ground with an agate mortar and pestle and can be ground directly with mineral oil to create the mull to be spread on the window. Otherwise, a small amount (about 10 mg) of the solid is placed on one window along with one small drop of mineral oil. A gentle rubbing of the two windows together with a circular or back-and-forth motion creates the mull and distributes it evenly between the windows. The windows are placed in the demountable cell fixture and placed in the path of the light as described previously.

A problem with this method is the fact that mineral oil is a mixture of covalent substances (high-molecular-weight hydrocarbons) and its characteristic absorption spectrum will be found superimposed in the spectrum of the solid analyte, as with the solvents used for liquid solutions discussed previously. However, the spectrum is a simple one (Figure 8.31) and often does not cause a significant problem, especially if the solid is not a hydrocarbon.

8.9.5 Reflectance Methods

Rather than measure the light *transmitted through* a sample, as in all of the IR sampling methods discussed thus far, some methods measure the light *reflected from* a sample. There are various methods employed to measure this reflectance, depending on the properties of the sample.

8.9.5.1 Specular Reflectance

If the sample can be cast as a thin smooth mirror-like film on a flat surface, the IR spectrum can be measured by measuring an IR light beam made to reflect from this surface (see Figure 8.32). The sample is mounted in air and the light beam travels through air before and after striking the sample.

8.9.5.2 Internal Reflectance

In this method, the sample (such as a polymer film) is pressed against a transparent material having a high refractive index (called the **internal reflection element**, or **IRE**). The IR light beam passes through this material, rather than air, before and after reflecting from the sample, hence the reason

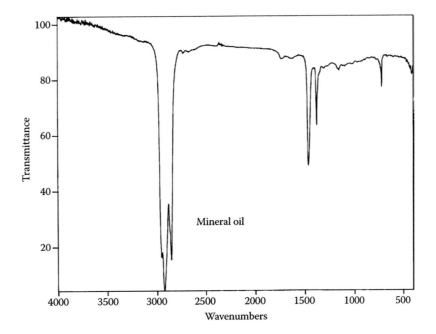

FIGURE 8.31   IR spectrum of mineral oil.

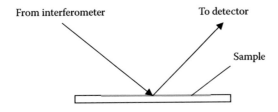

FIGURE 8.32   Specular reflectance.

for describing the technique as "internal." Devices that utilize multiple such reflections, with the sample pressed against both the upper and lower edges of the IRE have been fabricated and the technique dubbed **Multiple Internal Reflectance**. A large number of such reflections (due to a very long IRE and sample) results in the intensity of the light beam being "attenuated," or systematically reduced, and thus multiple internal reflectance is sometimes called **attenuated internal reflectance** (see Figure 8.33).

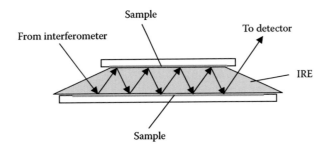

FIGURE 8.33   Multiple internal reflectance.

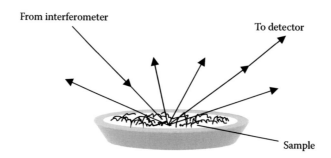

From interferometer

To detector

Sample

FIGURE 8.34 Diffuse reflectance.

8.9.5.3 Diffuse Reflectance

Powdered samples that cannot be cast as films can be measured by this technique. Quite simply, the IR light beam shines on a quantity of the powdered sample held in a "cup." In this case, the light scatters, or reflects in all directions. Some of this light is captured by the detector. A disadvantage is that the light intensity is decreased substantially by the scattering, hence the term "diffuse" (see Figure 8.34).

8.9.6 GAS SAMPLING

Gases, such as automobile exhausts and polluted air, can be measured by IR Spectrometry. A special cell is used for containing the gas. This cell is cylindrically shaped and has inlet and outlet stopcocks for introducing the sample. The side windows of the cylinder are nonabsorbing inorganic salt crystals.

8.10 BASIC IR SPECTRA INTERPRETATION

Once the spectrum is obtained, the question remains as to how to interpret it. We stated previously (Section 8.6) that IR spectra are characterized by rather sharp absorption bands ("peaks"), and each such peak is characteristic of a particular covalent bond in a molecule in the path of the light. Thus, for qualitative analysis (identification and purity characterization), the analyst correlates the location (the wavenumber), the shape, and the relative intensity of a given peak observed in a spectrum with a particular type of bond.

The region of the electromagnetic spectrum involved here, of course, is the infrared region. This spans the wavelength region from about 2.5 to about 17 μm, or, in terms of wavenumber, from about 4000 to about 600 cm^{-1}. Infrared spectra are usually displayed as transmission spectra in which the x-axis is wavenumber. Thus, we will be depicting the spectra with a 100%T baseline at the top of the chart (as the maximum y-axis value) and the peaks deflecting toward the 0%T level when an absorption occurs, as in the spectra shown previously (Figures 8.19, 8.26, and 8.31).

The region from 4000 to approximately 1500 cm^{-1} is especially useful for correlating peak location with bonds. For this reason, we will refer to this region as the "peak ID region." The region from 1500 to 600 cm^{-1} is typically very "busy" (refer to Figures 8.19, 8.26, and 8.31 as examples) and is not as useful for such correlation. We will refer to this region as the "fingerprint region" because, while not generally useful for peak correlation, it remains very useful as the molecular fingerprint. This means that we can still use this region, as we can the peak ID region, for peak-for-peak matching with a known spectrum from a "library" of known spectra (see Figure 8.35). Thus, we look for characteristic bands in the 4000- to 1500-cm^{-1} region to perhaps assign the unknown to a particular class of compounds, i.e., to narrow down the possible structures, and then look to the fingerprint region and the overall spectrum to make the final determination, matching peak for

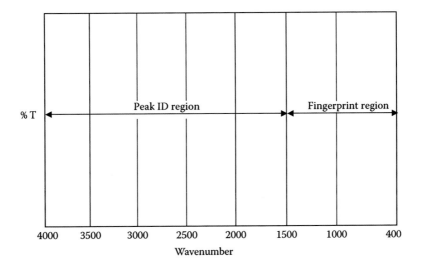

FIGURE 8.35 Peak ID and fingerprint regions of the IR spectrum.

peak. In addition to the location (i.e., wavenumber) of the peaks, it can also be useful to examine their width and depth (strength). The descriptions "broad" and "sharp" are often used to describe the width of the peaks and "weak," "medium," and "strong" are used to describe the depth or strength of the peaks. To understand these descriptions, refer to Figures 8.36, 8.37, and 8.38. Figures 8.36 and 8.38 show broad, strong peaks centered around 3300 cm^{-1}. Figure 8.37 shows a sharp, strong peak at 1685 cm^{-1}. Figure 8.31 shows a sharp, medium peak at 1220 cm^{-1} and a series of weak peaks between 1700 and 2000 cm^{-1}.

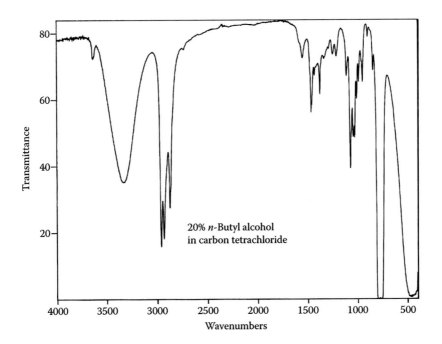

FIGURE 8.36 IR spectrum of an alcohol with no benzene rings.

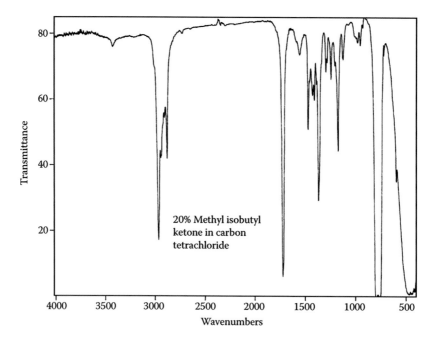

FIGURE 8.37   IR spectrum of a compound with a carbonyl group and no benzene ring.

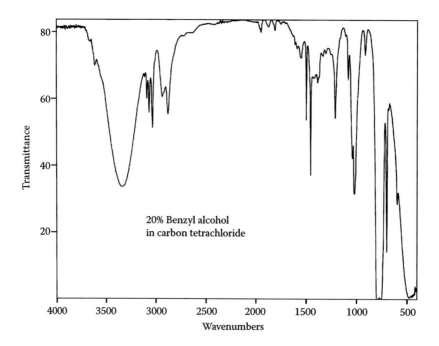

FIGURE 8.38   IR spectrum of an alcohol with a benzene ring.

TABLE 8.1
Some Easily Recognizable IR Absorption Patterns

Bond	Description of Peak
–C-H, where C is not part of a benzene ring	Sharp, strong peak on the low side of 3000 cm^{-1}, between about 2850 and 3000 cm^{-1}
–C-H, where C is part of a benzene ring or double bond	Sharp, medium peak on the high side of 3000 cm^{-1}, between about 3000 and 3100 cm^{-1}
–C-H, where C is part of a triple bond	Sharp, medium peak on the high side of 3000 cm^{-1}, between about 3250 and 3300 cm^{-1}
–O-H, in alcohols, phenols, and water, for example	Broad, strong peak centered at about 3300 cm^{-1}
–C=O, in aldehydes, ketones, etc.	Sharp, strong peak at about 1700 cm^{-1}
–C-C-, in benzene rings	Two sharp, strong peaks near 1500 and 1600 cm^{-1} and a series of weak peaks (called "overtones") between 1600 and 2000 cm^{-1}, the latter in the case of a monosubstituted benzene ring

Table 8.1 correlates the location, width, and depth or strength of various IR absorption patterns for some common kinds of bonds. By comparing Figures 8.29, 8.30, and 8.31 with Table 8.1, we can surmise that the spectrum in Figure 8.36 is that of an alcohol with no benzene rings. We can also surmise that the spectrum in Figure 8.37 is the spectrum of a compound containing a carbonyl group (such as an aldehyde) with no benzene ring. We can surmise that the spectrum in Figure 8.38 is the spectrum of an alcohol with a benzene ring. The so-called IR correlation charts, which are graphical displays showing the location (wavenumber ranges) of the IR absorption peaks of most bonds, can be helpful in the interpretation of spectra. These are found, in the *CRC Handbook of Chemistry and Physics*, for example, and some organic chemistry textbooks.

8.11   QUANTITATIVE ANALYSIS

Quantitative analysis procedures using infrared spectrometry utilize Beer's law. Thus, only sampling cells with a constant pathlength can be used. Once the %T or absorbance measurements are made, the data reduction procedures are identical with those outlined previously in Chapter 7 (preparation of standard curve, etc.).

Reading the %T from the recorded IR spectrum for quantitative analysis can be a challenge. In the first place, there can be no interference from a nearby peak due to the solvent or other component. One must choose a peak to read that is at least nearly, if not completely, isolated (see Application Note 8.3).

Second, the baseline for the peak must be well defined. The baseline for the entire spectrum is the portion of the trace where there are no peaks. It does not usually correspond to the 100%T line. Thus, two %T readings must be taken, one for the baseline (corresponding to where the baseline would be if the peak were absent—a "blank" reading) and one for the minimum %T, the tip of the peak. The two readings are then converted to absorbance and the absorbance of the baseline subtracted from the absorbance of the peak. The result is the absorbance of the sample.

It should also be mentioned again that the pathlength of the sample cell used must be constant for all standards and the unknown. Using care when filling and cleaning cells is also important to avoid alterations in pathlength due to excessive pressure or leaking.

APPLICATION NOTE 8.3: SIMETHICONE ANALYSIS BY FTIR

Simethicone is an antigas ingredient in many liquid and solid pharmaceutical preparations, and FTIR is used in quality assurance laboratories to determine whether its concentration is at the specified level. A sample of the product is dispersed in an HCl solution and the simethicone extracted from this solution with toluene. The toluene solutions are then analyzed by FTIR using a liquid sampling cell. For the quantitative analysis, a simethicone absorption band that is free from interference from the toluene absorption bands is used in a manner similar to the isopropyl alcohol band in Experiment 24 in this chapter. The wavelength for simethicone is 1266 cm^{-1}, which is in the fingerprint region.

(*Reference:* USP 24 NF 19, U.S. Pharmacopeia and National Formulary, Rockville, MD [January 2000], pp. 1518.)

EXPERIMENTS

EXPERIMENT 19: SPECTROPHOTOMETRIC ANALYSIS OF A PREPARED SAMPLE FOR TOLUENE

Note: The slightest contamination with water or other solvent will result in gross error. Remember to wear safety glasses.

1. Prepare one 100-mL volumetric flask and five 25-mL volumetric flasks. These flasks should be clean and dry. They may be prepared by washing with soapy water and a brush, rinsing with water, removing the water with several rinses with acetone, and removing the acetone with several rinses with the alkane solvent to be used (e.g., hexane, heptane, or cyclohexane).
2. Prepare 100 mL of a stock standard solution of toluene in an alkane solvent by pipetting 0.1 mL of toluene into the flask and diluting to the mark with the alkane. Use the 100-mL flask prepared in Step 1. Shake well. This solution has a concentration of 0.870 g/L.
3. Prepare four calibration standards with concentrations of 0.0435, 0.0870, 0.174, and 0.261 g/L in the alkane solvent by diluting the stock solution prepared in Step 2. Use the 25-mL volumetric flasks prepared in Step 1.
4. Obtain an unknown from your instructor and dilute to the mark with the alkane solvent. Rinse the remaining 25-mL flask from Step 1 with the control sample several times before filling the flask nearly to the mark with this sample.
5. Obtain an absorption spectrum of each of your standards, the unknown, and the control sample on a scanning UV/Vis spectrophotometer interfaced to a computer for data acquisition. Follow the instructions provided for your instrument and software.
6. Obtain the maximum absorbance value for each spectrum and prepare the standard curve (Beer's law plot) as in Experiment 13. Obtain the concentration of the unknown and the control sample.
7. Maintain the logbook for the instrument used in this experiment. Record the date, your name(s), the experiment name or number, the correlation coefficient, and the results for the control sample. Also, plot the control sample results on a control chart for this experiment posted in the laboratory.
8. Dispose of all solutions as directed by your instructor.

EXPERIMENT 20: EXTRACTION OF IODINE WITH CYCLOHEXANE

Important note: Be cautious with iodine crystals, being careful not to spill any crystals on the floor or benchtop. Wear gloves. Remember to wear safety glasses.

1. Prepare 100.0 mL of a 0.0075-M stock solution of iodine in cyclohexane. Use a top-loading balance, since iodine fumes are corrosive and can damage an analytical balance.

2. Prepare a series of standard solutions of iodine in cyclohexane, 25.00 mL each, by diluting the stock solution from Step 1. The concentrations should be 0.00015, 0.00030, 0.00045, 0.00060, and 0.00075 M.

3. Prepare 500.0 mL of a 0.00025-M stock solution of iodine in water. An ultrasonic cleaner bath can be used to hasten the dissolution. Pipet 25.00 mL of this solution into a clean, dry, 125-mL separatory funnel. Extract this solution with 10.00 mL of cyclohexane. After extraction, let stand, stopper in place, for at least 5 min. Save the remainder of the stock solution for other students.

4. If a scanning visible spectrophotometer is available, scan the visible region, 400 to 700 nm, for each of the standards and record the maximum absorbance of each. Alternatively, determine the wavelength of maximum absorbance manually using a Spectronic-20 or a similar instrument using one of the standards and measure the absorbance at this wavelength for all the standards. Do the same for the extract by drawing a portion of it (enough to fill the cuvette) from the separatory funnel using a dropper.

5. Create the standard curve, absorbance vs. concentration, using Experiment 13 or other program and obtain the concentration of iodine in the extract and the correlation coefficient.

6. From the data, calculate the distribution coefficient and the percent extracted according to the calculations studied in Chapter 2.

EXPERIMENT 21: DETERMINATION OF NITRATE IN DRINKING WATER BY UV SPECTROPHOTOMETRY

Remember to wear safety glasses.

1. Your instructor has prepared a 50-ppm stock solution of nitrogen in water by dissolving 0.361 g of KNO_3 in 1000 mL of solution.

2. Prepare calibration standards that are 0.5, 1.0, 3.0, 5.0, and 10.0 ppm N in 50-mL volumetric flasks by diluting the stock standard with distilled water. After diluting to the mark, pipette 1.0 mL of 1 N HCl into each flask and shake well.

3. Filter the water samples if necessary. Then, prepare them for measurement by adding 1 mL of 1 N HCl to 50 mL of each. This can be done by filling (rinse first!) 50-mL volumetric flasks with the samples to the 50-mL mark and then pipetting 1.0 mL of the HCl into these flasks. Shake well. Prepare the control sample and a blank (distilled water) in the same way as the samples.

4. Zero the UV spectrophotometer with the blank at 220 nm. Measure all solutions (calibration standards, unknowns, and control) as follows using quartz cuvettes. Measure the absorbances at 220 and at 275 nm for each solution. Since dissolved organic matter may also absorb at 220 nm, it is necessary to correct for interference. Subtracting two times the absorbance at 275 from the absorbance at 220 is a sufficient correction procedure for this interference. Thus, subtract two times the absorbance at 275 from the absorbance at 220 for each.

5. If any samples are outside the range of the standards (high), dilute them by an amount that you feel will bring them into range and measure them again.

6. Create the standard curve as practiced in Experiment 13 by plotting the resulting absorbance vs. concentration and obtain the concentration of the samples and the control. If any water samples were diluted prior to the measurement, apply the dilution factor before reporting the results.

7. Maintain the logbook for the instrument used in this experiment. Record the date, your name(s), the experiment name or number, the correlation coefficient, and the results for the control sample. Also, plot the control sample results on a control chart posted in the laboratory for this experiment.

EXPERIMENT 22: FLUOROMETRIC ANALYSIS OF A PREPARED SAMPLE FOR RIBOFLAVIN

Note: Riboflavin solutions are light sensitive. Store them in the dark as you prepare them. Remember to wear safety glasses.

1. Your instructor has prepared a 50-ppm riboflavin stock solution in 5% (by volume) acetic acid. Prepare a series of calibration standard solutions that are 0.2, 0.4, 0.6, 0.8, and 1.0 ppm from the 10-ppm stock solution and again use 5% acetic acid for the dilution. Use 25-mL volumetric flasks for these standards and shake well.

2. Obtain the unknown from your instructor and again dilute with 5% acetic acid to the mark and shake well.

3. Obtain fluorescence measurements (F) on all of your standards, the unknown, and control. Use 5% acetic acid for a blank. Your instructor will demonstrate the use of the instrument and explain how it was set up to measure riboflavin.

4. Create the standard curve, using the procedure practiced in Experiment 13, by plotting fluorescence intensity vs. concentration and determine the concentration of the unknown and the control sample.

5. Maintain the logbook for the instrument used in this experiment. Record the date, your name(s), the experiment name or number, the correlation coefficient, and the results for the control sample. Also, plot the control sample results on a control chart for this experiment posted in the laboratory.

EXPERIMENT 23: DESIGN AN EXPERIMENT: DETERMINATION OF RIBOFLAVIN IN A VITAMIN TABLET USING FLUOROMETRY

Redesign Experiment 22 such that the amount of riboflavin in a vitamin tablet is determined rather than in a solution prepared by your instructor. Write a step-by-step procedure as in other experiments in this book for preparing standards, preparing the sample, using the fluorometer to measure the standards and sample, and calculating the milligrams of riboflavin in one tablet. After your instructor approves your procedure, perform the experiment in the laboratory.

EXPERIMENT 24: QUANTITATIVE INFRARED ANALYSIS OF ISOPROPYL ALCOHOL IN TOLUENE

DISCUSSION

Isopropyl alcohol exhibits an infrared absorption peak at 817 cm^{-1}. This peak is well-isolated from any other peak either due to the analyte or toluene, the solvent. Thus, this peak is appropriate for quantitative analysis of isopropyl alcohol dissolved in toluene. Your instructor may ask you to disassemble and reassemble the cell so as to have the appropriate spacer in place. Remember to wear safety glasses.

1. In dry 25-mL volumetric flasks, prepare four calibration standards of isopropyl alcohol in toluene solvent that are 20%, 30%, 40%, and 50% in alcohol concentration. Obtain an unknown from your instructor and dilute to the mark with toluene. Shake well.

2. Obtain a liquid sampling cell from your instructor and fill it with the 20% solution. Place this cell in the FTIR instrument and obtain the transmittance spectrum according to the instructions specific to your instrument. Record the %T value at the tip of the 817-cm^{-1} peak and at the baseline. Repeat with the other three standards, the unknown, and the control sample.

3. Convert all %T readings to absorbance. Subtract the absorbance at the baseline from the absorbance at the tip of the 817-cm^{-1} peak in each spectrum.

4. Create the standard curve, using the procedure practiced in Experiment 13, by plotting absorbance vs. concentration and determine the concentration of the unknown and control.

5. Maintain the logbook for the instrument used in this experiment. Record the date, your name(s), the experiment name or number, the correlation coefficient, and the results for the control sample. Also, plot the control sample results on a control chart for this experiment posted in the laboratory.

QUESTIONS AND PROBLEMS

UV-Vis Instrumentation

1. Draw a diagram showing all the components of a basic spectrophotometer.
2. Define *colorimeter*, *UV spectrophotometer*, and *UV-Vis spectrophotometer*.
3. Briefly describe the light sources used for both visible and UV work.

4. Is the intensity of light emitted by the various light sources used for visible and ultraviolet work the same at all emitted wavelengths? How does intensity vary with wavelength for a tungsten filament lamp?

5. How is the wavelength called for in a procedure isolated from the many wavelengths present in the light from the light source in a spectrophotometer?

6. Define *bandwidth* and *resolution*.

7. What is an absorption filter? What is a monochromator?

8. What are the components of a monochromator? What is the function of each component?

9. Consider an experiment in which an analyst wants to change the wavelength used in a given colorimetry experiment from 392 nm (blue light) to 728 nm (red light). He/she turns a knob on the face of the instrument to do this. Tell exactly what is happening inside the instrument when he/she does this.

10. What is a diffraction grating?

11. Why is the sample compartment a "box" isolated from the rest of the instrument?

12. What features of a single-beam spectrophotometer differentiate it from a double-beam spectrophotometer? In what situations is it used over a double-beam instrument?

13. What is meant by the "precalibration" of a single-beam spectrophotometer?

14. For what kind of experiment is the use of a single-beam spectrophotometer a slow, tedious process? Why is it slow and tedious?

15. Can a single-beam spectrophotometer be used for rapid scanning? Explain.

16. What problem exists if you want to use a single-beam instrument for precise work, even if it has a rapid scanning capability?

17. What is a light chopper? What is a beamsplitter?

18. What is a double-beam spectrophotometer?

19. What are two designs of a double-beam spectrophotometer and what are the advantages and disadvantages of each?

20. What are the advantages of a double beam instrument over a single-beam instrument?

21. Why is a double-beam spectrophotometer preferred for rapid scanning?

22. Imagine an experiment in which the molecular absorption spectrum of a particular chemical species is needed. Which instrument is preferred—a single-beam instrument or a double beam instrument? Why?

23. What is a photomultiplier tube? Describe what it does and how it works.

24. Define *photocathode*, *dynode*, and *photodiode*.

25. What is a diode array spectrophotometer? What advantages does it have?

Cuvette Selection and Handling

26. Can cuvettes used for visible spectrophotometry be made of plastic? Explain.

27. What does it mean to say that the cuvettes used must be "matched"? Explain why the cuvettes need to be "matched."

28. A certain pair of cuvettes is not a matched pair. Tell what exactly may be different about them—things over which the analyst does not have any control. Name at least two things.

Interferences, Deviations, Maintenance, and Troubleshooting

29. What does it mean to say that a given substance "interferes" with a spectrochemical analysis?

30. What does it mean to say that there is a "deviation" from Beer's law?

31. What %T range and what absorbance range are considered to be the optimum working ranges for spectrochemical measurements?

32. Elaborate on the maintenance procedures required for spectrophotometers.
33. How can the wavelength calibration be checked on a spectrophotometer?
34. Discuss troubleshooting procedures for spectrophotometers for (1) failure of electrical components, (2) unexpectedly high absorbance readings, (3) an unexpected "peak" in an absorption spectrum.

Fluorometry

35. Is the wavelength of fluorescence longer or shorter than the wavelength of absorption? Explain your answer with the help of an energy level diagram.
36. Is the energy of absorption more or less than the energy of fluorescence? Explain your answer with the help of an energy level diagram.
37. Why are there two wavelength selectors in a fluorometer?
38. What is meant by a "right angle configuration" in a fluorometer and why is this instrument constructed this way?
39. A fluorometer differs in basic design from an absorption spectrophotometer in two major ways.
    (a) What are they?
    (b) Explain the need for these design differences.
40. Draw a diagram of a fluorescence instrument and point out the differences between it and the basic single-beam absorption spectrophotometer.
41. When performing a quantitative analysis procedure using fluorometry, what parameter is measured by the instrument and plotted vs. concentration?
42. Why is it that benzene ring systems such as riboflavin can be analyzed by fluorometry, while uncomplexed metal ions cannot?
43. Fluorometry is more selective and more sensitive than absorption spectrometry. Tell what is meant by selectivity and sensitivity.
44. Under what circumstances would you want to use a fluorometric procedure rather than an absorption spectrometry procedure. Explain briefly.
45. What two advantages does fluorometry have over absorption spectrophotometry?
46. The fact that very few chemical species fluoresce works both to the advantage and disadvantage of fluorometry as a quantitative technique. Explain this.
47. How do absorption spectrophotometry and fluorometry compare in terms of
    (a) Instrument design
    (b) Sensitivity
    (c) Applicability

Introduction to IR Spectrometry

48. Compare the energy transitions caused by infrared light absorption to those caused by UV-Vis light absorption.
49. Compare the cuvettes used for UV-Vis spectrophotometry with the liquid sampling cells used for IR spectrometry.
50. Compare typical IR spectra with typical UV-Vis spectra. What is similar and what is different?
51. Why is it that the infrared spectrum of an organic compound is more useful than the UV-Vis spectrum for qualitative analysis?

IR Instrumentation

52. What do the letters FTIR stand for?
53. Fill in the blanks with either "double-beam dispersive" or "FTIR," whichever correctly completes the statement:

(a) Uses an interferometer ⎯⎯⎯⎯⎯⎯⎯⎯⎯⎯⎯
(b) Uses a moveable mirror ⎯⎯⎯⎯⎯⎯⎯⎯⎯
(c) Utilizes the slit/dispersing element/slit arrangement ⎯⎯⎯⎯⎯⎯⎯⎯⎯⎯
(d) Measures the IR spectrum in a matter of seconds
(e) Disperses IR light into a "spray" of wavelengths, a narrow band of which is selected
⎯⎯⎯⎯⎯⎯⎯⎯⎯
(f) A Fourier transform is performed on the data in order to obtain the absorption data at all wavelengths ⎯⎯⎯⎯⎯⎯⎯⎯⎯⎯

54. What is an interferometer and what is its function in an FTIR instrument?
55. Answer with either "double-beam dispersive" or "FTIR" in order to indicate which instrument design is described by each statement.
   (a) Utilizes a moveable mirror to create an interference pattern.
   (b) Records the IR spectrum in a matter of seconds.
   (c) Is an older design that is nearly obsolete.
   (d) Utilizes an "interferometer."

Sampling

56. Why are inorganic compounds useful as sample "windows" and matrix material for infrared analysis?
57. What is a "neat" liquid?
58. Name three methods for mounting a liquid sample in the path of the light in an infrared spectrometer.
59. What are the differences between a "sealed" cell, a "demountable" cell, and a "sealed demountable cell" for liquid sampling?
60. What defines the pathlength in a sealed demountable IR cell?
61. Describe the pressure/vacuum method of filling an IR cell equipped with inlet and outlet ports.
62. Name two separate problems that occur when a sample for infrared analysis is contaminated with water.
63. Why is a water-contaminated sample a problem for the IR cells?
64. What problem is encountered with spectra interpretation when a solution of the analyte in a particular solvent is analyzed? Why does the use of carbon tetrachloride as the solvent minimize this problem?

Solid Sampling

65. Name at least five methods by which the IR spectrum of a solid can be obtained.
66. What is a KBr pellet?
67. Give two reasons why the potassium bromide used to make the KBr pellet must be dry.
68. Explain the use of a hydraulic press for IR laboratory work.
69. What is a Nujol mull?
70. What advantage do the KBr pellet method and the reflectance methods have over the solution and mineral oil mull methods for solids?
71. Why is it that the presence of mineral oil in the mineral oil mull usually does not cause a problem with spectral interpretation?
72. Name three "reflectance" methods for solid sampling and tell what the differences are.
73. Briefly describe the diffuse reflectance method for solids.
74. Can gases be measured by IR spectrometry? Explain.

Basic IR Spectra Interpretation

75. For each of the four IR spectra shown in Figures 8.39, 8.40, 8.41, and 8.42, tell which of the six bonds in Box 8.1 are present and which are absent. When finished, suggest a total structure for each that fits your observations in each case.

76. Look at the three infrared spectra in Figures 8.43, 8.44, and 8.45 and answer the following questions.

   (a) Are any of the spectra that of an alcohol? If so, which? What absorption pattern(s) at what wavelength(s) identifies an alcohol?

   (b) Are any of the spectra that of a compound containing a benzene ring? If so, which? What three absorption patterns at what wavelengths show that a compound has a benzene ring?

   (c) Are any of the spectra that of a compound containing only carbons and hydrogens? If so, which? Benzene rings contain only carbons and hydrogens. Might the spectrum or spectra you chose for your answer above indicate a benzene ring? (Tell what absorption patterns are present or not present that would support your answer.)

   (d) Do any of the three compounds have a carbonyl group (–C=O)? If so, which? What absorption pattern, or lack thereof, at what wavelength supports your answer?

   (e) The three compounds whose spectra appear in these figures are *n*-heptane, 3-methylphenol, and benzophenone. Which is which?

Quantitative Analysis

77. Describe an experiment in which the quantity of an analyte is determined by infrared spectrometry.

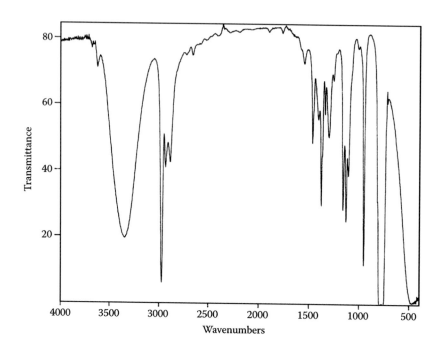

FIGURE 8.39   Infrared spectrum for question 75.

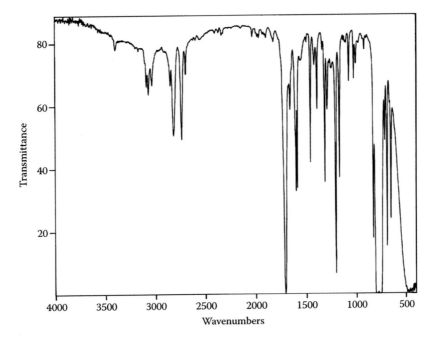

FIGURE 8.40   Infrared spectrum for question 75.

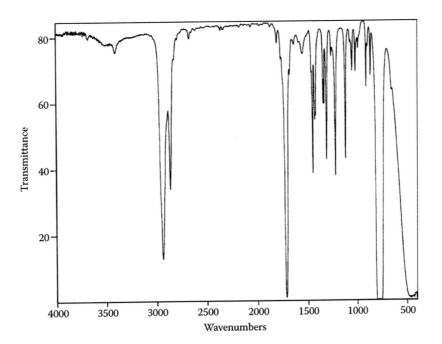

FIGURE 8.41   Infrared spectrum for question 75.

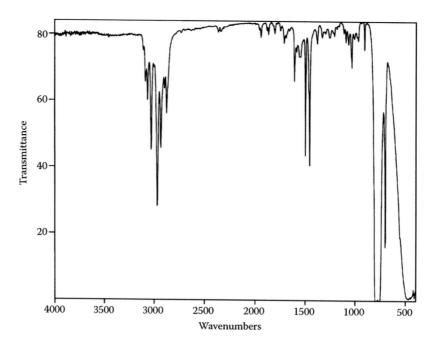

FIGURE 8.42   Infrared spectrum for question 75.

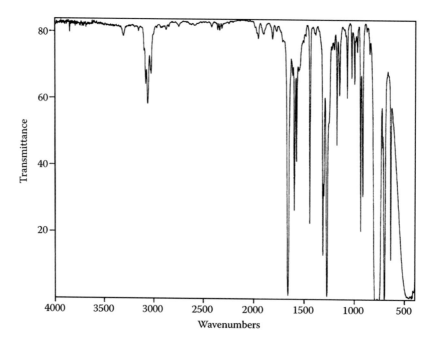

FIGURE 8.43   Infrared spectrum for question 76.

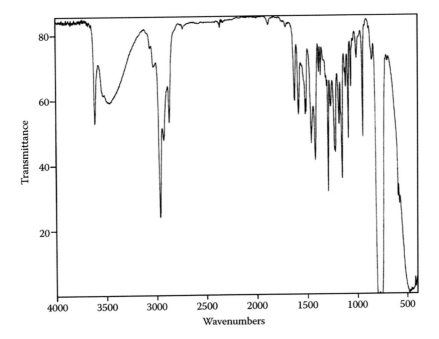

FIGURE 8.44   Infrared spectrum for question 76.

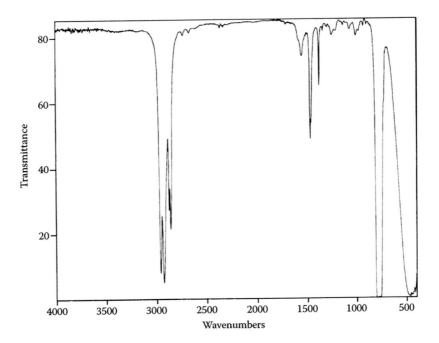

FIGURE 8.45   Infrared spectrum for question 76.

Report

78. For this exercise, your instructor will select a "real-world" UV or IR spectrometric analysis method, such as an application note from an instrument vendor or from a methods book, a journal article, or Web site, and will give it to you as a handout. Write a report giving the details of the method according to the following scheme. On your paper, write (or type) "A," "B," etc. to clearly present your response to each item.

(a) Method Title: Give a more descriptive title than what is given on the handout.

(b) Type of Material Examined: Is it water, soil, food, a pharmaceutical preparation, or what?

(c) Analyte: Give both the name and the formula of the analyte.

(d) Sampling Procedure: This refers to how to obtain a sample of the material. If there is no specific information given as to obtaining a sample, make something up. If you want to be correct in what *you write, do a library or other search to discover a reasonable response. Be brief.*

(e) Sample Preparation Procedure: This refers to the steps taken to prepare the sample for the analysis. It does *not* refer to the preparation of standards or any associated solutions.

(f) Standard Curve: Is a standard curve suggested? If so, what is plotted on the *y*-axis and what is plotted (give units) on the *x*-axis?

(g) Added Reagents: Are any reagents added to the sample and standards prior to measurement? If so, what are they and what is their purpose?

(h) Reactions Involved to Enhance the Absorption: If a chemical equation is given, write it down here. If no chemical equation is given, state in words what the reaction is. If this information is not given, write "not given."

(i) What Wavelength Is Used?

(j) How Are the Data Gathered?
The "data" are the absorbance measurements. In other words, you've got your blank, your standards, and your samples ready to go. Now what? *Be brief.*

(k) Concentration Levels for Standards: This refers to what the concentration range is for the standard curve. Be sure to include the units. If no standard curve is used, state this.

(l) How Are the Standards Prepared? Are you diluting a stock standard? If so, how is the stock prepared and what is its concentration? If not, how are the standards prepared?

(m) Potential Problems: Is there anything mentioned in the method document that might present a problem? *Be brief.*

(n) Data Handling and Reporting: Once you have the data for the standard curve, then what? Be sure to present any post-run calculations that might be required.

(o) References: Did you look at any other references/sources to help answer any of the above? If so, write them here.

9 Atomic Spectroscopy

9.1 REVIEW AND COMPARISONS

Atomic spectroscopy refers to the absorption and emission of UV-Vis light by atoms and monoatomic ions and is conceptually similar to the absorption and emission of UV-Vis light by molecules as discussed in Chapter 8. Some differences have already been discussed in Chapter 7 (see Section 7.5). In brief summary, absorption spectra for atoms are due to atoms in the gas phase absorbing UV-Vis light from a light source and are characterized by very narrow wavelength absorption bands called **spectral lines**. These very narrow bands result from energy transitions between electronic energy levels in which there are no vibrational levels superimposed (because covalent bonds are required for vibrational levels to exist and unbound gas phase atoms have no covalent bonds).

Similarly, emission spectra for atoms and monoatomic ions are due to gas phase atoms or monoatomic ions in excited states emitting UV-Vis light because they are returning to the ground state. Emission spectra are also characterized by very narrow wavelength bands (also called spectral lines) because they too involve electronic energy levels in which there are no vibrational levels superimposed. In fact, for a given kind of atom, the emission lines occur at the same wavelengths as the absorption lines because they involve the same energy levels (see Figure 9.1).

Besides these differences in electronic energy levels and spectra, atomic spectroscopy differs from UV-Vis molecular spectroscopy in the following ways:

1. Analysis for atoms means that atomic spectroscopy is limited to the elements. In fact, the key word for atomic spectroscopy is *metals*. The vast majority of methods involving atomic spectroscopy are methods for determining metals.
2. Sample preparation schemes for atomic spectroscopy usually place the metals in water solution. Since metals are present as ions in water solution, atomic spectroscopy methods must have a means for converting metal ions into free gas phase ground state atoms (a process called **atomization**) in order to measure them. Most of these methods involve a large amount of thermal energy.
3. Perhaps the most noticeable difference in instrumentation is the sample "container" used for atomic spectroscopy. This container is the source of the thermal energy needed for the conversion of ions in solution to atoms in the gas phase (and hence is called an **atomizer**), and in no way resembles a simple cuvette. Recall, for example, the brief discussion and photograph of the flame "container" in Chapter 7, Section 7.6.
4. The need for and use of thermal energy as outlined above has resulted in the invention of a number of separate and distinctly different atomizer and instrument designs, albeit based on the same theory, under the heading of atomic spectroscopy.
5. **Spectral line sources** are used as light sources in atomic absorption instruments rather than the **continuum sources** used for UV-Vis molecular absorption instruments and several atomic emission techniques require no light source at all apart from the thermal energy source.
6. Since the analytes for atomic spectroscopy are severely limited (elements only) compared with the large number of molecular and complex ion analytes for UV-Vis molecular absorption spectrometry, the wavelengths used for quantitation are well known and do not require the analyst to ever first measure the absorption or emission spectra. Hence, although atomic absorption and emission spectra are atomic "fingerprints" in the same

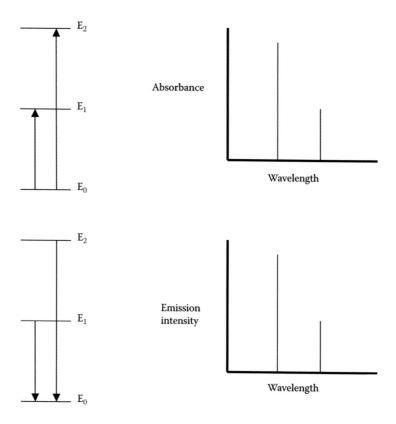

FIGURE 9.1 (Top left) Energy level diagram depicting energy transitions for atoms from the ground state to two excited states. (Top right) Line absorption spectrum with two spectral lines resulting from the transitions in the top left. (Bottom left) Energy level diagram depicting energy transitions for atoms from two excited states to the ground state. (Bottom right) Line emission spectrum with two spectral lines resulting from the transitions in the bottom left. Note what is plotted on the y-axis in the two spectra.

context that molecular absorption and emission spectra are molecular "fingerprints" and can be used for identification purposes, they are not needed for quantitative work. The wavelengths needed are simply looked up in tables of spectral lines available in instrument manufacturer's literature or handbooks.

9.2 BRIEF SUMMARY OF TECHNIQUES AND INSTRUMENT DESIGNS

The most important of the techniques and instrument designs requiring thermal energy (Item 4 above) are **flame atomic absorption**, which we will be referring to as "flame AA," **graphite furnace atomic absorption**, or "graphite furnace AA," also known as **electrothermal atomic absorption** and **inductively coupled plasma atomic emission**, or "ICP." Each of these will be discussed in detail in this chapter. Less important techniques and instrument designs that require thermal energy are **flame emission** and **atomic fluorescence**. Two that do not require thermal energy, or minimal thermal energy, are the **cold vapor mercury system**, and the **metal hydride generation** technique. One that requires electrical energy is **arc and spark emission**. Each of these techniques will be briefly discussed.

FIGURE 9.2 Photograph of the flame atomizer.

Flame AA utilizes a large flame as the atomizer. A photograph of this atomizer was first shown in Chapter 7, Figure 7.29, and is reproduced here in Figure 9.2. The sample (a solution) is drawn into the flame by a vacuum mechanism that will be described. The atomization occurs immediately. The light beam for the absorption measurements is directed through the width of the flame, right to left or left to right depending on the instrument orientation, and measured.

The atomizer for graphite furnace AA is, as the name implies, a furnace. It is actually a small graphite tube that can be quickly electrically heated to a very high temperature. A relatively small volume of the sample solution is placed in the tube either manually (with a micropipette) or drawn in with a vacuum mechanism. This "furnace" is then electrically brought to a very high temperature in order to atomize the sample inside. The light beam is directed through the tube, in which there is a "cloud" of atoms, and measured. Instruments that utilize a flame for the atomizer can also utilize the graphite furnace. It is a matter of replacing the flame module with the furnace module and lining it up with the light beam. Figure 9.3 is a photograph of a graphite furnace module.

ICP is an emission technique, which means it does not use a light source. The light measured is the light emitted by the atoms and monatomic ions of the sample in the atomizer. The ICP atomizer is an extremely hot plasma, which is a high-temperature partially ionized gas composed of atoms electrons, and positive ions. This gaseous mixture is directed through an induced magnetic field, which causes the elevation of the mixture to the very high temperature. The very high temperature means that the atoms and monatomic ions of a sample undergo sufficient excitation (and de-excitation) such that relatively intense emission spectra result. The sample is drawn in with a vacuum mechanism that will be described. The intensity of an emission line is measured and related to concentration.

FIGURE 9.3   Photograph of a graphite furnace module. This module is approximately 6 inches across.

9.3 FLAME ATOMIC ABSORPTION

Let us begin with flame AA. It is the oldest of the three techniques and has been widely used for many years. The instruments are relatively inexpensive and methods have been well-tested and are well-understood.

9.3.1 FLAMES AND FLAME PROCESSES

When solutions of metal ions are introduced into a flame, several processes occur in extremely rapid succession (refer to Figure 9.4 throughout the following discussion). First, the solvent evaporates leaving behind formula units of the formerly dissolved salt. Next, dissociation of the formula units of salt into atoms occurs—the metal ions "atomize" or are transformed into atoms. Then, if the atoms are easily raised to excited states by the thermal energy of the flame, a **resonance** process occurs in which the atoms resonate back and forth between the ground state and the excited states. Actually, only a small percentage of the atoms (less than 0.1%, depending on the temperature of

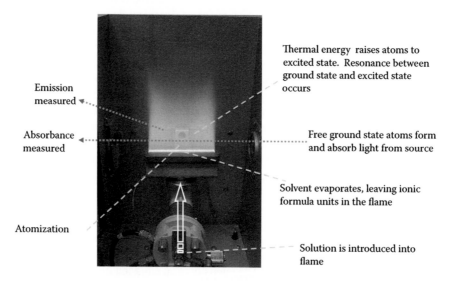

FIGURE 9.4   Photograph of a flame atomizer and illustrations of the processes that occur when a solution containing a metal ion is aspirated into it (see text for full discussion).

TABLE 9.1

Oxidants Recommended for Various Metals and Nonmetals Analyzed by Flame AA

Oxidant	Metals and Nonmetals
Air	Antimony, arsenic, bismuth, cadmium, calcium, cesium, chromium, cobalt, copper, gold, indium, iridium, iron, lead, lithium, magnesium, manganese, mercury, nickel, palladium, platinum, potassium, rhodium, rubidium, ruthenium, selenium, silver, sodium, tellurium, thallium zinc
Nitrous oxide	Aluminum, barium, beryllium, boron, dysprosium, erbium, europium, gadolinium, gallium, germanium, hafnium, holmium, lanthanum, molybdenum, neodymium, niobium, phosphorus, praseodymium, rhenium, samarium, scandium, silicon, strontium, tantalum, terbium, thulium, tin, titanium, tungsten, uranium, vanadium, ytterbium, yttrium, zirconium

the flame) are found in the excited state at any particular moment. As these atoms drop back to the ground state (a natural process) the emission spectrum is emitted. For those atoms that are easily excited under these conditions, the emitted wavelengths are in the visible region of the spectrum and the entire flame takes on a color characteristic of the element that is in the flame. It is a characteristic color because each element has its own characteristic line spectrum in the visible region of the spectrum. This line spectrum is the atomic "fingerprint," resulting from the particular energy transitions that element has. Easily excited elements include sodium (yellow), calcium (orange), lithium (red), potassium (violet), and strontium (red). It is possible to quantitate these elements using the flame emission technique mentioned in Section 9.2, and this will be discussed later.

The unexcited atoms in the flame (the 99.9%) are available to be excited by a light beam. Thus, as shown in Figure 9.4, a light source is used and a light beam is directed through the flame as shown. The experiment is a Beer's law experiment, the width of the flame being the pathlength.

We implied in the discussion above that the flame temperature is important. This is true for both the atomization and excitation processes.

All flames require both a fuel and an oxidant in order to exist. Bunsen burners and Meker (Fisher) burners utilize natural gas for the fuel and air for the oxidant. The maximum temperature of an air–natural gas flame is 1800 K. In order to sufficiently atomize most metal ions and excite a sufficient number of atoms of the more easily excited elements for quantitation purposes and therefore achieve a desirable sensitivity for quantitative analysis by atomic spectroscopy, however, a hotter flame is desirable. Most flames used for flame AA are air/acetylene flames—oxidant, air; fuel, acetylene. A maximum temperature of 2300 K is achieved in such a flame. Ideally, pure oxygen with acetylene would produce the highest temperature (3100 K), but such a flame suffers from the disadvantage of a high burning velocity, which decreases the completeness of the atomization and therefore actually lowers the sensitivity. Nitrous oxide (N_2O) used as the oxidant, however, produces a higher flame temperature (2900 K) while burning at a low rate. Thus, N_2O-acetylene flames are fairly popular. The choice is made based on which flame temperature/burning velocity combination works best with a given element. Since all elements have been studied extensively, the recommendations for any given element are available. Table 9.1 lists most metals and the recommended flame for each air/acetylene flames are the most commonly used.

9.3.2 SPECTRAL LINE SOURCES

As mentioned in Item 5 of Section 9.1, the light sources used in atomic absorption instruments are sources that emit spectral lines. Specifically, the spectral lines used are the lines in the line spectrum of the analyte being measured. These lines are preferred because they represent the precise wavelengths that are needed for the absorption in the flame, since the flame contains this analyte.

Spectral line sources emit these wavelengths because they themselves contain the analyte to be measured and when the lamp is on, these internal atoms are raised to the excited state and emit their line spectrum when they return to the ground state. It is this emitted light, the same light that will be absorbed in the flame, which is directed at the flame.

9.3.2.1 Hollow Cathode Lamp

The most widely used spectral line source for atomic absorption spectroscopy is the hollow cathode lamp. An illustration of this lamp is shown in Figure 9.5. The internal atoms mentioned above are contained in a **cathode**, a negative electrode. This cathode is a hollowed "cup" pictured with a "C" shape in the figure. The internal excitation/emission process occurs inside this cup when the lamp is on and the anode (positive electrode) and cathode are connected to a high-voltage source. The light is emitted as shown.

The lamp itself is a sealed glass tube and is filled with an inert gas, such as neon or argon, at a low pressure. This inert gas plays a role in the emission process as shown in Figure 9.6. Argon (Ar)

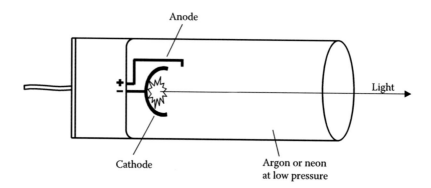

FIGURE 9.5   Drawing of a hollow cathode lamp.

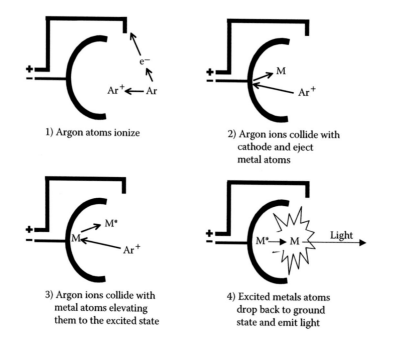

1) Argon atoms ionize

2) Argon ions collide with cathode and eject metal atoms

3) Argon ions collide with metal atoms elevating them to the excited state

4) Excited metals atoms drop back to ground state and emit light

FIGURE 9.6   Process occurring in a hollow cathode lamp providing the light used in an atomic absorption instrument (see text for discussion and definitions of symbols).

is the inert gas shown in this figure. When the lamp is on, the argon atoms undergo ionization, as shown. The positively charged argon ions (Ar^+ in the figure) then crash into the interior surface of the cathode because of the strong attraction of their positive charge for the negative electrode. The violent contact of the argon ions with the interior surface of the cathode causes **sputtering** (transfer of surface atoms in the solid phase to the gas phase due to the collisions). Additional collisions of the argon ions with the gas phase metal atoms ("M" in the figure) cause the metal atoms to be raised to the excited state (M* in the figure) and when they drop back to the ground state, the light is emitted.

The hollow cathode lamp must contain the element being measured. A typical atomic absorption laboratory has a number of different lamps in stock that can be interchanged in the instrument, depending on what metal is being determined. Some lamps are "multielement," which means that several different specified kinds of atoms are present in the lamp and are all raised to the excited state when the lamp is on. The light emitted by such a lamp consists of the line spectra of all the kinds of atoms present. One may think that the lines of the elements other than the analyte might interfere with the measurement of the analyte. This is not usually a problem because a monochromator is used following the flame to isolate the spectral line of the analyte.

9.3.2.2 Electrodeless Discharge Lamp

A light source known as the electrodeless discharge lamp (EDL) is sometimes used. In this lamp, there is no anode or cathode. Rather, a small, sealed quartz tube containing the metal or metal salt, and some argon at low pressure is wrapped with a coil for the purpose of creating a radiofrequency (RF) field. The tube is thus inductively coupled to an RF field and the coupled energy ionizes the argon. The generated electrons collide with the metal atoms, raising them to the excited state. The characteristic line spectrum of the metal is thus generated and is directed at the flame just as with the hollow cathode tube. EDL's are available for 17 different elements. Their advantage lies in the fact that they are capable of producing a much more intense spectrum and thus are useful for those elements whose hollow cathode lamps can produce only a weak spectrum.

Regardless of what source is used, the line spectra of the elements present in the source are passed through the flame. The intensities of the spectral lines emerging from the flame are therefore diminished because they have all been absorbed due to the energy transitions occurring in the flame (see Figure 9.7).

9.3.3 PREMIX BURNER

The burner used for flame AA is a "premix burner." It is called that because all the components of the flame (fuel, oxidant, and sample solution) are "premixed" as they take a common path to the

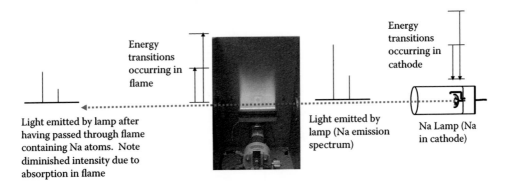

Energy transitions occurring in flame

Energy transitions occurring in cathode

Light emitted by lamp after having passed through flame containing Na atoms. Note diminished intensity due to absorption in flame

Light emitted by lamp (Na emission spectrum)

Na Lamp (Na in cathode)

FIGURE 9.7 Illustration of the complete flame AA experiment, including energy level diagrams and line spectra using sodium metal as the example. The energy level diagrams and line spectra are not intended to match that of sodium exactly.

flame. The fuel and oxidant originate from pressurized sources, such as compressed gas cylinders, and their flow to the burner is controlled at an optimum rate by flow control mechanisms that are part of the overall instrument unit.

The sample solution is aspirated (drawn by vacuum) from its original container through a small tube and converted to an aerosol, or fine mist, prior to the mixing. These steps (aspiration and conversion to an aerosol) are accomplished with the use of a **nebulizer** at the head of the mixing chamber. The nebulizer is a small (3 cm long and 1 cm diameter) adjustable device resembling the nozzle one places on the end of a garden hose to create a water spray. There are two inlets to the nebulizer. One inlet is a small plastic tube (the "garden hose") protruding from the back and the other is a sidearm entering at right angle. With the oxidant flowing, the small plastic tube is dipped into the sample solution by the operator in order to initiate the aspiration. The oxidant, due to its rapid flow through the sidearm, creates suction in the sample line (called the Venturi effect) and draws the sample solution from its container and into the mixing chamber (see Figure 9.8).

The aerosol spray, as it emerges from the nebulizer, contains variable-sized solution droplets. The larger droplets are undesirable because solvent evaporation from such droplets in the flame is inefficient. Hence, they should be removed. In addition, there must be some sort of mechanism for the thorough mixing of the aerosol spray, the oxidant, and the fuel in the mixing chamber. For these reasons, an impact device is used in the mixing chamber. This impact device is typically a flow spoiler, or series of baffles, placed in the path of the flow. It may also include a glass bead positioned near the tip of the nozzle (see Figure 9.9 for an illustration of the complete system: nebulizer, glass

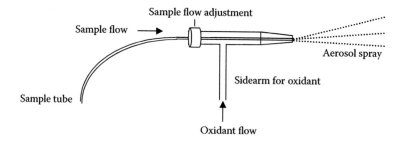

FIGURE 9.8   Illustration of a nebulizer.

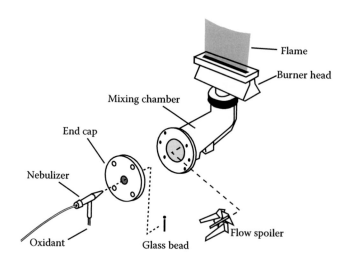

FIGURE 9.9   Illustration of the complete premix system.

FIGURE 9.10 Drawing of the top of a burner head. The slot is either 10 cm long (for air/acetylene flames) or 5 cm long (for nitrous oxide/acetylene flames).

bead, flow spoiler, and burner head). With this design, the larger droplets in the nebulizer spray fall to the bottom of the mixing chamber. A drain hole (shown in the figure) allows the accumulated solution to drain out to a waste receptacle. Approximately 90% of aspirated aqueous solution is actually eliminated through the drain and never reaches the flame.

The burner head itself is interesting. The flame must be "wide" (for a long pathlength) but does not have to be "deep." In other words, the burner has a long slot cut in it (5 or 10 cm long) to shape the flame in a wide but shallow contour. Typically, the flame is 10 cm wide for air/acetylene and 5 cm wide for N_2O/acetylene (see Figure 9.10).

9.3.4 OPTICAL PATH

The optical path for flame AA is arranged in this order: light source, flame (sample container), monochromator, detector. Compared with UV-Vis molecular spectrometry, the sample container and monochromator are switched. The reason for this is that the flame is, of necessity, positioned in an open area of the instrument surrounded by room light. Hence, the light from the room can leak to the detector and therefore must be eliminated. In addition, any light emitted by the flame must be eliminated. Placing the monochromator between the flame and the detector accomplishes both. However, any flame emissions that are the same wavelength as the source wavelength are not blocked by the monochromator but also must be eliminated. These emissions exist because the analyte atoms are in the flame and may be emitting the same wavelength as the wavelength from the source that the analyst wishes to measure. Modulating the light from the source either through electronic pulsing or by placing a light chopper (no mirror) in the optical path ahead of the flame solves this dilemma.

The optical path for flame AA can be either single- or double-beam. However, as we explained in Section 9.1 Item 6, the absorption spectra are seldom measured. Thus, the disadvantages of a single-beam instrument are not as severe. In a given experiment, the wavelength is seldom changed, and so there is no need to recalibrate with the blank in the tedious manner described for UV-Vis molecular absorption instruments in the previous chapter. The optical path for a single-beam instrument is shown in Figure 9.11a. With the light from the source modulated, the detector sees alternating signals of (1) source light combined with flame emissions and room light and (2) flame emissions and room light only. Knowing the frequency of the modulation, the detector then eliminates the flame emissions and room light by subtraction (see Figure 9.11b). However, the problems of source drift and fluctuations still exist, and this is handled with the double-beam design.

The double-beam design uses a beamsplitter or a chopper to divert the light from the source around the flame. The two beams are then joined again before entering the monochromator. The second beam does not pass through a blank. If it did, it would require a second flame matched to the first. As stated previously, atomic absorption analyses seldom require wavelength scans and so there would be no advantage to such a design. The primary function of a double-beam design is to eliminate problems due to the source drift and fluctuation. A side advantage is that source warm-up time is eliminated since changes in intensity during warm-up are immediately compensated and thus very rapid changeover of lamps in automated instruments is possible. Both the sample beam and the reference beam are chopped so that the detector can differentiate and eliminate emissions from the flame and room light (see Figure 9.12a and 9.12b).

(a)

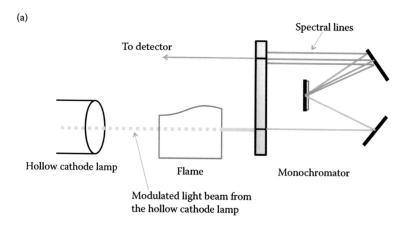

(b)

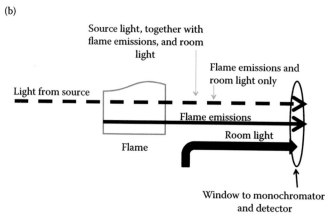

FIGURE 9.11 (a) Illustration of a single-beam flame atomic absorption spectrometer. The modulated light from the hollow cathode lamp is passed through the flame where it is joined to emissions from the flame and room light. This light then enters a monochromator that is set so that one of the lines from the line spectrum of the analyte metal passes through the exit slit to the detector. The source light is differentiated from the others because it is modulated. (b) Closer view of the joining of the light from the source to the flame emissions and the room light.

9.3.5 PRACTICAL MATTERS AND APPLICATIONS

9.3.5.1 Slits and Spectral Lines

There is more than one spectral line in the line spectrum of an element and therefore more than one line to choose from when setting the monochromator. For each element, there is one line that gives the optimum absorptivity for that element, and this line is therefore the most sensitive and most useful. This line is called the **primary line** and is the line most often chosen—the monochromator control is set at that wavelength. Other lines are called **secondary lines**. A secondary line may be chosen if, for some reason, the primary line is inappropriate, such as if there is another element in the sample with a line too close to the primary line of the analyte (see Application Note 9.1; primary and secondary lines are listed in manufacturers' literature and in handbooks). The more automated instruments automatically sense what lamp is installed and set the monochromator control to the primary line as part of the instrument setup process during power-up. The on-board software does allow for a secondary line to be chosen if desired.

The potential problem of one line being too close to another line is solved by a slit control. The wider a slit, the greater the bandpass and the more apt an interfering line would be allowed to

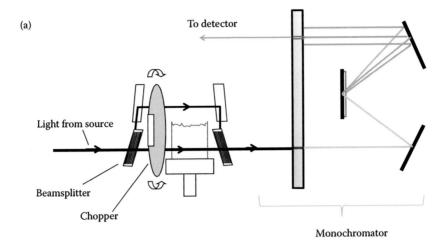

(a)

To detector

Light from source

Beamsplitter

Chopper

Monochromator

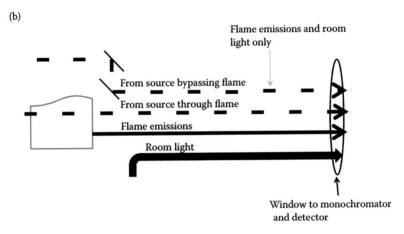

(b)

Flame emissions and room light only

From source bypassing flame

From source through flame

Flame emissions

Room light

Window to monochromator and detector

FIGURE 9.12 (a) Illustration of a double-beam flame atomic absorption spectrophotometer. A beamsplitter creates a second beam that bypasses the flame. A chopper chops both the sample beam and the reference beam using a small "window" in a rotating wheel. In this way, room light and flame emissions can be subtracted out, since they are not chopped, and the light through the flame and the light bypassing the flame can be differentiated. (b) Closer view of the joining of the split light from the source to the flame emissions and the room light.

enter and exit the monochromator and be measured along with the primary line. Atomic absorption monochromators are typically operated with bandpass values ranging from 0.2 to 2.0 nm. The required bandpass for a given element is selected by choosing the appropriate slit width setting. The slit width settings, like the wavelength settings, are listed in manufacturers' literature. For a given element, the operator sets the slit value using a slit control on the instrument. Again, the more automated instruments, having sensed what lamp is installed and what line has been selected, automatically set the slit to the optimum value as part of the instrument set-up during the power-up process. The slit value represents the bandpass for both the entrance and exits slits. If there is an interfering line at the optimum setting, either the slit width is narrowed or a secondary line is chosen. Both of these choices result in less desirable sensitivity.

9.3.5.2 Linear and Nonlinear Standard Curves

In atomic absorption spectrometry, the standard curve is a plot of absorbance vs. concentration. In discussing standard curve up to now, the usual procedure is to prepare a series of standard solutions

APPLICATION NOTE 9.1: THE ANALYSIS OF PAINT CHIPS FOR LEAD

Paint chips scraped from older buildings can be analyzed for their lead content by flame AA. Samples are crushed to a powder and then weighed into test tubes, 16 × 125 mm. Samples weights are between 10.0 and 20.0 mg. Sample preparation includes digesting the samples for 30 min in 1.0 mL of concentrated HNO_3 on a hot plate. After evaporating the resulting extracts to dryness, the residue is reconstituted with 10.0 mL of 0.5 M HCl by placing the test tubes in a boiling water bath for 30 min. The resulting solutions, or dilutions of these, are aspirated into the flame and the absorbances measured.

Standards are prepared by first preparing a 1000-ppm stock solution of lead using solid lead nitrate. The reference standards for the standard curve have concentrations in the 0.0- to 40.0-ppm range. The flame is an air/acetylene flame. The wavelength used is a secondary line, 282.3 nm, and the slit width is 0.5 nm. The secondary line is used due to interfering substances present in the paint.

(*Reference:* Markow, P., Determining the Lead Content of Paint Chips, *Journal of Chemical Education*, 73(2) (February 1996) pp. 178–179.)

over a concentration range suitable for the samples being analyzed, i.e., such that the expected sample concentrations are within the linear range established by the standards. The standards and the samples are then aspirated into the flame and the absorbances read from the instrument.

Nonlinear standard curves, such as the hypothetical curve drawn in Figure 9.13, are common in atomic absorption spectroscopy. This does not necessarily present a problem for the analyst, however. Calibration algorithms that draw a smooth curve through the points despite their nonlinear nature have been incorporated into modern signal processing software and accurate determinations of unknown concentrations are routine. One can select a linear or nonlinear curve as part of the instrument setup so that the proper algorithms are used.

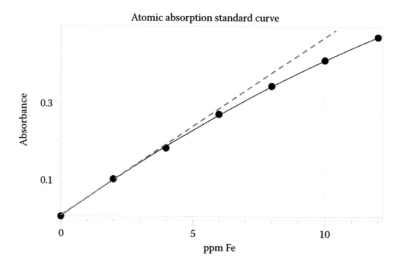

FIGURE 9.13 Example of a standard curve for flame atomic absorption that is not linear.

9.3.5.3 Hollow Cathode Lamp Current

The electrical power or current the hollow cathode lamp receives is adjustable. There is an optimum setting (the "operating current") representing the most intense light without significantly shortening the life of the lamp. The operator sets the value of the current when setting up the instrument. In the more automated instruments, the lamp current is set automatically during the power-up process. Setting the lamp current above the designated operating current shortens the life of the lamp, an item of significant expense. Setting the lamp current below the operating current may produce satisfactory results and extends the life of the lamp. It is recommended that a logbook be maintained for each lamp in the laboratory so that intensity, sensitivity, and hours used records are maintained. Again, the modern instrument, having sensed what lamp is installed, will set the lamp current automatically.

9.3.5.4 Lamp Alignment

The hollow cathode lamp must be aligned properly to provide the optimum intensity throughout the optical path. This is done during setup by monitoring the light energy reaching the detector while moving the lamp horizontally and vertically. Each lamp is slightly different in shape and internal alignment, and thus this external alignment procedure must be performed each time the lamp is changed. Highly automated instruments can perform this function automatically.

9.3.5.5 Aspiration Rate

There is a sample flow adjustment nut on the nebulizer (see Figure 9.7). Turning this nut in or out changes the flow rate of the sample solution through the nebulizer and can even result in oxidant being drawn back and bubbled through the sample solution if it is not adjusted properly. The setting is optimized by monitoring the absorbance of an analyte standard during the adjustment. A maximum absorbance would indicate an optimum flow rate.

9.3.5.6 Burner Head Position

The burner head can be adjusted vertically, horizontally (toward and away from the operator as well as rotationally). Initially, it should be adjusted vertically so that the light beam should pass approximately 1 cm above the center slot and horizontally so that the light beam passes directly through the center of the flame, end to end. All three positions can be optimized by monitoring the absorbance of an analyte standard while making the adjustments. A maximum absorbance would indicate the optimum position.

9.3.5.7 Fuel and Oxidant Sources and Flow Rates

The sources of acetylene, nitrous oxide, and sometimes air, are usually steel cylinders of the compressed gases purchased from specialty gas or welders' gas suppliers. Thus, several compressed gas cylinders are usually found next to atomic absorption instrumentation and the analyst becomes involved in replacing empty cylinders with full ones periodically. Safety issues relating to storage, transportation, and use of these cylinders will be addressed in Section 9.5. The acetylene required for atomic absorption is a purer grade of acetylene than that which welders use.

There is an optimum fuel and oxidant flow rate to the flame or, more precisely, an optimum fuel/oxidant flow rate ratio. If the flame is "oxidant-rich," it is too cool. If it is "fuel-rich," it is too hot. Again, monitoring the absorbance of an analyte standard while varying the flow rates helps find the optimum ratio. Instrument manufacturers' literature will also provide assistance. Safety issues relating to the proper flow rate of these gases will be addressed in Section 9.3.7.

9.3.6 Interferences

Interferences can be a problem in the application of flame AA. Interferences can be caused by chemical sources (chemical components present in the sample matrix that affect the chemistry of the analyte in the flame), or spectral sources (substances present in the flame other than the analyte that absorb the same wavelength as the analyte).

9.3.6.1 Chemical Interferences

Chemical interferences are the result of problems with the sample matrix. For example, viscosity and surface tension affect the aspiration rate and the nebulized droplet size, which, in turn, affect the measured absorbance. The most useful solution to the problem is **matrix matching**, matching the matrix (all components except the analyte) of standards (and blank) as these solutions are prepared with that of the sample (eg., same solvent, same components at the same concentrations, etc.) so that the same matrix effect is seen in all solutions measured and therefore becomes inconsequential. Of course, the complete qualitative and quantitative composition of the sample matrix must be known in order to match it by measuring out individual components for the preparation of the standards and blank. Unfortunately, the analyst usually does not have such sweeping information about the sample.

An alternative is the **standard additions method**. In this method, a certain volume of the sample solution itself is present in the same proportion in all standard solutions. It is equivalent to adding standard amounts of analyte to the sample solution, hence the term "standard additions." This solves the interference problem because the sample matrix is always present at the same component concentrations as in the sample—the matrix is matched. In addition, the sample solution components need not be identified.

The best way to accomplish this is to prepare standards in the usual way—add increasing volumes of a standard analyte solution to a series of volumetric flasks (include zero added)—but also add a volume of the sample solution to each before diluting to the mark with solvent (see Figure 9.14). Thus, you would have a series of standards in which the concentration of analyte *added* would be known, the smallest concentration added being zero. Exactly how much sample solution is used and what "concentration added" values would be prepared would be dictated by what concentration levels, with additions, would produce the desired standard curve. In any case, a diluted sample matrix is present in each standard and the matrices are matched (see Figure 9.15). A disadvantage is that it is impossible to prepare a blank with a matched matrix. Thus, a pure solvent blank, or other approximation, must be used.

The standard curve takes on a slightly different look in the standard additions method. It is a plot of absorbance vs. concentration *added* (to the sample) rather than just concentration. The y-axis in such a plot is not at its true position. It is offset to the right by the concentration in the solution with zero-added concentration, which is the sample solution. The concentration in this solution is the concentration sought. To show it on the graph, the standard curve is extrapolated to intersect with the x-axis. This intersection point is the true position of the y-axis and the concentration of the unknown is represented by the length of x-axis between the two y-axes (see Figure 9.16). The precise concentration in the zero-added solution is found using the equation of the straight line.

The extrapolation means that the curve is extended lower than the lowest standard into a range not covered by the standard curve. This is another disadvantage of this method.

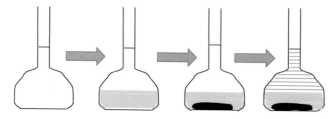

FIGURE 9.14   When preparing a standard for the standard additions technique, volumes of the stock standard and the sample are each pipetted into the flask before it is diluted to the mark with the solvent.

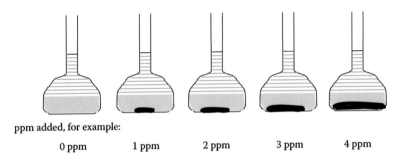

ppm added, for example:

    0 ppm   1 ppm   2 ppm   3 ppm   4 ppm

FIGURE 9.15 In the standard additions technique, increasing volumes of the stock standard and a constant amount of the sample are pipetted into the flasks prior to being diluted to the mark with the solvent. In this way, each of the standards and the sample (0.0 ppm added) have the same matrix.

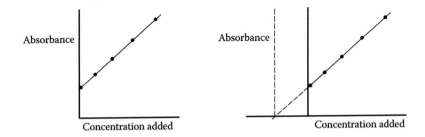

FIGURE 9.16 Illustration of the manipulation of the standard curve in the standard additions method. (Left) Standard curve before extrapolation. Note that the line does not pass through the origin. (Right) After extrapolation. The vertical dashed line represents the true y-axis.

One notable chemical interference occurs when atomization is hindered due to an unusually strong ionic bond between the ions in the ionic formula unit. A well-known example occurs in the analysis of a sample for calcium. The presence of sulfate or phosphate in the sample matrix along with the calcium suppresses the reading for calcium because of limited atomization due to the strong ionic bond between calcium and the sulfate and phosphate ions. This results in a low reading for the calcium in the sample in which this interference exists. The usual solution to this problem is to add a substance to the sample that would chemically free the element being analyzed, calcium in our example, from the interference. With our calcium example, the substance that accomplishes this is lanthanum. Lanthanum sulfates and phosphates are more stable than the corresponding calcium salts, and thus the calcium is free to atomize when lanthanum is present. Thus, analyses for calcium usually include addition of a lanthanum salt to the sample and standards.

Other chemical interferences may be overcome by changing oxidants (air or nitrous oxide are the choices) or by changing the fuel/oxidant flow ratio, giving a lower or higher flame temperature (see Application Note 9.2 for an interesting application of the standard additions method).

9.3.6.2 Spectral Interferences

Spectral interferences are due to substances in the flame that absorb the same wavelength as the analyte causing the absorbance measurement to be high. The interfering substance is rarely an element, however, because it is rare for another element to have a spectral line at exactly the same wavelength, or near the same wavelength, as the primary line of the analyte. However, if such an interference is suspected, the analyst can tune the monochromator to a secondary line of the analyte to solve the problem, but with a less desirable sensitivity.

Absorption due to the presence of light-absorbing molecules in the flame and light dimming due to the presence of small particles in the flame are much more common spectral interferences.

APPLICATION NOTE 9.2: ANALYSIS OF INTRAVENOUS INFUSION SOLUTIONS FOR KCl AND NaCl BY THE METHOD OF STANDARD ADDITIONS

Intravenous infusion solutions can be analyzed for KCl and NaCl by atomic spectroscopy and the method of standard additions. Stock standard solutions of KCl and NaCl are prepared. Then a portion of the KCl stock standard and the NaCl stock standard are combined into one diluted solution at a particular concentration level for each. The varying volume aliquots of this diluted stock standard are pipetted into 100-mL volumetric flasks along with 5 mL of a diluted infusion solution before diluting to the mark, in the same manner as indicated in Figure 9.15. Readings for both sodium and potassium are then recorded for each standard.

Two standard curves are created, one for potassium and one for sodium, each appearing as in Figure 9.16, and the concentration determined for each "0-added" solution. The concentrations of each in the undiluted infusion solutions are then calculated.

(*Reference:* Watson, D., *Pharmaceutical Analysis: A Textbook for Pharmacy Students and Pharmaceutical Chemists*, 2nd edition (2005), pp. 140–141.)

Such phenomena are referred to as **background absorption**. To solve the problem, **background correction** techniques have been employed. These techniques are designed to isolate the background absorption and subtract it out. A common technique is to use a deuterium arc lamp, a light source that we indicated previously is used in UV-Vis molecular spectrophotometers, to direct a continuum light beam through the optical path simultaneous with the light from the spectral line source (the hollow cathode lamp or electrodeless discharge lamp). Both light beams are absorbed by the sample in the flame, but the continuum source absorption is measured separately (with the use of chopping frequencies) and subtracted from the total in a manner similar to flame emissions.

9.3.7 SAFETY AND MAINTENANCE

There are a number of important safety considerations regarding the use of AA equipment. These center around the use of highly flammable acetylene, as well as the use of a large flame, and the possible contamination of laboratory air by combustion products.

All precautions relating to compressed gas cylinders must be enforced—the cylinders must be secured to an immovable object, such as a wall; they must have approved pressure regulators in place; they must be transported on approved carts; etc. Tubing and connectors must be free of gas leaks. There must be an independently vented fume hood in place over the flame to take care of toxic combustion products. Volatile flammable organic solvents and their vapors, such as ether and acetone, must not be present in the laboratory when the flame is lit.

Precautions should be taken to avoid flashbacks. Flashbacks result from improperly mixed fuel and air, such as when the flow regulators on the instrument are improperly set or when air is drawn back through the drain line of the premix burner. Manuals supplied with the instruments when they are purchased give more detailed information on the subject of safety.

Finally, periodic cleaning of the burner head and nebulizer is needed to ensure minimal noise level due to impurities in the flame. Scraping the slot in the burner head with a sharp knife or razor blade to remove carbon deposits and/or removing the burner head for the purpose of cleaning in an ultrasonic cleaner bath are two commonplace maintenance chores.

The nebulizer should be dismantled and inspected and cleaned periodically to remove impurities that may be collected there. See Application Note 9.1 for an interesting application of flame AA.

9.4   GRAPHITE FURNACE ATOMIC ABSORPTION

9.4.1   GENERAL DESCRIPTION

Another very important atomic absorption technique that we briefly mentioned in Section 9.2 is graphite furnace atomic absorption, or graphite furnace AA, also called electrothermal atomic absorption. As stated in the earlier section, the atomizer for graphite furnace AA is a small hollow graphite cylinder (see Figure 9.17) that can be quickly electrically heated to a very high temperature. A small volume of the sample solution is placed inside the cylinder. This "furnace" is electrically brought to a very high temperature in order to atomize the sample inside. The cylinder becomes filled with an atomic "cloud" and the light beam is directed through it, absorbed, and the absorption measured.

Instruments that utilize a flame for the atomizer can also utilize the graphite furnace. It is a matter of replacing the flame module with the furnace module and lining it up with the light beam. The same light source (usually a hollow cathode lamp) and the same optical path are used (see Figure 9.18 and compare with Figure 9.11a).

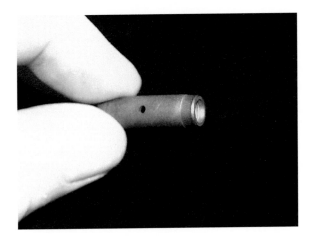

FIGURE 9.17   Small graphite cylinder that is the graphite furnace. The small hole is for introducing samples to the interior of the cylinder.

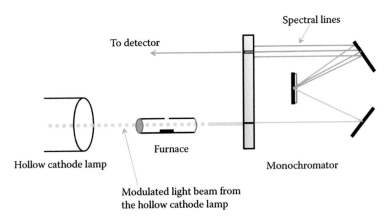

FIGURE 9.18   Illustration of the light path for a single-beam atomic absorption spectrophotometer with a graphite furnace as the atomizer.

The furnace module is much more than a small graphite cylinder, however (refer back to Figure 9.3). The furnace must be protected from thermal oxidation (conversion of graphite carbon to oxides of carbon in the presence of air at the high temperatures). Thus, an inert gas (argon) must be circulated inside and outside the cylinder to protect it from contact with air. A source of argon and purge controls must be provided. Flowing water must be circulated through an enclosed housing for the purpose of rapidly cooling the furnace between runs. A source of cold water and the appropriate plumbing must be provided. In addition, a relatively high-voltage power source is used and electrical contacts to the tube must be provided.

The graphite cylinder must therefore be encased in a rather complex maze of argon flow channels for flow both internal, and external to the furnace, channels for water flow, and electrical contacts. Despite the apparent clutter, the unit must still be open on each end for the light beam to pass through. Quartz "windows" cap each end of the assembly and a hole for introducing the sample to an internal platform through the hole in the cylinder is included (see Figure 9.19 for an illustration). The internal platform is shown in Figure 9.20.

FIGURE 9.19 Photograph looking inside the graphite cylinder. The sample platform is visible.

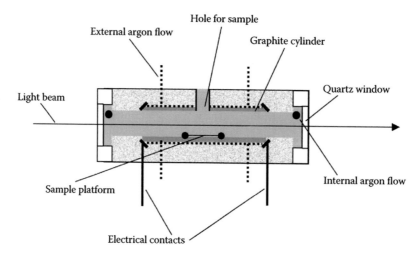

FIGURE 9.20 Drawing of the graphite furnace assembly. Channels for the cooling water are not shown.

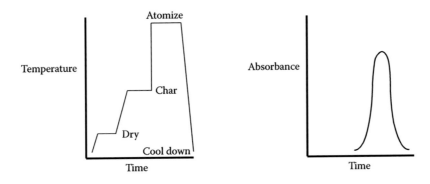

FIGURE 9.21 Illustration of a temperature program for a graphite furnace experiment (left), and the absorbance signal that results (right). The absorbance signal corresponds to the third temperature plateau (see text for a more detailed explanation).

The temperature of the furnace is programmed to take into account several effects of the heating so as to optimize the measurement. First, since the evaporation of the solvent occurs at a much lower temperature than the atomization, the temperature is initially ramped to the solvent boiling point for a short period. After this, the temperature is ramped still further to a temperature at which organic matter is charred, vaporized, and swept away. It is important to note that the flowing argon continuously sweeps through the interior of the furnace such that solvent molecules and the "smoke" from the charring are swept from the furnace.

Finally, the temperature is stepped to the atomization temperature (approximately 2500 K). It is at this temperature that the atomic cloud is present in the furnace and the absorbance is measured. This absorbance is monitored over the period of time while the atomic cloud is in residence and usually recorded as an absorbance vs. time peak. The flow of argon continues to sweep through the interior of the furnace and the absorbance signal is therefore a transient one. With some units, it is possible to momentarily stop the argon flow so as to increase the residence time of the atomic cloud (see Figure 9.21 for an illustration of a temperature vs. time program and the absorbance vs. time signal that results). A final step (not illustrated in the figure) to a clean-out temperature prepares the furnace for the next sample.

9.4.2 ADVANTAGES AND DISADVANTAGES

The flame atomizer continuously sweeps a given analyte concentration across the light path such that the number of ground state atoms present in the light path at a given moment is comparatively small. This means that this atomizer is not highly efficient, does not produce a result that is as sensitive as may be desired and, at a minimum, requires several milliliters of sample solution to obtain a reliable absorbance reading. The graphite furnace, on the other hand, confines the atomic vapor produced from a small sample volume (microliters) to a very small space in the path of the light resulting in a much larger concentration of ground state atoms in this space compared to the flame. This results in a much more sensitive measurement, meaning that much smaller concentrations in the sample solution can be detected and measured. The detection limit is also much improved over flame AA (see Section 9.7). While the excellent sensitivity and the use of extremely small sample volumes are the major advantages, see Application Note 9.3 for yet another reason that the graphic furnace technique may be chosen for some analyses.

As for disadvantages, absorbance readings from the graphite furnace are less reproducible due to a less-controlled pathlength and physical presence in the furnace. In addition, interferences (background absorptions) are more common and more severe. So-called **matrix modifiers** are often used to control interferences. These are chemicals added to the sample to assist in the

APPLICATION NOTE 9.3: TRACE METALS IN ENGINE OIL

It can be important to determine the concentration of metals in engine lubricating oils. Oil samples cannot be easily dissolved in water, so organic solvents are used. Since organic solvents are flammable, the use of a flame atomizer can result in problems. The electrothermal (graphite furnace) atomization, on the other hand, is not adversely affected by this situation. This, combined with the advantages of greater sensitivity and small sample sizes, make the graphite furnace technique ideal. Standards are prepared using the same solvent used for dissolving the oil. A "blank oil" is added to all the standards to match the matrix of the samples. Both the standards and samples have a high viscosity as a result, and this can present problems pipetting and diluting to the mark of a flask. These can be solved by weighing the oil instead of measuring the volume and also weighing the solution as you add the solvent instead of diluting to the mark of a flask.

(*Reference:* McKenzie, T., *The Determination of Trace Metals in Engine Oil Using the GTA-95 Tube Atomizer*, Agilent Technologies, www.agilent.com/chem [2010].)

separation of the analyte from the matrix prior to atomization such that the interference is not present at the time of atomization. Background absorptions can also be eliminated with the use of **background correction** techniques such as the use of a continuum source as we discussed for flame AA. However, a more common background correction procedure for graphite furnace AA is the **Zeeman background correction** technique. In this technique, a powerful magnetic field is used to shift the energy levels of atoms and molecules and therefore to shift the wavelengths that are absorbed. Pulsing the magnetic field on and off allows subtraction of the background absorption.

9.5   INDUCTIVELY COUPLED PLASMA

Besides flame AA and graphite furnace AA, there is a third atomic spectroscopic technique that enjoys widespread use. It is called the **inductively coupled plasma spectroscopy**, or **ICP**. Unlike flame AA and graphite furnace AA, the ICP technique measures the emissions from an atomization/ionization/excitation source rather than the absorption of a light beam passing through an atomizer.

Recalling Figure 9.4, we know that thermal energy sources, such as a flame, atomize metal ions. But we also know that a small number of metals have atoms that are raised to the excited state at the temperature of a flame. These atoms experience resonance between the excited state and ground state such that the emissions that occur when the atoms drop from the excited state back to the ground state can be observed and measured. Thus, a sodium solution aspirated into a flame will turn the flame yellow, calcium turns a flame orange, etc. Only a handful of elements will do this because the flame is not hot enough, i.e., does not provide enough energy, for other metals. The ICP emission source, on the other hand, is much hotter and does provide the required energy for the other metals (see Figure 9.22). Thus, ICP is the most important among emission techniques because the excitation source is much hotter, resulting not only in atomization, but also in ionization and many more emissions and more intense emissions, than from a flame.

So, ICP does not use a flame. Rather, as briefly mentioned in Section 9.2, this atomization/ionization/excitation source is a high-temperature plasma. A plasma, in this context, is a gaseous mixture of atoms, cations, and electrons that is directed though an induced magnetic field, causing "coupling" of the ions in the mixture with the magnetic field, hence the name "inductively coupled plasma." The interaction with the magnetic field causes atoms and ions to undergo extreme resistive

The 2500 K temperature of a flame can excite atoms of sodium, calcium, lithium, etc., which do not require much energy

But to excite atoms (and even ions) of other elements, more energy is needed – the energy supplied from the very hot plasma at temperatures exceeding 6000 K

FIGURE 9.22 Only those metals with a low excitation energy are excited by a flame. However, due to its very high temperature, the ICP torch excites all metals.

heating. The ICP "torch" resembles a flame as the plasma emerges from the magnetic field (see Figure 9.23). The temperature of the ICP torch is in excess of 6000 K.

In the ICP system, as shown in Figure 9.23, a quartz tube serves as the conduit for the plasma through the magnetic field. The quartz tube is actually three concentric tubes in one. The plasma (and sample) path is the center tube. The other tubes contain flowing auxiliary and coolant gases. The plasma itself consists of a flowing stream of argon gas that has been partially ionized with the use of an in-line electrical discharge device (not shown in the figure) prior to reaching the magnetic field. The magnetic field is induced with the use of an RF generator (not shown) and an induction

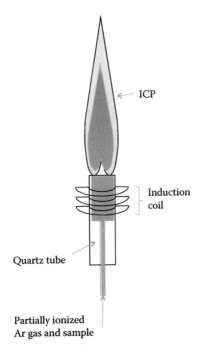

FIGURE 9.23 Drawing of the ICP torch (see text for a full description).

coil by passing an alternating RF current through the coil. This coil consists of a three- or four-turn heavy copper wire, as drawn in Figure 9.23.

The sample is drawn from its container with a peristaltic pump to a nebulizer. There are several designs for the nebulizer, but it performs the same function as the flame AA nebulizer, converting the sample solution into fine droplets (with the larger droplets flowing to a drain) that flow with the argon to the torch. The emissions are measured by the spectrometer at a particular "zone" in the torch, often called the "viewing zone" or "analytical zone."

As stated, unlike a flame, in which only a very limited number of metals emit light because of the low temperature, virtually all metals present in a sample emit their line spectrum from the ICP torch. Not only does this make for a very broad application for ICP, but it also means that a given sample may undergo very rapid and simultaneous multielement analysis. With this in mind, it is interesting to consider the options for the optical path for the ICP instrument.

One is the sequential optical path in which a monochromator is used and the emissions are measured one wavelength after the other sequentially by rotating the dispersing element while the individual lines emerge from the exit slit. In this case, one phototube measures all emission lines (see Figure 9.24a). The advantages are (1) all wavelengths of all elements can be utilized, and (2) spectral interferences can be avoided by careful wavelength selection. The disadvantage is that it is slower than the alternative.

The other option is comparable to the diode array design used with UV-Vis molecular absorption instruments previously described in Chapter 8. This design is known as the "simultaneous

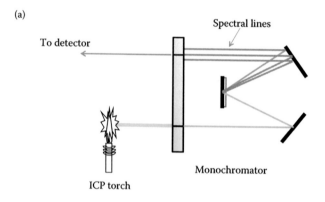

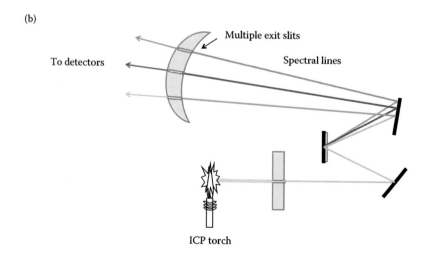

FIGURE 9.24 Drawings of the two options for ICP instruments. (a) The sequential design. (b) The simultaneous design.

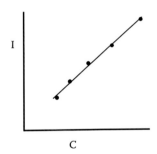

FIGURE 9.25   Standard curve for ICP quantitative analysis: intensity vs. concentration.

direct reading polychromator" design in which there are a number of exit slits and phototubes for measuring a number of lines at once (see Figure 9.24b). The wavelengths to be measured are set by the manufacturer. The advantages of this design are (1) it is fast, since all the desired wavelengths are measured simultaneously, and (2) it is good for samples with known matrices because it can be preset to the optimum wavelengths. Disadvantages are (1) that the wavelength choices are changeable, but it is expensive, (2) spectral interferences can be a problem.

To repeat a statement in our opening paragraph of this section, ICP measures the emission of light from the atomization/ionization/excitation source (ICP torch) rather than absorption of light by atoms in this source, as was the case with flame AA and graphite furnace AA. For this reason, it should be stressed that quantitative analysis is not based on Beer's law. Rather, the intensity of the emissions is proportional to concentration—the higher the concentration, the more intense the emission. Thus, for quantitative analysis, a series of standard solutions that bracket the expected concentration in the unknown is prepared and the intensity for each measured. The standard curve is then a plot of intensity vs. concentration as shown in Figure 9.25.

The advantages of ICP are that the emissions are of such intensity that it is usually more sensitive than flame AA (but less sensitive than graphite furnace AA). In addition, the concentration range over which the emission intensity is linear is broader. These two advantages, coupled with the possibility of simultaneous multielement analysis offered by the direct reader polychromator design make ICP a very powerful technique. The only real disadvantage is that the instruments are more expensive.

ICP has been coupled with another analytical technique known as **mass spectrometry**. This technique, as well as the technique coupling it with ICP, will be described in Chapter 13.

9.6   MISCELLANEOUS ATOMIC TECHNIQUES

9.6.1   FLAME PHOTOMETRY

Flame photometry is the name given to the technique that measures the intensity of the light emitted by analyte atoms in a flame. It is the oldest of all the atomic techniques. It is not highly applicable because of the low temperature of the flame. Only a handful of elements can be measured with this technique, including sodium, potassium, lithium, calcium, strontium, and barium. The technique was formerly used by hospital and clinical laboratories for measuring these elements in body fluids but has been supplanted by the use of ion-selective electrodes (see Chapter 14). Instruments designed for flame atomic absorption can be used as flame photometers. It is a mode of operation in which the hollow cathode is not used and the detector measures the flame emissions. Advantages are that the instruments are simple and inexpensive. Also, the technique provides adequate sensitivities for the metals listed. Disadvantages include lower sensitivity compared with other techniques. The standard curve for quantitative analysis is intensity vs. concentration similar to what is shown in Figure 9.25 for ICP.

9.6.2 COLD VAPOR MERCURY

Mercury is the only metal that is a liquid at ordinary temperatures. It is therefore also the only metal that has a significant vapor pressure at ordinary temperatures. For this reason, it is possible to obtain mercury atoms in the gas phase for measurement by atomic absorption without the use of thermal energy. It is a matter of chemically converting mercury ions in the sample into elemental mercury, getting it in the gas phase and channeling it into the path of the light of an atomic absorption instrument.

Reducing agents typically used for the conversion of mercury ions in solution to elemental mercury include stannous chloride ($SnCl_2$) and sodium borohydride ($NaBH_4$). Elemental mercury, the product of the reduction, is converted to the gas phase by simply bubbling air through the solution. The mercury/air mixture is channeled to a quartz absorption cell, a long-pathlength tube positioned in the path of the light from a mercury hollow cathode lamp, and the absorption measured. The gas phase mixture is ultimately channeled to waste. The sensitivity is far greater than what can be achieved with flame AA.

9.6.3 HYDRIDE GENERATION

A useful technique for the otherwise difficult elements arsenic, bismuth, germanium, lead, antimony, selenium, tin, and tellurium is the hydride generation technique. This technique is similar to the cold vapor mercury technique in that a reducing agent, in this case sodium borohydride, is used to convert ions of these metals to a more useful state—in this case, to hydrides, such as AsH_3. Hydrides of the metals listed above are volatile and can be channeled to an absorption cell using moving air as in the cold vapor mercury technique. Once in the absorption cell, however, the gaseous mixture must be heated in order to achieve atomization. Heating can be accomplished with a low-temperature flame or with an electrical heating system. As with cold vapor mercury, the vapor in the cell is ultimately channeled to waste. Absorbance is measured and Beer's law applies.

Advantages include a very acceptable method for the elements listed when other techniques for these elements fail. Sensitivity for these elements is also very good.

9.6.4 SPARK EMISSION

A technique that utilizes a solid sample for light emission is spark emission spectroscopy. In this technique, a high voltage is used to excite a solid sample held in an electrode cup in such a way that when a spark is created with a nearby electrode, atomization, excitation, and emission occurs and the emitted light is measured. Detection of what lines are emitted allows for qualitative analysis of the solid material. Detection of the intensity of the lines allows for quantitative analysis.

9.6.5 ATOMIC FLUORESCENCE

Molecular fluorescence, described in Chapter 8, has a counterpart in atomic spectroscopy called atomic fluorescence. In order for atoms to emit light of a different wavelength than was absorbed, they must detour to an intermediate excited state before returning to the ground state. This can occur when the light absorbed elevates the atoms to a higher excited state than first excited state, so that the route back to the ground state includes the first excited state and thus a fluorescence—light of less energy being emitted. The light must be measured at right angles, as with the molecular instruments. This technique is not popularly used.

9.7 SUMMARY OF ATOMIC TECHNIQUES

We have spoken frequently in this chapter about sensitivity and detection limit in reference to advantages and disadvantages of the various techniques. Sensitivity and detection limit have specific definitions in AA. **Sensitivity** is defined as the concentration of an element that will produce an absorption of 1% (%T of 99%). It is the smallest concentration that can be determined with a reasonable degree of precision. **Detection limit** is the concentration that gives a readout level that is double the electrical noise level inherent in the baseline. It is a qualitative parameter in the sense that it is the minimum concentration that can be detected, but not precisely determined, like a "blip" that is barely seen compared with the electrical noise on the baseline. It would tell the analyst that the element is present, but not necessarily at a precisely determinable concentration level. A comparison of detection limits for several elements for the more popular techniques is given in Table 9.2.

The description and application of the techniques discussed in this chapter are given in Table 9.3.

TABLE 9.2
Detection Limits for Selected Elements (in µg/L)

Element	Flame AA	Graphite Furnace AA	ICP	Cold Vapor Hg	Hydride
Arsenic	150	0.05	2	—	0.03
Bismuth	30	0.05	1	—	0.03
Calcium	1.5	0.01	0.05	—	—
Copper	1.5	0.014	0.4	—	—
Iron	5	0.06	0.1	—	—
Mercury	300	0.6	1	0.009	—
Potassium	3	0.005	1	—	—
Zinc	1.5	0.02	0.2	—	—

Source: Perkin Elmer Instruments literature. With permission.

TABLE 9.3
A Summary of the Techniques Described in This Chapter

Technique	Principle	Comments
Flame AA	Light absorbed by atoms in a flame is measured	A well-established technique that remains very popular for a wide variety of samples and analytes
Graphite furnace AA	Light absorbed by atoms in a small graphite cylinder furnace is measured	A popular and very sensitive technique that is applicable to small volumes of sample solution. Precision not as good as other techniques
ICP	Light emitted by atoms and monoatomic ions in an inductively coupled plasma is measured	A popular technique useful over a broad concentration range. Multielement analysis is possible. Instruments are costly
Hydride generation	Light absorbed by atoms derived from hydrides that are generated by chemical reaction are measured	An excellent technique for a limited number of analytes that are difficult to measure otherwise
Cold vapor mercury	Light absorbed by atoms of mercury generated by chemical reaction at room temperature is measured	An excellent technique for mercury analysis
Flame photometry	Light emitted by atoms in a flame is measured	A well-established technique for a limited number of elements. Not used much due to emergence of other techniques
Spark emission	Light emitted by atoms generated from a powdered solid sample in a spark source is measured	Useful for qualitative analysis of solid materials
Atomic fluorescence	Light emitted by atoms excited by the absorption of light in a flame is measured	Not a popular technique but can offer sensitivity advantages

EXPERIMENTS

EXPERIMENT 25: QUANTITATIVE FLAME ATOMIC
ABSORPTION ANALYSIS OF A PREPARED SAMPLE

Note: Your instructor will select the metal to be the analyte in this experiment.

Remember to wear safety glasses.

1. Prepare calibration standards of the selected metal, with concentrations suggested by your instructor, from a 1000-ppm stock solution. Dilute each to the mark with distilled water and shake well.
2. Obtain an unknown from your instructor and dilute to the mark with distilled water and shake well.
3. Select the hollow cathode lamp to be used and install it in the instrument. Your instructor may wish to demonstrate the lamp installation first. Also check the mixing chamber drain system to be sure it is ready.
4. A number of instrument variables need to be set prior to making measurements. These include slit, wavelength, lamp current, lamp alignment, amplifier gain, aspiration rate, burner head position, acetylene pressure, air pressure, acetylene flow rate, and air flow rate. Some instruments are rather automated in the set up process, whereas others are not. Your instructor will provide detailed instructions for the particular instrument you are using. Be sure to turn on the fume hood above the flame.
5. The blank for this experiment is distilled water. Follow the instructions provided for your instrument for zeroing with the blank and reading the standards, the unknown, and control.
6. Follow the instructions provided for instrument shutdown.
7. Create the standard curve, using the procedure practiced in Experiment 13, by plotting absorbance vs. concentration and determine the concentration of the unknown and control.
8. Maintain the logbook for the instrument used in this experiment. Record the date, your name(s), the experiment name or number, the correlation coefficient, and the results for the control sample. Also plot the control sample results on a control chart for this experiment posted in the laboratory.

EXPERIMENT 26: THE ANALYSIS OF SOIL SAMPLES FOR
IRON USING ATOMIC ABSORPTION

Note: Soil samples may be dried and crushed ahead of time. If not, you may report the results on an "as-received" basis.

It is presumed that you have previously performed Experiment 25 or other introductory experiment on the AA equipment. If you have not done this, carefully read through the Experiment 25 procedure so that you are aware of certain precautions and instrument set-up requirements. Remember to wear safety glasses.

1. Prepare 500 mL of one extracting solution with two solutes. Diluting 25 mL of 1 N HCl and 12.5 mL of 1 N H_2SO_4 to 500 mL with distilled water.
2. Weigh 5.00 g of each soil sample to be tested into dry 50-mL Erlenmeyer flasks. Pipet 20 mL of extracting solution into each flask. Stopper and shake on a shaker for at least 20 min.
3. Filter all samples simultaneously into small dry beakers using dry Whatman #42 filter paper, transferring only the supernatant. Do not rinse the soil samples in the filters, since that would dilute the sample extracts by an inexact amount.
4. While waiting for the samples to be shaken and/or filtered, prepare standards that are 0.5, 1.0, 3.0, and 5.0 ppm iron. Use a 1000-ppm iron solution, and prepare one 100-ppm intermediate stock solution. Use 50-mL volumetric flasks, and use the extracting solution as

the diluent. A control may also be provided. Dilute it also to the mark with the extracting solution.

5. Set up the atomic absorption instrument and obtain absorbance readings for all solutions using the extracting solution for the blank. Follow the instructions provided for instrument shutdown.

6. Create the standard curve, using the procedure practiced in Experiment 13, by plotting absorbance vs. concentration and determine the concentration of the unknown and control. Calculate the ppm Fe in the soil by the following calculation:

$$\text{ppm Fe (extractable)} = \text{ppm (in solution)} \times 4$$

7. Maintain the logbook for the instrument used in this experiment. Record the date, your name(s), the experiment name or number, the correlation coefficient, and the results for the control sample. Also plot the control sample results on a control chart for this experiment posted in the laboratory.

EXPERIMENT 27: DESIGN AN EXPERIMENT: A STUDY TO DETERMINE THE OPTIMUM pH FOR THE EXTRACTION SOLUTION FOR EXPERIMENT 26

The procedure for Experiment 26 calls for a very acidic solution, a combination of HCl and H_2SO_4, to be used for the extraction solution for the soil. And this makes sense based on the discussion in Chapter 2 where HCl is indicated as the acid of choice for dissolving iron and iron oxides that may be found in any sample. Design an experiment that would prove that such an acidic solution is indeed what is needed. The end goal would be a plot of absorbance vs. pH covering a very broad pH range, indeed even into the basic range, for the extraction solution. Write a step-by-step procedure as in other experiments in this book for preparing the needed solutions and using the spectrophotometer to measure the absorbances of each. After your instructor approves your procedure, perform the experiment in the laboratory.

EXPERIMENT 28: THE DETERMINATION OF SODIUM IN SODA POP

Note: It is presumed that you have previously performed Experiment 25 or other introductory experiment on the AA equipment. If you have not done this, carefully read through the Experiment 25 procedure so that you are aware of certain precautions and instrument setup requirements. Remember to wear safety glasses.

1. Prepare 100 mL of a 100-ppm sodium solution from the available 1000 ppm. Obtain soda pop samples and degas approximately 5–10 mL of each.

2. From the 100-ppm Na stock, prepare standards of 1, 3, 5, and 7 ppm in 25-mL volumetric flasks. A control sample may also be provided. Dilute to the mark with distilled H_2O.

3. Pipet 1 mL of each soda pop sample into separate 25-mL volumetric flasks and dilute to the mark with distilled water. Shake.

4. Obtain absorbance values for all standards and samples and control using an AA instrument.

5. Follow the instructions provided for instrument shutdown.

6. Create the standard curve, using the procedure practiced in Experiment 13, by plotting absorbance vs. concentration and determine the concentration of the unknown and control. Multiply the concentrations by 25 to get the ppm Na in the soda pop. Also calculate the milligrams of sodium in one 12-oz can of the soda pop.

7. Maintain the logbook for the instrument used in this experiment. Record the date, your name(s), the experiment name or number, the correlation coefficient, and the results for the control sample. Also plot the control sample results on a control chart for this experiment posted in the laboratory.

EXPERIMENT 29: DESIGN AN EXPERIMENT: SODIUM IN SODA POP BY THE STANDARD ADDITIONS METHOD

In Experiment 28, you determined the ppm Na in soda pop using the normal standard curve procedure. The analysis can also be done using the standard additions method. Review the standard
additions method as presented in this chapter and then write a step-by-step procedure as in other
experiments in this book for preparing the needed solutions and using the AA spectrophotometer
to measure the absorbances of each. You should assume that the standard concentrations used in
Experiment 28 represent the linear range. Prepare your stock standard using solid NaCl. After your
instructor approves your procedure, perform the experiment in the laboratory.

EXPERIMENT 30: DESIGN AN EXPERIMENT: ANALYSIS OF FERTILIZER FOR POTASSIUM

The potassium in solid fertilizer granules can be extracted using a simple liquid–solid extraction
procedure and the potassium in the extract determined by flame AA. With an eye on the minimal information given below, design an experiment for this analysis. Then write a step-by-step
procedure as in other experiments in this book for preparing the needed solutions and using the
AA spectrophotometer to measure the absorbance of each. After your instructor approves your
procedure, perform the experiment in the laboratory.

Sample: Dry multicolored granules.
Sample preparation: Place 2.5 g in a 250-mL volumetric flask. Add 150 mL of distilled water
and boil for 30 min. Cool, dilute to volume with distilled water and shake. Filter through a
dry Whatman #42 filter paper, or equivalent. For samples containing less than 20% K_2O,
transfer a 25-mL aliquot to a 100-mL flask, dilute to volume, and shake thoroughly. For
samples containing more than 20% K_2O, use a smaller aliquot.
Wavelength: 404.4 nm. This is a secondary, less sensitive, line. It is used due to the high concentration of potassium.
Standards preparation: Prepare a stock standard with a concentration of 1000 ppm using solid
KCl.
Linear range: Up to 600 ppm.

QUESTIONS AND PROBLEMS
Review and Comparisons

1. Absorption spectra for atoms are characterized by spectral lines. Explain this statement.
2. (a) With help of an energy level diagram, tell what is the source of the "lines" in the line
spectra of metals.
   (b) We speak of line spectra often as both emission spectra and as absorption spectra.
Name two common sources of atomic emission spectra.
3. Explain why emission lines occur at the same wavelengths as absorption lines.
4. What are three major differences between UV-Vis molecular spectrophotometry and
atomic absorption spectrophotometry?
5. What is an "atomizer?" Identify at least four atomizers used in atomic spectroscopy.

Brief Summary of Techniques and Instrument Designs

6. Name two techniques that are atomic absorption techniques and two techniques that are
atomic emission techniques.

Flame Atomic Absorption

7. Atomization occurs in a flame as one of several processes after a solution of a metal ion is aspirated. What other processes occur and in what order?

8. (a) The percentage of atoms in a typical air/acetylene flame that are in the excited state at any point in time is less than 0.01%. What does this have to do with the fact that flame AA is a useful technique?

   (b) Define resonance as it pertains to atomic spectroscopy.

9. What temperature can be achieved by each of the following flames?
   (a) air/natural gas
   (b) air/acetylene
   (c) N_2O/acetylene
   (d) oxygen/acetylene

10. Which of the flames listed in question 9 are most commonly used? Why?

11. Even though an oxygen/acetylene flame can produce the hottest temperature, what disadvantage does it possess that limits its usefulness in practice?

12. There is no monochromator placed between the light source and the flame in an AA experiment. Why is this?

13. The hollow cathode lamp must contain the metal to be analyzed in the cathode. Explain.

14. A typical AA laboratory has many hollow cathode lamps available and interchanges them in the instrument frequently. Explain why this is in terms of the processes occurring in the lamp and in the flame.

15. What is an EDL?

16. What is a nebulizer? Describe its use in conjunction with a premix burner.

17. Describe the design and use of the premix burner in flame AA.

18. Why must a premix burner have a drain line attached? What safety hazard exists because of this drain line and how do we deal with it?

19. What purpose does the light chopper serve in an AA instrument?

20. How does a single-beam atomic absorption instrument differ from a double-beam instrument? What advantages does one offer over the other?

21. How does a double-beam atomic instrument differ from a double-beam molecular instrument?

22. Since the sample "cuvette" is the flame located in an open area of the AA instrument, rather than a glass container held in the light tight box in the case of molecular instruments, how is room light prevented from reaching the detector and causing an interference?

23. Define what is meant by the "primary line" and the "secondary lines" in a line spectrum.

24. Name five instrument controls that must be optimized when setting up an atomic absorption spectrophotometer for measurement. Explain why each must be optimized.

25. Does Beer's law apply in the case of flame AA? Explain.

26. What is the difference between a chemical interference and a spectral interference?

27. What is meant by matrix matching? Why is this important in a flame AA experiment?

28. What is the standard additions method and why does it help with the problem of chemical interferences?

29. What are two disadvantages of the standard additions method?

30. Why is lanthanum used in an analysis for calcium?

31. Define "background absorption" and "background correction." How does the use of a continuum light source help with background correction?

32. What are three safety issues with regard to flame AA and how are they dealt with?

Graphite Furnace Atomic Absorption

33. Give a brief description of the graphite furnace method of atomization.
34. Describe exactly how the atomic vapor is produced in a graphite furnace.
35. What is the difference in the optical path of graphite furnace AA and flame AA?
36. The graphite furnace assembly utilizes argon gas, cold water, and a source of high voltage. Explain.
37. Describe the temperature program applied to a graphite furnace and explain each of the processes involved.
38. Why must the graphite furnace be protected from air?
39. Why is the absorbance signal developed in the case of the graphite furnace AA technique said to be "transient"?
40 What are the advantages and disadvantages of the graphite furnace atomizer?
41. What is meant by Zeeman background correction?

Inductively Coupled Plasma

42. What do the letters ICP stand for? Is the ICP technique more closely related to AA or flame photometry? Explain.
43. Describe the ICP analysis method in detail.
44. What are the advantages of the ICP technique?
45. What are the advantages of the following atomic techniques over the standard flame atomic absorption?
    (a) Graphite furnace AA
    (b) ICP

Miscellaneous Atomic Techniques

46. Why is the cold vapor mercury technique good only for mercury?
47. Describe the hydride generation technique. Why is it useful?
48. Compare atomic absorption (both flame and graphite furnace), ICP, flame photometry, cold vapor mercury, hydride generation, atomic fluorescence and spark emission in terms of (a) the process measured, (b) instrumental components and design, and (c) data obtained.
49. Match each statement with a choice or choices from the following list:
    flame AA, flame photometry, atomic fluorescence, ICP, graphite furnace AA, Spark emission
    (a) The technique that uses a partially ionized stream of argon gas called a plasma.
    (b) The technique in which a light source is used to excite atoms, the emission from which is then measured.
    (c) The three techniques that require light to be directed at the atomized ions.
    (d) The technique that utilizes a solid sample.
    (e) The four techniques that measure some form of light emission by excited atoms present in an atomizer.
    (f) The absorption technique in which the atoms are in the path of the light for a relatively short time.
    (g) The technique in which the emitting source may reach a temperature of 6000 K.
50. Answer the following questions true or false:
    (a) In flame atomic absorption, the flame serves solely as an atomizer (in addition to being the sample "container").
    (b) Line spectra are emitted by atoms in a flame.
    (c) In ICP, one needs a light source to excite the atoms in a flame.

(d) In ICP and flame photometry, the standard curve is a plot of emission intensity vs. concentration.

(e) In flame atomic absorption, the monochromator is placed between the light source and the flame.

(f) The population of atoms excited by the flame is what is measured in atomic absorption.

(g) In flame AA, the pathlength, b, is the width of the flame.

(h) The hollow cathode lamp has a tungsten filament.

(i) Atoms are raised to the excited state within a hollow cathode lamp.

(j) Detection limits are generally better with graphite furnace AA than with any other atomic technique.

(k) A "premix" burner has a drain line attached.

(l) In a "premix" burner, the fuel, oxidant, and sample solution meet at the base of the flame.

(m) The graphite furnace is an example of a nonflame atomizer.

(n) A "flashback" results from an improperly mixed sample solution.

(o) The fuel used in a flame AA unit is typically natural gas.

(p) Lanthanum is frequently used to prevent chemical interferences when analyzing for calcium.

(q) In the method of standard additions, standard quantities of the analyte are added to the blank in increasing amounts.

(r) The method of standard additions utilizes Beer's law.

(s) The atomic fluorescence technique requires a light source.

(t) The spark emission technique requires a dispersing element, but not a monochromator.

Summary of Atomic Techniques

51. Differentiate between detection limit and sensitivity. Compare flame AA, graphite furnace AA, ICP, cold vapor mercury, and hydride generation as to applicability and detection limit.

Report

52. For this exercise, your instructor will select a "real-world" atomic spectroscopic analysis method, such an application note from an instrument vendor or from a methods book, a journal article, or Web site, and will give it to you as a handout. Write a report giving the details of the method according to the following scheme. On your paper, write (or type) "(a)," "(b)," etc. to clearly present your response to each item.

(a) Method Title: Give a more descriptive title than what is given on the handout.

(b) Specific Technique Used: Is it flame AA, graphite furnace A, ICP, or what?

(c) Type of Material Examined: Is it water, soil, food, a pharmaceutical preparation, or what?

(d) Analyte: Give both the name and the symbol or formula of the analyte.

(e) Sampling Procedure: This refers to how to obtain a sample of the material. If there is no specific information given as to obtaining a sample, make something up. If you want to be correct in what you write, do a library or other search to discover a reasonable response. Be brief.

(f) Sample Preparation Procedure: This refers to the steps taken to prepare the sample for the analysis. It does NOT refer to the preparation of standards or any associated solutions.

(g) Series of Standard Solution or Standard Additions: Which of these two procedures is it?

(h) If flame AA, what fuel/oxidant combination is used? If graphite furnace AA, what furnace program is used? If ICP, which of the two instruments designs is used?

(i) What Wavelength Is Used? What Exit Slit Setting?

(j) How Are the Data Gathered? The "data" are the absorbance or emission measurements. In other words, you have got your blank, your standards, and your samples ready to go. Now what? Be brief.

(k) Concentration Levels for Standards: This refers to what the concentration range is for the standard curve. Be sure to include the units. If no standard curve is used, state this.

(l) How Are the Standards Prepared? Are you diluting a stock standard? If so, how is the stock prepared and what is its concentration? If not, how are the standards prepared?

(m) Potential Problems: Is there anything mentioned in the method document that might present a problem? Be brief.

(n) Data Handling and Reporting: Once you have the data for the standard curve, then what. Be sure to present any post-run calculations that might be required.

(o) References: Did you look at any other references sources to help answer any of the above? If so, write them here.

10 Introduction to Chromatography

10.1 INTRODUCTION

Modern-day chemical analysis can involve very complicated material samples—complicated in the sense that there can be many substances present in the sample, creating a myriad of problems with interferences when the laboratory worker attempts the analysis. These interferences can manifest themselves in a number of ways. The kind of interference that is most familiar is one in which substances other than the analyte generate an instrumental readout similar to the analyte, such that the interference adds to the readout of the analyte, creating an error. However, an interference can also suppress the readout for the analyte (e.g., by reacting with the analyte). An interference present in a chemical to be used as a standard (such as a primary standard) would cause an error, unless its presence and concentration were known. Analytical chemists must deal with these problems, and chemical procedures designed to effect separations are now commonplace.

This chapter, and also Chapters 11 and 12, describe modern analytical separation science.

10.2 CHROMATOGRAPHY

A myriad of techniques used to separate complex samples come under the general heading of "chromatography." The nature of chromatography allows much versatility, speed and applicability particularly when the modern instrumental techniques of gas chromatography (GC) and high-performance liquid chromatography (HPLC) are considered. These latter techniques are covered in detail in Chapters 11 and 12. In this chapter, we introduce the general concepts of chromatography and give a perspective on its scope. Since there are many different classifications, this will include an organizational scheme covering the different types and configurations that exist.

Chromatography is the separation of the components of a mixture based on the different degrees to which they interact with two separate material phases. The nature of the two phases and the kind of interaction can be varied, and this gives rise to the different "types" of chromatography, which will be described in the next section. One of the two phases is a moving phase (the "mobile" phase), whereas the other does not move (the "stationary" phase) (see Figure 10.1). The mixture to be separated is usually introduced into the mobile phase, which then is made to move or percolate through the stationary phase either by gravity or some other force. The components of the mixture are attracted to and slowed by the stationary phase to varying degrees, and as a result, they move along with the mobile phase at varying rates, and are thus separated. Figure 10.2 illustrates this concept.

The mobile phase can be either a gas or a liquid, whereas the stationary phase can be either a liquid or solid. One classification scheme is based on the nature of the two phases. All techniques that utilize a gas for the mobile phase come under the heading of "gas chromatography" (GC). All techniques that utilize a liquid mobile phase come under the heading of "liquid chromatography" (LC). Additionally, we have gas–liquid chromatography (GLC), gas–solid chromatography (GSC), liquid–liquid chromatography (LLC), and liquid–solid chromatography (LSC) if we wish to stipulate the nature of the stationary phase as well as the mobile phase. It is more useful, however, to classify the techniques according to the nature of the interaction of the

Stationary phase

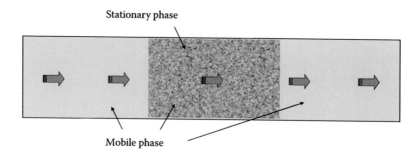

Mobile phase

FIGURE 10.1   In chromatography, a mobile (moving) phase moves through a stationary phase.

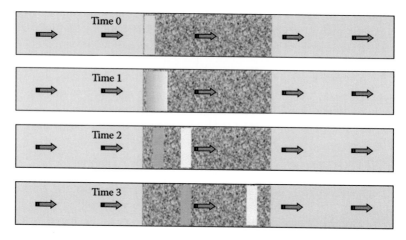

FIGURE 10.2   Illustration of the separation of a two-component mixture in a chromatography experiment. As the mobile phase moves through the stationary, the mixture components separate based on the way they interact with the two phases.

mixture components with the two phases. These classifications we refer to in this text as "types" of chromatography.

10.3   "TYPES" OF CHROMATOGRAPHY

10.3.1   PARTITION CHROMATOGRAPHY

In liquid–liquid partition chromatography, the mobile phase is a liquid that moves through a liquid stationary phase as the mixture components "partition" or distribute themselves between the two phases and become separated. The separation mechanism is thus one of dissolution of the mixture components to different degrees in the two phases according to their individual solubilities in each.

It may be difficult to imagine a liquid mobile phase used with a liquid stationary phase. What experimental setup allows one liquid to move through another liquid (immiscible in the first) and how can one expect partitioning of the mixture components to occur? The stationary phase actually consists of a thin liquid film chemically bonded to the surface of finely divided solid particles as shown in Figure 10.3. It is often referred to as **bonded phase chromatography**, or **BPC**. Such a stationary phase cannot be removed from the solid substrate by dissolving in the mobile phase, nor by heat, nor by reaction.

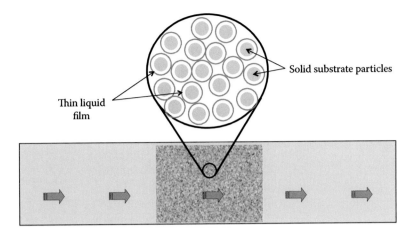

FIGURE 10.3   In partition chromatography, the stationary phase is a thin liquid film, such as one bonded to the surfaces of solid substrate particles as illustrated here. Mixture components separate in part based on their differing solubilities in this thin liquid film.

Since the separation depends on the relative solubilities of the components in the two phases, the polarities of the components and that of the stationary and mobile phases are important to consider. If the stationary phase is somewhat polar, it will retain polar components more than it will nonpolar components, and thus the nonpolar components will move more quickly through the stationary phase than the polar components. The reverse would be true if the stationary phase were nonpolar. Of course, the polarity of a liquid mobile phase plays a role as well.

The mobile phase for partition chromatography can also be a gas (GLC). In this case, however, the mixture components' solubility in the mobile phase is not an issue—rather their relative vapor pressures are important. This idea will be expanded in Chapter 11.

In summary, partition chromatography is a type of chromatography in which the stationary phase is a liquid chemically bonded to the surface of a solid substrate, whereas the mobile phase is either a liquid or gas. The mixture components dissolve in and out of the mobile and stationary phases as the mobile phase moves through the stationary phase and the separation occurs as a result. Examples of mobile and stationary phases will be discussed in Chapters 11 and 12.

10.3.2   ADSORPTION CHROMATOGRAPHY

Another chromatography "type" is **adsorption chromatography**. As the name implies, the separation mechanism is one of adsorption. The stationary phase consists of finely divided solid particles packed inside a tube but with no stationary liquid substance present to function as the stationary phase, as is the case with partition chromatography. Instead, the solid itself is the stationary phase, and the mixture components, rather than dissolve in a liquid stationary phase, adsorb or "stick" to the surface of the solid. Different mixture components adsorb to different degrees of strength, which also depends on the mobile phase, and thus again they become separated as the mobile phase moves. The nature of the adsorption involves the interaction of polar molecules, or molecules with polar groups, with a very polar solid stationary phase. Thus, hydrogen bonding, or similar molecule–molecule interactions, is involved.

This "very polar solid stationary phase" is typically silica gel or alumina. The polar mixture components can be organic acids, alcohols, etc. The mobile phase can be either a liquid or a gas. This type of chromatography is depicted in Figure 10.4.

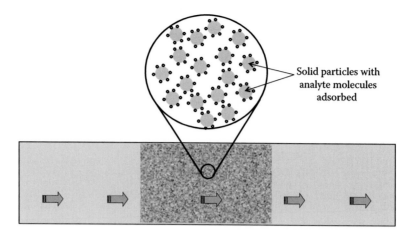

FIGURE 10.4 In adsorption chromatography, the stationary phase is a solid, such as the solid particles illustrated here. The mixture components separate based in part on their adsorption to differing degrees of strength on the surface of this solid.

10.3.3 Ion-Exchange Chromatography

A third chromatography "type" is **ion-exchange chromatography**, IEC, or simply "**ion chromatography**," IC. As the name implies, it is a method for separating mixtures of ions, both inorganic and organic. The stationary phase consists of very small polymer resin "beads," which have many ionic bonding sites on their surfaces. These sites selectively exchange ions with certain mobile phase compositions as the mobile phase moves. Ions that bond to the charged site on the resin beads are thus separated from ions that do not bond. Repeated changing of the mobile phase can create conditions that will further selectively dislodge and exchange bound ions that then are also separated. This stationary phase material can be either an **anion exchange resin**, which possesses positively charged sites to exchange negative ions, or a **cation exchange resin**, which possesses negatively charged sites to exchange positive ions. The mobile phase can only be a liquid. Further discussion of this type can be found in Chapter 12. Figure 10.5 depicts ion-exchange chromatography.

A special application of ion exchange resins is in the deionizaton of water. Wide columns packed with a mixture of an anion exchange resin that exchanges dissolved anions for hydroxide ions and

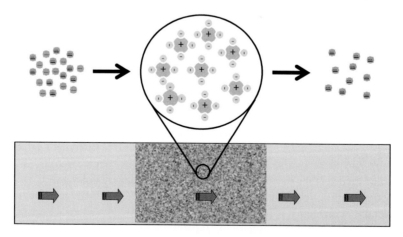

FIGURE 10.5 In ion exchange chromatography, certain ions ionically bond to sites of the opposite charge on the stationary phase structure and certain others do not, hence the separation. In this illustration, –1 ions bond but –2 ions do not.

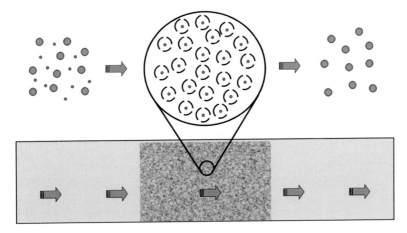

FIGURE 10.6 Illustration of size exclusion chromatography. Smaller particles fit in the pores on the stationary phase, whereas the larger particles do not, hence the separation.

a cation exchange resin that exchanges dissolved cations for hydrogen ions are used because water that is passed through such a column becomes free of ions (deionized) since the hydrogen and hydroxide ions combine to form more water.

10.3.4 SIZE EXCLUSION CHROMATOGRAPHY

Size exclusion chromatography, SEC, also called **gel permeation chromatography** (GPC) or **gel filtration chromatography** (GFC), is a technique for separating dissolved species on the basis of their size. The stationary phase consists of porous polymer resin particles. The components to be separated can enter the pores of these particles and be slowed from progressing through this stationary phase as a result. Thus, the separation depends on the sizes of the pores relative to the sizes of the molecules to be separated. Small particles are slowed to a greater extent than larger particles, some of which may not enter the pores at all, and thus the separation occurs. The mobile phase for this type can also only be a liquid, and it is discussed further in Chapter 12 as well. The separation mechanism is depicted in Figure 10.6.

10.4 CHROMATOGRAPHY CONFIGURATIONS

Chromatography techniques can be further classified according to "configuration"—how the stationary phase is "contained," how the mobile phase is configured with respect to the stationary phase in terms of physical state (gas or liquid), positioning, and how and in what direction the mobile phase travels in terms of gravity, capillary action, or other forces.

Configurations can be broadly classified into two categories: the "planar" methods and the "column" methods. The planar methods utilize a thin sheet of stationary phase material and the mobile phase moves across this sheet, either upward (**ascending chromatography**), downward (**descending chromatography**) or horizontally (**radial chromatography**). Column methods utilize a cylindrical tube to contain the stationary phase and the mobile phase moves through this tube either by gravity, with the use of a high-pressure pump, or by gas pressure. Additionally, with the exception of paper chromatography, those that utilize a liquid for the mobile phase are capable of utilizing all of the "types" reviewed above. Paper chromatography utilizing unmodified cellulose sheets is strictly partition chromatography (see the next section). If the mobile phase is a gas (**GC**), the "type" is limited to adsorption and partition methods. Table 10.1 summarizes the different configurations. Let us consider each individually.

TABLE 10.1

Summary of the Different Chromatography Configurations

Configuration	Geometry	Flow Direction	Applicable Types
Paper	Planar	Ascending, descending, or radial	Partition
Thin layer	Planar	Ascending, descending, or radial	Adsorption, partition, ion exchange, size exclusion
Open column	Column	Descending	Adsorption, partition, ion exchange, size exclusion
GC	Column	N/A	Adsorption, partition
HPLC	Column	N/A	Adsorption, partition, ion exchange, size exclusion

10.4.1 PAPER AND THIN-LAYER CHROMATOGRAPHY

Paper chromatography and thin-layer chromatography (TLC) constitute the planar methods mentioned above. Paper chromatography makes use of a sheet of paper having the consistency of filter paper (cellulose) for the stationary phase. Since such paper is hydrophilic, the stationary phase is actually a thin film of water unintentionally adsorbed on the surface of the paper. Thus, paper chromatography represents a form of partition chromatography only. The mobile phase is always a liquid. This configuration is largely obsolete today due to the emergence of thin-layer chromatography.

With thin-layer chromatography, the stationary phase is a thin layer of material spread across a plastic sheet or glass or metal plate. Such plates or sheets may either be purchased commercially already prepared or they may be prepared in the laboratory. The thin-layer material can be any of the stationary phases described earlier and thus TLC can be any of the four types, including adsorption, partition, ion-exchange, and size exclusion. Perhaps the most common stationary phase for TLC, however, is silica gel, a highly polar stationary phase for adsorption chromatography, as mentioned earlier. Also common is pure cellulose, the same material for paper chromatography, and here also we would have partition chromatography. The mobile phase for TLC is always a liquid.

The most common method of configuring a paper or thin-layer experiment is the ascending configuration shown in Figure 10.7. The mixture to be separated is first "spotted" (applied as a small "spot") within one inch of one edge of a 10-in.^{2} rectangular paper sheet or TLC plate. A typical experiment may

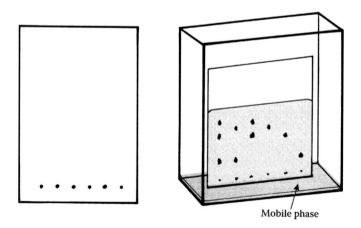

Mobile phase

FIGURE 10.7 Paper or thin-layer chromatography configuration. The drawing on the left shows the paper or TLC plate with spots applied. The drawing on the right shows the chromatogram in the developing chamber nearing complete development.

be an attempt to separate several spots representing different samples and standards on the same sheet or plate. Thus, as many as eight or more spots may be applied on one sheet or plate. So that all spots are aligned parallel to the bottom edge, a light pencil mark can be drawn prior to spotting. The size of the spots must be such that the mobile phase will carry the mixture components without streaking. This means that they must be rather small—they must be applied with a very small diameter capillary tube or micropipette. An injection syringe with a 25-μL maximum capacity is usually satisfactory.

Following spotting, the sheet or plate is placed spotted edge down in a "developing chamber" that has the liquid mobile phase in the bottom to a depth lower than the bottom edge of the spots. The spots must not contact the mobile phase. The mobile phase proceeds upward by capillary action (or downward by both capillary action and gravity if in the descending mode) and sweeps the spots along with it. At this point, chromatography is in progress, and the mixture components will move with the mobile phase at different rates through the stationary phase and, if the mixture components are colored, evidence of the beginning of a separation is visible on the sheet or plate. The result, if the separation is successful, is a series of spots along a path immediately above the original spot locations each representing one of the components of the mixture spotted there (see Figure 10.8).

If the mixture components are not colored, any of a number of techniques designed to make the spots visible may be employed. These include iodine staining, in which iodine vapor is allowed to contact the plate. Iodine will absorb on most spots rendering them visible. Alternatively, a fluorescent substance may be added to the stationary phase prior to the separation (available with commercially prepared plates), such that the spots, viewed under an ultraviolet light, will be visible because they do not fluoresce while the stationary phase surrounding the spots does fluoresce.

The visual examination of the chromatogram can reveal the identities of the components, especially if standards were spotted on the same paper or plate. Retardation factors (so-called R_f factors) can also be calculated and used for qualitative analysis. These factors are based on the distance the mobile phase has traveled on the paper (measured from the original spot of the mixture) relative to the distances the components have traveled, each measured from either the center or leading edge of the original spot to the center or leading edge of the migrated spot:

$$R_f = \frac{\text{distance mixture component has traveled}}{\text{distance mobile phase has traveled}}$$

These factors, which are fractions less than or equal to 1, are compared with those of standards to reveal the identities of the components (see Application Note 10.1).

Quantitative analysis is also possible. The spot representing the component of interest can be cut (in the case of paper chromatography) or scraped from the surface (TLC), dissolved and quantitated

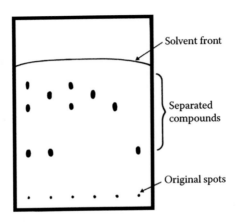

FIGURE 10.8   Developed paper or thin-layer chromatogram.

APPLICATION NOTE 10.1: DRUG SCREENING BY TLC

Drugs of abuse, including opiates, amphetamines, and barbiturates, can be detected in urine by a TLC procedure. The drugs are first removed from the urine by a liquid–liquid extraction procedure. The solvent containing the extracted drugs is then spotted on a TLC plate next to drug standards. The plates are dipped into a mobile phase in a developing chamber and the chromatography accomplished as shown in Figure 10.8. The resulting spots are often not visible unless they are fluorescent, or unless a colored derivative of the drug can form easily when the plate is sprayed with a reacting chemical. If they are not fluorescent, they may also be detected using a plate to which a fluorescent chemical has been added to the stationary phase, such that they then appear under an ultraviolet light, as described in Section 10.4.1. If the R_f factor of a spot from the urine sample matches that of one of the standards, then the drug has been identified.

(*Reference:* Kaplan, A. et. al., *Clinical Chemistry, Interpretation and Techniques, Fourth Edition*, Williams & Wilkins, Baltimore, MD [1995].)

by some other technique, such as spectrophotometry. Alternatively, modern scanning densitometers, which utilize the measurement of the absorbance or reflectance of ultraviolet or visible light at the spot location, may be used to measure quantity.

Using the TLC concept to prepare pure substances for use in other experiments, such as standards preparation, or synthesis experiments, is possible. This is called preparatory TLC and involves a thicker layer of stationary phase so that larger quantities of the mixture can be spotted and a larger quantity of pure component obtained.

Additional details of planar chromatography, such as methods of descending and radial development, how to prepare TLC plates, tips on how to apply the sample, what to do if the spots are not visible, and the details of preparatory TLC, etc., are beyond our scope.

To summarize, the most common form of planar chromatography is the ascending, meaning that the direction of mobile phase flow is "up" by the force of capillary action (see Figure 10.9).

10.4.2 CLASSICAL OPEN-COLUMN CHROMATOGRAPHY

Another configuration for chromatography consists of a vertically positioned glass or plastic tube in which the stationary phase is placed. It is typical for this tube to be open at the top (hence the name **open-column chromatography**), to have an inner diameter on the order of a centimeter or more, and to have a stopcock at the bottom, making it similar to a buret in appearance. With this configuration, the mixture to be separated is placed at the top of the column and allowed to pass onto the stationary phase by opening the stopcock. The mobile phase is then added and continuously fed into the top of the column and flushed through by gravity flow. The mixture components separate on the stationary phase as they travel downward and, unlike the planar methods, are then collected as they elute from the column. In the classical experiment, a "fraction collector" is used to collect the eluting solution. A typical fraction collector consists of a rotating carousel of test tubes positioned under the column such that fractions of eluate are collected over a period of time, such as overnight or a period of days, in individual test tubes (see Figure 10.10). This makes qualitative or quantitative analysis possible through the examination of these fractions by some other technique, such as UV-Vis spectrophotometry.

The length of the column is determined according to the degree to which the mixture components separate on the stationary phase chosen. Difficult separations would require more contact with the stationary phase and thus may require longer columns. Again, all four types (adsorption, partition, ion exchange, and size exclusion) can be used with this technique.

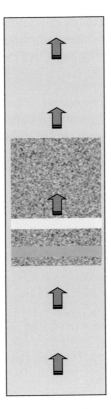

FIGURE 10.9   Direction of the mobile phase flow in most planar chromatography configurations is "up" and the force that moves the mobile phase is capillary action.

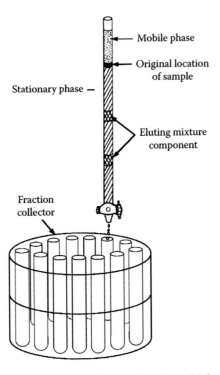

FIGURE 10.10   Classic open-column chromatography configuration with fraction collector.

It is well known that classical open-column chromatography has been largely displaced by the modern instrumental techniques of liquid chromatography. However, open columns are still used where extensive sample "clean-up" in preparation for the instrumental method is necessary. One can imagine that "dirty" samples originating, for example, from animal feed extractions or soil extractions, etc., may have large concentrations of undesirable components present. Since only very small samples (on the order of 1–20 μL) are needed for the instrumental method, the time required for obtaining a clean sample by this method, assuming the components of interest are not retained, is the time it takes for an initial amount of mobile phase to pass through from top to bottom. Compared with the "overnight" time frame, such a clean-up time is quite minimal and does not diminish the speed of the instrumental methods (see Application Note 10.2).

APPLICATION NOTE 10.2: PURIFYING URINE SAMPLES

The urine of people who are heavy smokers contains mutagenic chemicals, chemicals that cause mutations in biological cells. Bioanalytical laboratories can analyze urine samples for these chemicals, but the samples must be "cleaned up" first prior to extraction with methylene chloride. The procedure for this "cleanup" utilizes an open-column chromatography procedure. Columns several inches tall and about an inch wide are prepared by packing them with an adsorbing resin that has been treated with methyl alcohol. The urine samples are passed through these columns and interfering substances are removed as part of the sample preparation scheme.

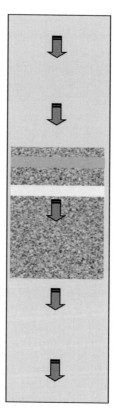

FIGURE 10.11   Direction of the mobile phase flow in the case of the open-column chromatography is "down" and the force that moves the mobile phase is gravity.

In summary, the direction of mobile phase flow for open-column chromatography is "down" and the force that moves the mobile phase is gravity (see Figure 10.11).

10.4.3 INSTRUMENTAL CHROMATOGRAPHY

The concept of the finely divided stationary phase packed inside a column allowing the collection of the individual components as they elute as discussed in the last section presents a useful, more practical alternative. One can imagine such a column along with a continuous mobile phase flow system (that does not use gravity for the flow), a device for introducing the mixture to the flowing mobile phase and an electronic sensor at the end of the column—all of this incorporated into a single unit (instrument) used for repeated, routine laboratory applications. There are two such chromatography configurations that are in common use today, known as GC and HPLC. These techniques essentially can incorporate all types of column chromatography discussed thus far (HPLC) as well as those types in which the mobile phase can be a gas (GC). Both add a degree of efficiency and speed to the chromatography concept. HPLC, for example, is such a "high-performance" technique for liquid mobile phase systems that a procedure that might normally take hours or days with open columns can usually be accomplished in a few minutes. The full details of these instrumental techniques are discussed in Chapters 11 and 12.

10.4.4 INSTRUMENTAL CHROMATOGRAM

Common to both GC and HPLC is the fact that they utilize an electronic sensor for detecting mixture components as they elute from the column. These sensors (more commonly called "detectors") generate electronic signals to produce a "chromatogram." A **chromatogram** is the graphical representation of the separation, a plot of the electronic signal vs. time. The chromatogram is traced on a computer screen or other recording device as the experiment proceeds. For the following discussion, refer to Figure 10.12 throughout.

Initially, in the time immediately after the sample is introduced to the flowing mobile phase and before any mixture component elutes ("Time 1" in the Figure 10.12), the electronic sensor sees only the flowing mobile phase. The mobile phase is used as a blank, such that the signal generated by the sensor is set to zero. Thus, when only mobile phase is eluting from the column, a baseline, at the zero electronic signal level, is traced on the chromatogram.

When a mixture component emerges and passes into the sensor's detection field ("Time 2"), an electronic signal different from zero is generated and the chromatogram trace deflects from zero. When the concentration of the mixture component is a maximum in the sensor's detection field ("Time 3"), the deflection reaches its maximum. Then, when the mixture component, continuing to move with the mobile phase, has cleared the sensor's detection field ("Time 4"), the sensor sees only mobile phase once again and the electronic signal generated is back to zero. When the second mixture component emerges, the process occurs again ("Time 5"), and the third, again, etc. The result is a "peak" observed on the chromatogram trace for each mixture component to elute.

The terminology associated with the chromatogram is the same for both GC and HPLC. For example, the time that passes from the time the sample is first introduced into the flowing mobile phase until the apex of the peak is seen on the chromatogram is called the **retention time** of that component, or the time that that mixture is retained by the column. The retention time is useful for a qualitative analysis of the mixture, as we shall see in the next chapter. It is given the symbol t_R (see Figure 10.13). Typically, retention times vary from a small fraction of 1 min to about 20 min, although much longer retention times are possible.

Another parameter often measured is the **adjusted retention time**, t'_R. This is the difference between the retention time of a given component and the retention time, t_M, of an unretained substance, which is often air for GC and the sample solvent for HPLC. Thus, the adjusted retention time is a measure of the exact time a mixture component spends in the stationary phase. Figure 10.14 shows

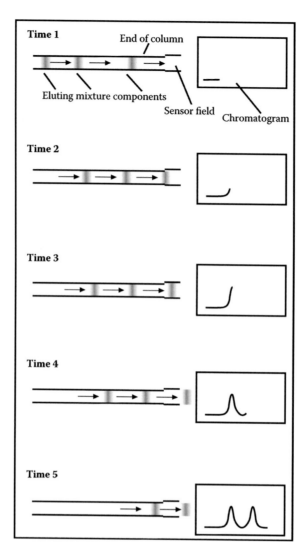

FIGURE 10.12 Illustration of the tracing of an instrumental chromatogram as separated mixture components elute from the column and pass through an electronic sensor.

how this measurement is made. The most important use of this retention time information is in peak identification, or qualitative analysis. This subject will be discussed in more detail in Chapter 11.

Other parameters sometimes obtained from the chromatogram, which are mostly measures of the degree of separation and column efficiency, are "resolution" (R), the number of "theoretical plates" (N), and the "height equivalent to a theoretical plate" (HETP or H). These require the measurement of the width of a peak at the peak base, W_B. This measurement is made by first drawing the tangents to the sides of the peaks and extending these to below the baseline, as shown for the two peaks in Figure 10.15. The width at peak base, W_B, is then the distance between the intersections of the tangents with the baseline, as shown. **Resolution** is defined as the difference in the retention times of two closely spaced peaks divided by the average widths of these peaks.

$$R = \frac{[t_R(B) - t_R(A)]}{\dfrac{[W_B(A) + W_B(B)]}{2}}$$ (10.1)

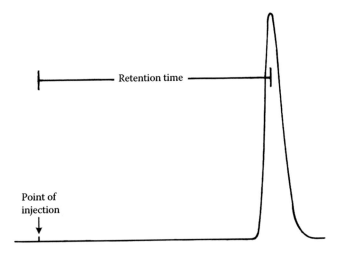

FIGURE 10.13 Retention time is the time from when the sample is injected into the flowing mobile phase to the time the apex of the peak appears on the chromatogram.

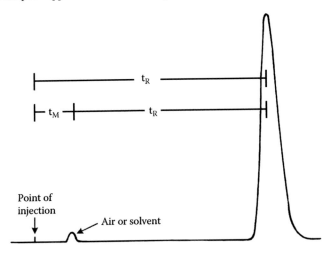

FIGURE 10.14 Chromatogram showing the definitions of t_R, t'_R, and t_M.

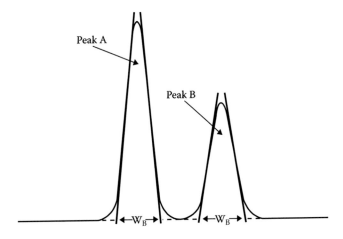

FIGURE 10.15 Measurement of the width at base, W_B, used in resolution and theoretical plate calculations.

where $t_R(B)$ is the retention time for peak B, $t_R(A)$ is the retention time for peak A, $W_B(A)$ is the width at base of peak A, and $W_B(B)$ is the width at base for peak B. R values of 1.5 or higher would indicate complete separation. Peaks that are not well resolved would inhibit satisfactory qualitative and quantitative analysis. An example of a chromatogram showing unsatisfactory resolution of two peaks is shown in Figure 10.16.

The **number of theoretical plates**, N, is also of interest. The concept of theoretical plates was discussed briefly in Section 10.2 for distillation. For distillation, one theoretical plate was defined as one evaporation/condensation step for the distilling liquid as it passes up a fractionating column. In chromatography, one theoretical plate is one "extraction" step along the path from injector to detector. Chromatography is analogous to a series of many extractions, but with one solvent (the mobile phase) constantly moving through the other solvent (the stationary phase), rather than being passed along through a series of separatory funnels. The equilibration that would occur in the fictional separatory funnel is one theoretical plate in chromatography. The number of theoretical plates can be calculated from the chromatogram using one of the peaks as follows.

$$N = 16 \left(\frac{t_R}{W_B} \right)^2$$

(10.2)

The **height equivalent to a theoretical plate**, H, is that length of column that represents one theoretical plate, or one equilibration step. Obviously, the smaller the value of this parameter, the more efficient the column. The more theoretical plates packed into a length of column the better the resolution. It is calculated by dividing the column length by the number of theoretical plates.

$$H = \frac{\text{length}}{N}$$

(10.3)

A parameter defined as the **relative retention**, or **selectivity**, is often reported for a given instrumental chromatography system as a number that ought to be able to be reproduced from instrument to instrument and laboratory to laboratory regardless of slight differences that might exist in the systems (column lengths, temperature, etc.). This parameter compares the retention of one component (1) with another (2) and is given the symbol alpha (α). It is defined as follows:

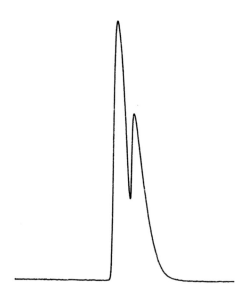

FIGURE 10.16   Unresolved peaks.

$$\alpha = \frac{t'_R(1)}{t'_R(2)} \qquad (10.4)$$

A selectivity equal to 1 would mean that the two retention times are equal, which means no separation at all.

Finally, a parameter known as the **capacity factor** may be determined. The capacity factor is a measure of the retention of a component per column volume, because the retention time is referred to the time for the unretained component. The capacity factor, k' (k prime), is calculated by dividing the adjusted retention time by the retention time, t_M, of an unretained substance, such as air in GC or the sample solvent in HPLC.

$$k' = t'_R / t_M \qquad (10.5)$$

The greater the capacity factor, the longer that component is retained and the better the chances for good resolution. An optimum range for k' values is between 2 and 6.

10.4.5 QUANTITATIVE ANALYSIS WITH GC AND HPLC

The physical size of a peak, or area under the peak, traced on the chromatogram is directly proportional to the amount of that particular component passing through the detector.* This, in turn, is proportional to the concentration of that mixture component in the sample solution. It is also proportional to the amount of solution injected, since this too dictates how much passes through the detector. The more material being detected, the larger the peak. Thus, for quantitative analysis, it is important that we have an accurate method for determining the areas of the peaks.

The most popular method of measuring peak area is by integration. Integration is a method in which the series of digital values acquired by the data system as the peak is being traced are summed. The sum is thus a number generated and presented by the data system and is taken to be the peak area (see Figure 10.17). We will discuss in Chapters 11 and 12 exactly how this area is converted to the quantity of analyte in GC and HPLC and the issues involved. For a hint of the process, see Figure 10.18. The analyte peak increases in size as the concentration increases as seen in Figure 10.18a. The standard curve can then be a plot of peak area vs. concentration, as shown in Figure 10.18b.

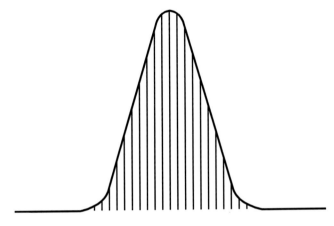

FIGURE 10.17 Illustration of how the area of an instrumental chromatography peak is determined by integration. The series of digital values acquired by the data system, represented by the vertical lines, are summed.

---

* A given amount of one component will not produce a peak the same size as an equal amount of another component, however. This will be discussed further in Chapter 11, Section 11.8.2.

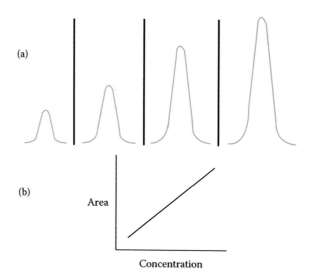

FIGURE 10.18 (a) Instrumental chromatography peak for an analyte gets larger as the concentration of that analyte increases. (b) Standard curve can then be a plot of peak area vs. concentration.

10.5 ELECTROPHORESIS

The final analytical separation technique we discuss is **electrophoresis**. Electrophoresis is a separation technique in which mixture components separate based on their differing migration rates in an electric field. A very popular electrophoresis configuration today is the instrumental method called capillary electrophoresis. Capillary electrophoresis closely resembles instrumental liquid chromatography, especially in the way mixture components are detected. For this reason, we defer the full discussion of this technique to Chapter 12.

EXPERIMENTS

EXPERIMENT 31: THIN-LAYER CHROMATOGRAPHY ANALYSIS OF COUGH SYRUPS FOR DYES

Note: Many cough medicine formulations exhibit the color they have because a food colorant has been added. Such colorants are standard FD&C dyes, typically blue #1, red #3, #33, and #40, or yellow #5, #6, and #10. In this experiment, identification of such dyes is accomplished by TLC. Remember to wear safety glasses.

1. Weigh 10 mg of each dye standard into 50-mL volumetric flasks, dilute to the mark with distilled water, and shake. If the dye is not water soluble ("Lake" dyes are insoluble dyes), dissolve it directly in n-butanol in Step 3.
2. Pipet 5 mL of each standard and 5 mL of each sample into separate 50-mL Erlenmeyer flasks.
3. Add 5 drops of diluted HCl (prepared by diluting 23.6 mL of concentrated HCl to 100 mL in the distilled water) and 5 mL of n-butanol to each flask and shake on a shaker for 30 min.
4. Pour each into test tubes, and allow the layers to separate. Spot the butanol (top) layer onto a TLC plate (consult instructor).
5. Place the development solvent (1-propanol/ethyl acetate/concentrated NH₄OH, 1:1:1) in the developing tank and develop the chromatography plate so that the solvent migrates for 12 cm.
6. Allow the plate to dry. Observe and measure R_f values, if possible, and identify the dyes in the cough syrups.

EXPERIMENT 32: THIN-LAYER CHROMATOGRAPHY ANALYSIS
OF JELLY BEANS FOR FOOD COLORING

Remember to wear safety glasses.

1. Add about 25 mL of warm water to three to six jellybeans of each color. Allow to stand until they look white (5–10 min). If allowed to stand longer, too much sugar will be extracted with the colors.
2. Remove the candy from the water and filter the extractant through Whatman #1 or equivalent-grade filter paper.
3. Transfer the liquid filtrate into a 125-mL separatory funnel. Add 25 mL of butanol and about 1 mL of 10% phosphoric acid. Shake for about 45 s, then allow the layers to separate.
4. Transfer the organic (top) layer to a small shaker using a transfer pipette. Evaporate the solvent in a steam bath.
5. Take up each dye residue in one or two drops of ammonia solution (prepared by diluting 1 mL of concentrated ammonium hydroxide to 100 mL). Using open-ended capillary tubes, spot each color onto both a cellulose TLC plate and a silica gel TLC plate at about 3 cm from the bottom edge.
6. Spot solutions of commercial food colorings on the same plates. Allow all the spots to dry completely and spot again any which are not dark enough in intensity.
7. Place the plates into a developing chamber and develop the plates using the developing solvent: butanol/ethanol/2% ammonia (3:1:2). Run for 1.5 h or until the solvent has progressed about 15 cm.
8. After drying the plates, identify the food dyes by matching the spot colors and positions with those of the commercial colorants. Your instructor may request that you measure the R_f factors for this identification.

QUESTIONS AND PROBLEMS

Introduction

1. Why is a study of modern separation science important in analytical chemistry?

Chromatography

2. Give a general definition of chromatography that would apply to all types and configurations. (To say that it is a separation technique is important but that alone is not a sufficient answer.)
3. Tell what each of the following abbreviations refer to: GC, LC, GSC, LSC, GLC, LLC.

Types of Chromatography

4. Name the four types of chromatography described in this chapter and give the details of the separation mechanism of each.
5. How does partition chromatography differ from adsorption chromatography?
6. Consider a mixture of compound A, a somewhat nonpolar liquid, and compound B, a somewhat polar liquid. Tell which liquid, A or B, would emerge from a chromatography column first under the following conditions and why.
   (a) A polar liquid mobile phase and a nonpolar liquid stationary phase.
   (b) A nonpolar liquid mobile phase and a polar liquid stationary phase.

7. We have studied four chromatography "types." One of these is partition chromatography. Answer the following questions concerning partition chromatography "yes" or "no":
   (a) Can the mobile phase be a solid?
   (b) Can the mobile phase be a liquid?
   (c) Can the mobile phase be a gas?
   (d) Can the stationary phase be a solid?
   (e) Can the stationary phase be a liquid?
   (f) Can the stationary phase be a gas?

8. (a) List all chromatography "types" that utilize a liquid stationary phase. If there are none, so state.
   (b) List all chromatography "types" that utilize a gaseous stationary phase. If there are none, so state.
   (c) List all chromatography "types" that utilize a solid stationary phase. If there are none, so state.

9. One type of chromatography separates small molecules from large ones. Name this type and tell how such a separation occurs.

10. Differentiate between the use of a cation exchange resin and an anion exchange resin in terms of whether the charged sites are positive or negative and whether cations or anions are exchanged.

11. Tell what each of the following abbreviations refer to: IEC, IC, SEC, GPC, and GFC.

Chromatography Configurations

12. (a) Name four chromatography "configurations."
    (b) Choose one of your answers to (a) and tell how the stationary phase is configured relative to the mobile phase and what force is used to move the mobile phase through the stationary phase.

13. Compare HPLC with open-column chromatography in terms of
    (a) How the stationary phase is contained.
    (b) The force that moves the mobile phase through the stationary phase.
    (c) How quantitative analysis takes place once the separation is completed.

14. Compare thin-layer chromatography to GC in terms of
    (a) What force moves the mobile phase through the stationary phase.
    (b) Whether it is described as planar or column.
    (c) How qualitative analysis is performed once the separation is completed.

15. What does the abbreviation LSC stand for? Give two examples of chromatography "types" that can be abbreviated as LSC.

16. Look at Figure 10.12 and explain how it is that a peak is traced on a chromatogram as separated mixture components elute from an instrumental chromatography column.

17. Match each item in Column A to a single item in Column B that most closely associates with it.

Column A	Column B
(a) Paper chromatography	(h) Thin-layer chromatography
(b) Ion-exchange chromatography	(i) Size-exclusion chromatography
(c) Gas chromatography	(j) Stationary Phase is Water
(d) Adsorption chromatography	(k) Column effluent is collected and analyzed
(e) Gel permeation chromatography	(l) Uses a stationary phase that trades ions with the mobile phase
(f) TLC	(m) An instrumental method
(g) Open-column chromatography	(n) Involves a mechanism in which the mixture components selectively "stick" to a solid surface

18. Fill in the blanks with a term from the following list:

stationary phase	mobile phase
adsorption	partition
paper	thin-layer
HPLC	ion exchange
size exclusion	electrophoresis
detector	

(a) In GC, the material packed within the column is usually a powdered solid that has a thin liquid film adsorbed on the surface. This thin liquid film is called the ____. This type of GC falls into the general classification of ____ chromatography.

(b) In one type of chromatography, the components of the mixture are separated on the basis of the relative sizes of the molecules. This is called ____ chromatography.

(c) A technique in which separation of charged species is effected by the use of an electric field is called ____.

(d) The fact that gravity flow of a liquid through a packed column is time consuming led to the development of ____.

(e) GSC is a type of ____ chromatography.

(f) A type of chromatography in which a layer of adsorbent is spread on a glass or plastic plate is called ____ chromatography.

19. Fill in the blanks in the following table.

Configuration	Type	Stationary Phase	To Separate Mixtures of
Paper	Partition		
HPLC			Different-size molecules
	Ion exchange		
	Partition		Gases and volatile liquids
	Adsorption	Layer of Adsorbent Spread on Glass Plate	

20. Answer the following questions with either true or false.

(a) The stationary phase percolates through a "bed" of finely divided solid particles in adsorption chromatography.

(b) The mobile phase can either be a liquid or a gas.

(c) The mobile phase is a "moving" phase.

(d) Partition chromatography can only be used when the mobile phase is a liquid.

(e) Adsorption includes LSC.

(f) In partition chromatography, the mobile phase "partitions" or distributes itself between the sample solution and the stationary phase.

(g) If the stationary phase is a polar liquid substance, nonpolar components will elute first.

(h) In GLC, the separation mechanism is partitioning.

(i) Size exclusion chromatography separates components on the basis of their charge.

(j) Gel permeation chromatography is another name for size exclusion chromatography.

(k) Ion-exchange chromatography is a technique for separating inorganic ions in a solution.

(l) Paper chromatography is a type of LLC.

(m) Thin-layer chromatography and open-column chromatography are two completely different configurations of GSC.

(n) It is useful to measure R_f values in open-column chromatography.

(o) R_f values are used for quantitative analysis.

(p) TLC refers to thin-layer chromatography.

(q) HPLC refers to high-performance liquid chromatography.

21. Match each statement to one of the terms given.

Partition chromatography	Adsorption chromatography
Ion exchange chromatography	Size exclusion chromatography
Paper chromatography	Thin-layer chromatography
Open-column chromatography	Gas chromatography
High-performance liquid chromatography	

(a) A chromatography configuration in which the stationary phase is spread across a glass or plastic plate.

(b) A chromatography type in which the stationary phase is a liquid.

(c) A chromatography type designed to separate dissolved ions.

(d) A chromatography configuration that utilizes a high-pressure pump to move the mobile phase through the stationary phase.

(e) One of two chromatography types that have an application in GC.

(f) A chromatogramphy configuration that utilizes a "fraction collector."

(g) The only chromatography type described by the letters GLC.

(h) One of two chromatography configurations in which the mobile phase moves by capillary action opposing gravity.

22. What is the difference between retention time and adjusted retention time?

23. Draw an example of an instrumental chromatogram showing one peak, label the x- and y-axes, and show clearly how the retention time and adjusted retention time are measured.

24. Draw an example of an instrumental chromatography peak and show in your drawing and describe in words the specific method by which peak area is measured by integration.

11 Gas Chromatography

11.1 OVERVIEW

Gas chromatography (GC) is an instrumental chromatography configuration, meaning we have electronic detection and peaks on a computer screen as discussed in Chapter 10. The mobile phase is a gas and that is the reason it is called "gas chromatography." The mobile phase is typically helium. Thus, a compressed gas cylinder of helium stands next to the bench, and this gas is plumbed into the interior of the instrument to connect with the column. The regulated pressure from the compressed gas cylinder is the force that pushes the helium through the column. The flow rate of the helium is controlled electronically by computer software or by a needle valve.

The mixture to be separated is usually a mixture of organic liquids, although gases and inorganic substances may also be separated. A liquid mixture is "injected" into the instrument with use of a microliter syringe that has a sharp tip for piercing a septum. The mixture must be in the gas phase when "carried" by the helium through the column. This means that the liquids must be rapidly evaporated immediately upon injection. Such a rapid evaporation requires that the injection port be heated. The column and detector must also be heated in order to sustain them in the gas phase. For this reason, the column is placed in an oven. Thus, the mixture to be separated is injected with a syringe into a heated port fitted with a rubber septum. The flowing helium carries the evaporated mixture onto the column where the separation then takes place.

The modern GC column has some unique features. Most often, it is a "capillary" column, meaning that it has a very small diameter, on the order of 0.1 to 0.25 mm. Also, the stationary phase liquid is not adsorbed on the surface of small particles packed inside, as was discussed generally for column configurations in Chapter 10, but rather it is bonded to the interior wall of this capillary tube. A capillary column can be up to 300 ft long and is coiled up to fit in the column oven. We will discuss this in more detail shortly.

The separated mixture components elute from the other end of the column and pass through the sensor field (more often referred to as a detector) one at a time, generating the peaks on the computer screen. A GC detector must detect the eluting compounds while in the gas phase, so detector designs are very unique, as we shall see. The entire setup is depicted in Figure 11.1, and photographs of a capillary column connected to the injection port and the detector are shown in Figure 11.2.

11.2 VAPOR PRESSURE AND SOLUBILITY

The separation depends on two things: (1) how well the mixture components dissolve in the liquid stationary phase and (2) the tendency of the mixture components to evaporate, i.e., to be in the mobile phase. Thus, it is their relative solubilities in the stationary phase and their relative vapor pressures that dictate the quality of the separation.

Vapor pressure is defined as the partial pressure exerted by a gas that is in equilibrium with its liquid phase in the same sealed container. It can be thought of as substance's tendency to be in the gas phase at a given temperature. If one compound has a higher vapor pressure than another, it will tend to be in the gas phase (the mobile phase) and move through the column more rapidly than one that has a lower vapor pressure. On the other hand, if a compound has a high solubility in the stationary phase, it will tend to dissolve in this phase and move through the column slowly. If a substance has a low solubility in the stationary phase, it will tend to not dissolve in this phase and move through the column more rapidly. In Figure 11.3, a separation of three mixture components

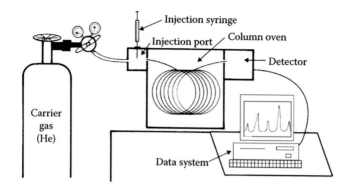

FIGURE 11.1   Drawing of a gas chromatography instrument with all of its parts.

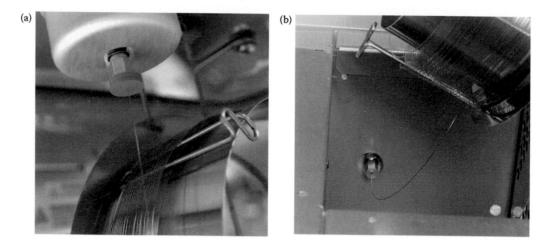

FIGURE 11.2   (a) Photograph showing a capillary GC column connected to the injection port. (b) Photograph of the other end of this same column connected to the detector.

is shown based on how they separate due to vapor pressure and solubility differences. This same information is conveyed in Table 11.1. Notice also that the figure depicts the column as a capillary column—the stationary phase is fixed to the column wall.

11.3   INSTRUMENT COMPONENTS

The helium moves from the compressed gas cylinder to the injection port, where it meets up with the injected mixture, then through the column where the mixture components separate due to their varying vapor pressures and solubilities. Finally, the helium and the separated mixture components emerge from the column and pass through the detector where the electronic signals are generated (see Figure 11.4). The three instrument components depicted in this figure are each at an elevated temperature. The reasons are as follows: (1) Liquid mixture components must be "flash vaporized" upon injection so that the flowing helium can immediately carry them into the column. Thus, the injection port must be heated. (2) Vapor pressures and solubilities can vary substantially with temperature. Because of this, it is useful to be able to use an elevated column temperature and also to be able to carefully control the temperature of the column. (3) Mixture components should not condense when they are inside the detector. For this reason, the detector must also be heated.

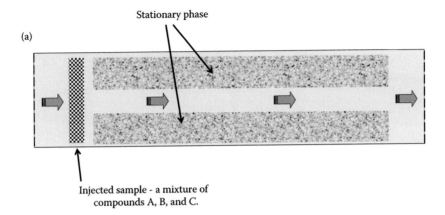

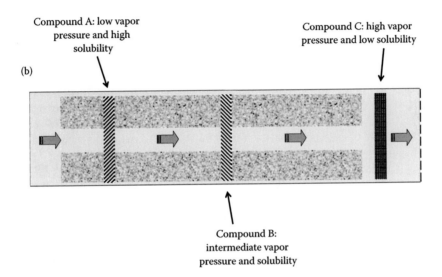

FIGURE 11.3 (a) Illustration of a capillary GC column showing the stationary phase bonded to the inside wall of the capillary tube, and an injected sample, a mixture of compounds A, B, and C. (b) This same column at a later time, showing the compounds A, B, and C having separated based on their differing vapor pressures and their differing solubilities in the stationary phase. In both (a) and (b), the mobile phase gas is flowing left to right as indicated by the wide arrows.

TABLE 11.1
Summary of Retention Concepts for GC

Component's Vapor Pressure	Component's Solubility in the Stationary Phase	Retention Time
High	Low	Short
High	High	Intermediate
Low	Low	Intermediate
Low	High	Long

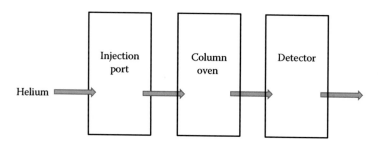

FIGURE 11.4 Three heated zones in a GC instrument are the injection port, the column oven, and the detector. Each is held at its own temperature independent of the others. The flowing mobile phase (helium is indicated here) carries the mixture components from one to the other (see the text for the reasons each is heated and what the typical temperatures are).

Column temperatures are typically in the 80°C–150°C range, but much higher temperatures are sometimes used. In addition, this temperature is often programmed to change while the components are in the column and the separation was taking place. More information on this is presented in Section 11.5.4. The temperature in the injection port depends on the volatility of the components but is typically in the 200°C–250°C range. The temperature of the detector is usually in the 200°C–250°C range also. You can see that all three, being at different temperatures and in different parts of the instrument, must have their own independent heating and temperature control system. Thus, the temperature of each is set and each must be allowed to reach this temperature before the experiment begins.

11.4 SAMPLE INJECTION

The most familiar design for the injection port consists of a small glass-lined or metal chamber equipped with a rubber septum to accommodate injection with a syringe. As the helium "blows" through the chamber, a small volume of injected liquid (typically on the order of 0.1 to 3 μL) is flash vaporized and immediately carried onto the column with the helium. The mobile phase is often referred to as a **carrier gas** for this reason. A variety of sizes of syringes and some additional features are available to make the injection easy and accurate. The rubber septum, after repeated sample introduction, becomes worn and can be replaced easily. Sample introduction systems for gases (gas sampling valves) and solids are also available.

The volume of injected liquid (appearing as an extremely small drop) may seem to be extraordinarily small. When vaporized, however, the sample volume is much larger and will occupy an appropriate volume in the column. Also, the detection system, as we will see later, is very sensitive and will detect very small concentrations even in such a small volume. As a matter of fact, too large a volume is a concern to the operator, since columns, especially capillary columns, can become overloaded even with volumes that are very small. Overloading means that the entire vaporized sample will not fit onto the column all at once and will be introduced over a period of time. A good separation would be in jeopardy in such an instance. To guard against overloading of capillary columns, **split injectors** have been developed. In these injectors, only a fraction of the liquid from the syringe actually is passed to the column. The remaining portion is split from the sample and vented to the air.

As implied above, the appropriate range of sample injection volume depends on column diameter. As we will see in the next section, column diameters vary from capillary size (0.1 mm) to 1/8 and 1/4 in. Table 11.2 gives the typical injection volumes suggested for these column diameters. The capillary columns are those in which the overloading problem mentioned above is most relevant. Injectors preceding the 1/8 in or larger columns are not split.

TABLE 11.2

Typical Injection Volumes for Different Column Diameters

Column Diameters	Maximum Injection Volumes (µL)
1/4 in. (packed column)	100
1/8 in. (packed column)	20
Capillary (open tubular)	0.1

Four different modes of injection are shown in Figure 11.5. In a **direct injection**, shown in Figure 11.5a, the sample is vaporized inside the injector and 100% of it is carried onto the column. In a **split injection**, shown in Figure 11.5b, the sample is vaporized inside the injector and is then split as mentioned above. A typical split has 1% entering the column and 99% vented through a "split valve" to the air. In a **splitless injection**, Figure 11.5c, the injector that is used is capable of split injection, but the split valve is closed such that 100% enters the column as in the "direct

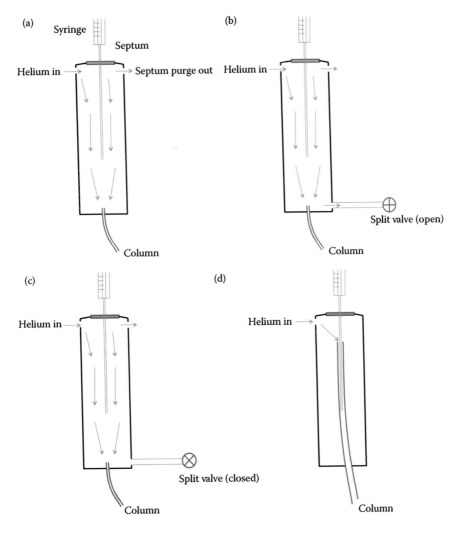

FIGURE 11.5 Illustrations of the GC injection port for (a) direct injection, (b) split injection, (c) splitless injection, and (d) on-column injection (see the text for descriptions).

APPLICATION NOTE 11.1: THE PURGE-AND-TRAP PROCEDURE

We mentioned in Chapter 2 (Section 2.9.1) that a **purge-and-trap** procedure sometimes precedes an analysis by gas chromatography. An example of this procedure is the analysis of water that has been chlorinated. When water is chlorinated, chlorine reacts with organic matter to form trihalomethanes (THMs), such as chloroform, bromoform, bromodichloromethane, and chlorodibromomethane. THMs in water are regulated by the Safe Drinking Water Act, and so a water analysis laboratory that is part of a water treatment plant must analyze the treated water to determine their concentration.

In the purge-and-trap technique, a procedure in which helium gas is vigorously bubbled through the sample solutions one-by-one, a technique known as helium "sparging," is carried out. Because the THMs are volatile, the helium sparging draws them out of the samples, in effect, "purging" them from the samples. The helium/THM gaseous mixture then flows through a "trap" in which the THMs are adsorbed and concentrated. This is followed by a desorption step in which the THMs are desorbed and then guided to the GC column in a single small microliter-size volume. No syringe is used.

injection." Finally, we have the **on-column injection** in which the end of the column actually extends into the injector such that the syringe needle is inside the column when the plunger is pushed as shown in Figure 11.5d.

The accuracy of the injection volume measurement can be very important for quantitation, since the amount of analyte measured by the detector depends on the concentration of the analyte in the sample as well as the amount injected. In Section 11.8.3, a technique known as the **internal standard** technique will be discussed. Use of this technique negates the need for superior accuracy with the injection volume as we will see. However, the internal standard is not always used. Very careful measurement of the volume with the syringe in that case is paramount for accurate quantitation. Of course, if a procedure calls only for identification (Section 11.7), then accuracy of injection volume is less important.

Another means of introducing the sample to the column is with what is termed the **purge-and-trap** procedure (see Application Note 11.1 for a summary of this technique).

11.5 COLUMN DETAILS

11.5.1 Instrument Logistics

GC columns, unlike any other type of chromatography column, are typically very long. Lengths varying from 2 to 300 ft or more are possible. Additionally, as mentioned previously, it is important for the column to be kept at an elevated temperature during the run in order to prevent condensation of the sample components. Indeed, maintaining an elevated temperature is very important for other reasons, as we shall see in the next subsection. The obvious logistical problem is how to contain a column of such length and be able to simultaneously control its temperature.

However long a column is, it is wound into a coil so that it fits nicely into a small oven, perhaps 1 to 3 ft^3 in size. This oven probably constitutes about half of the total size of the instrument. Connections are made through the oven wall to the injection port and the detector, as was seen in Figure 11.2.

GC instruments are designed so that columns can be replaced easily by undoing a pair of tubing joint fittings inside the oven. This not only facilitates changing to a different stationary phase altogether but also allows the operator to replace a given column with a longer one containing the same stationary phase. The idea of a longer column is to allow more interaction of the mixture components with the mobile and stationary phases, which in turn is bound to improve the separation. For

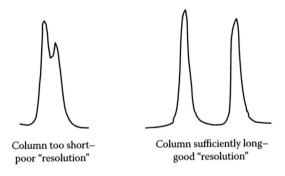

Column too short–
poor "resolution"

Column sufficiently long–
good "resolution"

FIGURE 11.6   Drawing showing the effect of column length on resolution.

example, a 12-ft column has more resolving power than a 6-ft column simply because the mixture components have more interaction with the two phases (see Figure 11.6).

11.5.2   PACKED, OPEN TUBULAR, AND PREPARATIVE COLUMNS

The capillary (also known as "open tubular") columns we have been discussing are the norm today. However, the so-called **packed columns**, those in which the stationary phase consists of the finely divided particles "packed" inside the tube as imagined throughout Chapter 10, can still find a use in the modern laboratory. One limitation of these columns is that they cannot be longer than about 20 ft, which means that they are severely limited in terms of resolving power. The force of the gas pressure from the helium bottle is not a sufficient force for moving the mobile phase through a packed column longer than such length. Compared with the 300-ft lengths that are possible with capillary columns, a 20-ft length simply does not work for most modern GC complex separations. The capillary column, being open down the middle, offers very little resistance to gas flow and can be made hundreds of feet long without demanding a high pressure. Such columns are extremely popular (see Figures 11.7 and 11.8).

In addition to the **analytical columns** (columns used mainly for analytical work), the so-called **preparative columns** may also be encountered. Preparative columns are used when the purpose of the experiment is to prepare a pure sample of a particular substance (from a mixture containing the substance) by GC for use in other laboratory work. The procedure for this involves the individual condensation of the mixture components of interest in a cold trap as they pass from the detector and as their peak is being traced on the computer screen. Although analytical columns can be suitable for this, the amount of pure substance generated is typically very small, since what is being

FIGURE 11.7   Photographs showing a packed column and a capillary column.

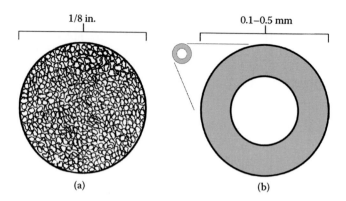

FIGURE 11.8 Drawings of cross sections of (a) a packed column and (b) a capillary column. The smaller diameter of the capillary column is indicated in the drawing on the right.

collected is only a fraction of the extremely small volume injected. Thus, columns manufactured with very large diameters (on the order of inches) and capable of very large injection volumes (on the order of milliliters), are manufactured for the preparative work. Also, the detector used must not destroy the sample, like the flame ionization detector (Section 11.6.1) does, for example. Thus, the thermal conductivity detector (Section 11.6.2) is used most often with preparative gas chromatography. We will mention this again in the later sections.

11.5.3   THE NATURE AND SELECTION OF THE STATIONARY PHASE

Most GC separations are of the partition type, meaning they utilize a liquid stationary phase. Since the interaction of the mixture components with the liquid stationary phase plays the key role in the separation process, the nature of this stationary phase is obviously important. In addition, most GC liquid phases today are those that are bonded to the inside wall of the capillary tube, as opposed to the surface of solid support particles. One might ask how it is that liquid stationary phases can chemically bond in this way. Also, what is the nature of the inside surface of the capillary and what sort of liquids are we talking about?

Most of the capillary tubes used for GC columns are made of "fused silica." Fused silica is a synthetic quartz material of very high purity. The outside of the tubing is coated with a polyimide material, which helps to prevent breakage and gives the capillary an amber tint. The inside surface is chemically treated and a "silylation" process takes place that activates the surface by exposing silanol groups (–Si-OH). Following this, the stationary phase material, a silane reagent consisting of a polymer chain with silicons bonded to oxygens (...–O-Si-O-Si-O-Si-O-Si—...) is added and a chemical reaction bonds this chain to the silicon of the silanol group. These silanes (silicon-based polymers) have organic groups on them, such as methyl groups, phenyl groups (benzene rings), and others, that, depending on their nature, give the stationary phase their polar/nonpolar properties. The results are polysilane liquids (polysiloxanes) that are fixed via chemical bond to the surface of the quartz. Most stationary phases are these polysiloxanes, but there are others as well.

Because there can be differences in the amount and nature of these organic groups along the polymer chain, several hundred different liquids useful as stationary phases are known. This means that the analyst has an awesome choice when it comes to selecting a stationary phase for a given separation. It is true, however, that relatively few such liquids are in actual common use. Their composition is frequently not obvious to the analyst because a variety of common abbreviations and trade names that have come to be popular for the names of some of them. V-101, OV-17, and SE-30 are examples.

The liquid stationary phase in a packed column is bonded to the surface of a solid substrate (also called the "support"). This material must be inert and finely divided (powdered). The typical diameter of a support particle is 125–250 μm, creating a 60- to 100-mesh material. This material is usually diatomaceous earth. **Diatomaceous earth**, the crushed, decayed silica skeletons of algae, is most commonly referred to by the manufacturer's (Johns Manville's) trade name **Chromosorb™**. Various types of Chromosorb™, which have had different pretreatment procedures applied, are available, such as Chromosorb™ P, Chromosorb™ W, and Chromosorb™ 101–104. The nature of the stationary phase and the nature of the substrate material are both usually specified in a chromatography literature procedure, and columns are tagged to indicate each of these as well.

11.5.4 COLUMN TEMPERATURE

As mentioned earlier the vapor pressure of a substance and the solubility of a substance in another substance change with temperature (see Figure 11.9). It should not be surprising then that the precise control of the temperature of a GLC column is very important, since, as we have indicated, the separation depends on both vapor pressure and solubility. Both **isothermal** (constant) and **programmed** (continuously changing) temperature experiments are possible. For simple separations, the isothermal mode may well be sufficient—there may be sufficient differences in the mixture components' vapor pressures and solubilities to effect a good separation at the chosen temperature. However, for more complicated mixtures, a complete separation is less likely in the isothermal mode and a temperature program, such as that shown in Figure 11.10 may be required.

For example, consider gasoline, which has a considerable number of highly volatile components as well as a significant number of less volatile components. It is possible that at a temperature of, say, 100°C, some of the less volatile components will be resolved, but the more volatile ones will pass through unresolved and have very short retention times. A lower temperature of, say, 80°C, may cause complete resolution of these more volatile components but would result in unwanted long retention times for the less volatile components and perhaps also result in poorly shaped peaks for these. If we could increase the temperature from 80°C to 120°C or higher in the middle of the run, however, we could have the best of both worlds—complete resolution and reasonable retention times for all peaks (see Figure 11.11). Thus, temperature-programmable ovens have been developed and are found on all modern GC units. Temperature programming can consist of simple programs, such as that suggested above—a single linear increase from a low temperature to a higher temperature—but it can also be more complex. For example, a chromatography researcher may find that several temperature increases, and perhaps even a decrease, must be used in some instances to effect

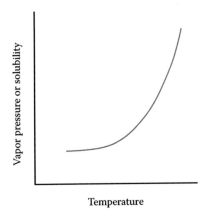

FIGURE 11.9 Both vapor pressure, the tendency of the analyte to be in the mobile phase, and solubility, the tendency of the analyte to be in the stationary phase, increase as temperature increases.

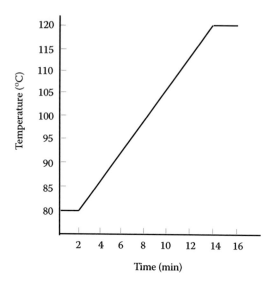

FIGURE 11.10   Temperature program that raises the temperature of the column oven from 80°C to 120°C in 12 min.

(a) Temperature too high: poor
     resolution of high vapor pressure
     components

(b) Temperature too low: retention
     times of low vapor pressures
     components are too long

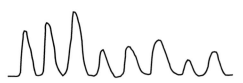

(c) The best of both worlds using
     temperature programming

FIGURE 11.11   Illustration of why temperature programming is useful.

an acceptable separation. Most modern GC units are capable of at least a slow temperature decrease in the middle of the run since they are equipped with venting fans that bring ambient air into the oven to cool it for the start of a second run, etc.

11.5.5   CARRIER GAS FLOW RATE

The rate of flow of the carrier gas affects resolution. A simple analogy here will make the point. Wet laundry hung out on a clothesline to dry will dry faster if it is a windy day. The components of the mixture will "blow" through the column more quickly (regardless of the degree of interaction with the stationary phase) if the carrier gas flow rate is increased. Thus, a minimum flow rate is needed for maximum resolution. It is known, however, that at extremely slow flow rates, resolution is dramatically reduced due to factors such as packing irregularities, particle size, column diameter, etc.

It is obvious that the flow rate must be precisely controlled. The pressure from the compressed gas cylinder of carrier gas, while sufficient to force the gas through a packed column, does not provide the needed flow control. Thus, a flow controller is built into the system. The flow rate of the carrier gas, as well as other gases used by some detectors, must be able to be carefully measured so that one can know what these flow rates are and be able to optimize them. Flow meters are commercially available.

11.6 DETECTORS

Detectors in gas chromatography are designed to generate an electronic signal when a gas other than the carrier gas elutes from the column. There have been a number of detectors invented to accomplish this. Not only do these detectors vary in design, but they also vary in sensitivity and selectivity. Sensitivity refers to the smallest quantity of mixture component that can generate an observable signal, and selectivity refers to the type of compound for which a signal can be generated. The flame ionization detector, for example, is a very sensitive detector, but does not detect everything, i.e., it is selective for only a certain class of compounds. The thermal conductivity detector, on the other hand, detects virtually everything, i.e., it is a "universal" detector, but is not very sensitive. What follows is a brief description of the designs of the detectors that are in common use, along with some indication of their sensitivity and selectivity.

11.6.1 FLAME IONIZATION DETECTOR (FID)

A very important GC detector design is the flame ionization detector, the FID. In this detector, the column effluent is swept into a hydrogen flame where the flammable components are burned. In the burning process, a very small fraction of the molecules becomes fragmented and the resulting positively charged ions are drawn to a "collector" (negatively charged) electrode, a metal cylinder above and encircling the flame, while electrons flow to the positively charged burner head. The negatively charged collector and the positively charged burner head are part of an electrical circuit in which the current changes when this process occurs and the change is amplified and seen as a peak on the chromatogram. Figure 11.12 shows a drawing of this detector. The design includes the hydrogen flame burner nozzle, the collector electrode, an inlet for air to surround the flame, and an igniter coil for igniting the hydrogen as it emerges from the nozzle.

Apparent near the bench on which the GC unit sits are pressure-regulated compressed gas cylinders of hydrogen and air (in addition to the carrier gas, helium, or nitrogen). Metal tubing, typically

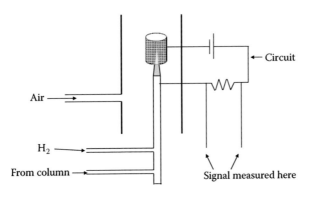

FIGURE 11.12 Drawing of a flame ionization detector.

**APPLICATION NOTE 11.2: THE ANALYSIS OF CLINICAL
SAMPLES BY GAS CHROMATOGRAPHY**

Clinical samples, such as whole blood, vitreous humor, and urine, have been analyzed by gas chromatography. The analytes are simple organic compounds: acetone, ethanol, acetaldehyde, and methanol. These compounds are volatile compounds that, if found in biological matrices such as these, are indicators of several diseases and intoxications. Standard solutions are prepared by diluting stock standards of these analytes with water to create concentrations of up to 240 ppm for acetone, methanol, and acetaldehyde, and up to 2400 ppm for ethanol. These solutions are further diluted with the addition of a surfactant and an acetonitrile solution. The internal standard is 1-propanol. Samples are also diluted with the surfactant and the acetonitrile solution. All solutions are vortexed and centrifuged. The fluids to be injected, samples and standards, are taken from the supernatant of the centrifuged solution.

A special glass-lined injection port is used. The column is a capillary column with a 0.25 internal diameter. The stationary phase has the polyethylene glycol type of structure mentioned in this text. The detector is a flame ionization detector.

(*Reference*: Pontes, H. et al., GC determination of acetone, acetaldehyde, ethanol, and methanol in biological matrices and cell culture, *Journal of Chromatographic Science*, 47 (2009).)

1/8-in. in diameter, connects the cylinders to the detector. A needle valve is used for flow control. These valves are located in the instrument for easy access and control by the operator.

The FID is very sensitive, but it is not universal and also destroys (burns) the sample. It only detects organic substances that burn and fragment in a hydrogen flame. These facts preclude its use for preparative GC or for inorganic substances that do not burn, such as water, carbon dioxide, etc. Still, it is a very popular detector, given its sensitivity, and given the fact that most analytical work involves flammable organic substances (see Application Note 11.2 for an example of a GC analysis that utilizes this detector).

11.6.2 Thermal Conductivity Detector (TCD)

The thermal conductivity detector (TCD) operates on the principle that gases eluting from the column have thermal conductivities different from that of the carrier gas, which is usually helium. Present in the flow channel at the end of the column is a hot filament, hot because it has an electrical current passing through it. This filament is cooled to an equilibrium temperature by the flowing helium, but is cooled differently by the mixture components as they elute, since their thermal conductivities are different from helium. This change in the cooling process causes the filament's electrical resistance to change and thus causes the current flowing through it and the voltage drop across it to change each time a mixture component elutes. Alongside this flow channel is a second channel with a second hot filament through which only pure helium flows (see Figure 11.13). Both filaments are part of a Wheatstone Bridge circuit, which allows a "comparison" between the two resistances and a voltage output to the data system. Such a design is intended to minimize effects of flow rate, pressure, and line voltage variations.

A flow-modulated design of the TCD has become popular. In this design, a single filament is used and the column effluent is alternated with the pure helium through the flow channel where the filament is located. This eliminates the need to use two matched filaments.

The thermal conductivity detector is universal (detects everything), it is nondestructive (can be used with preparative GC) but is less sensitive than other detectors.

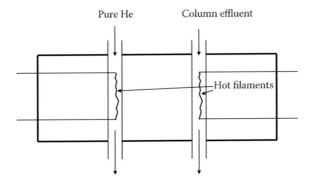

FIGURE 11.13   Drawing of a thermal conductivity detector.

11.6.3   ELECTRON CAPTURE DETECTOR (ECD)

A third type of detector, required for some environmental and biomedical applications, is the electron capture detector, or the ECD. This detector is especially useful for large halogenated hydrocarbon molecules since it is the only one that has an acceptable sensitivity for such molecules. Thus, it finds special utility in the analysis of halogenated pesticide residues found in environmental and biomedical samples.

The electron capture detector is another type of ionization detector. Specifically, it utilizes the beta emissions of a radioactive source, often nickel-63, to cause the ionization of the carrier gas molecules, thus generating electrons that constitute an electrical current. As an electrophilic component, such as a pesticide, from the separated mixture enters this detector, the electrons from the carrier gas ionization are "captured" creating an alteration in the current flow in an external circuit. This alteration is the source of the electrical signal that is amplified and sent on to the recorder. A diagram of this detector is shown in Figure 11.14. The carrier gas for this detector is either pure nitrogen or a mixture of argon and methane.

An additional consideration regarding pesticides warrants mentioning here. Most of these compounds decompose on contact with hot metal surfaces. This problem has, however, been adequately solved for most pesticides by constructing the entire path of the sample out of glass or glass-lined materials. Thus, glass or glass-lined injection ports and all-glass columns are used.

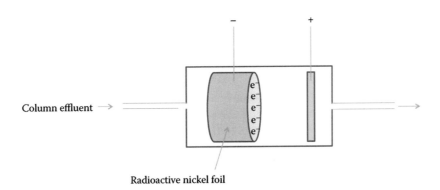

FIGURE 11.14   Drawing of an electron capture detector. The beta particles emitted by the nickel foil cause ionization of the carrier gas and the electrons thereby released are captured by eluting analyte molecules causing an alteration of the detector signal.

In terms of advantages and disadvantages, the ECD is extremely sensitive, but only for a very select group of compounds—halogenated hydrocarbons. Other gases will not give a peak. It does not destroy the sample and thus may be used for preparative work.

11.6.4   NITROGEN/PHOSPHORUS DETECTOR (NPD)

While the ECD is useful for chlorinated hydrocarbon pesticides, the NPD, also known as the "thermionic" detector, is useful for the phosphorus and nitrogen-containing pesticides, the organophosphates and carbamates. The design, however, represents a slight alteration of the design of the FID. In the NPD, we basically have an FID with a bead of alkali metal salt positioned just above the flame. The hydrogen and air flow rates are lower than in the ordinary FID, and this minimizes the fragmentation of other organic compounds. These changes result in a somewhat mysterious increase in both the selectivity and sensitivity for the pesticides.

11.6.5   FLAME PHOTOMETRIC DETECTOR (FPD)

A detector that is specific for organic compounds containing sulfur or phosphorus is the flame photometric detector (FPD). A flame photometer (Chapter 9) is an instrument in which a sample solution is aspirated into a flame and the resulting emissions from the flame are measured with a photomultiplier tube. The FPD is a flame photometer positioned to accept the effluent from the column rather than an aspirated sample. The flame in this case is a hydrogen flame as in the FID. The basic operating principle is that the sulfur or phosphorus compounds burn in the hydrogen flame and produce light-emitting species. A wavelength selector, typically a glass filter, makes this detector specific for these compounds. The signal for the recorder is the signal proportional to light intensity that is produced by the phototube.

The advantages are that it is a very selective detector and also very sensitive. Disadvantages include the problems associated with the need to carefully control the flame conditions so that the correct species are produced (S = S for the sulfur compounds and HPO for the phosphorus compounds). Such conditions include the gas flow rates and the flame temperature. It is a destructive detector.

11.6.6   ELECTROLYTIC CONDUCTIVITY (HALL DETECTOR)

The Hall detector converts the eluting gaseous components into ions in liquid solution and then measures the electrolytic conductivity of the solution in a conductivity cell. The solvent is continuously flowing through the cell and thus the conducting solution is in the cell for only a moment while the conductivity is measured and the peak recorded before it is swept away with fresh solvent. The conversion to ions is done by chemically oxidizing or reducing the components with a "reaction gas" in a small reaction chamber made of nickel positioned between the column and the cell. The nature of the reaction gas depends on what class of compounds is being determined. Organic halides, the most common application, use hydrogen gas at 850°C or higher as the reaction gas. The strong HX acids are produced, which give highly conductive liquid solutions.

The Hall detector has excellent sensitivity and selectivity, giving a peak for only those components that produce ions in the reaction chamber. It is a destructive detector.

11.6.7   GAS CHROMATOGRAPHY–MASS SPECTROMETRY (GC-MS)

Mass spectrometry (Chapter 13) has been adapted to and used with GC equipment as a detector with great success, and the so-called **gas chromatography–mass spectrometry**, or **GC-MS**, will be discussed in Chapter 13. The technique is "two-dimensional," meaning that in addition to a gas chromatogram, the combination also yields the mass spectrum of each mixture component as it

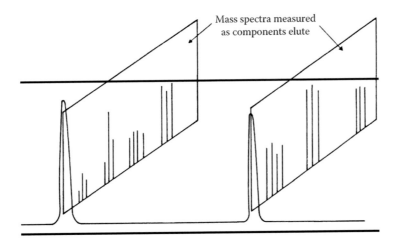

Mass spectra measured
as components elute

FIGURE 11.15   Illustration of the two-dimensional nature of the mass spectrometer detector. The mass spectrum of a mixture component is generated as the chromatography peak for that component is being traced.

elutes. It is like taking a photograph of each component as it elutes, as shown in Figure 11.15 (see Chapter 13 for the details).

11.6.8   PHOTOIONIZATION DETECTOR (PID)

The **photoionization detector (PID)**, as the name implies, involves the ionization of eluting mixture components by light, specifically, UV light. The UV source emits a wavelength characteristic of the gas (either helium or argon) inside. This light passes into an "ionization chamber" through a metal fluoride window and into the path of the column effluent there. This is where the mixture components absorb the light and ionize. The resulting ions are detected using a pair of electrodes in the ionization chamber, the current from which constitutes the signal to the recorder. The specific lamp and window are chosen according to the ionization energy needed for the compounds in the sample.

Since different lamps and windows are available, this detection method can often be selective for only some of the components present in the sample. Its sensitivity is especially good for aromatic hydrocarbons and inorganic substances. It is a very sensitive nondestructive detector.

11.7   QUALITATIVE ANALYSIS

As mentioned in Chapter 10, Section 10.8.4, the parameters that are most important for a qualitative analysis using most GC detectors are the retention time, t_R, and the adjusted retention time, t'_R. Their definitions were graphically presented in Figures 10.13 and 10.14. Under a given set of conditions (the nature of the stationary phase, the column temperature, the carrier flow rate, the column length and diameter, and the instrument dead volume), the retention time is a particular value for each component. It changes only when one or more of the above parameters changes. Thus, when the temperature changes, retention time will change; when the carrier gas flow rate changes, the retention time will change; and so forth.

Repeated injections into a given system under a given set of conditions should always yield a particular retention time for a given component. When one of the parameters changes, the retention time for that component will be slightly different. This is true, for example, when an analyst in another laboratory, trying to duplicate a given qualitative analysis, sets up with a different **dead volume**. The dead volume is the volume of the space between the injection port and the column packing and the space between the column packing and the detector. It also occurs in this scenario if the stationary phase has a slightly different composition than the original. The adjusted retention time will correct for changes in the dead volume but will not correct for any other change. Selectivity, however, does adjust for other changes. The selectivity is thus an important parameter for qualitative analysis if the work involves other setups with other instruments and columns that do not exactly match the original.

The usual qualitative analysis procedure, then, is to establish the conditions for the experiment, perhaps by trial and error in one's own laboratory or by matching conditions outlined in a given procedure, that would resolve all compounds that may potentially be in the unknown. The idea is to match the retention time data, either ordinary retention time or the selectivity, whichever is appropriate, for standards (pure samples) with that for the unknown. The analyst can then proceed to match the retention time data for the unknown to those of the pure samples to determine which substances are present in the unknown (Experiment 34).

One important caution, however, is that there may be more than one component with the same retention time (no separation) and thus further experimentation may be required. For example, when working with a complex mixture whose components are perhaps not all known, it may be necessary to change the experimental conditions to determine whether a given peak is due to one component (known) or more (e.g., one known and one unknown). Changing the stationary phase may prove useful. Such a change would produce a chromatogram with completely different retention times and probably a different order of elution. Thus, two components that were co-eluted before may now be separated, evidence for which would be a different peak size for the known component.

11.8   QUANTITATIVE ANALYSIS

11.8.1   QUANTITATION METHODS

Several different approaches exist as to what peaks are measured and how the mixture component of interest is actually quantitated. We now discuss three of the more popular methods.

11.8.2   RESPONSE FACTOR METHOD

Consider a four-component mixture to be analyzed by GC. The chromatogram may look something like that shown in Figure 11.16. One might think it logical that in order to quantitate the mixture for,

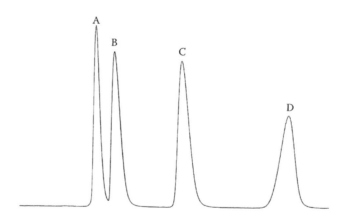

FIGURE 11.16   Chromatogram of a hypothetical four-component mixture of substances A, B, C, and D.

say, component B, all one would need to do is to measure the sizes of all four peaks and divide the size of the peak representing B by the total of all four:

$$\%B = \frac{Area_B}{Area_A + Area_B + Area_C + Area_D} \tag{11.1}$$

The problems with this approach are (1) without comparing the peaks to a standard or a set of standards, it is not known whether the result is a weight percent, volume percent, or mole percent; (2) the instrument detector does not respond to all components equally. For example, not all components will have the same thermal conductivity, and thus the thermal conductivity detector will not give equal sized peaks for equal concentrations of any two components. Thus, the sum of all four peaks would be a meaningless quantity, and the size of peak B by itself would not represent the correct fraction of the total.

It is possible, however, to measure a so-called response factor for the analyte, which is the area generated by a unit quantity injected, such as a microliter (μL) or microgram (μg). The procedure is to inject a known quantity of the analyte, measured by the position of the plunger in the syringe (μL) or by weighing the syringe before and after injection (μg). The peak size that results is measured and divided by this quantity:

$$Response\ factor = \frac{Size\ of\ peak}{Quantity\ of\ pure\ sample\ injected} \tag{11.2}$$

The quantity of analyte in an unknown sample is then determined by measuring the peak size of the analyte resulting from an injection of a known quantity of unknown sample and dividing by the analyte's response factor:

$$Quantity\ of\ analyte = \frac{Size\ of\ peak}{Response\ factor} \tag{11.3}$$

The percent of the analyte can then be calculated as follows:

$$\%\ of\ Analyte = \frac{Quantity\ of\ analyte\ (from\ Equation\ 11.3)}{Total\ quantity\ injected} \tag{11.4}$$

In this method, only the peak of the analyte need be measured in the four component mixture in order to quantitate this component.

11.8.3 Internal Standard Method

Since the peak size is directly proportional to concentration, one may think that one could prepare a series of standard solutions and obtain peak sizes to be used for a standard curve of peak size vs. concentration, a method similar to Beer's law in spectrophotometry, for example, as shown in Figure 11.17. But since peak size also varies with amount injected, there can be considerable error due to the difficulty in injecting consistent volumes as discussed above and in Section 11.4. A method that does away with this problem is the internal standard method. In this method, all standards and samples are spiked with a constant known amount of a substance to act as what is called an internal standard. The purpose of the internal standard is to serve as a reference for the peak size measurements, so that slight variations in injection technique and volume injected are compensated by the fact that the internal standard peak and the analyte peak are both affected by the slight variations.

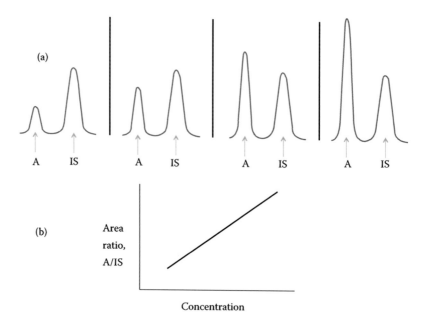

FIGURE 11.17 Illustration of the internal standard method in GC. (a) A gas chromatography peak for an analyte, A, gets larger as the concentration of that analyte increases. However, the peak for the internal standard, IS, in the same injected solution stays the same size, since its concentration does not change. (b) The standard curve can then be a plot of peak area ratio (A/IS) vs. concentration and the problem of the inconsistent injection volume goes away (compare with Figure 10.18).

The procedure is to measure the peak sizes of both the internal standard peak and the analyte peak and then to divide the analyte peak area by the internal standard peak area. The "area ratio" thus determined is then plotted vs. concentration of the analyte, as shown in Figure 11.17. The result is a method in which the volume injected is not as important and, in fact, can vary substantially from one injection to the next because this ratio does not change as the volume injected changes since both peaks are affected equally by the changes.

Can just any substance serve as an internal standard? There are certain characteristics that the internal standard should have, and these are listed below:

(a) Its peak, like the analyte's, must be completely resolved from all other peaks.
(b) Its retention time should be close to that of the analyte.
(c) It should be structurally similar to the analyte.

11.8.4 STANDARD ADDITIONS METHOD

Increasing standard amounts of analyte are added to the sample and the resulting peak areas, which should show an increase with concentration added, are measured. This method is not as useful in GC as it would be in AA (see Chapter 9), since the sample matrix is not an issue in GC as it is in AA due to the fact that matrix components become separated. However, standard additions may be useful for convenience sake, particularly when the sample to be analyzed already contains a component capable of serving as an internal standard. Thus, standard additions could be used in conjunction with the internal standard method (see Experiment 44 in Chapter 13) and the internal standard would not have to be independently added to the sample and to the series of standards—it is already present, a convenient circumstance. Area ratio would then be plotted vs. concentration added and the unknown concentration determined by extrapolation to zero area ratio (refer to Section 9.3.6 for other details of the method of standard additions).

11.9 TROUBLESHOOTING

Problems that arise during a GC experiment usually manifest themselves on the chromatogram. Examples of such manifestations are peak shapes being distorted, peak sizes diminishing for reasons other than quantity of analyte, the baseline drifting, or the retention times changing for no apparent reason, etc. These kinds of problems can usually be traced to injection problems, problems with the column, or problems with the detector. There can, of course, be problems associated with the electronics of the instrument. However, we will not be concerned with those here because of the large number of different instrument designs that have been manufactured over the years. The operator can usually find assistance for these in a troubleshooting section of the manuals that accompany the instrument.*

In the following paragraphs, we will address some of the most common problems encountered, pinpoint possible causes, and suggest methods of solving the problems.

11.9.1 DIMINISHED PEAK SIZE

We could also refer to this as reduced sensitivity. The peaks are smaller than expected based on previous observations when equal or greater quantities of a particular sample were injected. Such an observation usually means a problem with injection (less injected than assumed) or a problem with the detector such that a smaller electronic signal is sent to the recorder. One should check for a leaky or plugged syringe, a worn septum, a leak in the precolumn and postcolumn connections, or a contaminated detector. Of course, detector attenuation, recorder sensitivity settings, electrical connections, and other associated hardware problems are potential causes.

11.9.2 UNSYMMETRICAL PEAK SHAPES

Peak "fronting" or peak "tailing" (Figure 11.18) are typical examples of this problem. These could be indicators of poor injections, meaning too large an injection volume for the diameter of the column in use, too slow with the syringe manipulation during injection, or not fully penetrating the septum. It may indicate a decomposition of thermolabile components in contact with the hot system components such as the metal walls of the injection port and column. It may also mean contamination of the injection port and/or column.

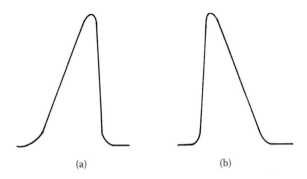

(a)   (b)

FIGURE 11.18   Peaks exhibiting (a) fronting and (b) tailing.

---

* See also the "GC Troubleshooting" column published regularly in the monthly journal *LC.GC, The Magazine of Separation Science.*

11.9.3 ALTERED RETENTION TIMES

This is usually caused by changes in the carrier gas flow rate or column temperature. Flow rate changes can be caused by leaks in the system upstream from the column inlet, such as in the injection port (e.g., the septum); by low pressure in the system due to an empty or nearly empty carrier supply; or by faulty hardware, such as the flow control valve or pressure regulator. Temperature changes can be caused by a faulty temperature controller, an improperly set temperature program, too short a cool-down period prior to the next injection in a temperature programmed experiment, etc. This could also be caused by overloading the column, or by diminished effectiveness (decomposition?) of the stationary phase.

11.9.4 BASELINE DRIFT

This occurs when a new column has not been sufficiently conditioned, when the detector temperature has not reached its equilibrium value, or when the detector is contaminated or otherwise faulty. New columns need to be conditioned, usually with an overnight "bakeout" at the highest recommended temperature for that column. Detector signals may very well change when the detector temperature changes. One should be sure that sufficient time has been given for the detector temperature to level off. The nature of detector problems depends, of course, on the detector. TCD filaments may become oxidized due to an air leak, ionization detectors may be leaking, or there may be a crack in the FID burner nozzle, etc.

Baseline drift may also occur in cases when a temperature program covers a large range.

11.9.5 BASELINE PERTURBATIONS

If the perturbations are in the form of spikes of an irregular nature, the problem is likely to be detector contamination. Such spikes are especially observed when dust particles have settled into the FID flame orifice. Of course, the problem may also be due to interference from electrical pulses from some other source nearby. Regular spikes can be due to condensation in the flow lines causing the carrier, or hydrogen (FID), to "pulse," or they can be due to a bubble flowmeter attached to the outlet of the TCD, as well as the electrical pulses referred to above. These can also be caused by pulses in the carrier flow due to a faulty flow valve or pressure regulator.

11.9.6 APPEARANCE OF UNEXPECTED PEAKS

Unexpected peaks can arise from components from a previous injection that moved slowly through the column, contamination from either the reagents used to prepare the sample or standards, or from a contaminated septum, carrier, or column. The solution to these problems include a rapid "bakeout" via temperature programming after the analyte peaks have eluted, using pure reagents, and replacing or cleaning septa, carrier, or column.

EXPERIMENTS

EXPERIMENT 33: A QUALITATIVE GAS CHROMATOGRAPHY
ANALYSIS OF A PREPARED SAMPLE

INTRODUCTION

This experiment is designed to be your first experience with a GC. The following conditions are suggested. Your instructor may choose other conditions.

Column: packed, 3% FFAP, 2 m in length
Carrier gas flow rate: 20 mL/min
Temperature program: 1.5 min at 80°C, then increase to 120°C at the rate of 50°/min, then hold
    at 120°C for 7 min

Detector: FID A TCD may be used if desired

Injector temperature: 250°C

Detector temperature: 200°C

Attenuation setting: probably either 128 or 256, depending on which gives the best size of peak without going off the scale as described in Step 6 below.

Under these conditions, a sample mixture consisting of benzene, toluene, ethylbenzene, chlorobenzene, bromobenzene, cyclohexane, and acetone should be well resolved.

Remember to wear safety glasses and gloves.

PROCEDURE

1. Examine the instrument to which you are assigned. Locate the source of the carrier gas and trace the line to the instrument. If a FID is to be used, also locate the source of the hydrogen and air and trace the lines of each to the instrument. Locate the injection port. Note any gauges and controls on the front of the instrument and try to identify their functions. Open the column oven and locate the column. Note the proximity of the inlet end of the column to the injection port. Note the outlet end of the column and locate the detector.

2. Locate the data system for your instrument. Turn on the computer and start the data acquisition software. Your instructor will provide instructions on the use of the software. Also turn on the printer and ensure that there is plenty of paper in the paper tray.

3. Open the valve on the pressure regulator of the carrier gas bottle and ensure that there is flow through the system. Turn on the instrument. Set the temperatures of the column, injector, and detector and also set up the temperature program indicated in the introduction above. Allow time for all components to come to the set temperatures.

4. If a FID is used, open the valves on the pressure regulators of the hydrogen and air bottles. After a minute or so, light the flame.

5. Obtain the unknown mixture contained in a vial from your instructor. It contains the mixture of organic liquids to be separated.

6. The purpose of the experiment is to identify the liquids in your unknown by retention time data. Your unknown may contain any number of the liquids listed in the introduction. Obtain chromatograms and determine the retention times of the pure liquids. Your instructor will demonstrate how to inject the sample (1 μL) and use the instrument and data system to determine the retention times of each. The pure liquids should also be contained in small vials. The temperature program may be interrupted and the temperature reset in any given run once the peak has been traced on the monitor. If any peak is so large that it goes off the scale (goes all the way to the top of the screen and displays a flat top), you should readjust your attenuation to a less sensitive setting to give a smaller peak.

7. Obtain a chromatogram of your unknown. The peaks will be smaller since the concentrations are not 100% as they were for the pure liquids. You may have to adjust the attenuation to get large well-defined peaks. Measure the retention times for each peak and compare these to your data for the pure liquids. Those liquids that have retention times matching a peak in your unknown are therefore contained in your unknown.

8. Report what organic liquids are in your unknowns and the evidence to support this conclusion.

9. Maintain the logbook for the instrument used in this experiment. Record the date, your name(s), and the experiment name or number.

EXPERIMENT 34: THE QUANTITATIVE GAS CHROMATOGRAPHY ANALYSIS OF A PREPARED SAMPLE FOR TOLUENE BY THE INTERNAL STANDARD METHOD

INTRODUCTION

The conditions for this experiment are the same as given in the introduction to Experiment 33. Perform Steps 1–4 of Experiment 33 to familiarize yourself with the instrument and prepare it for the experiment. All flasks and pipettes should be free of water and other solvents.

Remember to wear safety glasses and gloves.

PROCEDURE

1. Prepare a series of standard solutions of toluene in cyclohexane that are 0.5%, 1%, 2%, and 3% toluene by volume. Use 25-mL volumetric flasks. Add exactly 0.50 mL of ethylbenzene to each flask before diluting to mark with the cyclohexane. Shake well. The ethylbenzene is the internal standard.

2. The unknown is contained in a 25-mL volumetric flask and has been prepared by your instructor to have a concentration in the range of the standards when diluted to the mark with cyclohexane. Obtain the unknown, add the internal standard, and dilute with cyclohexane to the mark.

3. Inject 1 µL and obtain a good chromatogram of each solution. Since the conditions are the same as in Experiment 31, the retention times will be the same as you discovered in that experiment. You will have to adjust the attenuation setting to get the proper sized peaks for toluene and ethylbenzene. The cyclohexane peak will be much larger than the others, since it is the solvent and is present at a much larger concentration. This peak will not enter into the determination of the toluene concentration and can be allowed to go off the scale and ignored.

4. Determine the areas of the toluene and ethylbenzene peaks on each chromatogram. Your instructor will demonstrate how to obtain the peak areas with the software used. Calculate the ratio of the area of the toluene peak to the area of the ethylbenzene peak.

5. Plot the standard curve using the spreadsheet procedure used in Experiment 13. The y-axis is the area ratio and the x-axis is the toluene concentration. Obtain the correlation coefficient and the concentrations of the unknown and control if there is one.

6. Maintain the logbook for the instrument used in this experiment. Record the date, your name(s), the experiment name or number, the R^2 value, and the results for the control sample. Also, plot the control sample results on a control chart for this experiment posted in the laboratory.

EXPERIMENT 35: THE DETERMINATION OF ETHANOL IN WINE BY GAS CHROMATOGRAPHY AND THE INTERNAL STANDARD METHOD

INTRODUCTION

The conditions for this experiment are the same as given in the introduction to Experiment 33, except that the temperature is not programmed. Perform Steps 1–4 of Experiment 33 (except for the temperature program) to familiarize yourself with the instrument and prepare it for the experiment. All flasks and pipettes should be free of water and other solvents.

Remember safety to wear glasses and gloves.

PROCEDURE

1. Prepare four to six standard solutions of ethanol in water such that the alcohol content of the wine (as indicated on the wine label) is in the "middle." For example, if the wine is 15% ethanol, standards of 5%, 10%, 20%, and 25% are appropriate. Use 25-mL volumetric flasks and pipette the ethanol accurately. Dilute to the mark with water and then add 1.00 mL of acetone (the internal standard) above the mark. Shake well.

2. Prepare each sample by filling a 25-mL flask with the wine and adding 1.00 mL of acetone above the mark.

3. There are three solution components, ethanol, acetone, and water. If a flame ionization detector is used, there will be only two peaks, since water will not give a peak. Set up the instrument for isothermal operation at 100°C. The two peaks should be nicely resolved and each run should only take a few minutes.

4. Obtain chromatograms for all standards and all wine samples. Obtain the areas of the ethanol peaks and acetone peaks and calculate the area ratio, ethanol peak area to acetone peak area.

5. Plot the standard curve using the spreadsheet procedure used in Experiment 13. The y-axis is the area ratio and the x-axis is the ethanol concentration. Obtain the correlation coefficient and the concentrations of the unknown and control if there is one.

6. Maintain the logbook for the instrument used in this experiment. Record the date, your name(s), the experiment name or number, the R^2 value, and the results for the control sample. Also plot the control sample results on a control chart for this experiment posted in the laboratory.

EXPERIMENT 36: DESIGN AN EXPERIMENT: DETERMINATION OF ETHANOL IN COUGH MEDICINE OR OTHER PHARMACEUTICAL PREPARATION

Design an experiment based on your experience in Experiment 34 to determine the alcohol content in cough medicine or other pharmaceutical preparation by gas chromatography. Ask your instructor to approve your approach before beginning the work. Remember to wear safety glasses.

EXPERIMENT 37: A STUDY OF THE EFFECT OF THE CHANGING OF GC INSTRUMENT PARAMETERS ON RESOLUTION

Note: A mixture of two organic liquids is used for this study. The specific liquids will be selected by your instructor. They should be available in a vial equipped with a rubber septum. Your instructor has also selected the initial column packing and length. The two liquids should be in a ratio of approximately 1:1 by volume. Remember to wear safety glasses.

1. Set the carrier gas flow rate for your instrument to 20 mL/min. The method of measuring this flow rate will be demonstrated by your instructor. Set the column temperature to 70°C. Inject 1.0 μL of the mixture and observe the resolution. Now change the column temperature to 80°C, wait 5 min for the oven temperature to become stable and inject 1.0 μL again, observing the resolution. Repeat at 90°C, 100°C, 110°C, and 120°C, observing the effect on resolution.
2. Set the column oven temperature to 100°C and wait for the temperature to become stable. Set the carrier gas flow rate to 10 mL/min. Inject 1.0 μL of the mixture and observe the resolution. Now increase the carrier gas flow rate by 5 mL/min so as to observe the resolution at 15, 20, 25, 30, and 35 mL/min.
3. Set the carrier gas flow rate to 20 mL/min and obtain a series of chromatograms each at a different volume injected—0.5, 1.0, 1.5, 2.0, and 2.5 μL. The attenuation should be set so that the peaks from the larger injection volumes will not be off scale.
4. If a longer column with the same stationary phase is available, change columns. Set the temperature and carrier gas flow rate to some combination of values that gave poor resolution in one of the previous steps. Inject 1.0 μL and observe the effect of a longer column on resolution.
5. Change columns to one with some other stationary phase, perhaps one suggested by your instructor. Set the column temperature to 100°C and the carrier gas flow rate to 20 mL/min. Inject 1.0 μL of the mixture. Assuming good resolution, observe the order of elution. Is it different from that observed with the former stationary phase? If so, explain how that could be. If not, compare the resolution here with that of a previous injection in which all the parameters were equal. Comment on the difference.
6. As discussed in Chapter 10, a numerical value for resolution can be calculated as follows:

$$R = \frac{2(t_2 - t_1)}{(w_1 + w_2)}$$

where t_2 is the retention time for the component with the longest retention time (component 2), t_1, the other component, and w_1 and w_2 represent the respective widths at the base bisected by the tangents to the sides (see discussion in Chapter 10, Section 10.4.4). As directed by your instructor, calculate resolution for some or all of the above data and construct graphs of R vs. volume injected, R vs. flow rate, and R vs. temperature. Comment on the results.

QUESTIONS AND PROBLEMS
Overview

1. What are some unique features of the modern GC column?
2. Track a liquid from the injection syringe through the GC instrument to the detector and briefly explain what is happening to it at every stage.

Vapor Pressure and Solubility

3. Define vapor pressure.
4. What role does vapor pressure play in a GC separation?
5. In a GC experiment in which the liquid stationary phase is polar, which would have a shorter retention time—a nonpolar mixture component with a high vapor pressure or a polar mixture component with a low vapor pressure? Explain.
6. If a GC operator expects a given mixture component to have a very long retention time compared with other mixture components, what vapor pressure and solubility properties would this mixture component have?

Instrument Components

7. What are the three major components of a GC instrument?
8. Why are each of the three instrument components depicted in Figure 11.4 at an elevated temperature?

Sample Injection

9. Why must the size of a liquid sample injected into the GC be small? What would happen if it were too large?
10. What is a "split" injector? What is a "splitless" injection?
11. What are the four different modes of injection in GC?

Column Details

12. GC columns can be up to 300 ft long, yet they are contained in a space of 1 to 3 ft³. How is that possible?
13. What is meant by a GC open-tubular capillary column? Why has the development of such a column been useful?
14. Contrast the packed column and the open tubular capillary column in terms of design, diameter, length, how the stationary phase is held in place, ability to resolve complex mixtures, amount of sample injected.
15. What is the difference between "analytical GC" and "preparative GC"?
16. How is it that a liquid stationary phase can be chemically bonded to the inside surface of a capillary column?"
17. What is "Chromosorb™"? What is its use in GC?
18. What is meant by "temperature programming" in GC?
19. Tell how the temperature programming feature of most modern gas chromatographs can be useful in separating complex mixtures.
20. In a GC separation involving four components, it was discovered that all four components, A, B, C, and D, separate cleanly at 80°C. At this temperature, A and B have fairly

short retention times, but components C and D have very long retention times. It was also discovered that C and D separate cleanly at 150°C but that A and B do not. Suggest a temperature program that would separate all four in a reasonable time and explain why it would work.

21. Would higher carrier gas flow rates increase or decrease retention time?

Detectors

22. What does it mean to say that a GC detector is "universal"? What does it mean to say that one GC detector is "more sensitive" and is "more selective" than another?

23. Which of the different types of GC detectors
    (a) Requires the use of hydrogen gas?
    (b) Is well-suited for pesticide residue analysis?
    (c) Uses the abilities of the eluting gases to conduct heat?
    (d) Is the least sensitive?
    (e) Will not work for noncombustible mixture components?
    (f) Breaks the eluting molecules into charged fragments, which are then analyzed in terms of charge and mass?
    (g) Is part of an extremely powerful (and expensive) system abbreviated GC-MS?
    (h) Uses a radioactive source?
    (i) Uses an FID with a bead of alkali metal salt positioned just above the flame.
    (j) Use an FTIR instrument so as to take an IR spectrum "photograph" of the mixture components as they elute.
    (k) Is a detector that is specific for organic compounds of sulfur and phosphorus.
    (l) Is a detector that utilizes a UV light beam to ionize the component molecules.
    (m) Is a detector that converts eluting molecules into ions in solution so that they can then be detected by electrical conductivity measurements.

24. Which of the GC detectors we have studied
    (a) Burn(s) the mixture components in a hydrogen flame?
    (b) Is (are) not very sensitive?
    (c) Is (are) "universal"?
    (d) Is (are) good for pesticide residue analysis?
    (e) Utilize(s) a hot filament in the flow stream?
    (f) Utilize(s) a source of radioactivity?
    (g) Would not detect water?
    (h) Do (does) not destroy the mixture components?
    (i) Convert(s) eluting molecules into ions?
    (j) Make(s) a measurement like a photograph of the eluting molecules?

25. Match the detector with the items that follow.

Thermal conductivity detector	Flame ionization detector
Electron capture detector	Mass spectrometer detector

    (a) FID
    (b) A GC detector good especially for pesticides.
    (c) A GC detector that utilizes hydrogen gas.
    (d) A GC detector that utilizes a radioactive source.
    (e) A GC detector that analyzes fragmented molecules according to their mass and charge.
    (f) A powerful GC detector that can perform qualitative analysis even without the use of retention times.

(g) A GC detector that is based on the differences of the abilities of helium and the mixture components to conduct heat.

(h) A GC detector in that noncombustible materials will not give a peak.

(i) A GC detector especially good for electrophilic substances.

(j) A GC detector that is part of the GC-MS assembly.

(k) A universal but not very sensitive GC detector.

26. Of these three detectors, thermal conductivity, flame ionization, and electron capture, which

(a) Requires the use of hydrogen gas?

(b) Is universally applicable, but not very sensitive?

(c) Does not destroy the sample?

(d) Is good for pesticide analysis?

(e) Can only be used for samples that are able to be burned?

27. What advantages does

(a) A thermal conductivity detector have over a flame ionization detector?

(b) A flame ionization detector have over a thermal conductivity detector?

28. What do the letters GC-MS stand for? Briefly describe how the GC-MS detector works and tell why it is such a useful detector.

Qualitative Analysis

29. Explain the principles by which qualitative analysis can be performed in GC with the use of retention times.

30. A certain sample is a mixture of four organic liquids and these liquids exhibit the following retention times in a GC experiment.

Liquid	Retention Time (min)
A	1.6
B	2.2
C	4.7
D	9.8

Some known liquids were injected into the chromatograph and the following data were determined:

Liquid	Retention Time (min)
Benzene	0.5
Toluene	1.6
Ethylbenzene	3.4
n-Propylbenzene	4.7
Isopropylbenzene	5.8

(a) Can you tell what liquids from these five are definitely not present? If not, why not? If you can tell, which liquids are they?

(b) What liquids are possibly present?

(c) Why can one not tell with certainty what liquids are present, based on the information given here?

31. A laboratory technician tests a liquid mixture with gas chromatography for the purpose of identifying the components. She injects the mixture and four peaks are displayed on

the chromatogram. She then obtains four pure liquids from a stock room and injects them into the GC (same conditions) one at a time. The retention time of one of the pure liquids exactly matches one of the retention times on the mixture chromatogram. Do you think she now knows, with certainty, the identity of one of the components? Explain.

32. Fill in the blanks with terms chosen from the following list:

Retention time	Carrier gas
Injection port	Resolution
Preparative GC	Thermal conductivity detector
Flame ionization detector	Column
Electron capture detector	Mobile phase
Temperature programming	Stationary phase
Open-tubular column	Detector

(a) Two terms that describe the helium used in gas chromatography are _____ and _____.

(b) A measurement that is made when doing a qualitative analysis of a mixture in gas chromatography is _____.

(c) A type of detector in gas chromatography that is "universal" is the _____.

(d) A type of detector in gas chromatography that is useful for pesticide analysis is the _____.

(e) The three heated parts in a gas chromatography instrument are the _____, _____, and _____.

(f) Increasing the length of a GC column is a way of improving the _____.

(g) Sometimes, a difficult separation in gas chromatography can be accomplished easily by changing the temperature of the column during the run. This is called _____.

(h) A gas chromatography instrument can be used to obtain pure samples of the components of a mixture to be used whenever pure samples are needed for some other experiment. This is called _____.

(i) In an experiment in which the _____ is used, one needs to have a source of hydrogen gas.

(j) The _____ has the stationary phase adsorbed directly onto the wall of the column.

Quantitative Analysis

33. An injection of 3.0 μL of methylene chloride (density 1.327 g/mL) gave a peak size of 3.74 cm². The injection of 3.0 μL of an unknown sample (density = 1.174 g/mL) gave a methylene chloride peak size of 1.02 cm². Calculate a response factor for methylene chloride and the percent of methylene chloride in the sample.

34. Define "internal standard." Tell why an internal standard is important in a quantitative analysis by GC. Also tell what is plotted on the x- and y-axes when plotting the standard curve in internal standard procedures.

35. Compare and/or differentiate between the internal standard method and the standard additions method.

36. Compare the internal standard method with the standard additions method in terms of

(a) Solution preparation

(b) What the chromatograms look like

(c) What is plotted for the standard curve

37. Consider the quantitative gas chromatography analysis of alcohol-blended gasoline for ethyl alcohol by the internal standard method, using isopropyl alcohol as the internal

standard. The peaks for these two substances are well resolved from each other and from other components. Assume there is no isopropyl alcohol in the gasoline.
   (a) Describe how you would prepare a series of standard solutions for this analysis.
   (b) What would the chromatograms of these solutions look like?
   (c) What is plotted in an analysis of this type?
38. Select one of the quantitation procedures we have discussed (response factor method, internal standard method, or standard addition method) and describe:
   (a) Experimental details (solution preparation and/or sample preparation)
   (b) What the raw data are
   (c) What is plotted

Troubleshooting

39. What observation on a chromatogram would lead you to conclude that you injected too much sample for the diameter of column you are using?
40. If a GC operator observes that a given mixture component gives a smaller peak than in previous work for the same sample and the same injection volume, what might be the problem?
41. What might be the cause of a drifting chromatogram baseline?
42. What will occur if a GC operator makes another sample injection before the column has been cleared of mixture components from previous injections?
43. What can a GC analyst do to solve the problem of unexpected peaks on the chromatogram?

Report

44. For this exercise, your instructor will select a "real-world" gas chromatography analysis method, such as an application note from an instrument vendor or from a methods book, a journal article, or Web site, and will give it to you as a handout. Write a report giving the details of the method according to the following scheme. On your paper, write (or type) "a," "b," etc. to clearly present your response to each item.
   (a) Method Title: Give a more descriptive title than what is given on the handout.
   (b) Type of Material Examined: Is it water, soil, food, a pharmaceutical preparation, or what?
   (c) Analyte: Give both the name and the formula of the analyte.
   (d) Sampling Procedure: This refers to how to obtain a sample of the material. If there is no specific information given as to obtaining a sample, make something up. If you want to be correct in what you write, do a library or other search to discover a reasonable response. Be brief.
   (e) Sample Preparation Procedure: This refers to the steps taken to prepare the sample for the analysis. It does not refer to the preparation of standards or any associated solutions.
   (f) Is an internal standard used? If so, what is it and what is its concentration?
   (g) What is the stationary phase and the solid support? If there is no solid support (i.e., if a capillary column is used), indicate that in your answer, but tell what the stationary phase is.
   (h) Is temperature programming used? If so, tell what the temperature program is exactly. If not, indicate what the temperature of the column is.
   (i) What mobile phase is used and what is the flow rate of the mobile phase?
   (j) What detector is used?

(k) Standard Curve: Is a standard curve suggested? If so, what is plotted on the y-axis and what is plotted (give units) on the x-axis?

(l) How are the data gathered? In other words, you have got your blank, your standards, and your samples ready to go. Now what? Be brief.

(m) Concentration Levels for Standards: This refers to what the concentration range is for the standard curve. Be sure to include the units. If no standard curve is used, state this.

(n) How the standards are prepared? Are you diluting a stock standard? If so, how is the stock prepared and what is its concentration? If not, how are the standards prepared?

(o) Potential Problems: Is there anything mentioned in the method document that might present a problem? Be brief.

(p) Data Handling and Reporting: Once you have the data, then what? Be sure to present any post-run calculations that might be required.

(q) Reference: Did you look at any other reference sources to help answer any of the above? If so, write them here.

12 High-Performance Liquid Chromatography and Electrophoresis

12.1 INTRODUCTION

12.1.1 SUMMARY OF METHOD

High-performance liquid chromatography (HPLC) is an instrumental chromatography configuration in which the mobile phase is a liquid. The discussion of the principles of instrumental chromatography presented in Chapter 10 (Section 10.4.3) provides the basic background for this technique. A liquid mobile phase is made to move through a column containing the stationary phase. A mixture of compounds injected ahead of the column separates as the compounds pass through the column. The mixture components are then detected electronically one at a time as they elute from the column, resulting in the recording of the instrumental chromatogram by the data collection system (refer back to Figures 10.2 and 10.12).

All the "types" of chromatography discussed in Chapter 10 can be utilized in the HPLC configuration. Thus, we have partition (LLC), bonded phase (BPC), adsorption (LSC), ion exchange (IEC and IC), and size exclusion (SEC), including gel permeation (GPC) and gel filtration (GFC), all as commonly used types of HPLC. More specific information on the specific stationary phases for each will be given in Section 12.5.

The rise in popularity of HPLC is due in large part to its "high performance" nature and the advantages offered over the older, non-instrumental, "open-column" method also described in Chapter 10. Separation and quantitation procedures that require hours and sometimes days with the open-column method can be completed in a matter of minutes, or even seconds, with HPLC. Modern column technology and gradient solvent elution systems, which will be described, have contributed significantly to this advantage in that extremely complex samples can be resolved with ease in a very short time.

The basic HPLC system is diagrammed in Figure 12.1. It consists of solvent reservoirs for containing liquid mobile phases, a special high-pressure pump for pumping the mobile phases from the reservoirs and through the column, a specially designed injection device for sample introduction, the column where the separation takes place, the detector for electronic sensing of the eluting mixture components, and a data system for acquiring and displaying the chromatogram. Besides these basic components, an HPLC instrument setup may be equipped with a gradient programmer (Section 12.2), an auto-sampler (see Application Note 6.1), a "guard column," and various in-line filters (see Application Note 12.1 for an important use of HPLC).

12.1.2 COMPARISONS WITH GC

Because the HPLC mobile phase is a liquid, there are some very obvious differences between HPLC and GC. First, the mechanism of separation in HPLC involves the specific interaction of the mixture components with a specific mobile phase composition, whereas in GC, the vapor pressure of the components, and not their "interaction" with a specific carrier gas, is the most important consideration (see Chapter 11). Second, the force that sustains the flow of the mobile phase is that of a high-pressure pump, rather than the regulated pressure from a compressed gas cylinder. Third, the injection

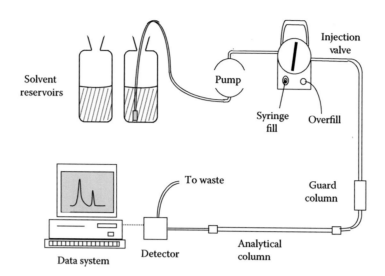

FIGURE 12.1   HPLC system.

APPLICATION NOTE 12.1: TRACKING DRUGS IN THE BODY

It is common for pharmaceutical companies and their contract laboratories to perform laboratory work on biological samples to track drugs in the body and to profile body fluids and excretions to determine exactly where the drugs end up in the body and at what concentrations. The drugs include over the counter medications as well as proprietary drugs that are still being researched. Samples analyzed in these kinds of studies include whole blood, plasma, and serum as well as urine and feces. Samples may also be parts of an animal's body, since the research is often performed on animals. In any case, the laboratory work begins by extracting the analyte from the sample and concludes with one of any number of methods to arrive at the concentration of the analyte drug. The analyst often uses HPLC, since there can be many other components of the sample that need to be separated before the analyte drug can be measured.

---

device requires a totally different design due to the high pressure of the system and the possibility that a liquid mobile phase may chemically attack a rubber septum. Fourth, the detector requires a totally different design because the mobile phase is a liquid. Finally, the injector, column, and detector need not be heated as in GC, although the mode of separation occurring in the column can be affected by temperature changes, and thus sometimes elevated column temperatures are used.

12.2   MOBILE PHASE CONSIDERATIONS

The mobile phase reservoir is made of an inert material, usually glass. There is usually a cap on the reservoir that is vented to allow air to enter as the fluid level drops. The purpose of the cap is to prevent particulate matter from falling into the reservoir. It is very important to prevent particulates from entering the flow stream. The tip of the tube immersed in the reservoir is fitted with a coarse metal filter. It functions as a filter in the event that particulates do find their way into the reservoir. It also serves as a "sinker" to keep the tip well under the surface of the liquid. In addition, in specially designed mobile phase reservoirs, this sinker/filter is placed into a "well" on the bottom of the reservoir so that it is completely immersed in solvent even when the reservoir is running low. This avoids drawing air into the line under those conditions. These details are shown in Figure 12.2.

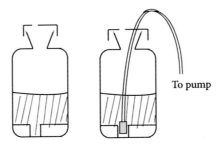

FIGURE 12.2   Drawings of mobile phase reservoirs. Note the vented caps. In addition, the reservoir on the right is shown with a coarse filter on the tip of the flow tube shown inside a "well" on the bottom of the reservoir.

The HPLC pump draws the mobile phase from the reservoir via vacuum action. In the process, air dissolved in the mobile phase may withdraw from the liquid and form bubbles in the flow stream unless such air is removed from the liquid in advance. Air in the flow stream is undesirable because it can cause a wide variety of problems, such as poor pump performance or poor detector response. Removing air from the mobile phase, called **degassing**, in advance of the chromatography is a routine matter, however, and can be done in one of several ways. These are (1) helium sparging, (2) ultrasonic agitation, (3) drawing a vacuum over the surface of the liquid, (4) a combination of items 2 and 3.

Helium sparging refers to the vigorous bubbling of helium gas through the mobile phase. This can be done while it is contained in the reservoir using a metal bubbler with tubing attached to a cylinder of helium. The bubbling causes the air to be efficiently displaced from the liquid by the helium. Helium does not saturate the solution because it is so light in weight. At a helium flow rate of 300 mL/min, complete degassing takes just a few minutes.

Drawing a **vacuum** over the surface of the liquid mobile phase has the same effect as the vacuum action of the HPLC pump—the air withdraws from the solvent. Such an action is usually sufficient for the degassing step for most applications. **Ultrasonic agitation**, by itself, is helpful in removing high levels of gases in certain samples, such as the carbonation in beverages. To remove standard levels of dissolved air, however, it must be combined with the vacuum action. Thus, a laboratory analyst will often draw a vacuum over the surface while also sonicating.

If the mobile phase contains liquids that are not certified as HPLC-grade solvents, they must be filtered ahead of time as well as degassed. The reason is that the packed bed of finely divided stationary phase particles through which the mobile phase percolates is itself an excellent filter. Particles in the mobile phase as small as 0.5 μm in diameter can be filtered out. The result of this is a decreased effectiveness of the column with time and possibly a blocked flow path. Unfiltered samples also may contain particles and cause this problem.

The problem is solved by **pre-filtering** all mobile phases and samples before beginning the experiment. For mobile phases and large sample volumes, this involves utilizing a vacuum apparatus that draws the liquid through a 0.5-μm filter. Since such filtration involves a vacuum, the mobile phase is automatically degassed as well, so the filtration need not be a separate step. An efficient operation would be to filter a mobile phase with a vacuum apparatus while simultaneously sonicating.

For nonaqueous solvents and their water solutions, filters made of paper are not a good choice due to the possibility of chemical incompatibility with the paper, which would then cause contamination. For these, the filter material of choice is usually nylon. A Teflon-based material (designated PTFE) may also be used, however, if the mobile phase contains some water, the filter must be "wetted" first with some pure organic solvent in order to provide a reasonable filtration rate. Aqueous solutions are often impossible to filter unless the Teflon-based filter is first wetted in this manner. For filtering samples, and standard solutions as well, a small syringe-type filtering unit is used. The filter material may also be nylon, but if the Teflon-based material is used and the sample solvent contains some water, the filter must be wetted first with an organic solvent. Again, paper cannot be used if the sample or standard contains an organic liquid.

In addition to these prefiltering steps, in-line filters and a "guard column" are often used. The sinker in the mobile phase reservoir is a filter, as mentioned previously. Another metal in-line filter is in the flow path between the injector and the column, often immediately preceding the column. This filter would serve to remove particulates that entered via the sample injection. The **guard column** is usually placed just before the regular "analytical" column. Its function is to remove other contaminating substances—substances that perhaps have long retention times on the analytical column and that eventually interfere with the detection in later experiments. Guard columns are inexpensive and disposable and are changed frequently.

12.3 SOLVENT DELIVERY

12.3.1 PUMPS

The pump that is used in HPLC cannot be just any pump. It must a special pump that is capable of very high pressure (up to 5000 psi) in order to pump the mobile phase through the tightly packed stationary phase at a reasonable flow rate, usually between 0.5 and 4.0 mL/min. It also must be nearly free of pulsations so that the flow rate remains even and constant throughout. Only manufacturers of HPLC equipment manufacture such pumps.

The pump is a **reciprocating piston pump**. A drawing is shown in Figure 12.3. In this pump, a small piston (approximately 1/4 in. in diameter) is driven back and forth drawing liquid in through an inlet check valve during its backward stroke and expelling the liquid through an outlet check valve during the forward stroke. A **check valve** is a device that allows liquid flow in one direction only. It typically consists of a ruby ball that moves with the mobile phase and seals against a sapphire seat that allows flow around it in one flow direction but not in the other (see Figure 12.3). Liquid is drawn in from the reservoir when the piston is in its backstroke and pushed out toward the column when the piston is in its downstroke. The piston moves through a short flexible sleeve called the **pump seal**, which prevents liquid from leaking through to the pump head.

This design is sometimes a "twin piston" design in which a second piston is 180° out of phase with the first. This means that when one piston is in its forward stroke, the other is in its backward stroke. The result is a flow that is free of pulsations. With the single piston design, a pulse-damping device positioned in the flow path following the pump is used.

Modern pump designs also include a means for flushing the piston with solvent *behind* the pump seal (not shown in Figure 12.3). The solvent for this is drawn in from a separate reservoir and pumped back into this same reservoir. The purpose is to continuously rinse the piston free of mobile phase residue such that abrasive solute crystals resulting from a mobile phase that has dried out on the piston will not deposit there. These solutes, such as the salts dissolved in the buffered mobile phases

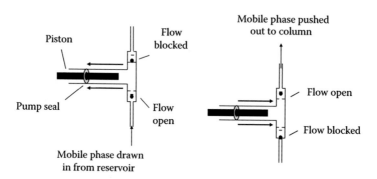

FIGURE 12.3 Illustration of a reciprocating piston pump with check valves. (Left) The piston is in its upstroke, drawing the mobile phase in from the reservoir. (Right) The piston is in its downstroke, pushing the mobile phase out to the column.

FIGURE 12.4 Photograph showing a pump piston (left), a pump seal (center), and a check valve (right). The pump seal is made to fit snugly over the piston, which is approximately 1/4 in. in diameter.

used in ion-exchange chromatography, may otherwise crystallize on the piston, damaging it or the pump seal when the piston moves back and forth. Mobile phases that contain such solutes must be flushed from the system after use so that there is also no crystallization on the front side of the seal.

Metal in-line filters and check valves may be removed and cleaned periodically, or replaced if they are damaged. This is done by dismantling the pump according to the manufacturer's instructions. In-line filters, including the sinker in the mobile phase reservoir, may be cleaned by soaking in a dilute nitric acid solution. Check valves may be cleaned by sonicating in an appropriate solvent. Photographs of a check valve, piston, and pump seal are presented in Figure 12.4.

12.3.2 GRADIENT VS. ISOCRATIC ELUTION

There are two mobile phase elution methods that are used to elute mixture components from the stationary phase. These are referred to as isocratic elution and gradient elution. **Isocratic elution** is a method in which a single mobile phase composition is in use for the entire separation experiment. A different mobile phase composition can be used, but the change to a new composition can only be accomplished by stopping the flow, changing the mobile phase reservoir and restarting the flow. **Gradient elution** is a method in which the mobile phase composition is changed, often gradually, in the middle of the run. It is analogous to temperature programming in GC.

In any liquid chromatography experiment, the composition of the mobile phase is very important in the total separation scheme. In Chapter 10, we discussed the role of a liquid mobile phase in terms of the solubility of the mixture components in both phases. Rapidly eluting components are highly soluble in the mobile phase and insoluble in the stationary phase. Slowly eluting components are less soluble in the mobile phase and more soluble in the stationary phase. Retention times, and therefore resolution, can be altered dramatically by a change in the mobile phase composition. The chromatographer takes advantage of this by sometimes changing the mobile phase composition in the middle of the run if that helps achieve a successful separation. The gradient elution method provides a convenient and automatic method for doing this.

The **gradient programmer** is a hardware module used for gradient elution. The gradient programmer is capable of drawing from at least two mobile phase reservoirs at once and gradually, in a sequence programmed by the operator in advance, changing the composition of the mobile phase delivered to the HPLC pump. A schematic diagram of this system is shown in Figure 12.5a and a sample "program" is shown in Figure 12.5b.

Solvent strength is a designation of the ability of a solvent (mobile phase) to elute mixture components. The greater the solvent strength, the shorter the retention times.

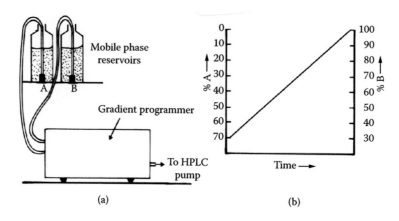

(a)   (b)

FIGURE 12.5   (a) Illustration of the gradient programmer hardware. (b) Example of a program. "A" and "B" represent two pure liquids being mixed in the programmer.

12.4   SAMPLE INJECTION

As mentioned previously, introducing the sample to the flowing mobile phase at the head of the column is a special problem in HPLC due to the high pressure of the system and the fact that the liquid mobile phase may chemically attack a rubber septum. For these reasons, the use of the so-called loop injector is the most common method for sample introduction.

The **loop injector** is a two-position valve that directs the flow of the mobile phase along one of two different paths. One path is a sample loop, which when filled with the sample causes the sample to be swept into the column by the flowing mobile phase. The other path bypasses this loop while continuing on to the column, leaving the loop vented to the atmosphere and able to be loaded with the sample free of a pressure differential. Figure 12.6 is a diagram of this injector, showing both the "load" and the

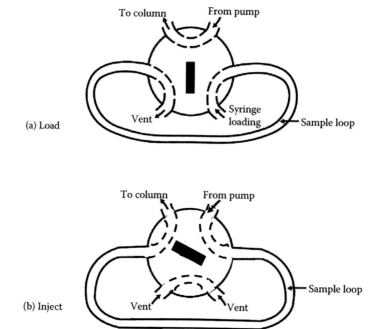

FIGURE 12.6   Loop injector for HPLC. (a) "Load" position—the sample is loaded into the loop via a syringe at atmospheric pressure. (b) "Inject" position—the mobile phase sweeps the contents of the loop into the column.

"inject" positions and the path of the mobile phase in both positions. Figure 12.7 shows two photographs of the loop injector, with Figure 12.7a being the front and Figure 12.7b being the back.

The sample loop has a particular volume such that a careful measuring of the sample volume using the syringe is unnecessary—the sample loop is simply filled to overflowing each time (from the inlet port, through the loop, and out the vent in Figure 12.6a) for a reproduced volume equal to the volume of the loop. If an injection volume smaller than the volume of the loop is desired, the loop may be changed to a smaller volume or a larger loop may be partially filled and the volume measured with the syringe. In some instruments, however, the dead volume between the injection port and the loop may be significant (and unknown), such that a partial filling of the loop with a known volume is impossible. In that case, for an experiment that depends on a reproduced injection volume (such as when injecting a series of standards in a quantitation experiment), the sample loop must be filled each time. If there is a need to inject a volume different from the volume of the sample loop installed on a given instrument, changing the loop to one of a different volume is an easy matter.

Automated injectors are often used when large numbers of samples are to be analyzed. Most designs involve the use of the loop injector coupled to a robotic needle that draws the samples from vials arranged in a carousel-type auto-sampler. Some designs even allow sample preparation schemes such as extraction and derivatization (chemical reactions) to occur prior to injection.

(a)

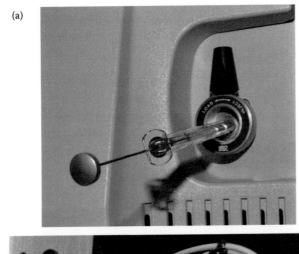

(b)

FIGURE 12.7 (a) Photograph of the loop injector, with syringe ready to load a sample solution, as seen looking at the front of an HPLC instrument. (b) Photograph of the back side of the loop injector. The ports that are connected via the internal channels are clearly seen. Notice that two of the ports are connected externally with a 20-μL sample loop.

12.5 COLUMN SELECTION

The stationary phases available for HPLC are as numerous as those available for GC. As mentioned previously, however, adsorption, partition, ion exchange, and size exclusion are all LC methods. We can therefore classify the stationary phases according to which of these four types of chromatography they represent. Additionally, partition HPLC, which is the most common, is further classified as "normal-phase" HPLC or "reverse-phase" HPLC. Both of these are "bonded-phase chromatography," which was described in Chapter 10. Let us begin with these.

12.5.1 NORMAL PHASE COLUMNS

Normal-phase HPLC consists of methods that utilize a nonpolar mobile phase in combination with a polar stationary phase. Adsorption HPLC actually fits this description as well, since the adsorbing solid stationary phase particles are very polar. (See discussion of adsorption columns in the next subsection.) Normal-phase partition chromatography makes use of a polar liquid stationary phase chemically bonded to these polar particles, which typically consists of silica, Si–O–, bonding sites. Typical examples of normal-phase bonded phases are those in which a cyano group (–CN), an amino group (–NH$_2$), or a diol group (–CHOH–CH$_2$OH) are parts of the structure of the bonded phase. Typical mobile phases for normal-phase HPLC are hexane, cyclohexane, carbon tetrachloride, chloroform, benzene, and toluene.

12.5.2 REVERSE-PHASE COLUMNS

Reverse-phase HPLC describes methods that utilize a polar mobile phase in combination with a nonpolar stationary phase. As stated above, the nonpolar stationary phase structure is a bonded phase—a structure that is chemically bonded to the silica particles. Here, typical column names often have the carbon number designation indicating the length of a carbon chain to which the nonpolar nature is attributed. Typical designations are C$_8$ or C$_{18}$ (or ODS, meaning "octadecyl silane"), etc. Common mobile phase liquids are water, methanol, acetonitrile (CH$_3$CN), and acetic acid buffered solutions (see Application Note 12.2).

**APPLICATION NOTE 12.2: SIBUTRAMINE IN
CAPSULES BY REVERSE-PHASE HPLC**

Reverse-phase HPLC has been used to analyze for sibutramine in capsules. Sibutramine is a drug used to treat obesity. Samples are prepared by weighing enough of the capsules to give 20 mg of sibutramine followed by dissolving in 50 mL of methanol. These solutions are then diluted to 100 mL with methanol, shaken, and filtered. Following centrifugation, 7.5 mL of the supernatant is diluted to 50 mL with methanol. This solution is then injected into the chromatograph using a 20-µL sample loop.

The stationary phase for this analysis is octadecyl silane (C$_{18}$). The mobile phase is a methanol-water-triethylamine solution prepared in the volume ratio 80:20:0.5, respectively, with the pH adjusted to 5.65 with concentrated phosphoric acid. The mobile phase flow rate is 1.0 mL/min. The detector is a UV-Vis absorption detector set at a wavelength of 223 nm. The standard solutions were prepared by diluting a 100.0-ppm stock solution to make concentrations in the range of 15–40 ppm.

(*Reference*: Diefenbach et al., *Journal of AOAC International*, Vol. 92, No. 1 (2009), pp. 148–151.)

12.5.3 ADSORPTION COLUMNS

Adsorption HPLC is the classification in which the highly polar silica particles are exposed (no adsorbed or bonded liquid phase). Aluminum oxide particles fit this description too and are also readily available as the stationary phase. As mentioned earlier, this classification can also be thought of as normal-phase chromatography, but LSC rather than LLC. Typical normal-phase mobile phases (nonpolar) are used here. The stationary phase particles can be irregular, regular, or "pellicular" in which a solid core, such as a glass bead, is used to support a solid porous material.

12.5.4 ION EXCHANGE AND SIZE EXCLUSION COLUMNS

As discussed in Chapter 10, ion exchange stationary phases consist of solid resin particles that have positive and/or negative ionic bonding sites on their surfaces at which ions are exchanged with the mobile phase (see Figure 10.10). Cation exchange resins (SCX) have negative sites so that cations are exchanged, while anion exchange resins (SAX) have positive sites at which anions are exchanged. A popular modern name for HPLC ion exchange is simply "ion chromatography." Detection of ions eluting from the HPLC column has posed special problems that are described in Section 12.6. The mobile phase for ion chromatography is always a pH-buffered water solution.

Size exclusion columns, as discussed in Chapter 10, separate mixture components on the basis of size by the interaction of the molecules with various pore sizes on the surfaces of porous polymeric particles. Size exclusion chromatography is subdivided into two classifications, gel permeation chromatography (GPC) and gel filtration chromatography (GFC). GPC utilizes nonpolar organic mobile phases, such as tetrahydrofuran (THF), trichlorobenzene, toluene, and chloroform, to analyze for organic polymers such as polystyrene. GFC utilizes mobile phases that are water-based solutions and is used to analyze for naturally occurring polymers, such as proteins and nucleic acids. GPC stationary phases are rigid gels, such as silica gel, whereas GFC stationary phases are soft gels. Neither technique utilizes gradient elution because the stationary phase pore sizes are sensitive to mobile phase changes.

12.5.5 THE SIZE OF THE STATIONARY PHASE PARTICLES

The typical diameter of the stationary phase particles is 5 μm. However, the smaller the size, the better the resolution. This stands to reason since the more finely divided the stationary phase material is, the more stationary phase surface is present in the column and the more exposure there is to the mixture to be separated. A rather recent development is to use particle diameters as small as 1.8 μm. Experiments utilizing stationary phase particles smaller than 2 μm has been referred to as **ultra-performance liquid chromatography**, or **UPLC**. The word "ultra" is appropriate because the resolution is greatly improved over the typical HPLC experiment.

12.5.6 COLUMN SELECTION

Since each type of HPLC just discussed utilizes a different separation mechanism, the selection of a specific column packing (stationary phase) depends on whether or not the planned separation is possible or logical with a given mechanism. For example, if a given mixture consists of different molecules all of approximately the same size, then size exclusion chromatography will not work. If a mixture consists only of ions, then ion chromatography is the logical choice. Although the cvonclusions drawn from these examples are obvious, others are less obvious and require a study of the variables and the mechanisms in order for a particular stationary phase to be chosen logically.

Table 12.1 presents some guidelines for column selection. Although these guidelines may prove helpful as a starting point, additional facts about the planned separation need to be determined in order to select the most appropriate chromatographic system, including facts that can only be discovered through

TABLE 12.1

Summary of Applications of the Different Types of HPLC

Type	Useful for Components That
Normal and reverse phase	Have a low formula weight (<2000)
	Are non-ionic
	Are either polar or nonpolar
	Are water or organic soluble
Adsorption	Have a low formula weight (<2000)
	Are nonpolar
	Are organic soluble
Ion exchange	Have a low formula weight (<2000)
	Are ionic
	Are water soluble
Size exclusion	Have a high or low formula weight
	Are non-ionic
	Are water or organic soluble

experimentation, or by searching the chemical literature. Several different mobile phase/stationary phase systems may work. Comparing reverse phase with normal phase, for example, one can see that there would only be a reversal in the order of elution. Polar components would elute first with reverse phase, whereas nonpolar components would elute first with normal phase. Experimenting with various mobile phase compositions, which may include a mixture of two or three solvents in various ratios, would be a logical starting point. Some considerations that would involve such experimentation are:

1. The mixture components should have a relatively high affinity for the stationary phase compared with the mobile phase. This would mean longer retention times and thus probably better resolution.
2. The various separation parameters should be adjusted to provide optimum resolution. These include mobile phase flow rate, stationary phase particle size, gradient elution, and column temperature (using an optional column oven).
3. Use partition chromatography for highly polar mixtures and adsorption chromatography for very nonpolar mixtures.

12.6 DETECTORS

The function of the HPLC detector is to examine the solution that elutes from the column and output an electronic signal proportional to the concentrations of individual components present there. In Chapter 11, we discussed a number of detector designs that serve this same purpose for gas chromatography. The design of the HPLC detectors, however, are more "conventional" in the sense that components present in a liquid solution can be determined with conventional instruments, including spectrophotometers, fluorometers, and refractometers.

12.6.1 UV ABSORPTION

The UV absorption HPLC detector is basically a UV spectrophotometer that measures a flowing solution rather than a static solution. It has a light source, a wavelength selector and a phototube like an ordinary spectrophotometer. The cuvette is a "flow cell," through which the column effluent flows. As the mobile phase elutes, the chromatogram traces a line at zero absorbance, but when a mixture component that absorbs the wavelength elutes, the absorbance changes and a peak is traced

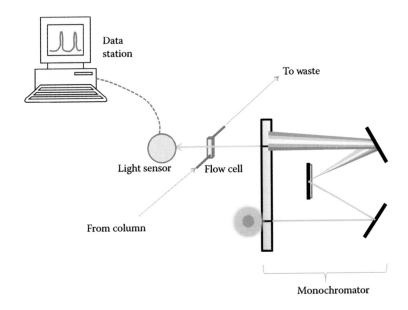

FIGURE 12.8 HPLC variable wavelength UV detector—a UV spectrophotometer with a flow cell. A peak appears when a mixture component that absorbs the set wavelength elutes from the column.

on the chromatogram (see Figure 12.8). The wavelength selector is a slit/dispersing element/slit monochromator as shown and the wavelength of maximum absorbance can be chosen in order to optimize sensitivity.

While a UV absorption detector is fairly sensitive, it is not universally applicable. The mixture components being measured must absorb light in the UV region in order for a peak to appear. Also, the mobile phase must not absorb an appreciable amount at the selected wavelength.

12.6.2 DIODE ARRAY

A diode array UV detector is a diode array UV spectrophotometer with a flow cell. Refer back to Chapter 8 (Section 8.2.3) for a description of the UV/VIS diode array spectrophotometer. The light from the source passes through the flow cell and is then dispersed via a grating. The dispersed light then sprays across an array of photodiodes, each of which detects only a narrow wavelength band. With the help of the data system, the entire UV absorption spectrum can be immediately measured as each individual component elutes (see Figure 12.9). With computer banks containing a library of UV absorption spectral information, a rapid, definitive qualitative analysis is possible in a manner similar to GC-MS or GC-IR (Chapters 11 and 13). In addition, the peak displayed on the chromatogram can be the result of a rapid change of the wavelength by the computer. Thus, the peaks displayed can represent the maximum possible sensitivity for each component. Finally, a diode array detector can be used to "clean up" a chromatogram so as to only display the peaks of interest. This is possible since we can rapidly change the wavelength giving rise to the peaks.

12.6.3 FLUORESCENCE

The basic theory, principles, sensitivity, and application of fluorescence spectrometry (fluorometry) were discussed in Chapter 8. Like the UV absorption detector described above, the HPLC fluorescence detector is based on the design and application of its parent instrument, in this case the fluorometer, and you should review Chapter 8 (Section 8.5) for more information about the fundamentals of the fluorescence technique.

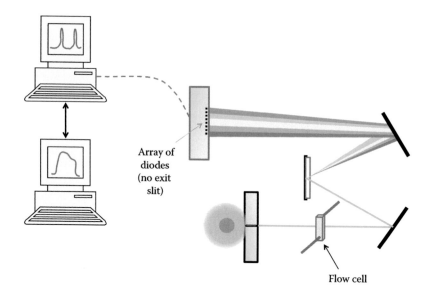

FIGURE 12.9   HPLC diode array UV absorbance detector. Notice that the flow cell is positioned before the dispersing element and there is no exit slit. When a mixture component elutes from the column, not only the chromatography peak but the entire UV absorption spectrum for that component can be recorded.

In summary, the basic fluorometer, and thus the basic fluorescence detector, consists of a light source and a wavelength selector (usually a filter) for creating and isolating a desired wavelength; a sample "compartment," and a second wavelength selector (another filter) with a phototube detector for isolating and measuring the fluorescence wavelength. The second monochromator and detector are lined up perpendicular to the light beam from the source (the so-called right angle configuration).

As with the UV absorption detector, the sample compartment consists of a special cell for measuring a flowing, rather than static, solution. The fluorescence detector thus individually measures the fluorescence intensities of the mixture components as they elute from the column (see Figure 12.10). The electronic signal generated at the phototube is recorded on the chromatogram.

The advantages and disadvantages of the fluorometry technique in general hold true here. The fluorescence detector is not universal (it will give a peak only for fluorescing species), but it is thus very selective (almost no possibility for interference) and very sensitive.

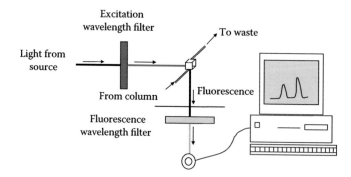

FIGURE 12.10   HPLC fluorescence detector. When a mixture component that exhibits fluorescence elutes from the column, the light is detected and a peak appears on the chromatogram.

12.6.4 REFRACTIVE INDEX

The refractometer instrument was introduced in Chapter 7. This instrument measures and displays the refractive index of a liquid or liquid solution. The refractive index depends on the degree of bending of a light beam in contact with the analyte or analyte solution compared with a reference. The refractive index HPLC detector is a refractometer with a flow cell. A drawing of the most common design is shown in Figure 12.11. In this design, both the column effluent and the pure mobile phase (acting as the reference) pass through adjacent flow cells in the detector. A light beam, passing through both cells, is focused onto an array of diodes and the location of the beam when both cells contain pure mobile phase is taken as the reference point and the signal is zeroed. When a mixture component elutes, the refractive index in one cell changes, the light beam is "bent" and becomes focused onto a different point of the diode array causing the signal to change and trace a peak.

The major advantage of this detector is that it is almost universal. All substances have their own characteristic refractive index (it is a physical property of the substance). Thus, the only time that a mixture component would not give a peak is when it has a refractive index equal to that of the mobile phase, a rare occurrence. The disadvantages are that it is not very sensitive and the output to the computer is subject to temperature effects. Also, it is difficult to use this detector with the gradient elution method because it is sensitive to changes in the mobile phase composition. A major application of this detector is the detection of sugars (see Application Note 12.3).

APPLICATION NOTE 12.3: ANALYSIS OF MILK FOR SUGARS

The amount of lactose, glucose, or galactose in milk can be determined using an HPLC instrument equipped with a refractive index detector. The sample preparation involves precipitating the proteins in the milk by mixing a known weight of the milk with isopropyl alcohol. Following this, the sample is filtered and/or centrifuged so as to obtain a clear solution. This clear solution is filtered through a 0.45-μm filter and injected into the instrument using a 25-μL sample loop. Standard solutions of the sugar, having concentrations of 2%–10%, are prepared and treated in the same way. The mobile phase is a solution of acetonitrile and water in a ratio of 70:30, respectively, by volume. The flow rate is 1.0 mL/min. The column is a reverse-phase column.

(*Reference*: James, C., *Analytical Chemistry of Foods*, Chapman & Hall (1996), pp. 132–133.)

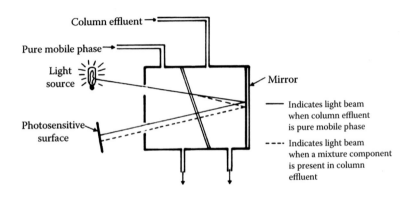

FIGURE 12.11 Illustration of a refractive index detector.

12.6.5 ELECTROCHEMICAL

Various detectors that utilize electrical current or conductivity measurements for detecting eluting mixture components have been invented. These are called electrochemical detectors. Let us now examine some of the basic designs.

12.6.5.1 Conductivity

Perhaps the most important of all electrochemical detection schemes currently in use is the electrical conductivity detector. This detector is specifically useful for ion-exchange, or ion, chromatography, in which the analyte is in ionic form. Such ions elute from the column and need to be detected as peaks on the computer screen.

A well-known fact of fundamental solution science is that the presence of ions in any solution gives the solution a low electrical resistance and the ability to conduct an electrical current. The absence of ions means that the solution would not be conductive. Thus, solutions of ionic compounds and acids, especially strong acids, have a low electrical resistance and are conductive. This means that if a pair of conductive surfaces (electrodes) is immersed into the solution and connected to an electrical power source, such as a simple battery, a current can be detected flowing in the circuit when there is a sufficient concentration of ions present. Conductivity cells based on this simple design are in common use in non-chromatography applications such as to determine the quality of deionized water. Deionized water should have no ions dissolved in it and thus should have a very low conductivity. The conductivity HPLC detector is based on this simple apparatus. When analyte ions elute from the column, a current flows. Otherwise, there would be no current and thus a baseline signal (zero). A peak is therefore seen when the analyte ions elute.

For many years, the concept of the conductivity detector for HPLC could not work, however. Ion chromatography experiments utilize solutions of high ion concentrations as the mobile phase. Thus, changes in conductivity due to eluting ions are not detectable above the already high conductivity of the mobile phase. This was true until the invention of so-called ion suppressors. Today, conductivity detectors are used extensively in HPLC ion chromatography instruments that also include suppressors. Analyte ions, in the absence of mobile phase ions, therefore generate the peaks like other detectors.

A suppressor is a short column (tube) that is inserted into the flow stream just after the analytical column. It is packed with an ion exchange resin itself; a resin that removes mobile phase ions from the effluent, much like a deionizing cartridge removes the ions in laboratory tap water, and replaces them with molecular species. A popular "mixed bed" ion exchange resin is used, for example, in deionizing cartridges, such that tap water ions (such as Ca^{2+} and CO_3^{2-}) are exchanged for H^+ and OH^- ions, which in turn react to form water. The resulting water is thus deionized. Of course, in the HPLC experiment, the analyte ions must not be removed in this process, and thus suppressors must be selective only for the mobile phase ions.

A typical design for a conductivity detector uses metal electrodes embedded into the wall of the postcolumn flow channel as illustrated in Figure 12.12.

12.6.5.2 Amperometric

A thorough discussion of electroanalytical techniques, including "polarography," "voltammetry," and "amperometry," is given in Chapter 14. An understanding of these would be useful for understanding the amperometric HPLC detector. For this reason, the description here is brief.

Electrochemical oxidation and/or reduction of eluting mixture components is the basis for amperometric electrochemical detectors. The three electrodes needed for the detection, the working ("indicator") electrode, reference electrode and auxiliary electrode, are either inserted into the flow stream or embedded in the wall of the flow stream (see Figure 12.13). The indicator electrode is typically glassy carbon, platinum, or gold; the reference electrode, a silver/silver chloride electrode; and the auxiliary, a stainless steel electrode. Most often, the indicator electrode is electrically

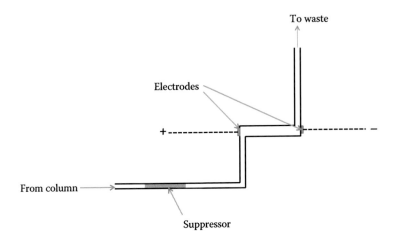

FIGURE 12.12   Illustration of a conductivity detector.

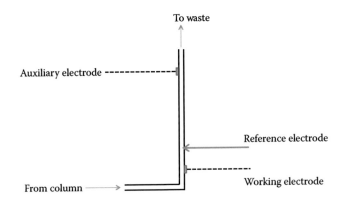

FIGURE 12.13   Illustration of an amperometric HPLC detector.

polarized so as to cause oxidation of the mixture components as they elute. The oxidation current is then measured and constitutes the signal sent to the computer.

Advantages of this detector include broad applicability to both ionic mixture components and molecular components, as long as they are able to be oxidized (or reduced) at fairly small voltage polarizations. Selectivity can be improved by varying the voltage. In addition, the sensitivity experienced with this detector is quite good—generally better than the UV detector, but not as good as the fluorescence detector. A disadvantage is that the indicator electrode can become fouled due to products of the electrochemical reaction coating the electrode surface. Thus, this detector must be able to be disassembled and cleaned with relative ease, since this may need to be done frequently.

12.7   QUALITATIVE AND QUANTITATIVE ANALYSIS

Qualitative and quantitative analysis with HPLC are very similar to that with GC (Chapter 11, Sections 11.7 and 11.8). In the absence of diode array, mass spectrometric, and FTIR detectors, which give additional identification information, qualitative analysis depends solely on retention time data, t_R and t_R'. (Remember that t_R' here is the time from when the solvent front is evident to the peak.) Under a given set of HPLC conditions, namely the mobile and stationary phase compositions, the mobile phase flow rate, the column length, temperature (when the optional column oven is used), and instrument dead volume, the retention time is a particular value for each component. It changes only when one of the of the above parameters changes. Refer to Section 11.7 for further discussion of qualitative analysis.

Peak size measurement and quantitation methods were outlined in Section 11.8. The reproducibility of the amount injected is not nearly the problem with HPLC as it is with GC. Roughly 10 times more sample is typically injected (5–20 µL), and there is no loss during the injection since the sample is not loaded into a higher pressure system through a septum. In addition, the sample loop is manufactured to have a particular volume and is often the means by which a consistent amount is injected, which means reproducibility is maximized through the consistent "overfilling" of the loop via the injection syringe. In this way, the loop is assured of being filled at each injection and a reproducible volume is always introduced. Sometimes, however, the analyst chooses to inject varying volumes of a single standard to generate the standard curve rather than equal volumes of a series of standard solutions. In this case, the injection syringe is used to measure the volumes—a less accurate method but better than an identical method with GC, since the sample volume is larger and there is less chance for sample loss. With this type of quantitation, the standard curve is a plot of peak size vs. amount injected, rather than concentration. As noted previously, however, some injectors have significant dead volume, which renders the method based on amount injected impossible.

The most popular quantitation "method" then is the series of standard solutions method with no internal standard or the variable injection of a single standard solution as outlined above.

12.8   TROUBLESHOOTING

Problems that arise with HPLC experiments are usually associated with abnormally high or low pressures, system leaks, worn injectors parts, air bubbles, and/or blocked in-line filters. Sometimes, these manifest themselves on the chromatogram and sometimes they do not. In the following paragraphs, we will address some of the most common problems encountered, pinpoint possible causes, and suggest methods of solving the problems.

12.8.1   UNUSUALLY HIGH PRESSURE

A common cause of unusually high pressure is a plugged in-line filter. In-line filters are found at the very beginning of the flow line in the mobile phase reservoir, immediately before and/or after the injector, and just ahead of the column. With time, they can become plugged due to particles that are filtered out (particles can appear in the mobile phase and sample even if they were filtered ahead of time), and thus the pressure required to sustain a given flow rate can become quite high. The solution to this problem is to backflush the filters with solvent and/or clean them with a nitric acid solution in an ultrasonic bath.

Other causes of unusually high pressure are an injector blockage, mismatched mobile and stationary phases, and a flow rate that is simply too high. An injector that is left in a position between "load" and "inject" can also cause a high pressure, since the pump is pumping but there can be no flow.

12.8.2   UNUSUALLY LOW PRESSURE

A sustained flow that is accompanied by low pressure may be indicative of a leak in the system. All joints should be checked for leaks (see below).

12.8.2.1   System Leaks

Leaks can occur within the pump, at the injector, at various fittings and joints, such as at the column, and in the detector. Leaks within the pump can be due to failure of pump seals and diaphragms, and loose fittings, such as at the check valves, etc. Leaky fittings should be checked for mismatched or stripped ferrules and threads, or perhaps they simply need tightening. Leaks in the injector can be due to a plugged internal line or other system blockage, gasket failure, loose connections, or use of the wrong size syringe if the leak occurs as the sample is loaded. Detector leaks are most often due to a bad gasket seal or a broken flow cell. Of course, loose or damaged fittings and a blockage in the flow line beyond the detector are possible causes.

12.8.2.2 Air Bubbles

An air pocket in the pump can cause low or no pressure or flow, erratic pressure, and changes in retention time data. It may be necessary to bleed air from the pump, or prime the pump according to system start-up procedures. Air pockets in the column will mean decreased contact with the stationary phase and thus shorter retention times and decreased resolution. Tailing and peak splitting on the chromatogram may also occur due to air in the column. Air bubbles in the detector flow cell are usually manifested on the chromatogram as small spikes due to the periodic interruption of the light beam (e.g., in a UV absorbance detector). Increasing the flow rate, or restricting and then releasing the postdetector flow, so as to increase the pressure, should cause such bubbles to be "blown" out.

12.8.2.3 Column "Channeling"

If the column packing becomes separated and a channel is formed in the stationary phase, the tailing and splitting of peaks will be observed on the chromatogram. In this case, the column needs to be replaced.

12.8.2.4 Decreased Retention Time

When retention times of mixture components decrease, there may be problems with either the mobile or stationary phases. It may be that the mobile phase composition was not restored after a gradient elution, or it may be that the stationary phase was altered due to unreversed adsorption of mixture components, or simply chemical decomposition. The use of guard columns (see previous discussion) may avoid stationary phase problems.

12.8.2.5 Baseline Drift

A common cause of baseline drift is a slow elution of substances previously adsorbed on the column. A column cleanup procedure may be in order, or the column may need to be replaced. This problem may also be caused by temperature effects in the detector. Refractive index detectors are especially vulnerable to this. In addition, a contaminated detector can cause drift. The solution here may be to disassemble and clean the detector.

You are also referred to the troubleshooting guide in Chapter 11 (GC) for possible solutions to problems.

12.9 ELECTROPHORESIS

12.9.1 INTRODUCTION

In our previous discussion of the HPLC conductivity detector (Section 12.6.5), we stated that when analyte ions elute from the column (and from the suppressor), the eluting mobile phase solution becomes electrically conductive due to the presence of these ions. This is the source of the electrical signal giving rise to the chromatography peaks on the computer screen. The design includes a pair of electrodes in contact with the eluting mobile phase solution, as illustrated in Figure 12.12. These electrodes are connected to the positive and negative poles of a power source, such as a battery, as illustrated by the "+" and "−" symbols in Figure 12.12. Thus, one of the electrodes is positively charged and the other is negatively charged. We say that an electric field exists between the electrodes.

In this process, anions (negatively charged ions) in the mobile phase are attracted to and migrate toward the positive electrode and cations (positively charged ions) in the mobile phase are attracted to and migrate toward the negative electrode. The technique known as electrophoresis makes use of this separation of ions that occurs in an electric field. There is a separation of ions based on the difference in their charges, i.e., positive ions separate from negative ions (see Figure 12.14). In addition,

Power supply

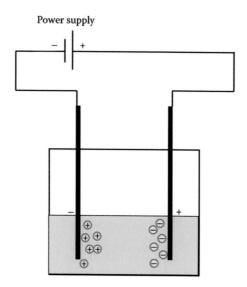

FIGURE 12.14 When an electric field exists between a pair of electrodes, the negative ions are attracted to the positive electrode (the electrode connected to the positive pole of the power supply) and the positive ions are attracted to the negative electrode (the electrode connected to the negative pole of the power supply).

in electrophoresis, ions that migrate in the same direction, i.e., the different anions that migrate toward the positive electrode or the different cations migrating toward the negative electrode, separate from each other because they migrate at different rates (see Figure 12.15). The separation that occurs because of these differences in ionic charge and migration rate is the basis for an analytical technique known as electrophoresis.

Electrophoresis is a technique that separates and measures dissolved analyte ions based on the fact that different ions dissolved in a migration medium migrate in an electric field in different directions and at different rates. The migration rates vary due to the ions having different "mobilities" in the particular medium. These mobilities depend on such things as different like charges (+1 ions separate from +2 ions, for example), different sizes and shapes (large organic cations migrate at different rates than monatomic metal or nonmetal ions, for example), and various properties of the migration medium, such as viscosity and pH. Migration rate is also affected by the strength of the electric field. In "noninstrumental" electrophoresis, the experiment is terminated after a time so that a given ion may be viewed and measured at a particular location, or zone, on the

Original location of
ion mixture

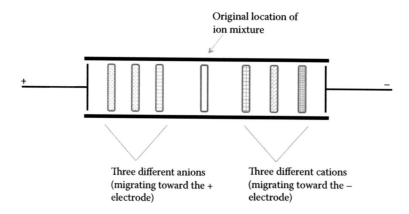

Three different anions
(migrating toward the +
electrode)

Three different cations
(migrating toward the −
electrode)

FIGURE 12.15 Illustration of the concept of electrophoresis.

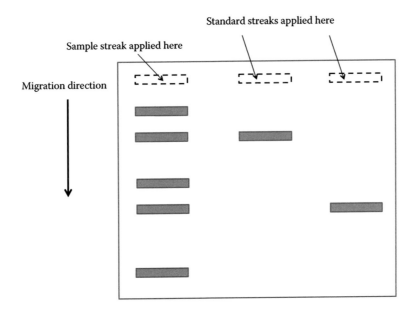

FIGURE 12.16 Illustration of a gel electrophoresis experiment (see text for discussion).

medium, in a way similar to paper and thin-layer chromatography. This type of electrophoresis is known as **zone electrophoresis**.

One common type of zone electrophoresis is known as **gel electrophoresis**. In this case, the migration medium is a "gel slab" sandwiched between two plastic plates. This gel slab is often made of polyacrylamide, a synthetic and transparent polymer. In any case, the sample is applied to ("streaked on") one end of the gel slab and the analyte ions migrate through to the other end. The experiment is terminated after a time, and the location of the each ion in the mixture is observed as a streak on the gel slab. DNA sequencing is carried out in this way through the detection of proteins and amino acids on the gel slab (see Figure 12.16).

The analyte ions are almost always present at a small concentration—too small to support an electric field for the electrical current that is required. Therefore, any electrophoresis medium must have a supporting electrolyte dissolved. This electrolyte is most often a buffer solution, such as an acetate buffer or a tris buffer (recall the discussion of buffers in Chapter 5) and is referred to as the "running buffer." The ions of the running buffer, while also migrating toward the two electrodes, must not interfere with the detection of the analyte ions.

12.9.2 Capillary Electrophoresis

Refer to Figure 12.17 throughout the following discussion. As the name implies, **capillary electrophoresis** is electrophoresis that is made to occur inside a piece (50–100 cm in length) of small diameter capillary tubing, similar to the tubing used for capillary GC columns. The capillary is made of glass with a polyamide coating to prevent breakage. The tubing contains the running buffer solution and the ends of the tube are dipped into reservoirs also containing the running buffer solution. Thus, we have an "inlet" reservoir and an "outlet" reservoir. Electrodes (connected to a high-voltage power supply) in these reservoirs create the required electric field across the capillary tube. The running buffer ions move from the inlet reservoir through the capillary to the outlet reservoir. The sample is introduced near the inlet end and electronic detection occurs near the outlet end. The electronic detector, such as those described for HPLC in this chapter, is used for the detection and quantitative analysis of mixture components. An electronic signal is generated in the detector when an analyte ion passes by, and peaks appear on a computer when this occurs, just as in instrumental chromatography.

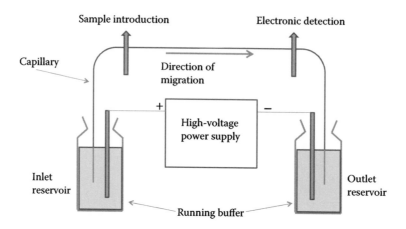

FIGURE 12.17   Process of capillary electrophoresis (see text for full discussion).

A very small volume of sample (5–50 nL) is introduced at the inlet end, typically the end dipped into the reservoir containing the positive electrode. When the experiment begins, the positive ions migrate quickly through the capillary tube toward the negative electrode in the opposite reservoir and separate on the basis of their mobilities, as in the other electrophoresis techniques. There are several types of capillary electrophoresis. The most common can be called **free solution capillary electrophoresis**, or **capillary zone electrophoresis**, and this is the type that is discussed in this text.

The basic equipment, consisting of the capillary, the inlet and outlet reservoirs, the high-voltage power supply, the electrodes, the sample introduction system, and the detection system, is shown in Figure 12.18.

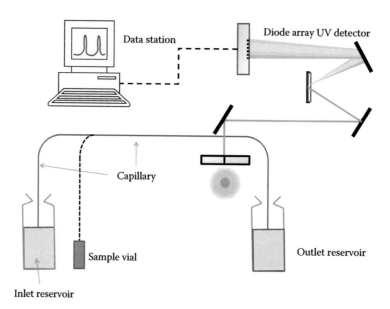

FIGURE 12.18   Capillary electrophoresis showing the sample vial and a diode array detector. The electrodes in the reservoir are not shown.

12.9.2.1   Electroosmotic Flow

The capillary is typically made of silica (glass) with silanol groups (Si-OH) on the surface of the inner wall. At pH values greater than 2, the hydrogen of the silanol group is stripped off. This imparts negative charge to the inner wall and the wall thus attracts solvated cations (cations surrounded by polar solvent molecules). When the electric field is switched on, these cations migrate, with solvent molecules in tow, toward the negatively charged electrode. This means that there will be a natural flow of solvent toward this electrode. Such electrically induced flow is called electroosmotic flow (EO flow). As one might expect, EO flow significantly affects the analyte ion migration. It will speed up the flow of analyte cations toward the negatively charged electrode and inhibit, or even reverse, the flow of analyte anions toward the positively charged electrode. Neutral molecules will also flow toward the negatively charged electrode as a result. Ultimately, all three (cations, anions, and neutral molecules) can then be detected and measured. That is why the sample solution is always introduced near the inlet reservoir and why the detector is situated near the outlet reservoir. No matter what the charge state is on an analyte species (negative, positive, or neutral), it will flow through the capillary (left to right in Figures 12.17 and 12.18), be separated from other mixture components, and be detected.

12.9.2.2   Sample Introduction

Samples are held in vials positioned in a carousel (i.e., an auto-sampler). Here we describe four common methods for getting samples out of the vials and into the capillary. (1) Nitrogen pressure: the capillary is moved to the sample vial and nitrogen gas pressure is applied to the air space above the sample solution. This forces the sample into the capillary. The capillary is then returned to the reservoir. (2) Vacuum: the capillary is moved to the sample vial and a vacuum is applied at the outlet reservoir, thus sucking the sample solution into the capillary. The capillary is then returned to the inlet reservoir. (3) Siphoning: the capillary is moved to the sample vial, and the sample vial and capillary are elevated above the level of the buffer in the outlet reservoir. The sample thus moves into the capillary via siphoning. The capillary is then returned to the inlet reservoir. (4) Electrokinetic injection: both the capillary and the electrode are moved to the sample vial. The sample is drawn into the capillary by applying a small electric field for a short time before the capillary and electrode are returned to the inlet reservoir.

   Regardless of the sample introduction method, it should be pointed out that the volume of sample used must be very small. As with capillary GC columns, we need to introduce extremely small sample volumes so that the capillary is not overloaded. The capillary has an extremely small diameter (like the diameter of a human hair). The timing of the methods described above must be such that this small volume introduction is accomplished.

12.9.2.3   Analyte Detection

Here we describe four common detectors for capillary electrophoresis. All four methods have been previously mentioned for HPLC. (1) Variable-wavelength UV: the capillary is designed so as to have a few millimeters of the polyamide coating removed from the exterior of the capillary where the beam of UV light of appropriate wavelength is passed through. If the analyte species is present and absorbs this wavelength, then this wavelength will be absorbed and a peak will appear on the computer screen. (2) Diode array UV: an illustration of this arrangement is shown in Figure 12.18. Again, a few millimeters of polyamide coating are removed from the capillary and the UV light passes through. As has been described for the diode array measurement technique in the previous sections and chapters, the light is dispersed into the component wavelengths after it has passed through the sample. This means that the entire UV spectrum of the analyte may be measured in addition to the electrophoresis peak. (3) Fluorescence: just as has been described in previous discussions in this text, analyte species that exhibit the property of fluorescence can be measured in the fluorometer design that uses two wavelength selectors and a right angle configuration. Again, a portion of the capillary has the polyamide coating removed for this detection process. (4) Mass spectrometry: this is known as capillary electrophoresis–mass spectrometry, or CE-MS. Mass spectrometry (MS) coupled to other instruments is discussed in Chapter 13.

EXPERIMENTS

EXPERIMENT 38: THE QUANTITATIVE DETERMINATION OF METHYL PARABEN IN A PREPARED SAMPLE BY HPLC

INTRODUCTION

In this experiment, a mixture of methyl, propyl, and butyl paraben (structures shown in Figure 12.19) in methanol solvent will be separated by reverse-phase HPLC. Mobile phase compositions of varying polarities will first be tested to see which one gives the optimum resolution of this mixture, and following this, a standard curve for methyl paraben will be constructed and its concentration in this solution determined. Remember to wear safety glasses.

PART 1: DETERMINATION OF THE OPTIMUM MOBILE PHASE COMPOSITION

1. Mobile phase compositions for this experiment are polar methanol/water mixtures in the ratios 90/10, 80/20, and 70/30 by volume. The stationary phase is C18. Prepare 200 mL of each mobile phase and then filter and degas each through 0.45-μm filters with the aid of a vacuum. (Instructor will demonstrate.) Slowly pour each (so as to avoid re-aeration by splashing) into individual mobile phase reservoirs that are labeled appropriately.
2. Obtain the sample from your instructor and filter it into a small vial using a syringe filter (instructor may choose to demonstrate.) The sample is a solution of methyl, propyl, and butyl paraben in methanol. Record any identifying label in your notebook.
3. Examine the HPLC instrument to which you are assigned. Find the inlet line to the pump and place the free end of this line in the reservoir containing the mobile phase with the 90/10 composition. Trace the path of the mobile phase from the reservoir through the pump, injection valve, column, and detector, to the waste container so that you identify and recognize all components of the flow path. Turn on the pump and detector and begin pumping the mobile phase at a rate between 1.0 and 1.5 mL/min. Allow plenty of time for the 90/10 mobile to completely flush the system before making the first injection (Step 5).
4. While the mobile phase is flushing the system, locate the data system for the instrument and turn it on. Start the software for the data acquisition. Your instructor will discuss how the data system works and how to set it up.

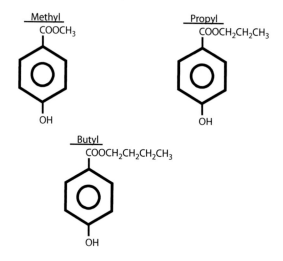

FIGURE 12.19   Structures of methyl, propyl, and butyl paraben.

5. Your instructor will discuss and/or demonstrate how to make an injection. With the injector valve in the LOAD position, flush the injector's sample loop with the sample and then make certain that the sample loop is completely filled. Turn the valve to inject and simultaneously activate the data system. The chromatogram should begin to be traced on the monitor.

6. After three peaks (perhaps poorly resolved, depending on the effect of the mobile phase used) appear in the chromatogram, stop the pump, and move the free end of the inlet line to the 80/20 mobile phase. Start up the pump again and allow time to flush the system once again.

7. Repeat Steps 5 and 6 to obtain a second chromatogram of the mixture, but with the 80/20 mobile phase and again with the 70/30 mobile phase, observing the effect of the mobile phase composition on resolution. Be sure to obtain peak area information from the data system for the experiment that gave the best resolution. This information will be used in Part 2.

8. Optional—calculate R, k, and α values according to your instructor's directions and interpret the results.

Part 2: Methyl Paraben Quantitation

9. Select the mobile phase from Part 1 that gave the best resolution and, if it is not already in the system, allow plenty of time for it to purge the system.

10. Prepare a stock standard solution of methyl paraben in methyl alcohol that has a concentration of 2 mg/mL (your instructor may choose to prepare this so that he/she can prepare unknowns and/or controls from it).

11. From this stock solution, prepare 4 standard solutions, using methanol as the solvent, having concentrations of 0.05, 0.1, 0.15, and 0.2 mg methyl paraben/mL, in 50-mL volumetric flasks. Filter these solutions into vials as you did the sample in Part 1.

12. Inject full sample loop volumes of each standard solution and the unknown and obtain a chromatogram for each. Each will show only one peak, because the propyl and butyl paraben compounds are absent from the standards. Obtain the peak areas for each standard. From the retention time of the lone peak (methyl paraben), identify the methyl paraben peak on the chromatogram from Part 1 and note its area. Also obtain these data for any additional unknowns or a control.

13. Plot the standard curve using the spreadsheet procedure used in Experiment 13 and obtain the correlation coefficient and the concentrations of the unknown and control.

14. Maintain the logbook for the instrument used in this experiment. Record the date, your name(s), the experiment name or number, the R^2 value, and the results for the control sample. Also plot the control sample results on a control chart for this experiment posted in the laboratory.

15. Dispose of all solutions as directed by your instructor.

EXPERIMENT 39: HPLC DETERMINATION OF CAFFEINE IN SODA POP

1. Prepare 500 mL of mobile phase (1.0 M acetic acid in 10% acetonitrile) as follows. Dilute 50 mL of acetonitrile and 28.5 mL of glacial acetic acid to 500 mL with water. Pour into a large beaker and place on a magnetic stirrer for pH adjustment. With a pH meter and a magnetic stirrer, adjust the pH to 3.0 by adding successive small amounts of a saturated solution of sodium acetate. Filter and degas as in previous experiments or as directed by your instructor. Prepare the instrument for use by flushing the HPLC system to be used with this mobile phase for 8 min at 2.0 mL/min. The system is a reverse-phase system with a C18 column.

2. Soda pop samples must be filtered and degassed. Prepare samples by vacuum filtering through 0.45-μm filter papers, then fill small labeled vials.

3. Prepare 50 mL of a stock standard solution that is 2.0 mg/mL in caffeine. Use an analytical balance to weigh the caffeine and a clean 50-mL volumetric flask for the solution. Use distilled water as the diluent. Shake well.

4. Prepare calibration standards that are 0.05, 0.10, 0.15, and 0.20 mg/mL in caffeine from the stock standard. Use 25-mL volumetric flasks and measuring pipettes or pipetters suggested by your instructor.
5. Filter each calibration standard, using the syringe filtering equipment, into small, labeled vials.
6. Obtain HPLC chromatograms of the calibration standards by injecting 20 μL of each. There should be two peaks. Record the peak size vs. concentration data.
7. Obtain chromatograms of the samples, injecting 20 μL of each. Allow the column to completely clear before making another injection. Also, obtain a chromatogram of the control sample if one is provided. Record the peak size data.
8. When finished, flush the system for 5 min with filtered and degassed water and then 5 additional minutes with filtered and degassed pure methanol or a methanol/water mixture.
9. Create the standard curve by plotting peak size vs. concentration. Use the spreadsheet procedure in Experiment 13. Obtain the concentrations of the unknowns and the control. Plot the results for the control sample, if one is provided on the control chart for this instrument posted in the laboratory.
10. Calculate the mg of caffeine present in one 12-oz can of the soda pop. There are 0.02957 L/fl oz.
11. Maintain the logbook for the instrument used in this experiment. Record the date, your name, the experiment name of number, the R^2 value, and the results of the control sample.

EXPERIMENT 40: DESIGN AN EXPERIMENT: DEPENDENCE OF CAFFEINE ANALYSIS ON THE pH OF THE MOBILE PHASE

The analysis of soda pop for caffeine in Experiment 39 calls for a pH of 3.00 for the mobile phase. Design an experiment in which a series of acidic pH values bracketing 3.00 are used for the mobile phase to see the effect that pH has on retention time and resolution. At least one of the goals should be a plot or retention time vs. pH to see if indeed the pH of the mobile phase as prepared in Experiment 39 is the optimum. Write a step-by-step procedure as in other experiments in this book for preparing the needed solutions. After your instructor approves your procedure, perform the experiment in the laboratory.

EXPERIMENT 41: THE ANALYSIS OF MOUTHWASH BY HPLC: A RESEARCH EXPERIMENT

INTRODUCTION

Various brands of mouthwash have a variety of components that should potentially be able to be determined by HPLC. In this experiment, you are "on your own" to determine what component to analyze for, what mobile and stationary phases to use, what flow rate to use, what concentrations to use, etc.

PROCEDURE

1. Visit your local pharmacy (outside of laboratory time) and examine the labels on various brands of mouthwash that are on the shelf. Purchase for laboratory analysis a brand that looks interesting and bring it into the laboratory.
2. Filter and degas a part of the sample. Prepare the instrument as you have done before, choosing a particular stationary and mobile phase system (such as a reverse-phase system using a methanol/water mixture for the mobile phase and a nonpolar stationary phase) and flow rate that you will use as a first trial.
3. Inject the sample and observe the separation of peaks. If there is at least one peak that is resolved, proceed to Step 4. Otherwise, try changing the composition of the mobile phase and the flow rate in order to achieve good resolution of at least one peak. If you are still unsuccessful, change the stationary phase and try again.
4. Prepare and filter standard solutions of several of the components shown on the label. The concentrations should be reasonable guesses of what might match what is in the sample. Inject each individually and observe the retention times for each component. Check to

see if one of these retention times matches the retention time for a resolved peak from the sample.

5. If you get a match of retention times, proceed to quantitate, comparing the peak size of the sample to that of the standard.

OPTIONAL

6. Just because the retention times match does not necessarily mean that you have identified the peak. To be sure, change the stationary phase and inject again. If a sample peak is again resolved from the others (and it is the same size), and if the retention time of the same standard peak matches that of the resolved peak, you can be sure that you have identified the peak and your quantitation in Step 5 is valid.

EXPERIMENT 42: THE DETERMINATION OF 4-HYDROXYACETOPHENONE BY CAPILLARY ELECTROPHORESIS

4-Hydroxyacetophenone is an organic acid with the following structure:

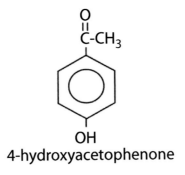

4-hydroxyacetophenone

It is a weak acid with a pK_a of 8.05. This means that in solutions having a pH greater than about 9.0, the hydrogen of the –OH group (the hydroxyl group) is stripped off creating an organic negative ion. In a capillary electrophoresis analysis, therefore, the running buffer is a borate buffer with a pH of 9.3. A diode array detector is an appropriate detector.

PROCEDURE

1. If not already available, prepare 100 mL of a 0.020 M solution of sodium borate using water as the solvent. This is the running buffer solution, which has a pH of approximately 9.3. Filter this solution through a 0.20 μm filter into four separate vials. Label these as "buf" using a permanent lab marker. Do not use labeling tape.
2. If not already available, prepare 100.0 mL of a 0.050 M solution of 4-hydroxyacetophenone using water as the solvent. This is the stock solution to be used in the preparation of the series of standards in step 3.
3. Prepare 50.00 mL of solutions of 4-hydroxyacetophenone that are 0.00010, 0.00050, 0.0010, and 0.0025 M from the stock solution prepared in step 2. Filter these as you did the buffer solution in step 1 into separate vials labeled with the concentrations.
4. Obtain a solution of unknown concentration of 4-hydroxyacetophenone from your instructor. Also filter this solution into a vial as you did above and label as "unk."
5. Fill two vials with distilled water, filtered as you did the solutions above. Label as H_2O.
6. Place all vials into position on the sample carousel and record which solution is in which vial. Perform capillary electrophoresis according to the instructions specific to your instrument. Obtain the peak areas. Plot peak area versus concentration, and obtain the concentration of the unknown using the spreadsheet procedure given in Experiment 13.

QUESTIONS AND PROBLEMS

Introduction

1. "HPLC" stands for ____.
2. Define *LC, LLC, LSC, BPC, IEC, IC, SEC, GPC*, and *GFC*.
3. We use the words "high performance" in the name for instrumental liquid chromatography. Why?
4. Give a simple but total definition of HPLC and describe with some detail the basic HPLC system.
5. Detail the path of the mobile phase through the HPLC system from the solvent reservoir to the waste receptacle, giving brief explanatory descriptions of instrument components along the way.
6. Why is HPLC an improvement over the "open column" technique?
7. Compare HPLC with GC in terms of (a) the force which moves the mobile phase through the stationary phase, (b) the nature of the mobile phase, (c) how the stationary phase is held in place, (d) what "types" of chromatography are applicable, (e) application of vapor pressure concepts, (f) sample injection, (g) mechanisms of separation, (h) detection systems, (i) recording systems, (j) data obtained.
8. The relative vapor pressures of the mixture components are important in GC but not in LC. Why is that?
9. Why do air bubbles form in the flow stream between the mobile phase reservoir and the pump when the mobile is not degassed?
10. Why must HPLC mobile phases and samples be degassed prior to use in the HPLC system?
11. Explain what "degassing" is and how it is accomplished.
12. What is "helium sparging?"
13. Why must mobile phases and all samples and standards be finely filtered before an HPLC experiment?
14. Why is it important to know whether the mobile phase and samples contain an organic solvent when preparing to filter them?
15. Paper filters are available to filter LC mobile phases but nylon filters are used more often. Why is that?
16. For what types of samples and mobile phases is Teflon-based filter material appropriate?
17. Explain the use of a "guard column" and "in-line filters."
18. Why is a metal in-line filter placed just ahead of the column when there already is a filter in the line dipped into the mobile phase reservoir? Why do we worry about filtering the mobile phase and sample in the first place?

Solvent Delivery

19. Why is it that no ordinary liquid pump can be used as the pump in an HPLC system?
20. Describe what is meant by a reciprocating piston pump.
21. What is a "check valve" and why is needed in a HPLC system?
22. Distinguish between isocratic elution and gradient elution.
23. What is a gradient programmer?
24. Define the gradient elution method for HPLC, tell what instrument component is needed for it, and tell how this method is useful.
25. How is gradient elution HPLC similar to temperature programming in GC?
26. Why is it that a change in the mobile phase in the middle of the run will change the retention and resolution of the mixture components?
27. What is meant by "solvent strength"?

Sample Injection

28. Give two reasons why an injection system similar to GC would not work for HPLC.
29. Describe in detail how the "loop injector" works and tell how it overcomes the problems that would be encountered with an injection port/septum system.
30. Show by means of a diagram the difference between the "load" and "inject" positions of the HPLC injection system.

Column Selection

31. What is meant by "bonded phase" chromatography? Would such a name describe normal phase, reverse phase, neither, or both? Explain.
32. Distinguish normal-phase HPLC from reverse-phase HPLC.
33. If a given HPLC system is using a methanol/water mixture for the mobile phase and a C18 column for the stationary phase, what classification of chromatography would be in use? Explain.
34. List some typical mobile and stationary phases for (a) reverse-phase HPLC and (b) normal-phase HPLC.
35. What is meant by "bonded phase" chromatography? Would such a name describe normal phase, reverse phase, neither, or both? Explain.
36. Answer the following with either "polar" or "nonpolar." How would you describe the mobile phase for normal-phase HPLC? How would you describe the stationary phase for reverse-phase HPLC?
37. Answer the following with "normal phase" or "reverse phase." For which type of LC is C18 column used? Which is similar to adsorption chromatography in terms of the polarity of the stationary phase?
38. Answer the following with either "cation exchange resin" or "anion exchange resin." Which has negatively charged bonding sites to which positive ions bond? Which separates positively charged ions?
39. Distinguish between cation exchange resins and anion exchange resins in terms of the nature of the charged sites on the surface of the particles and in terms of application.
40. What is "ion chromatography" and what is a typical mobile phase composition for ion chromatography?
41. Distinguish between gel permeation chromatography and gel filtration chromatography in terms of mobile phases that are used and application.
42. What type of HPLC should be chosen for each of the following separation applications?
    (a) All mixture components have formula weights less than 2000, are molecular and polar, and are soluble in nonpolar organic solvents.
    (b) Mixture components have formula weights varying from very large to rather small, and are nonionic.
    (c) Mixture components have formula weights less than 2000, are molecular and polar, and are water soluble.
43. Explain why the order of elution of polar and nonpolar mixture components would be reversed when switching from normal-phase to reverse phase.
44. What are some HPLC system parameters that can be altered in an attempt to improve resolution?
45. Compare and/or differentiate between the two items in each of the following (be complete, but concise):
    (a) the isocratic elution technique and the gradient elution technique.
    (b) normal-phase chromatography and reverse-phase chromatography.

Detectors

46. Distinguish between the variable-wavelength UV detector and the diode array UV detector in terms of design and use.
47. What is a diode array detector and what are its advantages?
48. In one or two words each, give one advantage and one disadvantage for each of the following:
    (a) UV absorbance detector
    (b) Refractive index detector
    (c) Fluorescence detector
    (d) Conductivity detector
    (e) Amperometric detector
    (f) LC-MS
    (g) LC-IR
49. Why is it that a UV absorption detector in HPLC is not universal?
50. Discuss the advantages and disadvantages of any three HPLC detectors discussed in the text.
51. Which of the four HPLC detectors (UV absorbance, refractive index, fluorescence, or conductivity)
    (a) Is the most sensitive?
    (b) Is the most universal?
    (c) Requires an ion suppressor to eliminate the ions present in the mobile phase?
    (d) Is a popular detector because it is "almost" universal and very sensitive?
    (e) Gives a signal based on the position of a light beam on the detector?
    (f) Has a right angle configuration design?
    (g) Has a single monochromator that can be either a glass filter or a slit/disperser/slit type?
    (h) Is used frequently for "ion chromatography?"
52. In Chapter 11, we discussed the need to calculate response factors specifically when a TCD detector is used (Section 11.8). Would response factors need to be calculated in HPLC when a UV absorbance detector is used? Explain.
53. What is a "suppressor" and why is one needed in an ion-exchange HPLC experiment in which the mobile phase contains ions?
54. Why is FTIR considered a "natural" as an HPLC detector? What problem does the presence of the mobile phase pose, especially if water is present?

Qualitative and Quantitative Analysis

55. Discuss qualitative and quantitative analysis methods for HPLC and how they are different from those of GC.

Troubleshooting

56. What is the most common cause of an unusually high pressure in an HPLC flow stream and how is the problem solved?
57. What symptoms appear if air bubbles enter the HPLC flow line?
58. The splitting of a peak into two peaks is a symptom of what problem in the column? How is it solved?
59. Name some causes of baseline drift in HPLC.

60. Consider the analysis of a soda pop sample for caffeine by the standard additions method (See Chapter 9, Section 9.3.6). Construct a graph from the following data and report the milligrams of caffeine in one 12-oz can of the soda pop.

C (ppm added)	Peak Size
0	1368
5	1919
10	2431
15	2997

Calculation hint: There are 0.02957 L/fl oz.

Electrophoresis

61. Describe the following: electrophoresis, zone electrophoresis, gel electrophoresis, and capillary electrophoresis.
62. Describe in detail the general mechanism of the separation of ions by electrophoresis.
63. What is "capillary electrophoresis" and what advantages does it have over other conventional electrophoresis techniques?
64. What is "electroosmotic flow" and what effects does it have in capillary electrophoresis?
65. What are some popular detectors in the design of a capillary electrophoresis instrument?

Report

66. For this exercise, your instructor will select a "real-world" HPLC analysis method, such as an application note from an instrument vendor or from a methods book, a journal article, or Web site, and will give it to you as a handout. Write a report giving the details of the method according to the following scheme. On your paper, write (or type) "(a)," "(b)," etc. to clearly present your response to each item.
    (a) Method Title: Give a more descriptive title than what is given on the handout.
    (b) Type of Material Examined: Is it water, soil, food, a pharmaceutical preparation, or what?
    (c) Analyte: Give both the name and the formula of the analyte.
    (d) Sampling Procedure: This refers to how to obtain a sample of the material. If there is no specific information given as to obtaining a sample, make something up. If you want to be correct in what you write, do a library or other search to discover a reasonable response. Be brief.
    (e) Sample Preparation Procedure: This refers to the steps taken to prepare the sample for the analysis. It does not refer to the preparation of standards or any associated solutions.
    (f) Quantitation procedure: Series of standards? Internal standard? Standard Additions? Concentrations of standards?
    (g) Type of Chromatography: Which of the four types mentioned in this chapter is it? If it is partition chromatography, is it normal phase or reverse phase? What specific stationary phase is used?
    (h) Mobile Phase Characteristics: What is the mobile phase composition? Is it gradient or isocratic elution? What is the flow rate?

(i) Detector: What detector is used? Are there any specific characteristics of the detector that are worth mentioning? For example, if it is a UV detector, is it variable wavelength UV, or diode array UV? If it is MS, is it tandem MS? Is it TIC or SIM?

(j) How Are the Data Gathered? In other words, you have got your blank, your standards, and your samples ready to go. Now what? Be brief.

(k) How the Standards are Prepared? Are you diluting a stock standard? If so, how is the stock prepared and what is its concentration? If not, how are the standards prepared?

(l) Potential Problems: Is there anything mentioned in the method document that might present a problem? Be brief.

(m) Data Handling and Reporting: Once you have the data, then what? Be sure to present any post-run calculations that might be required.

(n) References: Did you look at any other reference sources to help answer any of the above? If so, write them here.

13 Mass Spectrometry

13.1 BASIC PRINCIPLES

Mass spectrometry is an instrumental analytical technique that provides information about a molecule's mass. It generates ions from analyte atoms or molecules, separates these ions from each other on the basis of their mass and charge, and then generates an electronic signal for each. The series of sharp peaks, one for each ion, is then displayed as a so-called **mass spectrum**, which is a plot of the signal vs. mass-to-charge ratio. The magnitude of each signal is proportional to the number of each ion found.

Although the technique is named a "spectrometric" or "spectroscopic" technique, and the word "spectrum" is used to describe the data, it does not use light at all. The name likely results from the very early instruments in which photographic plates or phosphor screens were used to detect the ions. Also, the modern instruments use ion detectors that are similar in function to photomultiplier tubes used in spectroscopy, as we shall see in Section 13.4. One could consider that the ions are separated from each other in a manner that is analogous to how a monochromator functions in separating wavelengths and displaying an absorption spectrum. Instead of scanning wavelengths and developing an absorption signal for each wavelength, many mass spectrometer designs scan electric field strength, separate ions based on mass-to-charge ratio, and then produce a signal for each in the mass detector. In any case, the technique is called mass spectrometry; instruments are called mass spectrometers; and the data generated are referred to as mass spectra.

The mass spectrum consists of a series of sharp peaks resembling vertical lines, each due to an ion of a particular mass-to-charge ratio. The example given in Figure 13.1 is for a very simple molecule, methane, CH_4. The charged formulas indicated there next to each vertical line all result from the fragmentization of methane molecules. The CH_3^+ and CH_4^+ ions greatly outnumber the others as indicated by the significantly greater signal intensity for these two particles.

Qualitative analysis is possible by virtue of the fact that the mass spectrum is unique to the analyte. No two elements or compounds have exactly the same mass spectrum. It is a fingerprint. Quantitative analysis is possible by measuring the signal intensity of one ion or by summing the intensity for several or for all of the ions. This signal or signal summation is proportional to the concentration of the element or compound.

The components of a mass spectrometer are the **sample inlet**, the **ion source**, the **mass analyzer** (sometimes called the **mass filter**), the **ion detector**, and a powerful **vacuum system**. The sample is fed into the ion source where the ions are formed. The ions then move into the mass analyzer where they are separated from each other. Finally, the electronic signal is generated and amplified as the ions pass one at a time into the ion detector. Because the nature of the sample can vary, there are a number of different designs for the ion source and also a number different designs for the mass analyzer. In any case, the entire system (ion source, mass analyzer, and ion detector) must be evacuated to a very low pressure. The reasons for this are that (1) air molecules in the ion source would form ions along with sample molecules and would thus contaminate the sample ions and (2) air molecules in the mass analyzer would collide with the sample ions as they move and interfere with their detection. Figure 13.2 shows a block diagram of the entire system.

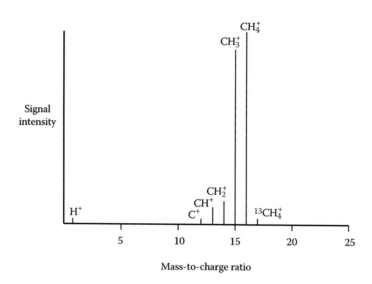

FIGURE 13.1 Mass spectrum of methane.

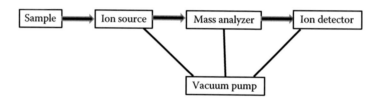

FIGURE 13.2 Block diagram showing the components of a mass spectrometer.

13.2 SAMPLE INLET SYSTEMS AND ION SOURCES

The analyte must be 100% pure when it enters the ion source, otherwise the results would show signal intensities for contaminant ions as well as the analyte ions. Sometimes the analyst has pure samples to run, but more often, the analyst has a sample that is a complex mixture. In addition, the sample may be a solid, a liquid, or a gas, but the ions must be in the gas phase when exiting the ion source. Sample inlet systems and the manner in which the ions are generated are thus important to consider.

The resolution of complex mixtures in both the liquid and gas phases are discussed in Chapters 10–12 (instrumental chromatography). Therefore, it should not be surprising that the sample inlet system can be a gas or liquid chromatography system. Indeed, it is very common today to find a gas chromatography instrument, or a liquid chromatography instrument, coupled to a mass spectrometer. Such setups have come to be known as gas chromatography–mass spectrometry (GC-MS) systems or liquid chromatography–mass spectrometry (LC-MS) systems. These "hyphenated techniques" will be discussed later in this chapter. But what about analytes that are already 100% pure? Pure gases are introduced via calibrated gas inlet systems. With volatile pure liquids and solids, ionization occurs when heated, and thus we have an ion source that is aptly referred to as the **thermal ionization (TI)** ion source. The primary use of a TI ion source is the analysis of radioactive metals for isotope distribution. In the case of isotopic distribution studies, the radioactive metal is electroplated onto a rhenium filament prior to being "desorbed" and vaporized as a result of the heating in the ion source.

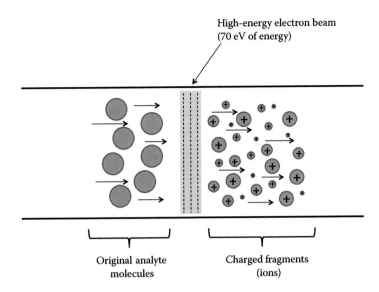

High-energy electron beam
(70 eV of energy)

Original analyte
molecules

Charged fragments
(ions)

FIGURE 13.3   Illustration of the EI ion source as described in the text. As the molecules move left to right through the high-energy electron beam, they fragment into positively charged ions.

The most common ionization source is the **electron impact (EI)** ion source. This is the ion source that is most often used when the analyte is a molecular compound, such as the compounds that elute from a GC column, and so are used in GC-MS systems. In this ion source, the molecules of the analyte move through a shower of high-energy electrons and will break apart into positively charged fragments (ions) as a result (see Figure 13.3 for an illustration). The electrons have an energy of around 70 eV, which is an energy sufficient to break the bonds for virtually all molecular compounds. At this energy, the distribution of the fragments is always the same, i.e., the same fragments always form in an exactly reproducible concentration profile for the same analyte such that the mass spectrum that results is always the same. Therefore, the mass spectrum that results is the fingerprint referred to earlier.

Besides the EI source, another that is popular with GC systems is **chemical ionization (CI)**, in which ions form by chemical reaction at low pressure. CI is a lower energy ionization procedure compared with EI, and so the spectra are simpler. We also have the ion sources used in conjunction with liquid chromatography, including **atmospheric pressure chemical ionization (APCI)** and **electrospray ionization (ESI)**. APCI, as the name implies, is chemical ionization that occurs at atmospheric pressure rather than at the low pressure of the CI source. ESI is currently the most popular ionization method when using mass spectrometry as a detector for HPLC, and this method will be covered in greater detail in Section 13.8.

There are still other ion sources, but these are beyond our scope.

13.3   MASS ANALYZERS

Following their formation in the ion source, the ions move forward to the mass analyzer where their separation takes place on the basis of their mass-to-charge ratio. Just as in the case of the ion sources, there are a number of different designs.

Probably the most common design is the **quadrupole mass analyzer**. In this design, four short parallel metal rods ("poles"), typically about 6 to 8 in. in length and with a diameter of about 0.8 cm each, are utilized. These rods are aligned parallel to and surrounding the ion path as illustrated in Figure 13.4 (see also the photographs in Figure 13.5). Direct current (DC) and radiofrequency

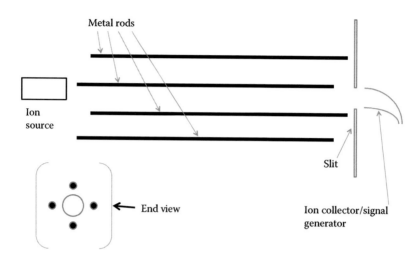

FIGURE 13.4 Drawing of the quadrupole mass analyzer as described in the text. The view from the ion source looking left to right down the center of the four rods toward the slit is shown in the inset on the lower left.

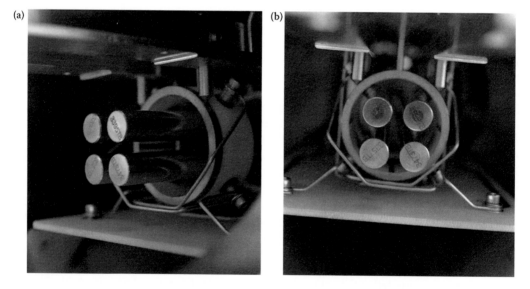

FIGURE 13.5 (a) Side view of the detector end of a quadrupole mass analyzer. (b) End view of a quadrupole mass analyzer looking toward the ion source.

alternating current (AC) voltages are applied to the rods such that the two opposite rods have the same voltages while the two adjacent rods have voltages of opposite signs. An electromagnetic field is thus created. The magnitude of these voltages can be altered such that the field strength is "scanned." As the ions enter the field and begin to pass down the center area, they deflect from their path according to how their charge and mass interact with the field. Varying the magnitude of the field creates the ability to "focus" the ions one at a time through the detector slit and into the detector where the electrical signal plotted on the y-axis of the mass spectrum is generated (see the illustration given in Figure 13.6a–f). The process is repeated (i.e., the field strength is scanned again) many times and, as long as the sample is present, a significant signal for each mass-to-charge ratio (i.e., for each ion) is generated. In this way, the mass spectrum is able to be plotted.

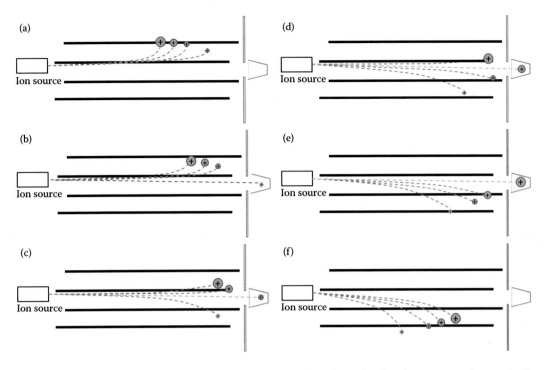

FIGURE 13.6 Illustration how a quadrupole mass analyzer functions, showing the sequence of events (a–f) that occurs in one complete scan of the electromagnetic field generated between the poles as discussed in the text.

Another popular design is the **time of flight mass analyzer (TOF)**. This design separates the ions based on how fast they move through a "drift tube" in which a strong electric field is present due to a high voltage metal grid positioned in the path. The lighter fragments travel through the tube faster than the heavier fragments, and the different masses are detected at different times, depending on their "time of flight" through the tube (see Figure 13.7). The signal is thus measured vs. time, but can be translated into the traditional mass spectrum.

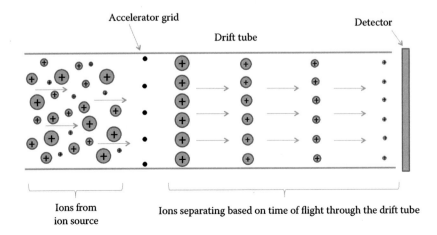

FIGURE 13.7 Illustration of a time-of-flight mass analyzer as described in the text.

A third design is known as the **magnetic sector mass analyzer**. This design is largely obsolete at this time. We mention it because of its historical value. In this design, a large electromagnet wraps around a curved tube leading away from the ion source. The ions moving through the tube deflect from their paths as a result of their interactions with the magnetic field. The degree to which they deflect depends on the mass-to-charge ratio. The ion detector is positioned at the end of the tube. As the magnetic field strength is scanned, the ions are detected one at a time at the detector and the mass spectrum of each is recorded. The quadrupole mass analyzer is an improvement over this design in that it is much more compact and can be used with a gas chromatograph with a much smaller footprint on the benchtop.

13.4   THE ION DETECTOR

As stated previously, the job of the ion detector is to generate a measurable electrical signal for each ion passing from the mass analyzer and through the detector slit. The most common design is an electron multiplier that is horned-shaped, as illustrated in Figure 13.8. The inside surface of this shape is coated with dynodes. A **dynode** is an electrode that is designed such that each electron impinging on its surface causes two or more electrons to be emitted from this surface. As the stream of ions enters from the mass analyzer through the slit and contacts the inside surface of the detector, a corresponding stream of electrons is ejected and these follow the path indicated in Figure 13.8. Because the inside surface consists of dynodes, the electron signal amplifies each time this stream of electrons bounces off the wall, such that, at the base of the horn, there is a significant signal, as illustrated. This signal then undergoes further amplification electronically resulting in a peak in the mass spectrum where the intensity of this signal is plotted on the y-axis. The different ions entering one at a time through the detector slit as a result of their separation in the mass analyzer create their own unique signal and each is found in the mass spectrum at a particular mass-to-charge ratio, which is what is plotted on the x-axis.

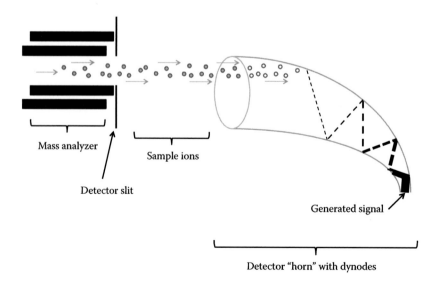

FIGURE 13.8   Illustration of the electron multiplier mass detector. Sample ions emerging from the mass analyzer enter the detector "horn" where an electronic signal is generated and amplified via collisions with dynodes lining the walls of the horn.

13.5 MASS SPECTRA

Mass spectra were first mentioned in Section 13.1 in conjunction with Figure 13.1. A **mass spectrum** is a recording of the "abundance" of a specific ion vs. mass-to-charge ratio, m/z, where "m" is the ion's mass and "z" is its charge. It is a plot of the detector signal vs. m/z and is that unique finger-print of, for example, a molecular organic compound. As such, it can be of help for identification purposes. Most ions formed have a charge of +1. The mass-to-charge ratio of any ion is therefore normally equal to the mass of that ion. Molecular fragments appear in a mass spectrum of such a compound and provide clues as to the identity and molecular structure of the analyte molecule simi-lar to the way pieces of a jigsaw puzzle provide clues to the structure of the intact puzzle. Imagine, for example, a molecule of chloroform, $CHCl_3$. When this molecule breaks apart in an EI source, major ions with mass-to-charge ratios of 83 and 35 form. The 83 ratio represents the sum of the atomic weights of one carbon atom (12), one hydrogen atom (1), and two chlorine atoms (70) and is therefore a $CHCl_2^+$ ion. The 35 ratio is a Cl^+ ion. There are other smaller (minor) signals representing other ions formed in the fragmentation process, including those ions that have the other chlorine isotopes in their structure. The largest peak in the spectrum is called the **base peak**.

Certain fragments are more prone to form from the analyte molecule than others, due to the pres-ence of functional groups in the molecule and their interconnection. The masses of these fragments can be used to deduce the structure of the analyte compound. The ionized analyte molecule, when seen as part of the mass spectrum, is referred to as the **molecular ion**. Occasionally, the molecule is so extensively fragmented by the ionizing process that little or no molecular ion is seen.

The effect of the presence of isotopes of elements in a sample is illustrated in Figure 13.9. A sample of elemental lead shows the presence of four isotopes as shown. In the mass spectra of compounds, the presence of isotopes in a sample is indicated when "clusters" of mass peaks around certain m/z values are observed. These clusters represent the naturally occurring isotopes that are present for carbon, nitrogen, sulfur, chlorine, bromine, and a few others. The relative percentages of these cluster ions provide more clues useful in unraveling the identity of an analyte molecule from its molecular fingerprint.

It is important to remember that the mass spectrum of a compound is a fingerprint of that com-pound. In modern GC-MS computerized data analysis, files of such mass spectral fingerprints can be stored on disk and scanned by the data system as the data are being acquired. This is a great tool then to help identify unknowns.

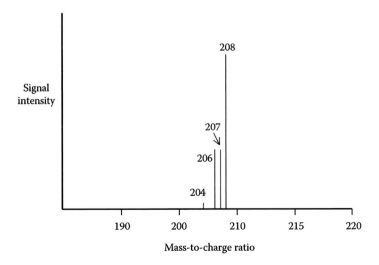

FIGURE 13.9 Mass spectrum of a sample of elemental lead showing the four isotopes of lead.

13.6   ICP-MS

The inductively coupled plasma atomizer (ICP) described in Chapter 9 is an ion source for mass spectrometry. And so, the inductively coupled plasma (ICP) technique has been coupled to mass spectrometry (ICP-MS) such that a mass spectrum is measured rather than the intensity of a spectral line in the atomic emission spectrum. The ICP source provides the mass spectrometer with a source of charged monatomic metal ions. The stream of extremely hot gas (the ICP gas flow) is fed into the mass spectrometer for analysis by the mass analyzer and detector portions of the mass spectrometer (Figure 13.10). Notice the "sampling cone" and the "skimmer cone" evident in the figure. The sampling cone has a hole approximately 1 mm in diameter and the skimmer cone has a hole approximately 0.4 mm in diameter. The vacuum draws the ions from the ICP through these holes and into the mass analyzer, which is usually a quadrupole as shown.

The intensity of the signal is proportional to the concentration of the analyte in the sample. In most cases, there will be more than one analyte ion for any given element (due to the presence of more than one naturally occurring isotope) and therefore more than one signal for each element. It may happen that more than one element in a sample will produce a signal with a particular mass-to-charge ratio. Therefore, there would be a spectral interference if that signal were to be measured. Nearly all elements have at least one isotope mass that no other element has. Thus, a signal with this spectral interference is ignored in favor of one that is due to that one element.

The detection limits for elements analyzed by ICP-MS are significantly lower in most case than the detection limits for other atomic techniques (see Table 13.1).

13.7   GC-MS

The GC-MS instrument consists of a gas chromatograph, an interface between the GC column and the MS, the electron-impact ion source, a vacuum system to reduce the pressure in the MS to approximately 10^{-5} Torr, a mass analyzer, the ion detector where the signal is generated, and a "data collection system." Modern GC-MS units are smaller than the original models, and they are often referred to as **benchtop GC-MS** units.

The **interface** must transfer the sample/mobile phase matrix from the positive pressure inside the GC column to the very low pressure inside the MS. It must reliably transfer the sample without compromising the performance of the GC or the MS. Most GC-MS applications utilize capillary

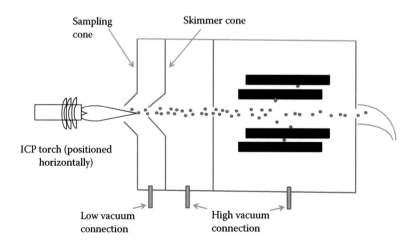

FIGURE 13.10   Illustration of an ICP-MS. Ions generated by the ICP torch are drawn into the mass analyzer through a sampling cone and a skimmer cone. These ions are then measured by the ion detector.

TABLE 13.1
Detection Limits for ICPMS Compared with the Best from Table 9.2

Element	ICPMS Detection Limit (μg/L)	Best Detection Limit from Table 9.2 (μg/L)
Arsenic	0.0006	0.03 (hydride)
Bismuth	0.0006	0.03 (hydride)
Calcium	0.0002	0.01 (graphite furnace)
Copper	0.0002	0.014 (graphite furnace)
Iron	0.0003	0.06 (graphite furnace)
Mercury	0.016	0.009 (cold vapor mercury)
Potassium	0.0002	0.005 (graphite furnace)
Zinc	0.0003	0.02 (graphite furnace)

columns through which the mobile phase flow rate is less than 5 mL/min. In that case, there is a direct transfer from the end of the column to the ion source, as shown in Figure 13.11.

The **data collection system** (computer with appropriate software) is responsible for the control of the GC-MS system. This includes GC temperatures, tuning the MS system, controlling the mass analyzer during data acquisition, detecting the abundance of each ion, and processing the acquired data. The quadrupole can be scanned quickly enough to measure signal intensities that appear over a useful mass range as each mixture component elutes from the column. Each peak in the GC chromatogram is usually acquired as the sum of the signals of all the ions detected for that component by mass spectrometry. For that reason, the chromatogram is called the **total ion chromatogram (TIC)**. The data system thus generates a complete mass spectrum for each mixture component as it elutes and the comparison to the library of spectra is immediate. So, in a GC-MS experiment, as each mixture component elutes producing the peak, the corresponding mass spectrum is obtained. GC-MS data are therefore often referred to as two-dimensional – the chromatography peak and the mass spectrum are both obtained and either, or both, can be displayed.

However, besides TIC, one can also acquire the data by setting up the mass analyzer such that intensities for only one or only a few ions are measured for the chromatography peak. This is called **selective ion monitoring (SIM)**. In cases where the sample is suspected to contain trace levels of the analyte, it is possible to monitor only two or three masses unique to that compound. In this mode, the mass spectrometer is a very specific detector, targeting only a particular compound. Selective ion monitoring is thus the method of choice for quantitation by GC-MS. Accurate peak areas can

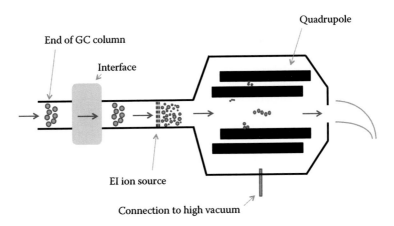

FIGURE 13.11 Illustration of a GC-MS. Sample molecules emerging from the GC column are fragmented by an EI ion source. These ions then pass through the mass analyzer and are measured by the ion detector.

**APPLICATION NOTE 13.1: ANALYSIS OF APPLES
FOR PESTICIDE RESIDUES BY GC-MS**

A GC-MS procedure has been developed for analyzing apples for pesticide residues. The apples to be tested are diced into 1-cm cubes and then frozen overnight. Prior to the extraction of the pesticides, a number of the cubes are ground up in a coffee grinder. A 10-g portion of the ground sample is then extracted with an acetonitrile/water mixture to which some $MgSO_4$ and NaCl has been added. A primary secondary amine is also added to remove fatty acids. Following vortexing and centrifuging, the internal standard (triphenyl phosphate) is added. A splitless injector for 1-µL injections and a complex temperature program are utilized. Selective ion monitoring is used to generate the GC peaks. Standard curves are created using concentrations of 10, 20, 50, 100, 250, and 400 ng/mL of the pesticide.

(*Reference*: Usher, K., and Majors, R., *Analysis of Pesticide Residues in Apple by GC-MS using Agilent SampliQ QuEChERS Kits for Pre-Injection Cleanup*, http://www.chem.agilent. com/Library/applications/5990-4468EN.pdf.)

be obtained using standard GC methods and quantitative analysis performed by calibration with standards and by plotting a standard curve as discussed in Chapter 11 (see Application Note 13.1).

In terms of advantages and disadvantages, compared with other GC dectectors and considering that a GC has the ability to resolve complex mixtures of organic materials quickly, the MS detector provides rapid, positive identification of the analytes even at very low concentrations (down to ppb, or µg/L, level). It can also provide reliable quantitative analysis results at these very low concentrations.

The major disadvantage is the expense: $60,000 for the minimum benchtop instrument. Also, they are more complicated than other GC detectors and require extensive operator training.

13.8   LC-MS

HPLC is an instrumental chromatography method similar to GC, as we have seen in previous chapters. However, connecting an HPLC system to a mass spectrometer is more complicated because we must deal with the huge amount of liquid mobile phase that would be incompatible with the very low pressure of the MS. To handle this situation, and also to accomplish the ionization, two interfacing/ionization schemes have been invented. These are **atmospheric pressure chemical ionization (APCI)** and **electrospray ionization (ESI)**.

As mentioned previously, ions are generated in the APCI by chemical reaction at atmospheric pressure. In this ion source, an aerosol cloud is created by directing the mobile phase emerging from the column to a nebulizer through which nitrogen gas also flows. A heating element heats the resulting "spray" to a very high temperature. The eluting analyte molecules ionize in a series of chemical reactions when the cloud is subjected to a special type of electrical discharge. Of course, the analyte molecules must be stable at the high temperature of the spray.

In the electrospray ionization method, the column effluent is routed through a metal capillary that has a very high electrical charge. Nitrogen gas also flows through the capillary. Solvent droplets that contain analyte ions emerge from this capillary and "fly apart" due to electrostatic repulsive forces. As they continue to move toward an electrode of opposite charge, the droplets become smaller and smaller due to solvent evaporation. The analyte ions then enter the mass analyzer free of solvent molecules (see Application Note 13.2 for an example of an LC-MS analysis that utilizes the electrospray ion source).

In either the APCI or the ESI designs, the analyte ions do not break apart into smaller particles as they do in the case of the EI ion source used in GC-MS. They are referred to as "soft" ion sources due to the fact that the energy used to create the ions is much less than with the EI source. So only

APPLICATION NOTE 13.2: PROCARBAZINE IN CLINICAL SAMPLES

Clinical pharmacological investigations of the effectiveness of anticancer drugs involve measuring the level of these drugs in blood, urine, feces, saliva, and other clinical samples after they have been administered to the patient. An example of such a drug is procarbazine. The concentration of procarbazine in human plasma, for example, can be measured using a reverse phase HPLC method with electrospray MS detection. Sample preparation includes precipitating the protein using trichloroacetic acid. The procarbazine remains dissolved, and this solution is treated with methyl *tert*-butyl ether to remove excess acid. It is a reverse phase experiment using a C18 column with a methanol–25-mM ammonium acetate buffer as the mobile phase. The flow rate is 1.0 mL/min. For the MS detection, selective ion monitoring is used. For the standard curve, concentrations of procarbazine in human plasma ranging between 0.5 and 50 ng/mL are used.

(*Reference*: http://ukpmc.ac.uk/abstract/MED/14670747.)

the ions forming as described above enter directly into the mass analyzer and are detected. The fragmentation patterns of GC-MS (using the EI source) spectra are not found in LC-MS (using the APCI or the ESI) spectra, so other means are employed to obtain structural information. These "other means" are the tandem mass spectrometry methods described below.

13.9   TANDEM MASS SPECTROMETRY

Tandem mass spectrometry refers to techniques in which two mass analyzers are used in tandem. This means that one mass analyzer is used to "select" a particular analyte ion from those that enter it. In this case, the magnetic field strength in the first mass analyzer is set (not scanned) such that only the selected ion is allowed through. This ion then undergoes fragmentation in a "collision cell." An ionized gas (Ar, He, or N_2) is pumped into this collision cell at low pressure. Analyte ions colliding with the charged gas particles (hence the name "collision cell") are energetic enough to generate fragments. The "product ions" so generated enter the second mass analyzer such that a complete mass spectrum of the analyte is generated in the usual way (see Figure 13.12). LC-MS/MS and GC-MS/MS are common designations for the complete procedure.

EXPERIMENTS

EXPERIMENT 43: THE QUANTITATIVE GC-MS ANALYSIS OF A PREPARED SAMPLE FOR CHLOROBENZENE BY THE INTERNAL STANDARD METHOD

1. Thoroughly rinse two clean 100-mL volumetric flasks with acetone.
2. Prepare a stock standard solution of chlorobenzene in acetone by pipetting 0.10 mL of chlorobenzene into one of the 100-mL flasks prepared in Step 1. Dilute to the mark with acetone and shake. This is the analyte stock solution. Calculate the volume percent concentration of this solution. Label accordingly.
3. Prepare a stock standard solution of bromobenzene in acetone by pipetting 0.10 mL of bromobenzene into the second 100-mL flask prepared in Step 1. Dilute to the mark with acetone and shake. This is the internal standard stock solution. Label accordingly.
4. Prepare four working standard solutions from the stock solution prepared in Step 2. The four solutions should have chlorobenzene concentrations of 0.01%, 0.02%, 0.03%, and 0.04%. Use 25-mL volumetric flasks that have been previously throroughly rinsed with acetone.

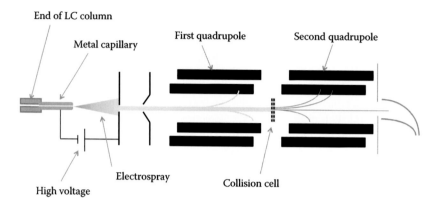

End of LC column

Metal capillary

First quadrupole

Second quadrupole

Electrospray

High voltage

Collision cell

FIGURE 13.12 Illustration of a LC-MS/MS. Sample molecules emerging from the LC column are ionized via an electrospray ion source. The first mass analyzer selects the analyte ion to be passed on to the collision cell. The ions from the collision cell then pass through the second mass analyzer/ion detector where the normal mass spectrum is measured.

Before diluting to the mark with acetone, pipette 5.00 mL of the internal standard stock solution into each. Shake each after diluting to the mark.

5. Obtain a 25-mL volumetric flask from your instructor containing an unknown amount of chlorobenzene. Add the 5.00 mL of internal standard stock solution and dilute to the mark with acetone.

6. Obtain chromatograms of each of the working standards and find the areas of the analyte and internal standard peaks. Prepare the standard curve by plotting the ratio of the analyte peak area to the internal standard peak area. Determine the chlorobenzene concentration of the unknown.

7. Optional: Obtain the mass spectra of the chlorobenzene and the bromobenzene and try to interpret the major peaks.

EXPERIMENT 44: GC-MS DETERMINATION OF ETHYLBENZENE IN GASOLINE BY COMBINED INTERNAL STANDARD AND STANDARD ADDITIONS METHODS

1. Prepare 100.0 mL of a 2% gasoline solution with acetone as the solvent. Also prepare 50.00 mL of a 1% ethylbenzene stock solution also with acetone as the solvent.

2. Into each of five 25-mL volumetric flasks, pipette 10.00 mL of the 2% gasoline solution. Stopper the flasks immediately to prevent loss by evaporation.

3. Pipet volumes of the ethylbenzene stock into four of the flasks such that the concentrations are 0.00500%, 0.0100%, 0.0150%, and 0.0200% added when diluted to the mark. Add nothing to the fifth flask. It will be the 0% added concentration. Dilute each of the five flasks to the mark with acetone and shake. The internal standard is the toluene that is already present in gasoline at the same concentration in all the standards.

4. With the column temperature set at 70°C, obtain a chromatogram of the 0% added solution. Allow the chromatogram to run for a period until you are fairly certain that there are no more peaks. Note the retention times of toluene and ethylbenzene. Now set the column temperature to be programmed from 70°C to 120°C to begin increasing at a time that will hasten the elution of the gasoline components that elute after ethylbenzene. This is so that you would not have to wait long between injections. Note the areas of the toluene and ethylbenzene peaks.

5. Inject the other four standards, again noting the areas of the toluene and ethylenebenzene peaks. Calculate the area ratio (ethylbenzene to toluene) for each solution.

6. Plot the standard curve using Excel spreadsheet, plotting area ratio vs. concentration added, including 0% added. For the unknown, type in a ratio of zero. Excel will give a negative value for the unknown concentration (because a zero ratio is on the negative side of the y-axis). The absolute value is the concentration of ethylbenzene in the 2% gasoline solution. Multiply by 50 to obtain the concentration in the undiluted gasoline.

QUESTIONS AND PROBLEMS

Basic Principles

1. Define *mass spectrometry* and *mass spectrum*.
2. What are two possible reasons why mass spectrometry is called spectrometry despite the fact that light is not used in the technique?
3. In the mass spectrum of methane shown in Figure 13.1, there are two peaks due to a CH_4^+ ion, and the one on the right is of much lower intensity. Why are there two and why is the one on the right of such low intensity?
4. Figure 13.13 shows the mass spectrum of acetone. Based on the mass-to-charge ratio of each peak, write a formula for as many of the fragments as you can as was done for methane in Figure 13.1.
5. Explain how both qualitative analysis and quantitative analysis are possible given the mass spectrum of a compound.
6. What are the five components of a mass spectrometer and what function does each have?
7. Track a sample through a mass spectrometer from the sample inlet to the ion detector stating exactly what happens in each of the instrument components.
8. Why is a powerful vacuum system needed for a mass spectrometer?

Sample Inlet Systems and Ion Sources

9. If the sample to be tested is a complex mixture in which the analyte is only one of many components, how can this be handled by mass spectrometry?
10. What is the function of the ion source in a mass spectrometer?
11. What ion source is used for the analysis of a metal for isotopic distribution? Exactly how does this ion source function for this application?

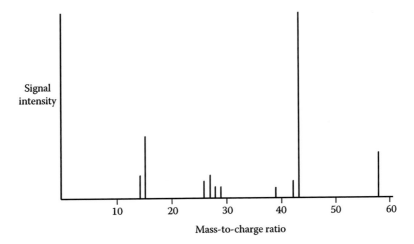

FIGURE 13.13   Mass spectrum of acetone.

12. The electron impact ion source is a popular ion source for analyzing molecular compounds. How does it work?
13. Five different ion sources are mentioned in Section 13.2. What are the names and the abbreviations?

Mass Analyzers

14. In general terms, what takes place in the mass analyzer component of a mass spectrometer?
15. Name three different designs of mass analyzers.
16. How is it that ions are separated from each other in a (a) quadrupole mass analyzer, (b) time of flight mass analyzer, (c) magnetic sector mass analyzer?

Ion Detector

17. What specifically is the function of the ion detector component of a mass spectrometer?
18. What is a dynode?
19. Compare the ion detector discussed in this chapter with the photomultiplier tube discussed in Chapter 8.

Mass Spectra

20. What is plotted on the y- and x-axis of a mass spectrum?
21. What is the "base peak" in a mass spectrum?
22. What is referred to as the "molecular ion" in a mass spectrum?
23. Why do the mass spectra of certain compounds show clusters of peaks?
24. What function can a computer perform that assists the operator in identifying an unknown compound?

ICP-MS

25. What is meant by ICP-MS?
26. How is it that the ICP source described in Chapter 8 can serve as an ion source for mass spectrometry?
27. What is the difference between ICP as an atomic spectroscopy technique and ICP-MS as a mass spectroscopy technique?
28. What is the advantage of ICP-MS compared with ICP as discussed in Chapter 8?

GC-MS

29. Why would a capillary GC column be a "natural fit" for a GC-MS unit?
30. Why is GC-MS often referred to as "two-dimensional?"
31. Under what circumstances is a GC chromatogram that has been generated by mass spectrometry detection often referred to as a TIC? Under what circumstances would a GC chromatogram *not* be referred to as a TIC?
32. What is meant by "selective ion monitoring" in the context of GC-MS?
33. Why is SIM the method of choice for quantitation in GC-MS?
34. What are the advantages and disadvantages of using mass spectrometry as a detector for gas chromatography?

LC-MS

35. In liquid chromatography, a huge amount of liquid, relatively speaking, emerges from the column and feeds into the MS detector. How is this handled in light of the very low pressure that must be maintained in a mass spectrometer?
36. Describe (a) atmospheric pressure chemical ionization and (b) electrospray ionization.
37. In a single stage LC-MS instrument, analyte molecules are not fragmented as they are in GC-MS. What difference does this make in the data that are generated?

Tandem Mass Spectrometry

38. What is "tandem" mass spectrometry?

14 Electroanalytical Methods

14.1 INTRODUCTION

In Chapter 12, we discussed the conductometric and amperometric HPLC detectors. We also discussed electrophoresis. In these techniques, we saw that a pair of electrodes dipped into a solution and connected to a power supply can be used to set up an electric field that affects the behavior of ions present in the solution. Anions are attracted to and migrate toward the electrode connected to the positive pole of the power supply while cations are attracted to and migrate toward the electrode that is connected to the negative pole of the power supply. This migration constitutes the flow of electrical current through a solution. The conclusion, in the case of the conductometric detector, was that anions and cations must be present in a solution in order for that solution to be able to conduct an electric current. And so, the conductometric detector generated an electric signal whenever analyte ions eluted from the HPLC column. In the case of electrophoresis, we saw that different ions migrated through the solution in different directions and at different rates such that they separated. These ions were then detected and measured independent of the other ions present in the solution. In capillary electrophoresis, the electroosmotic flow that occurs in the capillary tube assisted the detection and measurement of neutral molecules as well as ions. It was seen that the amperometric HPLC detector required further development of the theory of such behavior occurring in such an electric field in order to understand how it works. We said that this detector required the use of not two, but three electrodes—the working electrode, a reference electrode, and an auxiliary electrode. In fact, there is still much to discuss with regard to the use of electrodes in chemical analysis and our goal in this chapter is to do just that. We start at the beginning and define some important terms.

The first question to be answered is: "what is electrical current?" Electrical current is the flow of electrons through a conductor. When we put a battery or other electrical power source into service, such as to use a battery to light a flashlight or electrical power to light a lamp in our living room, electrons move from one pole of the power source, through the light bulb, and back to the other pole of the power source. There are, of course literally hundreds of examples of the use of batteries to produce this flow of electrons and to power battery-driven devices. There is a battery in our car. There is a battery in our computer. There is a battery in our calculator. There is a battery in our cell phone. The list is very long. The question to be answered in the context of solution conductivity is: "Where do the electrons come from and what happens when they move from the power source through a wire to an electrode dipped into a solution and encounter a solution with ions?" And also, "How is it that these electrons appear at the other electrode so that they can go back to the power source and complete the circuit?" The answer is that oxidation and reduction processes occur at the electrode surfaces that are in contact with the solution. An example is given in Figure 14.1. In this figure, the solution into which the electrodes are dipped is a solution of hydrochloric acid, HCl, and so hydrogen ions and chloride ions are present in the solution and make the solution conductive. As indicated, these ions migrate through the solution when the power source is connected. The hydrogen ions migrate toward the negative electrode and the chloride ions migrate toward the positive electrode. At the surface of the negative electrode, hydrogen ions are accepting electrons from the electrode and are reduced to hydrogen gas, as shown. At the surface of the positive electrode, chloride ions are releasing electrons to the electrode and are oxidized to chlorine gas, as shown. Oxidation and reductions reactions are involved. In this way, the circuit is completed—electrons flow from the power source to the negative electrode and from the positive electrode and back to the power source.

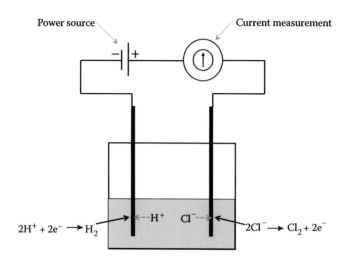

FIGURE 14.1 Illustration of the fact that oxidation–reduction reactions occur at the electrode surfaces when a solution conducts an electric current.

The so-called **electroanalytical methods** involve either the measurement of the electrical current flowing between a pair of electrodes immersed in the solution tested (**voltammetric and amperometric methods**) or the measurement of an electrical potential (voltage) developed between a pair of electrodes immersed in the solution tested (**potentiometric methods**). In either case, the measured parameter (current or potential) is related to the concentration of analyte.

Electroanalytical techniques are therefore an extension of classical oxidation–reduction chemistry, since oxidation and reduction processes occur at the surface of or within the two electrodes, oxidation at one and reduction at the other. Electrons are consumed by the reduction process at one electrode and generated by the oxidation process at the other. The electrode at which oxidation occurs is termed the **anode**. The electrode at which reduction occurs is termed the **cathode**. The complete system, with the anode connected to the cathode via an external conductor, is often called a **cell** (see Figure 14.2). The individual oxidation and reduction reactions are called

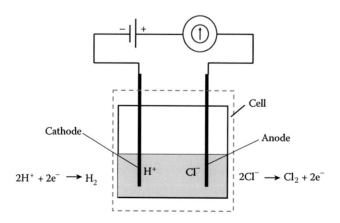

FIGURE 14.2 Illustrations of "anode," "cathode," and "cell."

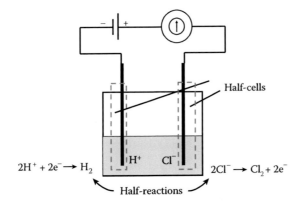

$$2H^+ + 2e^- \rightarrow H_2 \qquad\qquad H^+ \quad Cl^- \qquad\qquad 2Cl^- \rightarrow Cl_2 + 2e^-$$

FIGURE 14.3   Illustrations of "half-reaction" and "half-cell."

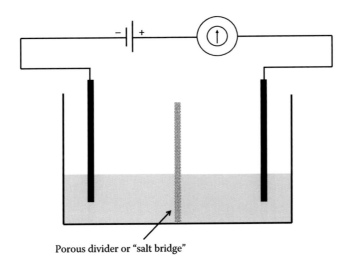

Porous divider or "salt bridge"

FIGURE 14.4   "Salt bridge" is a porous divider between two half-cells. This is also an illustration of an electrolytic cell.

the **half-reactions**.* The individual electrodes with their half-reactions are called **half-cells** (see Figure 14.3). As we shall see in this chapter, the half-cells are often in separate containers (mostly to prevent contamination) and are themselves often referred to as electrodes because they can be housed in portable glass or plastic tubes. In any case, there must be contact between the half-cells to facilitate the ionic diffusion. This contact is called the **salt bridge** and is shown in Figure 14.4 as a porous divider between the two half-cells.

A **galvanic cell** is one in which this current flows (and the redox reaction proceeds) spontaneously because of the strong tendency for the chemical species involved to give and take electrons. An **electrolytic cell** is one in which the current is not a spontaneous current, but rather is the result of incorporating an external power source, such as a battery, in the circuit to drive the reaction in

---

* The concept of oxidation and reduction half-reactions was first introduced in Chapter 5, Section 5.7.2.

one direction or the other. An electrolytic cell is represented in Figure 14.4 and a galvanic cell is represented in Figure 14.5. In Figure 14.6, we see a galvanic cell in which one of the half-cells is the portable design that is so prevalent in modern analytical laboratories. Notice the porous fiber salt bridge at the tip of this portable electrode. It is important to recognize that potentiometric methods involve galvanic cells and voltammetric and amperometric methods involve electrolytic cells.

Redox reactions, as discussed in Chapter 5, occur upon direct contact between the oxidizing agent and the reducing agent. Upon contact, the electrons jump from the reducing agent to the oxidizing agent. In the case of galvanic and electrolytic cells, the electrons transfer from one to the other by flowing through the external conductor from the anode to the cathode. The flow of

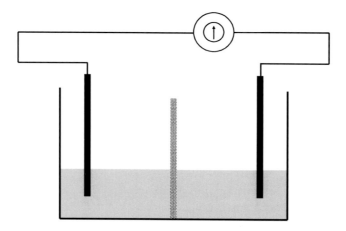

FIGURE 14.5   "Galvanic cell." The current in such a cell flows spontaneously without a power source in the circuit.

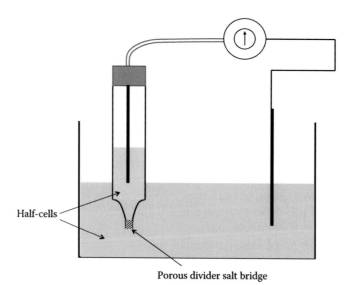

FIGURE 14.6   In modern electroanalytical chemistry, a half-cell is often a portable unit that can be dipped into a solution as shown in order to create a full cell. The other electrode and its solution constitute the other half-cell. The salt bridge in this portable electrode is at the tip as shown.

electrons through the external conductor in an electrolytic cell constitutes the current that is measured in the voltammetric and amperometric methods. The concentration of the analyte is related to this current. When one of the electrodes is a reference electrode (to be discussed later), the potential (voltage) difference between the two half-cells in a galvanic cell is measured. These are the potentiometric methods and the analyte concentration is related to this potential.

14.2   TRANSFER TENDENCIES: STANDARD REDUCTION POTENTIALS

Different redox half-reactions have different tendencies to occur. Consider the three metals sodium, magnesium, and aluminum, all in the same row of the periodic table. A sodium atom, as well as atoms of other alkali metals, has a very strong tendency to give up one electron to become a sodium ion. The tendency is so strong that the reaction of sodium with ordinary water borders on explosive!

$$2\,Na + 2\,H_2O \rightarrow 2\,Na^+ + 2\,OH^- + H_2 \text{ (explosive)}$$

Magnesium, the alkaline earth metal next to sodium in the periodic table, has a relatively strong tendency to give up its two outermost electrons, but this tendency is much less than that of sodium. When a fresh surface of magnesium metal is exposed to water, the reaction occurs very slowly.

$$Mg + 2\,H_2O \rightarrow Mg^{2+} + 2\,OH^- + H_2 \text{ (slow)}$$

Aluminum metal, the next element to the right in the same row as sodium and magnesium in the periodic table, has an even lesser tendency to lose electrons, even though it does lose electrons to become the aluminum ion, Al^{3+}. Its reactivity with water is almost nil.

$$Al + H_2O \rightarrow \text{no reaction}$$

A similar illustration can be made with the *nonmetals* and their tendency to *take on* electrons. The point is that different half-reactions have different tendencies to occur.

To understand **potentiometric methods**, those that measure electrical potentials and determine analyte concentrations from these potentials, it is necessary that numerical values for these tendencies be known under conventional standard modes and conditions. What are these standard modes and conditions? First, for convenience's sake, all half-reactions must be written as either reductions or oxidations. Scientists, by international agreement, have decided to write them as reductions. Second, the tendencies for half-reactions to proceed depend on the temperature, the concentrations of the chemical species involved,* and, if gases are involved, the pressure in the half-cell. Scientists have defined standard conditions to be a temperature of 25°C, a concentration of exactly 1 M for all dissolved chemical species involved, and a pressure of exactly 1 atm for any gases that may be involved. Third, because every cell consists of two half-cells, it is not possible to measure the value directly. However, if we were to assign the tendency of a certain half-reaction to be zero, then the tendencies of all other half-reactions can be determined relative to this "reference" half-reaction.

The result of all of this is what has come to be known as the table of **standard reduction potentials**. An abbreviated such table is given here as Table 14.1. In this table, the half-reactions are listed on the left and the numerical values for each is listed on the right. The heading for the right hand column is **E°**, the symbol for standard reduction potential. Notice that all half-reactions are written as reductions and that the half-reaction $2H^+ + 2e^- \rightleftharpoons H_2$ is the reference half-reaction (0.0000 V).

The half-reactions at the top in the table are those that have the strongest tendency to occur as written. For example, the fluorine to fluoride half-reaction, the first one listed, has a very strong

---

* To be thermodynamically correct, the tendencies for half-reactions to occur depend on the *activities* of the chemical species involved, not the concentrations (see Chapter 5, Section 5.5.2, for a brief discussion of activity).

TABLE 14.1

Some Standard Reduction Potentials at 25°C

Half-Reactions	E° (V)
$F_2 + 2e^- \rightleftharpoons 2F^-$	+2.866
$O_3 + 2H^+ + 2e^- \rightleftharpoons O_2 + H_2O$	+2.076
$H_2O_2 + 2H^+ + 2e^- \rightleftharpoons 2H_2O$	+1.776
$Ce^{4+} + 1e^- \rightleftharpoons Ce^{3+}$	+1.72
$MnO_4^- + 8H^+ + 5e^- \rightleftharpoons Mn^{2+} + 4H_2O$	+1.507
$Cl_2 + 2e^- \rightleftharpoons 2Cl^-$	+1.35827
$Cr_2O_7^{2-} + 14H^+ + 6e^- \rightleftharpoons 2Cr^{3+} + 7H_2O$	+1.232
$O_2 + 4H^+ + 4e^- \rightleftharpoons 2H_2O$	+1.229
$Hg^{2+} + 2e^- \rightleftharpoons Hg$	+0.851
$Ag^+ + 1e^- \rightleftharpoons Ag$	+0.7996
$Fe^{3+} + 1e^- \rightleftharpoons Fe^{+2}$	+0.771
$I_2 + 2e^- \rightleftharpoons 2I^-$	+0.5355
$Cu^{2+} + 2e^- \rightleftharpoons Cu$	+0.3419
$Hg_2Cl_2 + 2e^- \rightleftharpoons 2Hg + 2Cl^-$ (SCE)	+0.2412
$AgCl + 1e^- \rightleftharpoons Ag + Cl^-$ (Ag–AgCl reference)	+0.22233
$Sn^{4+} + 2e^- \rightleftharpoons Sn^{2+}$	+0.151
$2H^+ + 2e^- \rightleftharpoons H_2$	0.00000
$Fe^{3+} + 3e^- \rightleftharpoons Fe$	−0.037
$Sn^{2+} + 2e^- \rightleftharpoons Sn$	−0.1375
$Ni^{2+} + 2e^- \rightleftharpoons Ni$	−0.257
$Fe^{2+} + 2e^- \rightleftharpoons Fe$	−0.447
$Cr^{3+} + 3e^- \rightleftharpoons Cr$	−0.744
$Zn^{2+} + 2e^- \rightleftharpoons Zn$	−0.7618
$Mg^{2+} + 2e^- \rightleftharpoons Mg$	−2.372
$Na^+ + 1e^- \rightleftharpoons Na$	−2.71
$K^+ + 1e^- \rightleftharpoons K$	−2.931
$Li^+ + 1e^- \rightleftharpoons Li$	−3.0401

Source: Lide, D.R., *Handbook of Chemistry and Physics,* 82nd ed., CRC Press, Boca Raton, FL, 2001–2002. With permission.

---

tendency to occur as any student of fundamental chemistry would be able to conclude based on the position of fluorine in the periodic table. What may be confusing is that the half-reactions involving lithium, potassium, and sodium are found at the bottom of the table indicating that they do not have a tendency to occur. The explanation for this apparent contradiction is that the reactions are, by convention, reductions, meaning that the ions are being reduced to the metals. Thus, the reductions of lithium ions, potassium ions, and so forth would indeed not have a tendency to occur since the reverse reactions in each case (the metal to metal ion reactions) have the strong tendencies to occur.

Another observation is that the reactions near the bottom of the table have a negative sign for their E°, whereas the reactions near the top have a positive sign. The explanation for this is that, compared with the standard ($2H^+ + 2e^- \rightleftharpoons H_2$), those that have a positive sign tend to occur as written, whereas those that have a negative sign tend to occur in the opposite direction. The more

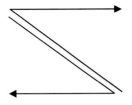

$Cl_2 + 2e^- \rightleftharpoons 2Cl^-$	+1.35827
$Cr_2O_7^{2-} + 14H^+ + 6e^- \rightleftharpoons 2Cr^{3+} + 7H_2O$	+1.232
$O_2 + 4H^+ + 4e^- \rightleftharpoons 2H_2O$	+1.229
$Hg^{2+} + 2e^- \rightleftharpoons Hg$	+0.851
$Ag^+ + 1e^- \rightleftharpoons Ag$	+0.7996
$Fe^{3+} + 1e^- \rightleftharpoons Fe^{+2}$	+0.771

FIGURE 14.7 Mnemonic devices, the arrows on the left, for predicting which chemicals will participate in a redox reaction and which will not. A segment of the table of standard reduction potentials (Table 14.1) is presented on the right as a help to understand the use of the arrows (see text for an example).

positive the E°, the stronger the tendency to occur as written. The less positive (or more negative) the E° value, the stronger the tendency to occur in the opposite direction to what is written.

A way to peruse the table is to understand that if the reduced form in a particular half-reaction in the table comes into contact with the oxidized form in a half-reaction *above* it in the table, a redox reaction between these two forms will occur. Stated in reverse, if the oxidized form in any half-reaction in the table comes into contact with the reduced form in a half-reaction *below* it in the table, a redox reaction between these two forms will occur. Redox reactions between other species, such as a reduced form in a given reaction in contact with an oxidized form in a half-reaction *below* it in the table, will *not* occur. Mnemonic devices (devices that aid in memorizing) summarizing these statements are the arrows shown in Figure 14.7. Thus, for example, as can be seen in Figure 14.7, Cl_2 will react with Fe^{2+} (to form Cl^- and Fe^{3+}), but Cl^- would *not* react with Fe^{3+}.

14.3   DETERMINATION OF OVERALL REDOX REACTION TENDENCY: $E°_{CELL}$

The more the two half-reactions are separated in the table, the greater is the tendency for the net reaction, the reaction that combines the two half-reactions, to occur. This tendency for this net redox reaction to occur, whether by direct contact between the oxidizing agent and the reducing agent, in the same container (no electrodes) or in an electrochemical cell in which the electrons move in the conductor outside the cell to get from one to the other, is determined from the standard reduction potentials, E° values, of the half-reactions involved. The value of this "net" potential is an indication of the tendency of the net redox reaction to occur. We will now present a scheme for determining this net potential, which is symbolized $E°_{cell}$.

In the following scheme, it is assumed that there is a proposed redox system given so that the half-reactions and standard reduction potentials can be found in Table 14.1 or other table of standard reduction potentials. An example follows:

Step 1: Write the equations representing the half-reactions as extracted from the overall reaction given and label as an "oxidation" and a "reduction."

Step 2: Locate the half-reactions in a table of standard reduction potentials, such as Table 14.1, and write the E° values adjacent to the respective half-reactions. For the oxidation half-reaction the sign of the E° must be changed, since the reaction is written in reverse.

Step 3: Balance charges, equalize electrons in both half-reactions, and add the two equations together (as in the scheme for equation balancing in Chapter 5), and also add the E° values together. Do not multiply E° values by the multiplying coefficients. The resulting E is the $E°_{cell}$.

Step 4: If $E°_{cell}$ is positive (+), the reaction will proceed spontaneously to the right as written. If it is negative (−), it will proceed spontaneously in the opposite direction, i.e., to the left.

Example 14.1

What is the $E°_{cell}$ for the following redox system and in which direction will the reaction spontaneously proceed?

$$Cu + Ag^+ \rightleftharpoons Ag + Cu^{2+}$$

Solution 14.1

Step 1: Oxidation: $Cu \rightleftharpoons Cu^{2+} + 2e^-$
    Reduction: $Ag^+ + 1e^- \rightleftharpoons Ag$
Step 2: Oxidation: $Cu \rightleftharpoons Cu^{2+} + 2e^-$ $E° = -0.3419$ V (note sign change)
    Reduction: $Ag^+ + 1e^- \rightleftharpoons Ag$ $E° = +0.7996$ V
Step 3: Oxidation: $Cu \rightleftharpoons Cu^{2+} + 2e^-$ $E° = -0.3419$ V
    Reduction: $2(Ag^+ + 1e^- \rightleftharpoons Ag)$ $E° = +0.7996$ V
    $Cu + 2Ag^+ \rightleftharpoons Cu^{2+} + 2Ag$ $E°_{cell} = +0.4577$ V
Step 4: Since $E°_{cell}$ is (+), the reaction is spontaneous to the right as written.

14.4   THE NERNST EQUATION

Both half- and overall reaction tendencies change with temperature, pressure (if gases are involved), and concentrations of the ions involved. Thus, far, we have only been concerned with standard conditions. Standard conditions, as stated previously, are 25°C, 1-atm pressure, and 1 M ion concentrations. An equation has been derived to calculate the cell potential when conditions other than standard conditions are present. This equation is called the **Nernst equation** and is used to calculate the true E (the so-called cell potential) from the E°, temperature, pressure, and ion concentrations. Consider the half-reaction in which "a" moles of chemical "Ox" react with "n" electrons ("e⁻") to give "b" moles of chemical "Red" (examples follow).

$$a\ Ox + ne^- \rightleftharpoons b\ Red \tag{14.1}$$

the Nernst equation* is the following:

$$E = E° - \frac{0.0592}{n} \log \frac{[Red]^b}{[Ox]^a} \tag{14.2}$$

For the general net reaction, in which "a" moles of chemical "A" react (with the transfer of electrons) with "b" moles of chemical "B" to give "c" moles of chemical "C" and "d" moles of chemical "D,"

$$aA + bB \rightleftharpoons cC + dD \tag{14.3}$$

the Nernst equation is the following:

$$E_{cell} = E°_{cell} - \frac{0.0592}{n} \log \frac{[C]^c[D]^d}{[A]^a[B]^b} \tag{14.4}$$

---

* Again, to be thermodynamically correct, activity should be used rather than concentration. Use of concentration is an approximation.

If any species involved is a gas, the partial pressure of the gas is substituted for the concentration. If the temperature is different from 25°C, the "constant" 0.0592 changes. The number of electrons, n, in Equation 14.4 is the total number of electrons transferred as discovered after equalizing the electrons in the two half-reactions in Step 3 of the scheme for determining E°_{cell} in Section 14.3.

As seen in Equations 14.2 and 14.4, the potential of cells and half-cells are dependent on the concentrations of the dissolved species involved. Clearly, the measurement of a potential (voltage) can lead to the determination of the concentration of an analyte. This therefore is the basis for all quantitative potentiometric techniques and measurements to be discussed in this chapter.

Example 14.2

What is the E for the Fe^{3+}/Fe^{2+} half-cell if $[Fe^{3+}] = 10^{-4}$ M and $[Fe^{2+}]$ is 10^{-1} M at 25°C?

Solution 14.2

$$Fe^{3+} + 1e^- \leftrightarrows Fe^{2+} \quad E^\circ = +0.771 \text{ V (from Table 14.1)}$$

$$
\begin{aligned}
E = E^\circ &- \frac{0.0592}{n} \log \frac{[Fe^{2+}]}{[Fe^{3+}]} \\
&= +0.771 - \frac{0.059}{n} \log \frac{10^{-1}}{10^{-4}} \\
&= 0.771 - 0.0592 \log 10^3 \\
&= 0.771 - 0.0592 \times 3 \\
&= 0.771 - 0.1776 \\
&= +0.593 \text{ V}
\end{aligned}
$$

Example 14.3

What is the E for the $Cu^{2+}/Cu//Ag^+/Ag$ cell if $[Cu^{2+}] = 0.10$ M and $[Ag^7] = 10^{-2}$ M at 25°C?

Solution 14.3

One way to write the equation for the reaction involved is the equation in Example 14.1. The E°_{cell} was calculated in Example 14.1.

$$Cu + 2Ag^+ \rightleftharpoons Cu^{2+} + 2Ag \quad E^\circ_{cell} = +0.4577 \text{ V}$$

The reaction could also be written as the reverse of this one—the way we write it is arbitrary since the sign of the E tells us what direction the reaction proceeds. Using the equation above, the Nernst equation is the following. Notice that there are two electrons involved as determined in Step 3 in the solution to Example 14.1.

$$E = E^\circ_{cell} - \frac{0.0592}{n} \log \frac{[Cu^{2+}]}{[Ag^+]^2}$$

Note: As with equilibrium constant problems, concentrations of pure undissolved solids (in this case Ag and Cu) do not appear in the expression.

$$E = +0.4577 - \frac{0.0592}{2} \log \frac{10^{-1}}{[10^{-2}]^2}$$

$$= +0.4577 - \frac{0.0592}{2} \log 10^3$$

$$= +0.4577 - \frac{0.0592}{2} \times 3$$

$$= +0.4577 - 0.0888 = +0.3689 \text{ V}$$

14.5 POTENTIOMETRY

As mentioned previously, electroanalytical techniques that measure or monitor electrode potential utilize the *galvanic* cell concept and come under the general heading of **potentiometry**. Examples include pH electrodes, ion-selective electrodes, and potentiometric titrations, each of which will be described in this section. In these techniques, a pair of electrodes is immersed, the potential (voltage), of one of the electrodes is measured relatively to the other and the concentration of an analyte in the solution into which the electrodes are dipped is determined. One of the immersed electrodes is called the **indicator electrode** and the other is called the **reference electrode**. Often, these two electrodes are housed together in one "probe." Such a probe is called a **combination electrode**.

14.5.1 Reference Electrodes

The measurement of any voltage is a relative measurement and requires an unchanging reference point. For voltage measurements in ordinary electronic circuitry, this reference is usually "ground." Ground is often a wire that is connected to the frame of the electronic unit and also to the third prong in an electrical outlet, which in turn is connected to a rod that is pushed into the earth, hence the name "ground." Thus, an electronics technician measures voltages relative to ground.

In electroanalytical chemistry, the unchanging reference is a half-cell that, at a given temperature, has an unchanging potential. There are two designs for this half-cell that are popular—the saturated calomel electrode and the silver-silver chloride electrode. These are described below.

14.5.1.1 The Saturated Calomel Reference Electrode (SCE)

The saturated calomel reference electrode (SCE) is one such constant-potential electrode. A drawing and two photographs of a typical SCE available commercially is shown in Figure 14.8. It consists of two concentric glass or tubes, each isolated from the other except for a small opening for electrical contact. The outer tube has a porous fiber tip, which acts as the salt bridge to the analyte solution and the other half-cell. A saturated solution of potassium chloride is in the outer tube. The saturation is evidenced by the fact that there is some undissolved KCl present. Within the inner tube is mercury metal and a paste-like material known as calomel. Calomel is made by thoroughly mixing mercury metal (Hg) with mercurous chloride (Hg_2Cl_2), a white solid. When in use, the following half-cell reaction occurs:

$$Hg_2Cl_2 + 2e^- \rightleftharpoons 2 \text{ Hg} + 2 \text{ Cl}^- \tag{14.5}$$

The Nernst equation for this reaction is

$$E = E° - \frac{0.0592}{2} \log[\text{Cl}^-]^2 \tag{14.6}$$

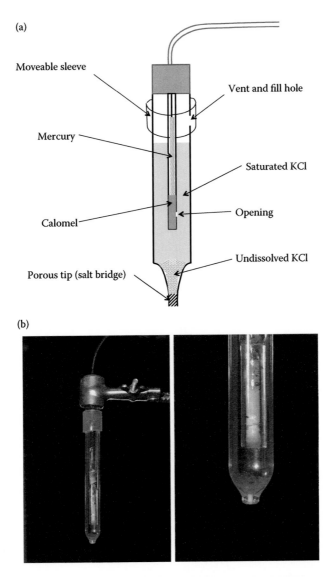

(a)

Moveable sleeve

Vent and fill hole

Mercury

Saturated KCl

Opening

Calomel

Undissolved KCl

Porous tip (salt bridge)

(b)

FIGURE 14.8   (a) Drawing of a commercial saturated calomel electrode. (b) Photographs of a commercial saturated calomel electrode.

or

$$E = E° - 0.0592 \log[Cl^-] \qquad (14.7)$$

Obviously, the only variable on which the potential depends is $[Cl^-]$. The saturated KCl present provides the $[Cl^-]$ for the reaction, and since it is a saturated solution, $[Cl^-]$ is a constant at a given temperature represented by the solubility of KCl at that temperature. If $[Cl^-]$ is constant, the potential of this half-cell, dependent only on the $[Cl^-]$, is therefore also a constant. As long as the KCl is kept saturated and the temperature kept constant, the SCE is useful as a reference against which all other potential measurements can be made. Its standard reduction potential at 25°C (see Table 14.1) is +0.2412 V.

The SCE is dipped it into the analyte solution along with the indicator electrode. A voltmeter is then externally connected across the lead wires leading to the two electrodes and the potential of the indicator electrode vs. that of the SCE is measured.

14.5.1.2 The Silver–Silver Chloride Electrode

The commercial silver–silver chloride electrode is similar to the SCE in that it is enclosed in glass, has nearly the same size and shape, and has a porous fiber tip for contact with the external solution. Internally, however, it is different. There is only one glass tube (unless it is a "double-junction" design; see Section 14.5.3) and a solution saturated in silver chloride and potassium chloride is inside. A silver wire coated at the end with a silver chloride "paste" extends into this solution from the external lead (see Figure 14.9). The half-reaction that occurs is

$$AgCl(s) + 1\ e^- \rightleftharpoons Ag(s) + Cl^- \tag{14.8}$$

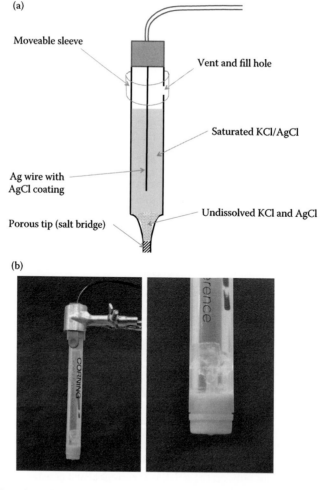

FIGURE 14.9 (a) Drawing of a commercial silver–silver chloride reference electrode. (b) Photographs of a commercial silver/silver chloride electrode.

and the Nernst equation for this is

$$E = E° - \frac{0.0592}{1} \log[Cl^-] \tag{14.9}$$

The standard reduction potential for this half-reaction (from Table 14.1) is +0.22233 V. The potential is only dependent on the $[Cl^-]$, as was the potential of the SCE, and once again the $[Cl^-]$ is constant because the solution is saturated. Thus, this electrode is also appropriate for use as a reference electrode.

14.5.2 INDICATOR ELECTRODES

As stated previously, the reference electrode represents half of the complete system for potentiometric measurements. The other half is the half at which the potential of analytical importance—the potential that is related to the concentration of the analyte—develops. There are a number of such **indicator electrodes** and analytical experiments that are of importance.

14.5.2.1 The pH Electrode

The measurement of pH is very important in many aspects of chemical analysis. Curiously, the measurement is based on the potential of a half-cell, the pH electrode.

The pH electrode consists of a closed-end glass tube that has a very thin fragile glass membrane at the tip. Inside the tube is a saturated solution of silver chloride that has a particular pH. It is typically a 1-M solution of HCl. A silver wire coated with silver chloride is dipped into this solution to just inside the thin membrane. Although this is almost the same design as the silver–silver chloride reference electrode, the presence of the HCl and the fact that the tip is fragile glass and does not have a porous fiber plug points out the difference (see Figure 14.10).

The purpose of the silver–silver chloride combination is to prevent the potential that develops from changing due to possible changes in the interior of the electrode. The potential that develops is a **membrane potential**. Since the glass membrane at the tip is thin, a potential develops due to the fact that the chemical composition inside is different from the chemical composition outside.

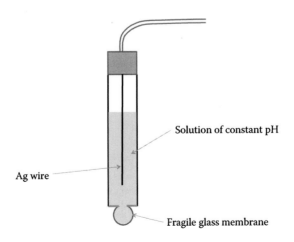

FIGURE 14.10 Drawing of a pH electrode.

Specifically, it is the difference in the concentration of the hydrogen ions on opposite sides of the membrane that causes the potential—the membrane potential—to develop. There is no half-cell reaction involved. The Nernst equation is

$$E = E° - 0.0592 \log \frac{[H^+](\text{internal})}{[H^+](\text{external})} \tag{14.10}$$

or, since the internal [H$^+$] is a constant, it can be combined with E°, which is also a constant, giving a modified E°, E*, and eliminating [H$^+$](internal):

$$E = E* + 0.0592 \log[H^+](\text{external}) \tag{14.11}$$

In addition, we can recognize that pH = −log[H$^+$] and substitute this into the above equation:

$$E = E* - 0.0592 \text{ pH} \tag{14.12}$$

The beauty of this electrode is that, as you can see from Equation 14.12, the measured potential (measured against a reference electrode) is thus directly proportional to the pH of the solution into which it is dipped. A specially designed voltmeter, called a pH meter, is used. A pH meter displays the pH directly rather the value of E.

The pH meter is standardized (calibrated) with the use of buffer solutions. Usually, two buffer solutions are used for maximum accuracy. The pH values for these solutions should "bracket" the pH value expected for the sample. For example, if the pH of a sample to be measured is expected to be 9.0, buffers of pH = 7.0 and pH = 10.0 are used. Buffers with pH values of 4.0, 7.0, and 10.0 are available commercially specifically for pH meter standardization. Alternatively, of course, homemade buffer solutions (see Chapter 5) may be used. In either case, when the pH electrode and reference electrode are immersed in the buffer solution being measured and the electrode leads are connected to the pH meter, the meter reading is electronically adjusted (refer to manufacturer's literature for specifics), to read the pH of this solution. The electrodes can then be immersed into the solution being tested and the pH directly determined.

14.5.3 Combination Electrodes

14.5.3.1 The Combination pH Electrode

In order to use the pH electrode described above, two half-cells (portable electrodes, or "probes") are needed—the pH electrode itself and a reference electrode, either the SCE or the silver–silver chloride electrode, and two connections are made to the pH meter. An alternative is **combination pH electrode**. This electrode incorporates both the reference probe and pH probe into a single probe and is usually made of epoxy plastic. It is by far the most popular electrode today for measuring pH. The reference portion is a silver–silver chloride reference. A drawing of the combination pH electrode is given in Figure 14.11a and a photograph in Figure 14.11b. The fragile glass membrane is protected with plastic as shown.

The pH electrode is found in the center of the probe as shown. It is identical to the pH electrode described above—a silver wire coated with silver chloride immersed in a solution saturated with silver chloride and having a [H$^+$] of 1.0 M. This solution is in contact with a thin glass membrane at the tip. The reference electrode is in an outer tube concentric with the inner pH electrode. It has a silver wire coated with silver chloride in contact with a solution saturated with silver chloride and potassium chloride. A porous fiber strand serves as the salt bridge to the outer tube with the solution tested. Figure 14.12 shows several photographs of the porous fiber strands on various combination probes. Notice the epoxy plastic surrounding the thin glass membranes in these photographs. The

(a)

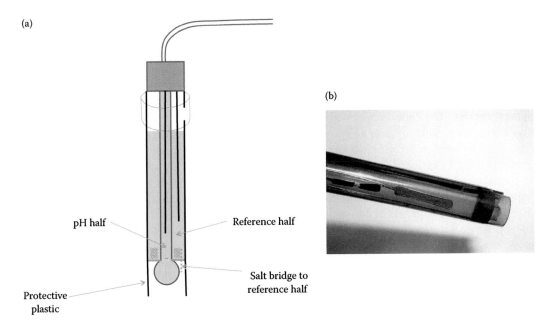

(b)

FIGURE 14.11 (a) Drawing of a commercial combination pH electrode. (b) Photograph of a commercial combination pH electrode. In the photograph, notice the silver electrodes in both the inner and outer tubes.

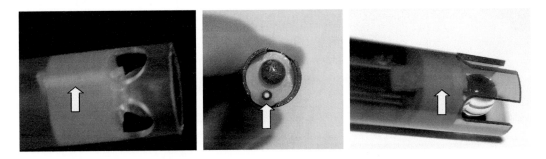

FIGURE 14.12 Photographs showing the porous fiber strands on various combination pH probes. The arrow points to the strand in each case. Also, notice the glass membranes and the protective plastic.

external end of the porous fiber strand must be in full contact with the solution being tested when the measurement is made.

14.5.3.2 Ion-Selective Electrodes

The concept of the pH electrode has been extended to include other ions as well. Considerable research has gone into the development of these **ion-selective electrodes** over the years, especially in studying the composition of the membrane that separates the internal solution from the analyte solution. The internal solution must contain a constant concentration of the analyte ion, as with the pH electrode. Today we utilize electrodes with (1) glass membranes of varying compositions, (2) crystalline membranes, (3) liquid membranes, and (4) gas-permeable membranes. In each case, the interior of the electrode has a silver–silver chloride wire immersed in a solution of the analyte ion.

Examples of electrodes that utilize a glass membrane are those for lithium ions, sodium ions, potassium ions and silver ions. Varying percentages of Al_2O_3, SiO_2, along with oxides of the metal analyte are often found in the membrane as well as other metal oxides. Selectivity and sensitivity of these electrodes vary.

With crystalline membranes, the membrane material is most often an insoluble ionic crystal cut to a round, flat shape, and having a thickness of 1 or 2 mm and a diameter of about 10 mm. This flat "disc" is mounted into the end of a Teflon or PVC tube. The most important of the electrodes with crystalline membranes is the fluoride electrode. The membrane material for this electrode is lanthanum fluoride. The fluoride electrode is capable of accurately sensing fluoride ion concentrations over a broad range and to levels as low as 10^{-6} M (see Figure 14.13). Other electrodes that utilize a crystalline membrane but with less impressive success records are chloride, bromide, iodide, cyanide, and sulfide electrodes. The main difficulty with these is problems with interferences.

Liquid membrane electrodes utilize porous polymer materials, such as PVC or other plastics. An organic liquid ion-exchanger immiscible with water contacts and saturates the membrane from a reservoir around the outside of the tube containing the water solution of the analyte and the silver–silver chloride wire. Important electrodes with this design are the calcium and nitrate ion-selective electrodes.

Finally, gas-permeable membranes are used in electrodes that are useful for dissolved gases, such as ammonia, carbon dioxide, and hydrogen cyanide. These membranes are permeated by the dissolved gases, but not by solvents or ionic solutes. Inside the electrode is a solution containing the reference wire as well as a pH probe, the latter positioned so as to create a thin liquid film between the glass membrane of the pH probe and the gas-permeable membrane. As the gases diffuse in, the pH of the solution constituting the thin film changes and thus the response of the pH electrode changes proportionally to the amount of gas diffusing in.

Calibration of ion-selective electrodes for use in quantitative analysis is usually done by preparing a series of standards as in most other instrumental analysis methods (see Chapter 6), since the measured potential is proportional to the logarithm of the concentration. The relationship is

$$E = E^* - \frac{0.0592}{z} \log[\text{Ion}] \tag{14.13}$$

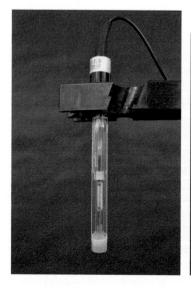 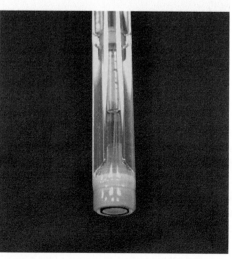

FIGURE 14.13   Photograph of a fluoride ion-selective electrode.

APPLICATION NOTE 14.1: THE DETERMINATION OF NITRATE IN SOILS

Soil can be analyzed for nitrate using a nitrate ion-selective electrode. A biological inhibitor is needed and is prepared by dissolving 0.1 g of phenylmercuric acetate in 20 mL of dioxane. This solution is then diluted to 100 mL with water. Soil samples are air-dried and 50-g portions of each are mixed with 100 mL of water containing 1 mL of the inhibitor. This mixture is homogenized, let stand for 1 h, and then filtered. The filtrate is the sample solution that it tested with the electrode.

Standards are prepared by diluting a 1000-ppm NO_3^- stock solution to make concentrations between 1 and 1000 ppm, such that the expected sample solution concentrations are bracketed. When all standards and sample solutions are ready, 1 mL of 2 M NH_4SO_4 is added to 50 mL of each and each of these solutions is then tested with the electrode. The standard curve is subsequently plotted and the sample solution concentrations calculated. To calculate the concentration of nitrate in the solid soil samples, the solution concentrations are each multiplied by 2.

(*Reference*: http://www.nico2000.net/analytical/nitrate.htm.)

in which z is the signed charge on the ion. The analyst can measure the potential of the electrode immersed in each of the standards and the sample (vs. the SCE or silver–silver chloride reference). The standard curve is a plot E vs. the −log[Ion]. Experiment 45 in this chapter is an analysis with a fluoride ion-selective electrode (see also Application Note 14.1).

14.5.4 OTHER DETAILS OF ELECTRODE DESIGN

The electrodes described in this section (Section 14.6) are commercially available. The body of these electrodes may be either glass or epoxy plastic, as we have discussed. Epoxy plastic electrodes are unbreakable. Some electrodes are **gel-filled electrodes** and are sealed. This means that the KCl solution has a gelatin mixed with it. There is no vent hole and they cannot be refilled with saturated KCl and solid KCl cannot be added.

Some electrodes are **double-junction electrodes**. Such electrodes are encased in another glass tube and therefore have two "junctions" or porous plugs. The purpose of such a design is to prevent contamination—the contamination of the electrode solution with the analyte solution, the contamination of the analyte solution with the electrode solution, or both, by the diffusion of either solution through the porous tip or plug (see the next section for other tips concerning these problems).

14.5.5 CARE AND MAINTENANCE OF ELECTRODES

Although the SCE or Ag/AgCl electrode is dipped into the solution, there will be a slight leakage of the solutions through the porous tip. In order for these electrodes to be used accurately, the measurement must not be adversely affected by the slight contamination from these ions. It is a good idea to slide the moveable sleeve (Figure 14.8, for example) downward so that the outer tube is vented while the electrode is in use so that the ions do indeed freely diffuse through the porous tip. Also, the electrode, under these circumstances, should not be immersed into the solution so deep that the level of solution in the external tube is lower than the level of the solution tested. This would cause the solution to diffuse into the reference electrode rather than the reverse, and thus would contaminate the solution inside and possibly damage the electrode.

The vent hole may also be used prior to the experiment to refill the outer tube with more fill solution (often supplied by the manufacturer) as this solution is lost with time or is sometimes drained out for long-term storage or cleaning. In addition, if the undissolved KCl in the SCE disappears, more solid KCl can be added through the vent hole.

Proper storage of electrodes is important. For short-term storage (up to 1 week), the tip of the electrode should be immersed in a solution to a level above the porous plug. Some manufacturers supply a solution to be used for this. Others recommend a particular solution, such as a pH 7 buffer solution. Other details for such storage solutions may be given in the manufacturer's literature. In any case, distilled water should *not* be used. For long-term storage (longer than 1 week), the protective plastic sleeve provided with the electrode should be placed back on the tip and a small bit of cotton moistened with the storage solution placed inside at the tip of the electrode. The manufacturer's literature usually contains the full recommendations for storage.

14.5.6 POTENTIOMETRIC TITRATIONS

It is possible to monitor the course of a titration using potentiometric measurements. The pH electrode, for example, is appropriate for monitoring an acid–base titration and determining an end point in lieu of an indicator, as in Experiment 11 in Chapter 5. The procedure has been called a **potentiometric titration** and the experimental setup is shown in Figure 14.14. The end point occurs when the measured pH undergoes a sharp change—when all the acid or base in the titration vessel is reacted. The same procedure can be used for any ion for which an ion-selective electrode has been fabricated and for which there exists an appropriate titrant (see, for example, Application Note 14.2).

In addition, potentiometric titration methods exist in which an electrode other than an ion-selective electrode is used. A simple platinum wire surface can be used as the indicator electrode when an oxidation–reduction reaction occurs in the titration vessel. An example is the reaction of Ce(IV) with Fe(II):

$$Ce^{4+} + Fe^{2+} \rightleftharpoons Ce^{3+} + Fe^{3+}$$

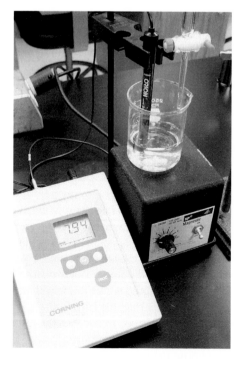

FIGURE 14.14   Setup for a potentiometric titration.

**APPLICATION NOTE 14.2: THE DETERMINATION
OF SALT IN ANIMAL FEEDS**

The chemical makeup of dry animal feed should be reflected on the labels found on the feed bags. Analytical technicians in an agriculture laboratory often perform the analyses involved. An example is the salt (sodium chloride) content in the feed. A common method for analyzing the feed for salt is a potentiometric titration using a chloride ion-selective electrode. After dissolving the feed sample, the chloride is titrated with a silver nitrate standard solution. The reaction involves the formation of the insoluble precipitate silver chloride:

$$Ag^+ + Cl^- \rightarrow AgCl$$

The electrode monitors the concentration of the chloride as the titration proceeds, ultimately detecting the end point when the chloride concentration is zero and the potential of the electrode remains constant.

If this reaction were to set be up as a titration, with Ce^{4+} as the titrant and the Fe^{2+} in the titration vessel and the potential of a platinum electrode dipped into the solution monitored (vs. a reference electrode) as the titrant is added, the potential would change with the volume of titrant added. This is because as the titrant is added, the measured E would change as the $[Fe^{2+}]$ is decreased, the $[Fe^{3+}]$ is increased, and the $[Ce^{3+}]$ is increased. At the end point and beyond, all the Fe^{2+} is consumed and the $[Fe^{3+}]$ and $[Ce^{3+}]$ change only by dilution, and thus, the E is dependent mostly on the change in the $[Ce^{4+}]$. At the end point, there would be a sharp change in the measured E.

Automatic titrators have been invented, which are based on these principles. A sharp change in a measured potential can be used as an electrical signal to activate a solenoid and stop a titration.

14.6 VOLTAMMETRY AND AMPEROMETRY

Electroanalytical techniques that measure or monitor the current flow between a pair of electrodes utilize the *electrolytic* cell concept. Such techniques were first mentioned in Chapter 12 (Section 12.6.5) in describing the amperometric detector for HPLC. An electrolytic cell, you will recall from Section 14.1, operates with a power source in the electrode circuit to force a current to flow through the electrode system. The voltammetric and amperometric techniques usually utilize three electrodes, rather than two. The three electrodes are the **working electrode**, the **auxiliary electrode** (sometimes called the **counter electrode**), and a **reference electrode** (such as those described in Section 14.5). A special electronic circuit for carefully controlling the power that the system receives utilizes the three-electrode system. The circuit applies a small, carefully controlled electrical polarization (an **"applied" potential**) to the working electrode and the resulting current is due to the analyte being either reduced or oxidized at this electrode. This current is proportional to the concentration of the analyte in the solution. **Voltammetric methods** measure the current that flows as the applied potential is varied. **Amperometric methods** measure the current that flows as a result of a *constant* applied potential—a potential that causes the desired electrode reaction to take place—while stirring the solution or while causing the solution to flow across the electrode.

14.6.1 VOLTAMMETRY

One classic design of a working electrode is the so-called **dropping mercury electrode** in which small drops of mercury dropping from the end of a capillary tube constitute the working electrode. This is a good electrode for this kind of work because the surface is continuously renewed while the measurement is made and the solution is also automatically stirred as the drops fall. A continuously renewed surface is helpful because the surface may become fouled from the products of the reduction or oxidation forming there. The voltammetric technique utilizing this electrode is called **polarography**. There are several variations of the basic polarographic method, but these are beyond our scope.

If a stationary electrode is used, such as platinum, gold, or glassy carbon, the technique is called **voltammetry**. One useful voltammetric technique is called **stripping voltammetry** in which the product of a reduction is deposited on the surface on purpose and then "stripped off" by an oxidizing potential—a potential at which the oxidation of the previously deposited material occurs. This technique can also use a mercury electrode, but one that is held stationary (i.e., not dropping).

14.6.2 AMPEROMETRY

We have already briefly described a popular application of amperometry in Chapter 12. This was the electrochemical detector used in HPLC methods. In this application, the eluting mobile phase flows across the working electrode embedded in the wall of the detector flow cell. With a constant potential applied to the electrode (one sufficient to cause oxidation or reduction of mixture components), a current is detected when a mixture component elutes. This current translates into the chromatography peak.

The concept of amperometry can also be applied to a titration experiment much like potential measurements were in Section 14.5 (potentiometric titration). Such an experiment is called an **amperometric titration**—a titration in which the end point is detected through the measurement of the current flowing at an electrode.

The polarization of the measuring (working) electrode, which is typically a rotating platinum disk embedded in a Teflon sheath, is held constant at some value at which the analyte reduces or oxidizes. The solution is stirred due to the rotation of the electrode. The resulting current is then measured as the titrant is added. The titrant reacts with the analyte, removing it from the solution, thus decreasing its concentration. The measured current therefore also decreases. When all of the analyte has reacted with the titrant, the decrease will stop and this signals the end point.

14.7   KARL FISCHER TITRATION

The Karl Fischer method is a titration to determine the water content in liquid and solid materials. The method utilizes a rather complex reaction in which the water in a sample is reacted with a solution of iodine, methanol, sulfur dioxide, and an organic base.

$$I_2 + CH_3OH + SO_2 + \text{organic base} + H_2O \rightarrow \text{products} \qquad (14.14)$$

There are two general ways by which the titration can take place. One is the volumetric method, in which the titrant is added to the sample via an automatic titrator. In this case, the titrant is either a mixture of all of the reactants above (a **composite** titrant) or an iodine solution (other components already in the titration vessel). The other is the coulometric method in which iodine is generated electrochemically. The coulometric method differs significantly from the classical titration. In either case, the end point is detected either potentiometrically or amperometrically as described below.

14.7.1   END POINT DETECTION

At the point where the last trace of water in the sample has reacted with the titrant, unreacted iodine appears in the titration vessel. The presence of unreacted iodine signals the end point. Unreacted iodine can be detected visually, since it imparts a dark yellow, or brown, color to most solutions. However, a visual detection scheme is not as reliable as an electrochemical detection scheme that now is in common use. The electrochemical scheme utilizes two platinum wire electrodes and no reference electrode. Either a constant potential or constant current is applied across these electrodes. When unreacted iodine

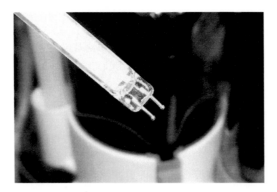

FIGURE 14.15   Photograph of the dual-platinum electrode probe.

appears in the titration vessel, either the current required to sustain the constant potential, or the potential required to sustain the constant current, changes sharply signaling the end point. If the potential is monitored (constant current), it is called a **bipotentiometric method**. If the current is monitored (constant potential), it is called a **biamperometric method**. The prefix "bi-" is used here to distinguish the use of two platinum electrodes and no reference electrode from the use of a single indicator electrode used with a reference electrode, such as in a normal potentiometric or amperometric titration. The two platinum electrode wires are available sealed in a single glass probe as shown in Figure 14.15.

14.7.2   ELIMINATION OF EXTRANEOUS WATER

All extraneous water present in the surrounding air and in the solvents used must be eliminated. Moisture from the air is prevented from entering the system by sealing off the cell from laboratory air. This is accomplished by the use of a rubber septum for sample injection and by the use of water absorbing material in the other openings to the system.

The elimination of water from the solvent used is accomplished by a solvent "conditioning" step in which the moisture in the solvent is actually "titrated" with iodine prior to introducing the sample. Once the solvent moisture is eliminated, the sample can be introduced and the titration begun.

14.7.3   THE VOLUMETRIC METHOD

In the volumetric method, the titrant can be a solution of iodine, methanol, sulfur dioxide, and an organic base as described previously. Such a mixture is commonly known as the **Karl Fischer reagent** and can be purchased from any chemical vendor. It can also be a solution of iodine in methanol solvent. In that case, a **Karl Fischer solvent** containing the other required components is needed for the titration vessel. There are two possibilities for the organic base. One is pyridine, which is the base in the traditional Karl Fischer recipe. The structure of pyridine is similar to that of benzene. The difference is that one of the carbons in the benzene ring is replaced by a nitrogen. Pyridine is toxic, has a disagreeable odor, and does not give the optimum pH for the determination. More recently, imidazole has been used for the base. The structures of pyridine and imidazole are shown in Figure 14.16.

The complete volumetric unit typically consists of a titrant reservoir, the automatic titrator, the titration vessel with dual pin platinum electrode, an automatic stirrer, and an electronics module to run the detection system and display the buret reading and results. The unit may also come equipped with a system for automatically emptying the titration vessel to a waste container and introducing fresh solvent (typically methanol) for the sample. Figure 14.17 shows the complete unit and a close-up view of the titration vessel. In operation, solvent is placed in the titration vessel and, with stirring, the reagent is added so that iodine reacts with the extraneous moisture to condition the solvent. The

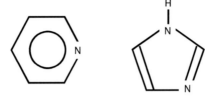

FIGURE 14.16   Structures of pyridine (left) and imidazole (right).

FIGURE 14.17   Photographs of a typical volumetric unit for the Karl Fischer analysis. The complete unit is shown on the left and a close-up of the titration vessel on the right. The dual platinum wire probe is visible in the right photograph as is the tube (dark in color) that introduces the titrant to the solution.

dual pin platinum electrode detects the excess iodine at the point when the extraneous moisture is eliminated and halts the addition of the reagent. The sample is then manually introduced and the process repeats, this time to titrate the moisture from the added sample. When the addition of titrant is again halted by the signal from the dual platinum electrode, the buret reading, and results are displayed.

Like other titrants, this titrant must be standardized. Standardization means that the titer of the titrant must be determined. Titer is the weight of analyte (usually in mg) consumed by 1 mL of titrant. To determine the titer, a known weight of water is introduced and titrated (after conditioning the solvent). The titer is then calculated by dividing the weight (mg) of water used by the milliliters of titrant used to titrate it. The typical volumetric unit calculates the titer automatically. This occurs after operator inputs the weight on a keypad (shown on the right of the complete unit photograph in Figure 14.17) and after the titration takes place. The percentage of water in a sample is then calculated as follows:

$$\%H_2O = \frac{\text{titrant (in mL)} \times \text{titer}}{\text{sample weight (in mg)}} \times 100$$

The volumetric method is used (rather than the coulometric method) when the water content is higher (greater than about 1%).

14.7.4 THE COULOMETRIC METHOD

As stated previously, the iodine titrant is generated electrochemically in the coulometric method. Electrochemical generation refers to the fact that a needed chemical is a product of either the oxidation half-reaction at an anode or the reduction half-reaction at a cathode. In the Karl Fischer coulometric method, iodine is generated at an anode via the oxidation of the iodide ion.

$$2\ I^- \rightarrow I_2 + 2e^-$$

An anode/cathode assembly is required in addition to the dual-pin platinum electrode used for the end point detection. The Karl Fischer reagent in this case contains iodide and no iodine. With the reagent in the cell, the current to the anode/cathode assembly is switched on to generate the iodine needed to eliminate the extraneous moisture. When this moisture is eliminated, unreacted iodine appears and the dual pin platinum electrode switches off the power to the anode/cathode assembly. The sample is then introduced and the current switched on again so that the iodine is again generated. When the dual pin platinum electrode detects unreacted iodine again, the iodine generation is halted again.

The critical data is not in the buret reading, as it was in the case of the volumetric method. Rather, the amount of iodine used is determined "coulometrically" by computing the **coulombs** (total current over time) needed to reach the end point. The coulombs are calculated by multiplying the current applied to the anode/cathode assembly (a constant value) by the total time (sec) required to reach the end point. The modern coulometric titrator computes the amount of moisture automatically from these data and displays it.

EXPERIMENTS

EXPERIMENT 45: DETERMINATION OF THE pH OF SOIL SAMPLES

Note: Safety glasses are required.

1. Obtain soil samples as directed by your instructor. These samples should be thoroughly dried and crushed to very small particles using a mortar and pestle.
2. Weigh 5 g of each soil sample into separate 50-mL beakers. An analytical balance is not necessary for this.
3. Prepare a pH meter and combination pH electrode with a pH = 7 buffer.
4. Perform this step one at a time for each sample. Pipet 5 mL of distilled water into one of the beakers. Swirl vigorously by hand for 5 s. Let stand 10 min. Swirl lightly, dip the pH electrode in, and measure the pH.
5. Repeat step 3 for each sample. During the 10-min waiting period, the next sample(s) may be prepared, but be sure to measure each sample immediately after the 10-min period.

EXPERIMENT 46: ANALYSIS OF A PREPARED UNKNOWN FOR FLUORIDE USING AN ION-SELECTIVE ELECTRODE

1. Prepare 100.0 mL of a 100.0-ppm F^- solution from the available 1000 ppm.
2. Prepare 100.0 mL of a 10.0-ppm F^- solution from the 100.0-ppm solution prepared in Step 1.
3. Prepare 100.0 mL of a 1.00-ppm F^- solution from the 10.0-ppm solution prepared in Step 2.
4. Prepare 100.0 mL of a 0.100-ppm F^- solution from the 1.00-ppm solution prepared in Step 3.

5. Obtain an unknown and dilute to the mark.

6. Pipet 50.00 mL of each of the five solutions (100, 10, 1, and 0.1 ppm and unknown) into 150-mL beakers. Then, pipette 5.00 mL of TISAB into each of these beakers. Stir each well and then measure the potential of a fluoride ion-selective electrode dipped in each.

7. Plot the standard curve (E vs. log [F⁻]) using spreadsheet software and determine the concentration of fluoride in the unknown.

EXPERIMENT 47: DESIGN AN EXPERIMENT: DETERMINATION OF FLUORIDE IN TOOTHPASTE USING AN ION-SELECTIVE ELECTRODE

Redesign Experiment 46 such that the amount of fluoride in a toothpaste is determined rather than in a solution prepared by your instructor. Write a step-by-step procedure as in other experiments in this book for preparing standards, preparing the sample, using the ion-selective electrode to measure the standards and sample, and calculating the percentage of fluoride in the toothpaste. Assume that you will use solid NaF to prepare to prepare a stock standard that is 1000 ppm F. After your instructor approves your procedure, perform the experiment in the laboratory.

EXPERIMENT 48: OPERATION OF METROHM MODEL 701 KARL FISCHER TITRATOR (FOR LIQUID SAMPLES)

1. If not already done, set up titration unit as indicated on page 4 of the Metrohm Model 703 Ti Stand *Instructions for Use* manual and the exchange unit as on pp. 72–73 of the Metrohm Model 701 KF Titrino *Instructions for Use* manual. Place titrant in bottle labeled <TITRANT>. Place methanol in the bottle labeled <SOLVENT>. Make sure both bottle caps are tight. Switch on the instrument.

2. Perform this step if the buret is not already filled. To fill the buret, press the <DOS> button until the piston is in the top end position. Then press the <FILL/STOP> button. Repeat so that all air bubbles are eliminated.

3. Press and hold the "IN" button on the top rear of the 703 Ti Stand in order to pump solvent into the titration vessel. Continue to add solvent until the electrode pins are completely immersed.

4. Press the <start> key (either on the keypad or on the exchange unit). The green <cond> light should blink as titrant is automatically added to the titration vessel to eliminate the residual water in the solvent. (The solvent is being "conditioned.") When this green light stops flashing and the display reads "conditioning," the solvent is conditioned and the cell is ready for titrations.

5. Determine the titer of the titrant as follows. Tare an empty 50-μL syringe on an analytical balance, then fill to the 40-μL line with distilled water and weigh again. This weight is the weight of the water in the syringe. Press <mode> on the keypad a few times until the display reads "Titer with H₂O or std." Press <enter>.

6. Press <Start>. Add the measured water to the titration vessel by piercing the septum in the sample inlet port with the needle of the syringe and pushing the plunger all the way in. Remove the syringe from the sample port. Type in the sample weight on the keypad and press <enter>. The water will now be titrated by the automatic addition of titrant. At the completion of the titration, read and record the titer on the display.

7. Repeat Steps 5 and 6 to obtain a second titer reading. The instrument will average the two readings automatically.

8. Determine the water in a liquid sample as follows. Tare an empty 50 μL syringe on an analytical balance, then fill to the 40 μL line with the sample and weigh again. This weight is the weight of the sample. Press <mode> on the keypad a few times until the display reads "KFT." Press <enter>.

9. Press <Start>. Add the sample to the titration vessel as you did the water in Step 6, typing in the sample weight and pressing <enter> as before. When finished, the percentage of water in the sample should be displayed. Repeat with a second sample if desired.

10. For more determinations, proceed with Step 8. When the titration vessel fills (after several runs), eliminate the solution in the titration vessel by pressing the <OUT> button on the 703 Ti Stand and holding it in. To perform more determinations after that, fresh methanol must be introduced and conditioned (Steps 3 and 4). The titer of the titrant should not change over a short period.

QUESTIONS AND PROBLEMS

Introduction

1. Differentiate between potentiometric methods and voltammetric and amperometric methods.
2. How are oxidation–reduction chemistry and electroanalytical chemistry related?
3. Define *cell, half-cell, anode, cathode, electrolytic cell*, and *galvanic cell*.
4. Distinguish between an electrolytic cell and a galvanic cell.
5. What is a "salt bridge?"
6. Does potentiometry utilize galvanic cells or electrolytic cells? Do voltammetry and amperometry utilize galvanic cells or electrolytic cells?

Transfer Tendencies—Standard Reduction Potentials

7. Define standard reduction potential, $E°$.
8. Without doing any calculations, look at the table of standard reduction potentials and indicate whether or not there would be a reaction between
   (a) Cu and Ni^{2+}
   (b) Hg and Sn^{2+}
   (c) Fe and Ag^+

Determination of Overall Redox Reaction Tendency: $E°_{cell}$

9. What is the $E°_{cell}$ for a cell where the following overall reaction occurs?

$$Ce^{3+} + Cu^{2+} \leftrightharpoons Ce^{4+} + Cu$$

10. Will the reaction in problem 9 proceed spontaneously to the right? Why or why not?
11. A certain voltaic cell is composed of a Ce^{4+}/Ce^{3+} half-cell and a Sn^{4+}/Sn^{2+} half-cell. The overall cell reaction is

$$Ce^{4+} + Sn^{2+} \leftrightharpoons Sn^{4+} + Ce^{3+}$$

   (a) What is $E°_{cell}$?
   (b) Is the reaction spontaneous as written, left to right? Why or why not?
12. What is $E°_{cell}$ for the following redox system and in which direction will the reaction spontaneously proceed? (Assume standard conditions.)

$$Zn + Cu^{2+} \leftrightharpoons Zn^{2+} + Cu$$

13. What is the $E°_{cell}$ for a cell where the following overall reaction occurs?

$$Cu^{2+} + Ni \leftrightharpoons Cu + Ni^{2+}$$

14. What is the E°_{cell} for a cell in which the following overall reaction occurs.

$$Ce^{3+} + Sn^{4+} \rightleftharpoons Ce^{4+} + Sn^{2+}$$

15. Compare question 11 with question 14. How are the answers different and why?
16. A voltaic cell is composed of a copper electrode (Cu) dipping into a solution of copper ions (Cu^{+2}) and a mercury electrode in a solution of Hg^{2+} ions. The cell reaction is

$$Cu + 2\,Hg^{2+} \rightleftharpoons Cu^{+2} + 2\,Hg$$

What is the E° for this cell? Is the reaction spontaneous? Explain your answer.

17. Under standard conditions, what is the E for a cell in which nickel ions react with magnesium metal to form nickel metal and magnesium ions?
18. Assuming standard conditions, what is the E for a cell in which iron metal reacts with manganous ion to form ferrous ion and manganese metal?
19. Using the information given in each row of the table below, calculate the items marked with a question mark (?).

	Reaction	E°_{cell}
(a)	$Ce^{3+} + Ag^+ \rightleftharpoons Ce^{4+} + Ag$	?
(b)	$Hg^{2+} + Na \rightleftharpoons Hg + Na^+$	?
(c)	$Fe^{3+} + Mg \rightleftharpoons Fe^{2+} + Mg^{2+}$	?
(d)	$Fe + Cr^{3+} \rightleftharpoons Fe^{2+} + Cr$	?
(e)	$Cl_2 + Cu \rightleftharpoons 2Cl^- + Cu^{2+}$	?

20. For a–e in question 19, tell whether the reaction will proceed to the right as written and explain your answer for each.

The Nernst Equation

21. What is the Nernst equation and what is its usefulness in electroanalytical chemistry?
22. The E°_{cell} for the following reaction is +0.46 V.

$$2\,Ag^+ + Cu \rightleftharpoons 2\,Ag + Cu^{2+}$$

If $[Ag^+] = 0.010$ M and $[Cu^{2+}] = 0.0010$ M initially, what is the true E for this cell at 25°C?

23. The concentration of Ag^+ ions in a cell is 0.010 M. The concentration of Mg^{2+} ions is 0.0010 M. The cell reaction is

$$Mg + Ag^+ \rightleftharpoons Mg^{2+} + Ag$$

The E° for the cell is +3.17 V. What is the E under the above concentration conditions?

24. What is the E for the Sn^{4+}/Sn^{2+} half-cell if $[Sn^{4+}] = 0.10$ M and $[Sn^{2+}] = 0.0010$ M?
25. What is the E for a Ce^{4+}/Ce^{3+} half-cell if $[Ce^{4+}] = 0.010$ and $[Ce^{3+}] = 0.0010$ M?
26. Using the information given in each row of the table below, calculate the items marked with a question mark (?).

	Half-Cell	[Red]	[Ox]	E
(a)	$Fe^{3+} + 1e^- \leftrightarrows Fe^{+2}$	0.0010 M	0.010 M	?
(b)	$Cr^{3+} + 3e^- \leftrightarrows Cr$	xxxxxx	0.10 M	?
(c)	$Hg^{2+} + 2e^- \leftrightarrows Hg$	xxxxxx	0.00010 M	?
(d)	$Sn^{4+} + 2e^- \leftrightarrows Sn^{2+}$	0.010 M	0.10 M	?
(e)	$I_2 + 2e^- \leftrightarrows 2I^-$	0.0010 M	xxxxxx	?

27. If the E (under standard conditions) for the reaction of zinc ions with iron metal (to give ferric ions) is −0.72 V, what is the E if the zinc ion concentration is 0.010 M and the ferric ion concentration is 0.0010 M?

28. What is the E for a cell in which tin(II) ions react with magnesium metal to give tin metal when the concentrations are the following: magnesium ions, 0.010 M; tin(II) ions, 0.0010 M?

29. What is the E for a $Cr^{3+}/Cr//Ni^{2+}/Ni$ cell if $[Cr^{3+}] = 0.000010$ M and $[Ni^{2+}] = 0.010$ M?

30. What is the E for a $Cu^{2+}/Cu//Zn^{2+}/Zn$ cell if $[Cu^{2+}] = 0.010$ M and $[Zn^{2+}] = 0.0010$ M?

$$Cu + Zn^{2+} \leftrightarrows Cu^{2+} + Zn$$

31. What is the E for a $Sn^{2+}/Sn^{4+}//Zn^{2+}/Zn$ cell if $[Sn^{2+}] = 0.010$ M, $[Sn^{4+}] = 0.00010$ M, and $[Zn^{2+}] = 0.0010$ M?

$$Sn^{2+} + Zn^{2+} \leftrightarrows Sn^{4+} + Zn$$

32. Using the information given in each row of the table below, calculate the items marked with a question mark (?).

	Reaction	E°_{cell}	Concentrations	n	E_{cell}
(a)	$Cu^{2+} + Fe \leftrightarrows Cu + Fe^{3+}$	?	$[Cu^{2+}] = 10^{-3}$ M, $[Fe^{3+}] = 10^{-2}$ M	?	?
(b)	$Hg^{2+} + Ni \leftrightarrows Ni^{2+} + Hg$	?	$[Hg^{2+}] = 10^{-2}$ M, $[Ni^{2+}] = 10^{-4}$ M	?	?
(c)	$Ag + Fe^{3+} \leftrightarrows Ag^+ + Fe$	?	$[Ag^+] = 10^{-1}$ M, $[Fe^{3+}] = 10^{-3}$ M	?	?
(d)	$Cu + Sn^{2+} \leftrightarrows Cu^{2+} + Sn$	?	$[Sn^{2+}] = 10^{-3}$ M, $[Cu^{2+}] = 10^{-4}$ M	?	?
(e)	$Ag^+ + Zn \leftrightarrows Ag + Zn^{2+}$	?	$[Ag^+] = 10^{-2}$ M, $[Zn^{2+}] = 10^{-1}$ M	?	?
(f)	$Fe + Na^+ \leftrightarrows Fe^{3+} + Na$	?	$[Na^+] = 10^{-4}$ M, $[Fe^{3+}] = 10^{-1}$ M	?	?

Potentiometry

33. Define *potentiometry, reference electrode, ground,* and *indicator electrode*.

34. Using the Nernst equation, tell how the SCE electrode works as a reference electrode.

35. Concerning the SCE, what does the fact that the KCl is saturated have to do with its usefulness as a reference electrode?

36. Tell how the concept of the salt bridge is put into practice with commercial reference electrodes.

37. Using the Nernst equation, tell how the silver–silver chloride electrode works as a reference electrode.
38. Using the Nernst equation, tell how the pH electrode works as an electrode for determining the pH of a solution.
39. Concerning the pH electrode, the following defines the potential that develops when it is immersed into a solution.

$$E = E° - 0.0592 \log \frac{[H^+](internal)}{[H^+](external)}$$

Given this, explain how we can then say that this potential is directly proportional to the pH of the solution into which it is immersed.
40. Briefly describe how a pH meter is standardized.
41. What is a combination pH electrode? Describe its construction.
42. Discuss the relationship between the salt bridge concept illustrated in Figure 14.2 and the fiber strands in Figure 14.8.
43. What is an ion-selective electrode? Identify and describe four different types.
44. The following data were obtained using a nitrate electrode for a series of standard solutions of nitrate:

$[NO_3^-]$ (M)	E (mV)
10^{-1}	85
10^{-2}	150
10^{-3}	209
10^{-4}	262

   (a) Plot the calibration curve for this analysis.
   (b) What is the nitrate ion concentration in a solution for which E = 184 mV?
45. What is special about gel-filled electrodes and double-junction electrodes?
46. What are two functions of the vent hole found near the top of commercial electrodes?
47. How should commercial electrodes be stored over the short term? How should they be stored over the long term?
48. Why is a platinum electrode needed in some half-cells?
49. What is a potentiometric titration?

Voltammetry and Amperometry

50. How do voltammetry and amperometry differ from potentiometry?
51. Define *working electrode, reference electrode, counterelectrode,* and *auxiliary electrode.*
52. What are two advantages of the dropping mercury electrode over some stationary electrode?
53. Differentiate between polarography and voltammetry.
54. Explain what is meant by amperometric titration.

Karl Fischer Titration

55. What is the objective of a Karl Fischer titration?
56. What chemicals are required in a Karl Fischer reaction? What is the reaction?

57. Explain the bipotentiometric method for detecting the end point in a Karl Fischer titration. Why is it called a "bipotentiometric" method?

58. In a Karl Fischer experiment, why is the titration vessel sealed off from the laboratory air?

59. What is meant by "conditioning" the solvent in a Karl Fischer experiment?

60. Distinguish between the volumetric and the coulometric methods for the Karl Fischer titration.

61. Explain how a Karl Fischer titrant is standardized in the volumetric method.

62. Define "titer." If 9.38 mg of water required 17.28 mL of Karl Fischer titrant, what is the titer of this titrant?

63. A sample weighing 0.091 g required 29.22 mL of a titrant with a titer of 0.692 mg/mL. What is the percent of water in this sample?

64. Why is an anode/cathode assembly as well as a dual pin platinum electrode needed in the coulometric Karl Fischer method?

65. There is no buret reading in the coulometric Karl Fischer method. How can the results be calculated without a buret reading?

Report

66. For this exercise, your instructor will select a "real-world" electroanalytical method, such an application note from an instrument vendor or from a methods book, a journal article, or Web site, and will give it to you as a handout. Write a report giving the details of the method according to the following scheme. On your paper, write (or type) "a," "b," etc., to clearly present your response to each item.

   (a) Method Title: Give a more descriptive title than what is given on the handout.

   (b) Type of Material Examined: Is it water, soil, food, a pharmaceutical preparation, or what?

   (c) Analyte: Give both the name and the formula of the analyte.

   (d) Sampling Procedure: This refers to how to obtain a sample of the material. If there is no specific information given as to obtaining a sample, make something up. If you want to be correct in what you write, do a library, or other search to discover a reasonable response. Be brief.

   (e) Sample Preparation Procedure: This refers to the steps taken to prepare the sample for the analysis. It does NOT refer to the preparation of standards or any associated solutions.

   (f) Specific Experiment: Is it a titration, a series of standard solutions, ion-selective electrode, Karl Fischer, potentiometric, voltammetric/amperometric, or what? Indicate all of these that apply, or write something not listed.

   (g) What is the titrant and how is it standardized? If it is not a titration, write N/A.

   (h) How is the standard curve (calibration curve) created? If a standard curve is not used, write N/A.

   (i) What is the quantitation procedure? Brief summary.

   (j) Standard Curve: Is a standard curve suggested? If so, what is plotted on the y-axis and what is plotted (give units) on the x-axis?

   (k) How are the Data Gathered? In other words, you have got your blank, your standards, and your samples ready to go. Now what? Be brief.

   (l) Concentration Levels for Standards. This refers to what the concentration range is for the standard curve. Be sure to include the units. If no standard curve is used, state this.

   (m) How the Standards are Prepared? Are you diluting a stock standard? If so, how is the stock prepared and what is its concentration? If not, how are the standards prepared?

(n) Potential Problems. Is there anything mentioned in the method document that might present a problem? Be brief.

(o) Data Handling and Reporting. Once you have the data for the standard curve, then what? Be sure to present any post-run calculations that might be required.

(p) References. Did you look at any other references sources to help answer any of the above? If so, write them here.

15 Miscellaneous Instrumental Techniques

15.1 X-RAY METHODS

15.1.1 INTRODUCTION

X-ray methods include x-ray diffraction, x-ray absorption, and x-ray fluorescence. X-ray diffraction is a technique for determining ultra-small spacings in materials, such as the spacings between the atoms or ions in a crystal structure, or the thickness of a thin electroplated material. An example of the former is in soil laboratories in which the minerals in various soils need to be characterized. X-ray absorption is limited in application, but has been used to determine heavy elements in a matrix of lighter elements, such as determining lead in gasoline. X-ray fluorescence is much more popular and is used to determine elements in a wide variety of solid materials.

X-rays have extremely short wavelengths and high energy. A source of a wide band of x-ray wavelengths is the **x-ray tube** in which x-rays are generated by bombarding a metal anode with high-energy electrons. These electrons are of sufficient energy as to cause inner shell electrons ($n = 1$, or K-shell) to be ejected from the atoms, which means the atoms are ionized. Subsequently, higher energy electrons drop into the vacancies thus created and a sort of domino effect follows with electrons from higher levels dropping into lower level vacancies created when electrons in those levels dropped to lower levels. The x-rays are generated in this process because the loss in energy that occurs when the electrons drop into the vacancies corresponds to the energy of x-rays.

Some applications utilize specific x-ray wavelengths generated by fluorescence (see also Section 15.1.3). Such x-rays are generated by irradiating a given elemental material, such as copper, with the x-rays emitted by the x-ray tube described above. These x-rays are of higher quality because they do not include background radiation. The higher energy x-rays from the x-ray tube are of sufficient energy to accomplish the ejection of electrons from the inner shells of the copper atoms, which in turn results in the domino effect again. However, now the wavelengths emitted constitute a high-quality x-ray line spectrum of copper and very specific x-ray wavelengths can be isolated.

In either the x-ray tube or the fluorescence source, the inner shell electrons involved are usually those in the K shell ($n = 1$), the L shell ($n = 2$), the M shell ($n = 3$), and the N shell ($n = 4$). The absorption-emission domino effect occurs specifically as follows: (1) the complete ejection of an inner shell electron (e.g., K or L shell) from an atom (ionization) occurs creating a vacancy in this inner shell; (2) an electron from an outer shell (e.g., L, M, or N) drops to the vacancy in the inner shell, thus losing energy—the energy of a low-energy x-ray; (3) electrons from outer shells (e.g., M or N) drop to the vacancies in the lower outer shells (e.g., L and M), also losing energy—also the energy of low energy x-rays (see Figure 15.1).

The x-ray emissions are categorized as K, L, M, etc., emissions and as alpha (α) and beta (β) emissions. It is a K emission if the electron drops from any higher level to the K shell. It is an L emission if it drops from any higher level to the L shell, etc. The α emissions are those that involve electrons that drop just one principle level, such as from the L shell to the K shell (K_α emissions) or from the M shell to the L shell (L_α emissions). The β emissions are those in which electrons drop two levels, such as from the M shell to the K shell (K_β emissions) or from the N shell to the L shell (L_β emissions). Emissions are further characterized by the subscript numbers 1, 2, 3, etc. to indicate the sublevel in the higher level (where the electron originated). Thus, we have, for example, $K_{\alpha 1}$, $K_{\alpha 2}$,

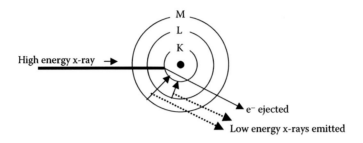

FIGURE 15.1 In x-ray fluorescence, high-energy x-rays are absorbed and low-energy x-rays are emitted, as discussed in the text.

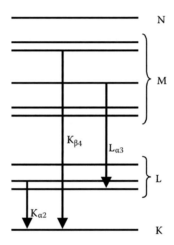

FIGURE 15.2 Energy level diagram showing examples of the different kinds of emissions and terminology.

etc. (see Figure 15.2). Beta emissions are higher-energy emissions (shorter wavelength) than alpha emissions because the energy jump is greater.

15.1.2 X-Ray Diffraction Spectroscopy

When a beam of light is directed at a structure that has very small regularly spaced lines or points in it, the light scatters forming many beams of light having a particular regular pattern that appears to be the original light beam expanded in space. This scattering has come to be known as **diffraction**. The effect occurs when the wavelength of the light and the spacings between the lines or points are equivalent and constructive and destructive interference takes place due to the reflection from the different planes in the structure.

The spacing between ions or atoms in crystal structures are approximately the same distance as the wavelengths of x-rays. These spacings then serve to diffract x-rays such that a particular pattern of diffraction results from shining x-rays on a given atomic or ionic structure. A particular structure can be expected to give a characteristic, reproducible diffraction pattern. The study of crystal structure by x-ray diffraction is an important spectroscopic technique in analytical chemistry.

In 1913, William and Lawrence Bragg, who were father and son, determined how the spacings of layers in crystal structures give different x-ray diffraction patterns. Today, it is possible to observe

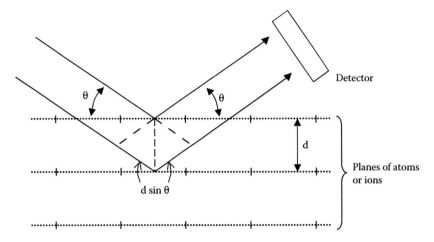

FIGURE 15.3 Illustration of d, θ, and d sinθ in Bragg's law. The distance traveled by the x-ray reflected from the second plane is greater than that reflected from the first plane by 2d sinθ in order for constructive interference to occur and a light intensity to be observed at the detector.

the diffraction pattern derived from a particular crystal and determine the structure of the crystal. The Braggs determined that the following factors are involved.

1. The distance between similar atomic planes in a crystalline structure. This distance is called the "d-spacing."
2. The angle of diffraction, also called the θ angle. A diffractometer measures an angle that is double the θ angle. We refer to it as "2θ."
3. The wavelength of the incident light.

These factors are combined in what has come to be known as **Bragg's law**.

$$n\lambda = 2d \sin\theta \tag{15.1}$$

In this equation, n is an integer (1, 2, 3, ...), the "order" of diffraction, λ is the wavelength in angstroms,[*] d is the interplanar spacing in angstroms, and θ is the diffraction angle in degrees. An illustration of these parameters is given in Figure 15.3.

In order to determine the interplanar spacing, which is critical to characterizing a crystal structure, the wavelength and the diffraction angle must be known (so that d can be calculated from Equation 15.1). The diffraction angle can be obtained from the instrument used (a **diffractometer**). In a diffractometer, in order to measure the angles, either the stage on which the sample is held is rotated or the source or detector is moved through an arc. The latter is illustrated in Figure 15.4. The data obtained are usually displayed in a plot of signal intensity vs. 2θ. A hypothetical example is given in Figure 15.5. Intensity peaks appear at certain 2θ values indicating constructive interference at that particular 2θ value. The pattern is specific to a particular interplanar spacing. Hence, crystal structures and film thicknesses can be determined from the pattern.

The wavelength used is dependent on the element used in the x-ray source. A common element for the source is copper and the wavelength isolated from the copper emission is 1.539 Å.

---

[*] 1 angstrom = 10^{-10} m.

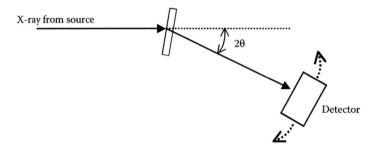

FIGURE 15.4   Illustration of a diffractometer in which the detector is moved through an arc in order to measure the angle. For convenience reasons, 2θ is measured as illustrated.

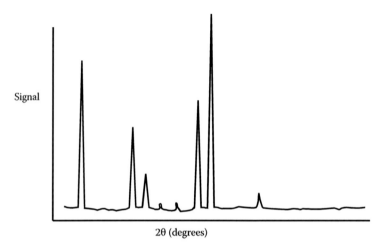

FIGURE 15.5   Hypothetical "diffractogram," or plot of signal intensity vs. 2θ, that results from an x-ray diffraction scan. The peaks represent θ angles at which constructive interference has occurred resulting in a significant x-ray intensity.

Example 15.1

The wavelength used in a particular x-ray diffraction analysis is 1.539 Å. If an intense peak occurs at a 2θ value of 39.38°, what is the d-spacing in this structure?

Solution 15.1

Since the 2θ value is given, we must divide by 2 to obtain θ.

$$\theta = \frac{39.38}{2} = 19.69°$$

Then, using Bragg's law, assuming n = 1, we have the following.

$$n\lambda = 2d \sin\theta$$

$$1.539 = 2d \sin(19.69)$$

$$d = \frac{1.539}{2\sin(19.69)} = 2.318 \text{ Å}$$

15.1.3 X-Ray Fluorescence Spectroscopy

The fluorescence process used in some x-ray sources as described in Section 15.1.1 can also be used as an analytical tool. One can direct either high energy electron beams or x-rays at an unknown sample and perform qualitative and quantitative analysis by making measurements on the lower energy x-ray emissions that occur. Let us first briefly review what we have discussed to this point concerning the concept of fluorescence.

In the past two chapters, we have considered fluorescence in two different contexts, molecular fluorescence and atomic fluorescence. In molecular fluorescence, ultraviolet light is absorbed by molecules or complex ions and outermost electrons are elevated to a higher level, an excited state. These electrons drop back to the ground electronic level, but because of vibrational energy losses in both electronic states, the energy lost (emitted) is less than the energy that was absorbed. (Refer back to Chapter 8, Figure 8.16.) In atomic fluorescence, ultraviolet light is absorbed by atoms, rather than molecules and complex ions, but again outermost electrons are elevated to a higher energy level, an excited state. However, in this case, the elevation is to an excited state higher in energy than one or more other excited states. The energy loss that subsequently occurs involves stops at the intermediate levels (before the final drop to the ground state) such that, once again, lower energy light is emitted (Chapter 9, Section 9.6.5). Thus, fluorescence, whether molecular or atomic, is the emission of light of a lower energy than the light that was absorbed due to outermost electrons losing less energy in returning to lower levels than they had gained. Thus, the x-ray emission process described in Section 15.1.1 is fluorescence too because the emissions are of lower energy then the absorbed energy. The emissions represent energy drops from the higher inner electron shells to lower ones following the absorption of the very high energy used to eject the K shell electron that starts the process.

X-ray fluorescence is a type of atomic spectroscopy since the energy transitions occur in atoms. However, it is distinguished from other atomic techniques in that it is nondestructive. Samples are not dissolved. They are analyzed as solids or liquids. If the sample is a solid material in the first place, it only needs to be polished well or pressed into a pellet with a smooth surface. If it is a liquid or a solution, it is often cast on the surface of a solid substrate. If it is a gas, it is drawn through a filter that captures the solid particulates and the filter is then tested. In any case, the solid or liquid material is positioned in the fluorescence spectrometer in such a way that the x-rays impinge on a sample surface and the emissions are measured. The fluorescence occurs on the surface and emissions originating from this surface are measured.

There are two different instrument designs. These are the **energy-dispersive system** and the **wavelength-dispersive system**. In the energy-dispersive system, x-rays from the source impinge on the surface of the sample, the fluorescence occurs and the intensity is measured by a detector. In the wavelength-dispersive system, x-rays impinge on the surface of the sample, the fluorescence occurs and these emissions are further dispersed by a diffraction crystal (known as the "analyzer crystal") of known interplanar spacing such that intensities at the 2θ angles of diffraction are measured (see Section 15.1.2). Illustrations of these two designs are shown in Figure 15.6.

Each of the two designs offers advantages. The energy-dispersive system measures all elements simultaneously, meaning that the analysis is fast. However, the resolution is not optimum since the emissions do not undergo the dispersion by the analyzer crystal. Also, the energy-dispersive analysis is more sensitive. The advantage of the wavelength-dispersive system is therefore higher resolution, but the analysis is slower and less sensitive.

15.1.4 Applications

Elemental qualitative analysis is a popular application of x-ray fluorescence spectroscopy. The values of the wavelengths reaching the detector are indicative of what elements are present in the sample. This is so because the inner shell transitions giving rise to the wavelengths are specific to

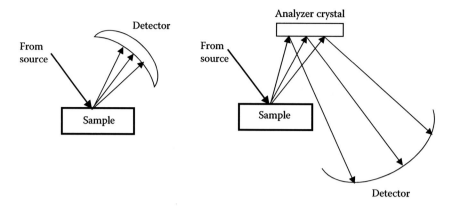

FIGURE 15.6 Two instrument designs for x-ray fluorescence spectroscopy. (Left) Energy-dispersive system. (Right) Wavelength-dispersive system.

the element. Qualitative analysis applications include detection of impurity metals in alloys, electroplated materials, and other products from industrial manufacturing processes. In addition, the detection and characterization of minerals in geological and soil samples by x-ray fluorescence is commonplace. Also important are the analyses of corrosion products for the purpose of identifying elements that may pinpoint the cause of the corrosion (and failure) of construction materials, and the analysis of rare or archeological objects for which a nondestructive method is important.

Quantitative analysis is possible by measuring the intensity of the x-ray emissions. Thus, if quantitative analysis is important in all the above cited qualitative applications, for example, then this intensity is measured and compared with standards.

15.1.5 Safety Issues Concerning X-Rays

The need for safety consciousness around x-ray equipment is well known. In health and dental care facilities, rooms in which x-ray equipment is used are shielded so that x-ray technicians need only to step out of the room behind a shielded wall in order to be protected. Patients are shielded with lead aprons.

Analytical laboratories housing x-ray instrumentation must follow strict safety rules. X-ray equipment must be constructed so as to minimize radiation leaks. Operators of the equipment must be properly trained and the training documented and updated. The person designated as the safety officer must be aware of the presence of the equipment and must follow state and federal guidelines relating to registration of the equipment, proper labeling of equipment, room, and hallway postings, training of employees, and communication of health risks, such as to pregnant employees. The equipment must be tested periodically to check the operation of safety devices and to check for radiation leaks.

15.2 NUCLEAR MAGNETIC RESONANCE SPECTROSCOPY

15.2.1 Introduction

Up to now, we have studied many instrumental analysis techniques that are based on the phenomenon of light absorption. We have discussed atomic and molecular techniques utilizing light in the ultraviolet and visible regions of the spectrum and involving electronic energy transitions—the elevation of electrons to higher energy states. We have also discussed techniques utilizing light in the infrared region involving molecular vibrational and rotational energy transitions. In the current chapter, we have looked at techniques utilizing the high energy of x-rays capable of exciting inner shell electrons in atoms. We now introduce the concepts of **nuclear magnetic resonance**

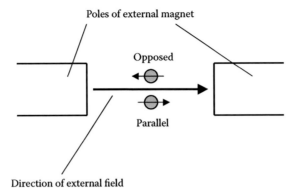

Poles of external magnet

Opposed

Parallel

Direction of external field

FIGURE 15.7 Illustration of the magnetic fields of spinning/precessing ¹H nuclei aligned parallel to or opposed to an external magnetic field. The opposed alignment is shown above the parallel alignment in order to represent it as a higher energy state.

spectrometry (NMR). This technique utilizes light in the very low-energy radio wave region of the spectrum and involves nuclear spin energy transitions that occur in a magnetic field.

The technique is based on the theory and experimental evidence that atomic nuclei with an odd number of either protons or neutrons behave as small magnets by virtue of the fact that they are spinning and precessing like a top or gyroscope. Since a nucleus is positively charged, the spinning/precession motion creates a small magnetic field around it. The nucleus most often studied and measured is the nucleus of the hydrogen atom, symbolized ¹H. If we were to position a compound with ¹H nuclei in its structure (true of the vast majority of organic compounds) in an external magnetic field, i.e., between the poles of a magnet, the magnetic fields of the spinning/precessing ¹H nuclei, representing smaller magnetic fields, will align themselves to the external field. They can align both parallel to the external field, or opposed to the external field. The opposed alignment represents a slightly higher energy state and so there are fewer nuclei in this state at a given point in time (see Figure 15.7). The two states are in resonance, meaning that they continually "flip" from one state to the other and back.

The energy difference between the two different alignments is on the order of approximately 6.6×10^{-26} joules. The frequency corresponding to this energy is 1×10^8 s⁻¹ Hz, or 100 MHz, which is a radio wave frequency (RF). Thus, with more nuclei in the lower energy state at any given moment, RF light, or light in the RF region of the electromagnetic spectrum, can be absorbed by the molecules in a magnetic field so as to cause the nuclear spin energy transition and more resonance, hence the name nuclear magnetic resonance (see Figure 15.8). Although this phenomenon applies to nuclei of certain other elements, NMR has found its most useful application in the measurement of the ¹H nucleus. For this reason, it is sometimes also referred to as proton magnetic resonance (PMR). The application lies mostly in the determination of the structure of organic compounds, and it is a powerful qualitative analysis tool for such compounds, as we shall see.

15.2.2 INSTRUMENTATION

A unique situation exists with NMR because the transition from one spin state to the other depends on both the energy of the RF and the strength of the external magnetic field. The greater the field strength, the larger the energy spread between the two energy levels, as shown in Figure 15.9, and therefore the greater RF required. Thus, one could say that the precise RF that will be absorbed depends on the strength of the magnetic field used. But since the strength of the magnetic field is variable, one can also say that the strength of the magnetic field required depends on the RF to be used. In the traditional instrument design, the RF is constant and the external magnetic field is

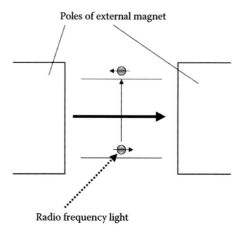

FIGURE 15.8   Those ^{1}H nuclei with a magnetic field parallel to the external field can absorb RF light and be promoted to the higher energy state in which their magnetic field is opposed to the external field.

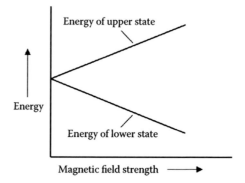

FIGURE 15.9   Energy difference between the two spin states depends on the magnetic field strength.

varied in order to generate the NMR spectrum. Fourier Transform instruments (FTNMR), which are similar in principle to FTIR instruments, are popular today. We will briefly describe this instrument later in this section.

The available NMR instruments are classified according to the RF employed. Instruments can utilize different RF. Examples are the 60- and 100-MHz instruments. Different frequencies require different field strengths. A 60-MHz instrument requires a field strength of 14,092 gauss (gauss is a unit of field strength), whereas a 100-MHz instrument requires 23,486 gauss, for example. In the traditional instruments, the field strength is varied over a very narrow range.

The basic components of the traditional NMR instrument therefore include a large-capacity superconducting magnet, a sample probe that includes the sample tube, an RF transmitter, and an RF receiver/detector (see Figure 15.10). The RF transmitter emits the RF, the RF receiver/detector receives the frequency and detects the absorption and a data system (computer) acquires the data and displays the NMR spectrum. The sample is held in a 5-mm OD glass tube containing less than half a milliliter of liquid, which in turn is held in a fixture called the "sample probe." This tube rotates rapidly in the sample probe as the spectrum is being measured. This rotation ensures that all nuclei in the sample are affected equally by the applied magnetic field.

As stated above, the modern NMR instrument uses Fourier transform technology similar to that of the FTIR instrument described in Chapter 8. It is therefore accurately described as Fourier transform nuclear magnetic resonance (FTNMR). In FTNMR instruments, the design is identical to the

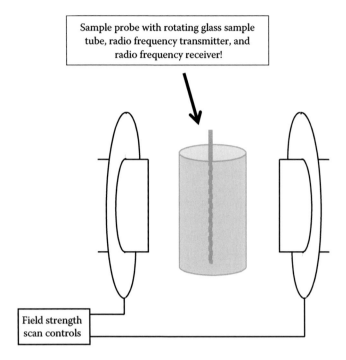

Sample probe with rotating glass sample tube, radio frequency transmitter, and radio frequency receiver!

Field strength scan controls

FIGURE 15.10 Illustration of an NMR spectrometer as described in the text.

older instruments, except that very brief repeating pulses of RF energy are applied to the sample. As with FTIR, the resulting detector signal contains the absorption information for all transitions occurring in the sample, which in this case are all the nuclear energy transitions of all ^{1}H nuclei. The Fourier transformation, performed by the computer, then sorts out and plots the absorption as a function of frequency, giving the same result, the NMR spectrum, as the traditional instruments but in much less time. In addition, it is more sensitive and gives better resolution.

15.2.3 THE NMR SPECTRUM

Absorption of various RF occurs at a constant field strength. Also, absorption of a particular RF occurs at various field strengths. The traditional NMR spectrum is a plot of the latter—absorption vs. field strength.

Intuitively, one may expect all hydrogen nuclei to absorb a particular RF at a particular field strength, reflecting an assumption that all hydrogen nuclei are identical. If this were true, the proton NMR spectrum of all compounds would be identical—hardly useful for any analytical characterization. Fortunately, this is not the case.

15.2.3.1 Chemical Shifts

Electrons, as well as nuclei (other than ^{1}H) with an odd number of protons or neutrons, are, of course, present around and near the hydrogen nuclei in a molecule, and these are generating individual magnetic fields too. These very small fields change the value of the applied field slightly giving an effective applied field somewhat smaller than expected and therefore a slightly shifted absorption pattern for the ^{1}H nuclei, the so-called chemical shift. If all hydrogen nuclei in a molecule were "shielded" by electrons and neighboring nuclei equally, we would still have the situation of just one major absorption peak. Due to the different environments surrounding the different hydrogens in an organic molecule, however, we find major absorption peaks at field strengths representing each of these environments.

In other words, all hydrogens with a given environment give one characteristic major absorption peak at a specific field value. Methyl alcohol, CH_3OH, for example, would give two major peaks, one representing the methyl hydrogens (all three hydrogens have the same environment) and one representing the hydroxyl hydrogen (a different environment from the other three and therefore not equivalent), while cyclohexane, which has 12 equivalent hydrogens, gives just one major peak. The importance of this information is that (1) from the number of different major peaks, we can tell how many different kinds of hydrogen there are, and (2) from the amount of shielding shown, we can determine the structure near each hydrogen.

Example 15.2

How many major peaks can be expected in the NMR spectrum of (a) ethyl alcohol, (b) *n*-propyl alcohol, (c) isopropyl alcohol, (d) acetone, and (e) benzene?

Solution 15.2

(a) In ethyl alcohol, CH_3CH_2OH, the three CH_3- hydrogens are equivalent and different from the others; the two $-CH_2-$ hydrogens equivalent and different from the others; and the hydroxyl hydrogen is different from the others. Therefore, there would be three major peaks.

(b) In *n*-propyl alcohol, $CH_3CH_2CH_2OH$, there are four different kinds of hydrogen, therefore four major peaks.

(c) In isopropyl alcohol,

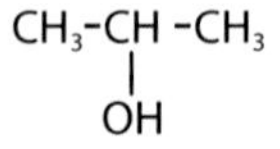

$$CH_3-CH-CH_3$$
$$|$$
$$OH$$

there are only three different kinds of hydrogen (the six hydrogens on the two methyl groups are equivalent) so only three major peaks.

(d) In acetone,

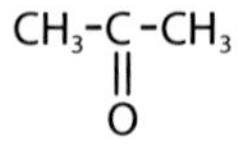

$$CH_3-C-CH_3$$
$$||$$
$$O$$

there is only one kind of hydrogen, so only one major peak.

(e) In benzene,

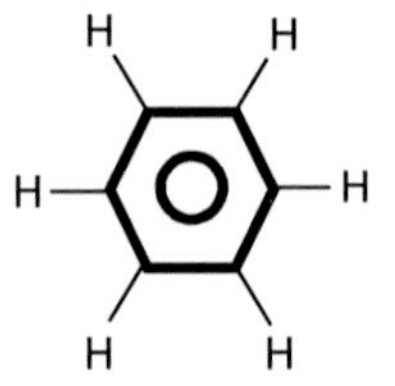

there are six hydrogens, all equivalent, so only one major peak.

Since the shifts can be extremely slight (on the order of parts per million of the external field strength), the actual field strength of the peak is difficult to measure precisely. For this reason, a reference compound, typically tetramethylsilane, TMS, is added to the sample. All the hydrogens in the TMS structure are equivalent (Figure 15.11). It is then convenient to modify the x-axis to show the difference between the single TMS peak and the compound's peaks. The x-axis thus typically

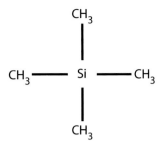

FIGURE 15.11 Structure of tetramethylsilane, TMS. All the hydrogens are equivalent, giving one NMR peak, which is used as a reference peak.

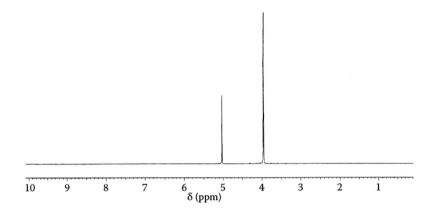

FIGURE 15.12 NMR spectrum of methyl alcohol.

represents a "difference factor" symbolized by the Greek letter, lower case delta (δ), in which the single TMS peak has a 0.0-ppm field strength difference and other peaks then are a certain parts per million away from the TMS peak. Figure 15.12 shows such an NMR spectrum for methyl alcohol.

15.2.3.2 Peak Splitting and Integration

Additional qualitative information is possible with NMR spectra. First, a given major absorption peak, while apparently arising from one particular kind of hydrogen, may appear to be split into two or more peaks, giving what are termed "doublets," "triplets," etc., a phenomenon called **spin–spin splitting**. The splitting is due to the effect that ^{1}H nuclei on adjacent carbons have on the nucleus in question. Splitting occurs because the magnetic fields that these different hydrogens generate become "magnetically coupled" to each other. In general, the number of peaks resulting from the splitting is predicted according to a rule called the **n + 1 rule**, which states that if there are "n" equivalent hydrogens on the adjacent carbon, then the number of peaks that appear will be n + 1. So, for example, if there are no hydrogens on the adjacent carbon, then 0 + 1, or only one, peak appears. If there is one hydrogen on the adjacent carbon, then 1 + 1, or two, peaks will appear as a result of the splitting. If two equivalent hydrogens, then three peaks, etc. For example, Figure 15.13 is the NMR spectrum of ethylbenzene. The triplet at a δ range of approximately 1.2 to 1.3 is due to the three hydrogens on the $-CH_3$ half of the ethyl group. It is a triplet because the adjacent carbon has two equivalent hydrogens and n + 1 = 3. The quartet at a δ range of approximately 2.6 to 2.7 is due to the $-CH_2-$ half of the ethyl group. It is a quartet because the adjacent carbon has three

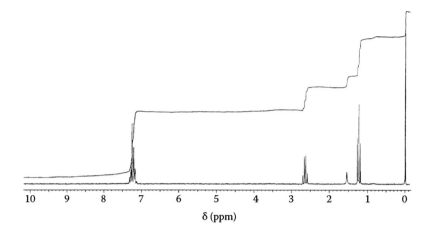

FIGURE 15.13 NMR spectrum of ethylbenzene with integration trace.

equivalent hydrogens and n + 1 = 4. An important point to remember is that if all the hydrogens on both carbons are equivalent, then there will be no splitting.

Second, the area under a peak is indicative of the number of hydrogens it represents. This, too, is important qualitative information, and so NMR spectrometer data systems are designed to determine the areas under the peaks to determine the number of hydrogens of each type. An integration trace is often displayed and can then used to determined this number of hydrogens of each type. Figure 15.13 (ethylbenzene) shows integration trace. Notice the rise in the trace above the –CH$_3$ half compared with the rise in the trace above the –CH$_2$– half of the ethyl group. They are in an apparent ratio of 3:2.

15.2.4 SOLVENTS AND SOLUTION CONCENTRATION

The 5-mm tube held in the sample probe is indicative of the sample size used in an NMR instrument—typically some fraction of a milliliter. Typical analyte quantities dissolved in this volume range from 10 to 50 mg.

A popular solvent for the sample is deuterated chloroform, CDCl$_3$. The "D" is the symbol for deuterium, the isotope of hydrogen that has one proton and one neutron. This solvent has no ^{1}H nuclei and does not absorb RF. It will therefore not interfere in the analysis.

15.2.5 ANALYTICAL USES

The primary use of NMR is in the determination of the structure of unknown organic compounds. It is often used in conjunction with other spectrometric techniques (FTIR and mass spectrometry, for example) in this determination. NMR spectra are molecular fingerprints, however, and by comparison with data files of known spectra, the structure of an unknown can be determined independent of other data.

15.3 VISCOSITY

15.3.1 INTRODUCTION

The quality of the ketchup that the American consumer uses on hamburgers and hotdogs is often judged by its resistance to flow. A ketchup that has a thick, flow-resistant consistency is generally considered to be of higher quality than one that flows readily from the bottle. However, this same consumer may become frustrated when the resistance to flow is so high that the ketchup does not flow in a timely manner from the bottle. The most desirable ketchup is thus the one that is judged to have a resistance to flow that is somewhere between the two extremes.

The same might be said of ordinary house paint. Paint that has a low resistance to flow is "runny" and will likely not provide satisfactory coverage of a wall with just one coat. It is therefore judged to be of poorer quality than one that does provide good one-coat coverage. On the other hand, paint that is thick and "gummy" may also be judged to be of poorer quality because it may not be possible to apply it uniformly with either a brush or roller.

There are dozens of other examples around the house for which we have similar quality standards in terms of their resistance to flow. Examples include other food products such as honey, pancake syrup, gravy, and ice cream toppings. Hygiene products include shampoo, hand lotion, and liquid soap. We might also cite pharmaceutical formulations, such as cough medicines, milk of magnesia, and liquid dietary supplements as well as home and car care products such as caulks, glues, and motor oils.

Just as a consumer may judge the quality of a product by its resistance to flow, so also quality assurance technicians in the chemical process industries may judge the quality of a fluid material or product by its resistance to flow. Examples include solutions of polymers (where a solution's resistance to flow is indicative of the quality of the undissolved polymer), asphalt formulations for roads and parking lots, lubricating oils, etc.

15.3.2 Definitions

The science that deals with the deformation and flow of matter is called **rheology**. An important rheological concept is the **shear force**, sometimes called the **shear stress**, or the force that causes a layer of a fluid material to flow over a layer of stationary material. The rate at which a layer of a fluid material flows over a layer of stationary material is called the **shear rate**. A fluid flowing through a tube, for example, would be the fluid material while the tube wall would be the stationary material. An important rheological measurement that is closely related to the resistance to flow is called **viscosity**. The technical definition of viscosity is the ratio of the shear stress to the shear rate, or

$$\text{viscosity} = \frac{\text{shear stress}}{\text{shear rate}} \tag{15.2}$$

If increasing (or decreasing) the shear stress increases (or decreases) the shear rate proportionally at a given temperature (such that the ratio does not change), the fluid is said to be a **Newtonian** fluid. If the shear rate does not increase proportionally with the shear stress, the fluid is said to be a **non-Newtonian** fluid. In the above example, if the flow rate of a fluid in a tube increases due to an increase in the force pushing it through, and the ratio of the force to the rate does not change, then the fluid is a Newtonian fluid. If it does change, then it is a non-Newtonian fluid. Some fluids exhibit Newtonian behavior and some do not. This has practical significance in that if a parameter relating to force or rate changes in the course of laboratory measurements of viscosity, then the results will vary.

Viscosity, as it was defined in Equation 15.2, is often called the **dynamic viscosity**. The most common unit of dynamic viscosity is the **centipoise**, a unit based on force per rate. The "base unit," the **poise** (100 centipoise per poise) is seldom used. Of more practical significance is **kinematic viscosity**, which is the dynamic viscosity divided by the density of the fluid.

$$\text{Kinematic viscosity} = \frac{\text{dynamic viscosity}}{\text{density}} \tag{15.3}$$

The most common unit of kinematic viscosity is the **centistokes**, cS, the centipoise per density unit. Again, the base unit, the **stokes**, S, (100 cS/S), is seldom used.

The laboratory technique for measuring viscosity is called **viscometry** (sometimes **viscosimetry**) and the device in the laboratory used to measure viscosity is called a **viscometer** (sometimes **viscosimeter**).

15.3.3   TEMPERATURE DEPENDENCE

Viscosity depends on temperature. The higher the temperature, the lower the viscosity. Pancake syrup, for example, flows more freely when heated. For reasonable accuracy when measuring viscosity, the temperature must be very carefully controlled. This means that the viscometer/sample must be immersed in a constant temperature bath and the temperature given time to equilibrate before the measurement is recorded. A calibrated thermometer must be used to measure the temperature.

15.3.4   CAPILLARY VISCOMETRY

Capillary viscometry measures viscosity by measuring the time it takes for the fluid test material to pass through a very small diameter tube—a capillary tube. While any capillary tube can conceivably be used, there are tubes that are commercially produced for this purpose. These tubes are called **capillary viscometers**. The basic design consists of a U-shaped glass tube with capillary portions in one arm of the "U." This portion has a large central bulb and two etched calibration lines. It opens up near the top so that suction can be conveniently applied. The other side has a larger opening at the top into which the test material is poured when preparing for a measurement. A simplified representation is given in Figure 15.14.

To measure the time of flow, the test fluid is poured into the appropriate side (Figure 15.15a) so that it occupies the bottom of the "U" and so that the meniscus on the capillary side is below the lower calibration mark (Figure 15.15b). The suction is then applied to the capillary side so that the

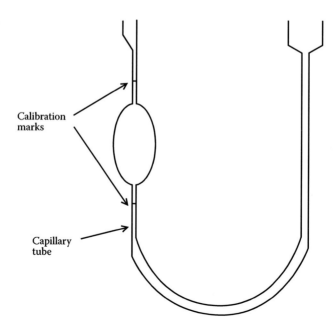

Calibration
marks

Capillary
tube

FIGURE 15.14   Simplified representation of a capillary viscometer.

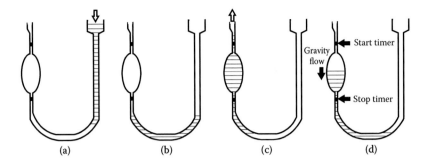

FIGURE 15.15 Concept of a capillary viscosity measurement (see text for full description).

fluid is drawn up to above the upper calibration mark (Figure 15.15c). When the fluid is released, it flows by gravity through the capillary (Figure 15.15d) so that the levels in the two sides will once again be at the same height (Figure 15.15b). The time it takes for the meniscus on the capillary side to pass the two calibration marks is the time that is measured (with a timer).

Capillary viscometers that have this design are called Ostwald viscometers. There are many specific designs of Ostwald viscometers. The most frequently used are the Cannon–Fenske viscometer and the Ubbelohde viscometer, and these are the two that will be described here.

The Cannon–Fenske viscometer is pictured in Figure 15.16. The U-tube has a bend in the center. There are several "bulbs" in the "U," which allows a greater volume of fluid to be tested and this also means a long and precise time measurement. The timing marks are clearly visible in Figure 15.16.

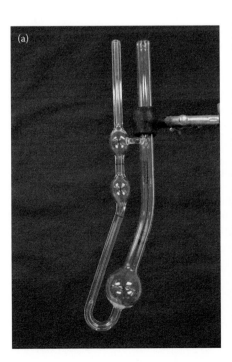
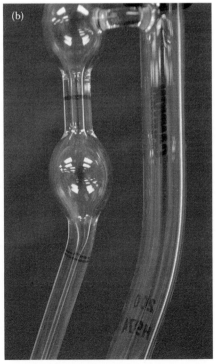

FIGURE 15.16 (a) Photograph of a Cannon–Fenske capillary viscometer. (b) Close-up view, showing the timing marks. Note the narrowing of the diameter near the lower timing mark.

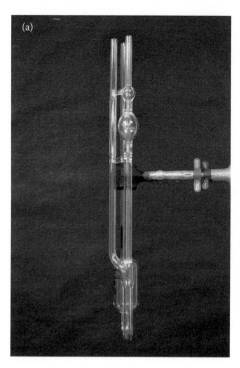

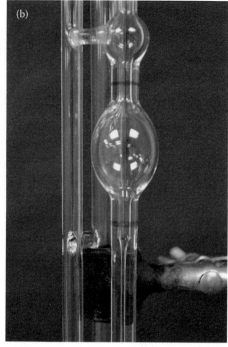

FIGURE 15.17 (a) Photograph of a Ubbelohde capillary viscometer. (b) Close-up view, showing the timing marks.

The Ubbelohde viscometer is pictured in Figure 15.17. In this design, the "U" is completely vertical and, like the Cannon–Fenske viscometer, also has several bulbs for containing the fluid. The timing marks are also clearly visible in Figure 15.17.

The Cannon–Fenske viscometer is used for measuring the kinematic viscosity of transparent Newtonian liquids, especially petroleum products and lubricants. The Ubbelohde viscometer is also used for the measurement of kinematic viscosity of transparent Newtonian liquids, but by the "suspended level" principle.

The time of flow is proportional to the kinematic viscosity as follows.

$$\text{kinematic viscosity} = \text{calibration constant} \times \text{time} \tag{15.4}$$

In order to calculate the kinematic viscosity, the calibration constant for the viscometer in question must be known. This calibration constant is often determined by the vendor before it is shipped, but it is also often checked by the user. In the calibration procedure, a fluid of known viscosity is tested so that the calibration constant can then be calculated.

$$\text{Calibration constant} = \frac{\text{kinematic viscosity}}{\text{time}} \tag{15.5}$$

Example 15.3

A given calibration liquid is known to have a kinematic viscosity of 15.61 cS at 25°C. Testing this liquid in a capillary viscometer gave a time of 139 s. An unknown liquid was then tested with the same viscometer and found to give a time of 238 s. What is the kinematic viscosity of the unknown liquid?

Solution 15.3

First, the calibration constant of the viscometer is determined from the calibration data.

$$\text{Calibration constant} = \frac{15.61 \text{ cS}}{139 \text{ s}} = 0.112 \text{ cS/s}$$

Then the viscosity of the unknown liquid can be calculated.

$$\text{kinematic viscosity} = \text{calibration constant} \times \text{time}$$

$$= 0.112 \text{ cS/s} \times 238 \text{ s} = 26.7 \text{ cS}$$

15.3.5 ROTATIONAL VISCOMETRY

Rotational viscosity methods measure viscosity by measuring the torque required to rotate a spindle immersed in the fluid sample. A motor rotates the spindle and the torque is proportional to the resulting deflection of a spring. This deflection is indicated by either a pointer/dial mechanism or via a digital display. In the digital model, the viscosity is displayed directly. A wide range of viscosities can be measured using interchangeable spindles and multiple rotation speeds as well as different motors (different rotational units). For a given spindle geometry and speed, an increase in viscosity is indicated by an increase in the deflection of the spring and thus the readout. A diagram of the rotational viscometer is shown in Figure 15.18.

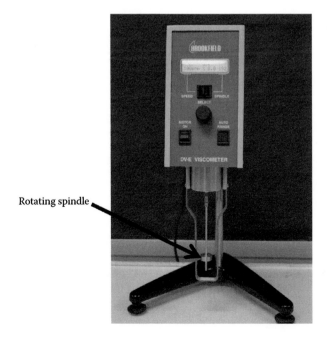

FIGURE 15.18 Photograph of a rotational viscometer. The arrow points to the rotating spindle.

15.4 THERMAL ANALYSIS

15.4.1 INTRODUCTION

Thermal analysis is the characterization of a sample of matter based on the properties it displays while subjected to a controlled temperature program. These properties include any property related to temperature changes (especially temperature increases). Examples include heat-related phase changes and degradations, crystallizations, heat capacities, heats of reaction, glass transitions, curing rates for adhesives, and weight changes. The properties are observed by monitoring either temperature or heat flow in and out of the sample or by monitoring the sample weight during the process.

Differential thermal analysis (DTA) is a technique in which the temperature difference between the sample tested and a reference material is measured while both are subjected to the controlled temperature program. **Differential scanning calorimetry (DSC)** is a technique in which the heat flow difference between the sample and the reference material is monitored while both are subjected to the controlled temperature program. **Thermogravimetric analysis (TGA)** is a technique in which the weight of a sample is monitored during the controlled temperature program.

15.4.2 DTA AND DSC

As a sample of a pure substance is heated, its temperature increases (Figure 15.19a). When a phase change begins to occur, however, such as when a solid sample begins to melt, the temperature does not increase, even though the heat continues to be added. This is because the heat added is used to change the phase of the sample (heat of fusion or heat of vaporization) rather than to raise its temperature. Once the phase transition is completed, its temperature "catches up" to the surroundings and increases again (Figure 15.19b).

Now consider the use of a reference material that does not melt in the temperature range used in the above scenario. Its temperature would match the temperature of the surroundings (T_E) for the entire temperature program. Consider plotting the difference ($\Delta T = T_S - T_R$) between the temperature of the sample (T_S) and the temperature of the reference material (T_R) vs. the temperature of the surroundings. Initially, there would be no difference, ΔT is zero, since the sample and surroundings are heated equally. However, when the sample melts, T_S lags behind T_R temporarily, making ΔT negative. After melting is complete, the sample "catches up" such that the two temperatures are again equal, $\Delta T = 0$. A plot of ΔT vs. T_E then results in the DTA curve in Figure 15.20. A negative peak occurs when the sample melts.

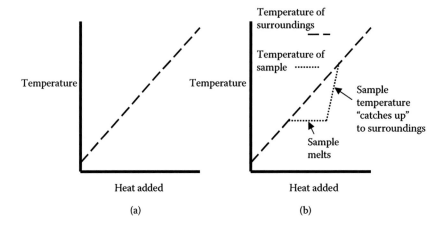

FIGURE 15.19   (a) Linear temperature program. (b) Applying the temperature program in (a) to a sample that melts at a particular temperature.

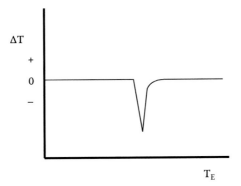

FIGURE 15.20 Representation of the results of the differential thermal analysis (a DTA curve) of a sample that melts at a particular temperature.

The negative peak shown in Figure 15.20 is a result of an endothermic process (a process that absorbs heat) such as melting. Other endothermic processes, other than melting, would also produce a negative peak. Examples include a chemical reaction or a decomposition. The particular characteristics of this peak (shape, width, sharpness, smoothness, etc.) provide clues concerning the sample composition and properties that are the object of a DTA.

Exothermic processes (processes that evolve heat) may also occur during the experiment. This would produce a surge in the sample temperature rather than the flattening observed in Figure 15.19a and would produce a positive peak in the ΔT vs. T_E plot (see Figure 15.21b). Again, the characteristics of this positive peak provide clues concerning the sample. Exothermic processes include crystallization as well as some chemical reactions and decomposition reactions.

Differential scanning calorimetry (DSC) is similar to DTA. However, rather than monitoring the temperature difference between the sample and the reference as the temperature of the surroundings is increased as in DTA, the energy required to keep the sample temperature equal to the reference temperature is monitored. This energy is either monitored according to the oven power required or according to the actual heat flow that occurs. In any case, separate electrical heating elements heat the sample and reference. When the sample temperature lags behind the reference temperature (when an endothermic process occurs), the sample heating element is given more power and there is greater heat flow to the sample. When the sample temperature surges ahead of the reference

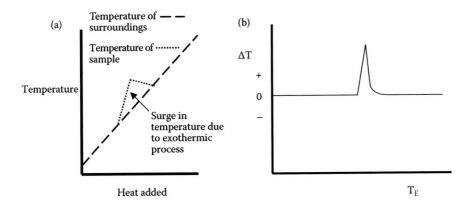

FIGURE 15.21 (a) Applying the temperature program in Figure 15.19a to a sample that undergoes an exothermic process. (b) Representation of the DTA curve of the exothermic process.

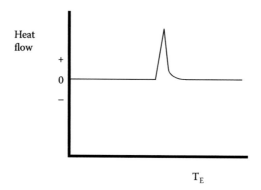

FIGURE 15.22   Representation of a DSC curve for an exothermic process. Note that the y-axis is heat flow.

temperature (when the exothermic process occurs), the reference heating element is given more power and there is greater heat flow to the reference. The difference in either the power or the heat flow (whichever is being monitored) is then plotted vs. the temperature of the surroundings and a DSC curve that appears very similar to the DTA curve is obtained (see Figure 15.22).

Some scientists describe DSC techniques as a subset of DTA. DTA can be considered a more global term covering all differential thermal techniques, whereas DSC is a DTA technique that gives calorimetric (heat transfer) information. This is the reason DSC has "calorimetry" as part of its name. Most thermal analysis work is DSC and Sections 15.4.3 and 15.4.4 provide information about the instrumentation and applications of this technique.

15.4.3   DSC Instrumentation

The instrumentation used for heat flow monitoring (rather than oven power monitoring) is described here. The DSC "oven" has a cylindrical shape and is approximately 2.5 cm in diameter and 3 cm high. Two tiny aluminum crucibles (approximately 20 μL capacity) are used, one to contain the sample and one to contain the reference material. These crucibles are placed on a platform that is positioned approximately 1 cm above the floor of the oven. The specific locations for the crucibles are designated (such as "S" for "sample" and "R" for "reference") (see Figure 15.23). The platform

FIGURE 15.23   Photograph of the interior of a small oven used for differential scanning calorimetry. The crucibles for the sample (S) and reference material (R) are placed in the small circles.

is made of a special material to facilitate thermoelectric heating and monitoring via thermocouples on the underside. Immediately below the sample and reference locations under the platform are small metal discs (not visible in Figure 15.23) that form the junction with the platform for the thermocouple.

The reference material is usually elemental indium, a soft metal (atomic number 49 in Group IIIA in the periodic table). One small "plug" of indium is placed in an aluminum crucible, which is then positioned on the reference location in the oven. A few milligrams of the sample are placed in another aluminum crucible, which in turn is positioned on the sample location in the oven. The oven is then closed and the temperature program begun.

It is common to provide a particular gaseous environment in the oven, i.e., to purge the oven with a gas such as air, helium, oxygen, nitrogen, etc. The oven module has an inlet for the purge gas. The purpose of a purge gas is to sweep out gases that form during the program and thus provide a constant environment for the sample and reference so that effects of environment changes are not measured. Also, a particular gas may create favorable conditions for a particular effect to be observed. For example, a chemical reaction caused by the presence of oxygen will be observed in the DSC curve if oxygen is used as the purge gas. This reaction will not be observed if helium or another inert gas is used. In addition, different gases have different thermal conductivities and this will affect the results.

The rate of heating is variable. The slower this rate the better, because fast heating will shift temperatures higher and cause the size of the peak to increase. The rate of heating during calibration should equal the rate of heating during sample analysis.

15.4.4 APPLICATIONS OF DSC

Two important applications of DSC are in the pharmaceutical industry and in the polymer industry. In the pharmaceutical industry, the purity of formulations and raw materials can be measured. Various levels of purity give different melting points and melting ranges. Very pure materials melt sharply (within 1–2°) and melt at expected temperatures. Impure materials have broader melting ranges and melt at lower temperatures than pure materials. Such phenomena can easily be detected with DSC.

In the polymer industry, the melting or degradation behaviors of polymers are important to determine. For example, when a polymer is extruded (i.e., when polymer "pellets" are converted to "film" by heating them and drawing them through an extruder), the thermal analysis of the polymer material determines the amount of heat needed in the extruder to make the material pliable. Given the large number of polymer formulations that have been developed and continue to be developed, thermal analysis procedures can be quite important.

15.5 OPTICAL ROTATION

Ordinary white light, such as has been dealt with in previous chapters, does not move in just one plane. In actuality, the light waves exist in all planes around the line of travel. There are an infinite number of such planes. Figure 15.24a is a simplified view of this idea, showing a "head-on" view of a beam of light with the planes depicted as double arrows. It is possible to "polarize" light so as to block all planes of light except for one (Figure 15.24b). This is done with the use of a polarizing filter. Light consisting of just a single plane of light such as this is called **plane-polarized light**.

Some compounds exhibit the property of being able to rotate the plane of polarized light. In other words, when a beam of plane-polarized light passes through a sample of such a compound, the plane is rotated to another position around the line of travel (Figure 15.24c). The property is called **optical rotation**, or **optical activity**. In order to be optically active, a compound must possess an asymmetric carbon atom in its molecular structure. **An asymmetric carbon atom** is one that has four different structural groups attached to it. An example of such a compound is 2-butanol.

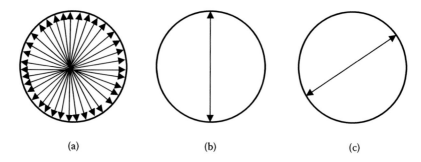

(a)   (b)   (c)

FIGURE 15.24   (a) Representation of a light beam with all wave planes shown as double arrows around the line of travel. (b) Representation of plane-polarized light—all planes are blocked but one. (c) Representation of plane polarized that has rotated as a result of passing through a solution of an optically active compound.

$$\underset{\text{CH}_3\text{-CH-CH}_2\text{-CH}_3}{\overset{\text{OH}}{|}} \qquad \text{2-butanol}$$

The second carbon from the left is an asymmetric carbon atom because it has four different groups attached to it—a methyl group, a hydrogen, a hydroxyl group, and an ethyl group.

Having an asymmetric carbon atom means that a structure will have a nonsuperimposable mirror image (like a right hand and a left hand), which in turn means that in drawing a structure such as 2-butanol above, we have actually represented two compounds with one drawing. There are actually two different structures in one and they are mirror images of each other. A pair of structures that are nonsuperimposable mirror images are called **enantiomers**. It turns out that one enantiomer rotates plane-polarized in one direction and the other rotates it through exactly the same angle, but in the opposite direction. Two compounds representing a pair of enantiomers are exactly identical in all other physical properties (color, odor, boiling point, solubility in water, refractive index, viscosity, etc.). They even have identical infrared spectra. The only property by which they differ is in the direction in which they rotate plane-polarized light.

Most of the time, enantiomers are found equally mixed together. Equally mixed enantiomers are *not* optically active because the rotation in one direction by one structure is cancelled by the rotation in the opposite direction by the other structure. Hence, a sample of 2-butanol, for example, as normally obtained from a chemical vendor, is not optically active. An equimolar mixture of two enantiomers is called a **racemic mixture** and is optically inactive. Separation of a racemic mixture is not possible by conventional methods because the enantiomers are identical with respect to properties that are used to effect the separation. However, it may be possible to separate them by *chemical* methods, meaning that one may undergo a chemical reaction that the other does not. Some biological reactions are such reactions, and hence, a single enantiomeric structure is sometimes found in nature.

The rotation of plane-polarized light can be measured with a laboratory instrument called a **polarimeter**. A simple polarimeter consists of a light source, a polarizing filter, a sample "tube" for containing the sample, and a device to observe/measure the rotation. The exact degree of rotation depends on how much of the enantiomer is in the path of the light. Hence, one can measure the concentration of an enantiomer with a polarimeter. The degree of rotation would also depend on the length of the sample tube. There is also an effect of temperature and wavelength. A temperature of 25°C and a wavelength of 589.3 nm (the primary sodium emission line) are the standard conditions for reporting this property in tables of physical properties. When reporting the optical rotation in such tables, it is referred to as the **specific rotation**. The measured angle of rotation, usually symbolized by the Greek letter theta (θ), is divided by the concentration (in g/mL) and the length of the sample tube (in decimeters) to calculate the specific rotation (see Application Note 15.1).

**APPLICATION NOTE 15.1: DETERMINATION OF
SUGAR IN MILK BY POLARIMETRY**

The sugar content in a solution can be determined by polarimetry. The rotation of the plane of polarized is related to concentration by the equation $\beta = \alpha lc$, where β is the observed rotation, α is the specific rotation, l is the length of the polarimeter tube, and c is the concentration of the sugar in grams/liter. The lactose concentration in milk can be determined in this way. Sample preparation involves weighing 50 mL of the milk held in a 100-mL volumetric flask and adding acidic mercury(II) nitrate (Millon's reagent) to the flask to precipitate the protein. Following this precipitation, the mixture is filtered and the clear filtrate tested in the polarimeter to determine the optical rotation. The specific rotation of lactose is +52.2° for the sodium D line.

(*Reference*: James, C., *Analytical Chemistry of Foods*, Chapman & Hall (1995), pp. 128–129.)

QUESTIONS AND PROBLEMS
X-Ray Methods

1. Name three analytical methods involving x-rays and give some information about the application of each.
2. Name two sources of x-rays and explain how the x-rays are generated in each case.
3. Define K_α emission, L_β emission, and $K_{\alpha 2}$ emission.
4. What is diffraction? How does constructive and destructive interference result in a diffraction pattern?
5. What is Bragg's law? Define each of the parameters in Bragg's law.
6. The wavelength used in a particular x-ray diffraction analysis is 1.539 Å. If an intense peak occurs at a 2θ value of 44.28°, what is the d-spacing in this structure?
7. The wavelength used in a particular x-ray diffraction analysis is 1.539 Å. If a standard crystal with a d-spacing of 0.844 Å is used, at what angle can a light intensity be expected to be observed?
8. Why is it that the x-ray emission depicted in Figure 15.1 is considered fluorescence?
9. Is x-ray fluorescence a molecular fluorescence, an atomic fluorescence, or neither of these? Explain.
10. What are the advantages and disadvantages to using an analyzer crystal in an x-ray fluorescence instrument as depicted in Figure 15.6?
11. What are some safety rules to be followed in laboratories in which x-ray equipment is used?

Nuclear Magnetic Resonance Spectroscopy

12. The letters "NMR" stand for _____ _____ _____.
13. Why is a large magnet needed as part of an NMR instrumentation?
14. What wavelength region of the electromagnetic spectrum is needed for an NMR experiment? Why?
15. What two energy states of spinning/precessing nuclei present in a magnetic field are involved in an NMR experiment? Which state represents the higher energy?
16. Why is NMR sometimes referred to as PMR?

17. The letters "FTNMR" stand for _____ _____
    _____ _____.

18. What is meant by the terms "hertz" and "megahertz"?

19. What is a typical value of the RF (in megahertz) used in an NMR experiment? What is the
    magnitude of the magnetic field (in gauss) needed for an instrument utilizing this RF?

20. What is a "gauss?"

21. Name seven components of the traditional NMR instrument and briefly describe the func-
    tion of each.

22. How does a modern FTNMR experiment differ from an experiment using a traditional
    instrument?

23. What parameters are represented on the y- and x-axes of an NMR spectrum?

24. What is the "chemical shift," and what causes it?

25. How is it that peaks representing different hydrogens in an organic molecule appear at dif-
    ferent locations in an NMR spectrum?

26. How many major peaks can be expected in the NMR spectrum of (a) dimethyl ether,
    (b) diethyl ether, (c) methyl ethyl ketone, (d) acetone, and (e) *t*-butyl alcohol?

27. What is it about the structure of methyl alcohol that would result in two major peaks in its
    NMR spectrum (Figure 15.12)?

28. Explain the use of tetramethylsilane (TMS) in an NMR experiment.

29. Look at the NMR spectrum in Figure 15.25. How many different kinds of hydrogen are
    represented? Explain your answer.

30. What does the integrator trace in Figure 15.25 tell you about the number of the different
    kinds of hydrogen present in the structure?

31. If you were told that the spectrum in Figure 15.25 was either ethyl alcohol or diethyl ether,
    what evidence would your cite that would lead you to conclude that it is ethyl alcohol and
    not diethyl ether?

32. If your were told that the spectrum in Figure 15.25 resulted from either ethyl alcohol or
    methyl ethyl ether, what evidence would your cite that would lead you to conclude that it is
    ethyl alcohol and not methyl ethyl ether?

33. In Figure 15.25, the major peak at 5 ppm is a singlet, the major peak at 4.2 ppm is a quartet,
    and the major peak at 1.8 is a triplet. How does the n + 1 rule confirm that the compound
    may be ethyl alcohol?

34. In Figure 15.26, the major peak at 5 ppm is a singlet, the major peak at 4.5 ppm is a septet
    (split into seven peaks), and the major peak at 1.5 is a doublet. The spectrum is either that
    of *n*-propyl alcohol, isopropyl alcohol, or acetone. Decide which and justify your decision.

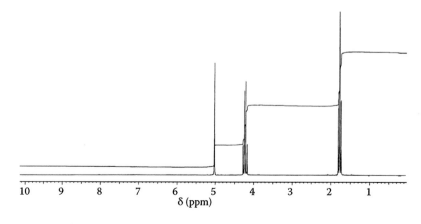

FIGURE 15.25   NMR spectrum for problems 29 to 33.

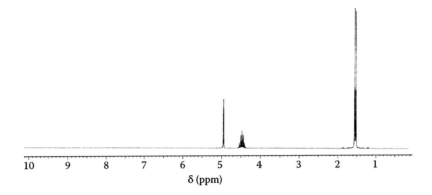

δ (ppm)

FIGURE 15.26 NMR spectrum for problems 34 and 35.

35. After answering question 34, describe what the integrator trace would look like on this spectrum.
36. What causes NMR peaks to be "split" into "doublets," "triplets," etc.? How does the presence of split peaks assist with qualitative analysis?

Viscosity

37. Other than those mentioned in Section 15.2.1, think of some other formulations the quality of which may be judged by their resistance to flow.
38. Define *rheology, shear force, shear stress, shear rate, Newtonian fluid, dynamic viscosity, centipoise, kinematic viscosity, centistokes, viscometry,* and *viscometer.*
39. How does capillary viscometry measure viscosity?
40. In capillary viscometry, the fluid being measured flows through a capillary tube. How is this helpful in measuring viscosity?
41. What is an "Ostwald viscometer?"
42. Is a Cannon–Fenske viscometer a capillary type or a rotational type? Explain.
43. Distinguish between a Cannon–Fenske viscometer and a Ubbelohde viscometer.
44. A given calibration liquid is known to have a kinematic viscosity of 12.72 cS at 25°C. What is the calibration constant for a viscometer if it takes this liquid 109 s to pass from the upper mark to the lower mark of a capillary viscometer?
45. The calibration constant of a capillary viscometer is 0.133 cS/s. If an unknown liquid flows between the marks in 98 s, what is the viscosity of the liquid?
46. Explain how a rotational viscometer measures viscosity.

Thermal Analysis

47. Define *thermal analysis, differential thermal analysis, differential scanning calorimetry,* and *thermogravimetric analysis.*
48. What do the following stand for: DTA, DSC, and TGA?
49. When you heat a sample that is melting, why does its temperature not increase?
50. How do the temperature of a sample and the temperature of its surroundings compare as the temperature of the surroundings is increased?
51. Explain the negative peak in Figure 15.20.
52. In a plot of the difference in temperature between a sample and a reference material vs. the temperature of the surroundings, why does an endothermic process produce a negative peak while an exothermic process produces a positive peak?
53. How does DSC differ from DTA? How are they similar? Why is DSC often considered a subset of DTA?

54. In Figure 15.23, what do the circles designated "S" and "R" refer to? Why must there be two locations like these in the oven?
55. For what purpose is indium often used in DSC?
56. What reactions other than melting produce results in DSC?
57. What are some applications of thermal analysis in the pharmaceutical and polymer industries?

Optical Rotation

58. Define *plane-polarized light, optical rotation, optical activity, asymmetric carbon atom, enantiomers, racemic mixture, polarimeter,* and *specific rotation.*
59. What exactly is the physical property known as "optical rotation?"
60. Write down the structures of at least five compounds that have an asymmetric carbon atom.
61. The compound 2-butanol has an asymmetric carbon atom, but a sample of 2-butanol out of the bottle is not optically active. Why is that?
62. If two structures are nonsuperimposable mirror images, what can you say about the direction that each will rotate the plane of polarized light? Would a mixture of the two be optically active? Explain.
63. Describe how a polarimeter measures optical rotation.

Appendix 1: Formulas for Solution Concentration and Preparation Calculations

SYMBOLISM:

FW = Formula weight
C_B = Concentration before dilution
C_A = Concentration after dilution
V_B = Volume before dilution
V_A = Volume after dilution
EW = Equivalent weight
L_D = Liters desired
M_D = Molarity desired
FW_{SOL} = Formula weight of the solute
N_D = Normality desired
EW_{SOL} = Equivalent weight of the solute
ppm_D = ppm desired

CALCULATION OF THE PERCENT CONCENTRATION OF A SOLUTION FROM SOLUTION PREPARATION DATA

$$\text{weight/weight \%} = \frac{\text{grams of solute}}{\text{grams of solution}} \times 100 \tag{A1.1}$$

$$\text{weight/volume \%} = \frac{\text{grams of solute}}{\text{milliliters of solution}} \times 100 \tag{A1.2}$$

$$\text{volume \%} = \frac{\text{milliliters of solute}}{\text{milliliters of solution}} \times 100 \tag{A1.3}$$

$$C_A = \frac{C_B \times V_B}{V_A} \tag{A1.4}$$

CALCULATION OF THE MOLARITY OF A SOLUTION FROM SOLUTION PREPARATION DATA

$$\text{molarity} = \frac{\text{moles of solute}}{\text{liters of solution}} \tag{A1.5}$$

$$\text{molarity} = \frac{\text{grams of solute/FW of solute}}{\text{liters of solution}} \tag{A1.6}$$

$$C_A = \frac{C_B \times V_B}{V_A} \tag{A1.7}$$

CALCULATION OF THE NORMALITY OF A SOLUTION FROM SOLUTION PREPARATION DATA

$$\text{normality} = \frac{\text{equivalents of solute}}{\text{liters of solution}} \tag{A1.8}$$

$$\text{normality} = \text{molarity} \times \text{equivalents per mole} \tag{A1.9}$$

$$\text{normality} = \frac{\text{grams of solute/EW}}{\text{liters of solution}} \tag{A1.10}$$

$$C_A = \frac{C_B \times V_B}{V_A} \tag{A1.11}$$

CALCULATION OF PPM OF A SOLUTION FROM SOLUTION PREPARATION DATA

$$\text{ppm} = \frac{\text{milligrams of solute}}{\text{liters of solution}} \tag{A1.12}$$

$$C_A = \frac{C_B \times V_B}{V_A} \tag{A1.13}$$

PREPARATION OF PERCENT SOLUTIONS

weight/weight %:

$$\text{grams to weigh} = \frac{\text{percent concentration}}{100} \times \text{solution weight} \tag{A1.14}$$

weight/volume %:

$$\text{grams to weigh} = \frac{\text{percent concentration}}{100} \times \text{solution volume} \tag{A1.15}$$

volume %:

$$\text{milliliters} = \frac{\text{percent concentration}}{100} \times \text{solution volume} \qquad \text{(A1.16)}$$

If the solute is already dissolved, but the solution is to be diluted,

$$V_B = \frac{C_A \times V_A}{C_B} \qquad \text{(A1.17)}$$

PREPARATION OF MOLAR SOLUTIONS

If the solute is a pure solid (or liquid) that will be weighed,

$$\text{grams to weigh} = L_D \times M_D \times FW_{SOL} \qquad \text{(A1.18)}$$

If the solute is already dissolved, but the solution is to be diluted,

$$V_B = \frac{C_A \times V_A}{C_B} \qquad \text{(A1.19)}$$

PREPARATION OF NORMAL SOLUTIONS

If the solute is a pure solid (or liquid) that will be weighed,

$$\text{grams to weigh} = L_D \times N_D \times EW_{SOL} \qquad \text{(A1.20)}$$

If the solute is already dissolved, but the solution is to be diluted,

$$V_B = \frac{C_A \times V_A}{C_B} \qquad \text{(A1.21)}$$

PREPARATION OF PPM SOLUTIONS

If the substance the ppm is expressed as is the same substance as that is to be weighed,

$$\text{ppm}_D \times L_D = \text{milligrams to be weighed} \qquad \text{(A1.22)}$$

If the substance the ppm is expressed as is not the same substance as that to be weighed,

$$\text{ppm}_D \times L_D \times \text{gravimetric factor} = \text{milligrams to be weighed} \qquad \text{(A1.23)}$$

If a more concentrated solution is to be diluted,

$$V_B = \frac{C_A \times V_A}{C_B} \qquad \text{(A1.24)}$$

Appendix 2: The Language of Quality Assurance and Good Laboratory Practice (GLP) Laws: A Glossary

accuracy: the degree to which the result obtained agrees with the correct answer.

action limits: horizontal lines on a control chart (see definition elsewhere in this glossary) that are designated to be the limits whereby a laboratory analyst is informed that a process or laboratory test result is not conforming to predetermined limits of accuracy.

calibration: a procedure by which an instrument or measuring device is tested to determine what its response is for an analyte in a test sample or samples for which the true response is either already known or needs to be established. If the true response is already known, we then make an adjustment, if possible, so that the known response is what the instrument or device reads for it. If we cannot adjust it to give the known response, the device is defective and is taken out of service and repaired. If the true response needs to be established, we establish it via a single standard or perhaps via a calibration curve or standard curve created using a series of standards and then correlate the response of unknowns with that of the known quantity or quantities.

certified reference material (CRM): a material that is shown by the vendor to be a reference material via a certificate that accompanies the material. In this way, the vendor certifies that the material has been made from or compared to a standard reference material (SRM). A CRM is often used by laboratory workers to prepare other reference materials to be used for calibration and standardization purposes. (Compare this definition to the definitions of "reference material" and "standard reference material" in this glossary.)

chain of custody: the documentation that accompanies the handling of a sample to be used in a laboratory. The documentation includes who has handled the sample and what responsibility each handler has at various junctures between the sampling site and the laboratory (see Chapter 2 for more discussion).

control: a material or solution of known composition (similar to the actual standards and samples analyzed) prepared and used expressly for the purpose of monitoring a measurement process for accuracy. The concentration of this material or solution is the "expected value" on a control chart (defined elsewhere in this glossary).

control chart: a plot of a measurement or analysis result on the y-axis vs. time (usually days) on the x-axis. The purpose is to provide a visual representation of a manufacturing process or laboratory test that will show if and when the process or test does not conform to what is expected.

current good manufacturing practices (cGMP): federal regulations (in the Code of Federal Regulations, CFR) specifically intended for the pharmaceutical industry to ensure that pharmaceutical products are produced and controlled according to the quality standards pertinent to their intended use.

detection limit: the smallest concentration of an analyte that is detectable with a given method; a concentration that produces a signal-to-noise ratio of 2:1.

good laboratory practices (GLP): federal regulations (in the Code of Federal Regulations, CFR) intended for the pharmaceutical industry (21 CFR 58) and the environmental industry

(40 CFR 160) to ensure that the laboratory tests of products created by and raw materials used by these industries are conducted according to quality standards pertinent to their intended use.

International Organization for Standardization (ISO): a worldwide federation of national standards bodies charged with promoting common standards to ensure that the products and services made available internationally are of the highest quality.

ISO: see International Organization for Standardization elsewhere in this glossary.

ISO 9000: a series of documents drafted by the International Organization for Standardization (ISO) that outline the requirements for ensuring that the exchange of goods between worldwide trading partners is of high and internationally acceptable quality.

measurement system: a system consisting of all the physical equipment, facilities, logistics, and processes that is configured in order to make the particular measurement that is needed.

nonclinical laboratory study: a term used in the FDA GLP regulations to mean a work assignment in which test articles are studied in a test system under laboratory conditions to determine their safety.

person: a term used in GLP regulations to mean any individual, partnership, corporation, association, scientific or academic establishment, government agency, or organizational unit thereof, and any other legal entity.

precision: the degree to which a series of measurements made on the same sample with the same measurement system agree with each other.

quality assurance: the system of operations which tests the product or service to ensure compliance with defined specifications.

quality assurance unit: a term used in GLP regulations to mean a person (see definition in this glossary) or organizational element, except the study director (see definition in this glossary), designated by testing facility management to perform the duties relating to quality assurance of the studies at hand.

quality control: the overall system of operations designed to control a process so that a product or service adequately meets the needs of the consumer.

quality system: the structure, responsibilities, procedures, and resources that are required within an organization for the implementation of the concept of Total Quality Management (defined elsewhere in this glossary).

raw data: any laboratory worksheets, records, memoranda, notes, or exact copies thereof, that are the result of original observations and activities of a (nonclinical laboratory) study (see definitions elsewhere in this glossary) and are necessary for the reconstruction and evaluation of the report of that study.

reference material (RM): a material of a known purity or concentration that has been either prepared from or compared to another RM by laboratory workers or purchased from a qualified vendor. As such, it can be a standard reference material (SRM), a certified reference material (CRM), or a material made from or compared to another RM. (See the definitions of standard reference material and certified reference material elsewhere in this glossary.)

reference substance (EPA): any chemical substance or mixture, or analytical standard, or material other than a test substance, feed, or water that is administered to or used in analyzing the test system in the course of a study for the purposes of establishing a basis for comparison with the test substance for known chemical or biological measurements.

sponsor: (1) a person who initiates and supports, by provision of financial or other resources, a (nonclinical laboratory) study; (2) a person who submits a (nonclinical laboratory) study to the EPA (FDA) in support of an application for a research or marketing permit; or (3) a testing facility if it both initiates and actually conducts the study (see definitions of terms used here elsewhere in this glossary).

standard operating procedures (SOP): the documented set of instructions a technician or chemist follows when carrying out an analysis or process. SOPs are very carefully considered instructions that become official only after thorough review and testing by the laboratory personnel. Careful attention to such written instructions is required in order for the work to be considered valid under GLP regulations. A SOP is intended to ensure the quality and integrity of the data generated in a laboratory and can only be changed by a study director.

standard reference material (SRM): the ultimate reference material from which another material is made or to which it is compared in order for it to become a certified reference material (CRM). All reference materials are ultimately traceable to a SRM in this way. SRMs are only available from NIST, the National Institute for Standards and Technology. (See also the definitions of reference material, certified reference material, and traceability elsewhere in this glossary.)

study: a term used in the EPA GLP regulations to mean a work assignment in which a test substance is studied in a test system under laboratory conditions or in the environment to determine or help predict its effects.

study director: the individual responsible for the overall conduct of a (nonclinical laboratory) study

system suitability: the validation of all components of an analysis scheme working together as a unit or "system." "All components" refers to all parts of the scheme from obtaining a sample to data handling (see also the definition of validation in this glossary).

test article (FDA): any food additive, color additive, drug, biological product, electronic product, medical device for human use, or any other article subject to regulation under the Federal Food, Drug, and Cosmetic Act (FFDCA) or the Public Health Service Act (PHSA).

test substance (EPA): a substance or mixture administered or added to a test system (see definition elsewhere in this glossary) in a study, in which the substance or mixture (1) is the subject, or contemplated subject, of an application for a research or marketing permit, or (2) is an ingredient or product of a substance as defined above.

test system (EPA): a term used in the EPA GLP regulations to mean any animal, plant, microorganism, chemical, or physical matrix, including but not limited to soil or water, or subparts thereof, to which the test, control, or reference substance is administered to or added for the study (see definitions of terms used here elsewhere in this glossary).

test system (FDA): a term used in the FPA GLP regulations to mean any animal, plant, microorganism, or subparts thereof, to which the test or control article is administered to or added for the study (see definitions of terms used here elsewhere in this glossary).

testing facility: a person who actually conducts a (nonclinical laboratory) study, i.e., who actually uses the test substance (article) in a test system (see definitions of terms used here elsewhere in this glossary).

Total Quality Management: a concept wherein all workers within an enterprise, from upper management to custodians and including laboratory workers, are managing their own particular piece of the puzzle with utmost concern and care for quality—quality in design, quality in development, quality in production, quality in installation, and quality in servicing.

traceability: a standardization chain in which one material is established as a standard via a second standard, which was established as a standard via a third standard, etc. All secondary standards can be traced to a primary standard, and this primary standard became a standard by comparison to an reference material (RM), ultimately being compared to an standard reference material (SRM).

validation: a study designed to ensure that a method, a procedure, an instrument, etc. provides accurate data and results for the work at hand.

warning limits: horizontal lines on a control chart (see definition elsewhere in this glossary) that are designated to be the limits whereby a laboratory analyst is warned that a process or laboratory test result may not be conforming to predetermined limits of accuracy.

Appendix 3: Significant Figure Rules

RULES FOR DETERMINING THE NUMBER OF SIGNIFICANT FIGURES IN A GIVEN NUMBER

1. Any nonzero digit is significant.

    Example: 916.3   four significant figures

2. Any zero located between two significant figures is significant.

    Example: 1208.4   five significant figures

3. Any zero to the left of nonzero digits is not significant unless it is also covered by Rule 2.

    Example: 0.00345   three significant figures

4. Any zero to the right of nonzero digits and also to the right of a decimal point is significant.

    Example: 34.10   four significant figures

5. Any zero to the right of nonzero digits and to the left of a decimal point and not covered by Rule 2 may or may not be significant, depending on whether the zero is a placeholder or was actually part of the measurement. Such a number should be expressed in scientific notation to avoid any confusion.

    Example: 430   do not know

    (4.3×10^2 or 4.30×10^2 would be better ways to express this number, depending on whether there are two or three significant figures in the number.)

RULES FOR DETERMINING SIGNIFICANT FIGURES IN THE ANSWER TO A CALCULATION

1. The answer to a multiplication or division has the same total number of significant figures as in the number with the least significant figures used in the calculation. Rounding to decrease the count of digits or the addition of zeros to increase the count of digits may be necessary.

    Example 1: $4.3 \times 0.882 = 3.7926$ (calculator answer)

    $= 3.8$ (answer with correct number of significant figures)

Example 2: $\dfrac{0.900}{0.2250} = 4$ (calculator answer)

$= 4.00$ (answer with correct number of significant figures)

2. The correct answer to an addition or subtraction has the same number of digits to the right of the decimal point as in the number with the least such digits that is used in the calculation. Again, rounding to decrease the count of digits or addition of zeros to increase the count of digits may be necessary.

Example 1: $24.992 + 3.2 = 28.192$ (calculator answer)

$= 28.2$ (answer with correct number of significant figures)

Example 2: $772.2490 - 0.049 = 772.2$ (calculator answer)

$= 772.200$ (answer with correct number of significant figures)

3. When several calculation steps are required, no rounding is done until the final answer is determined.

Example: $\dfrac{3.026 \times 4.7}{7.23} = 1.9363762$ (calculator answer with premature rounding)

$= 1.9$ (incorrect answer due to preamture rounding)

$= 1.9671093$ (calculator answer correctly determined)

$= 2.0$ (correct answer)

4. When both Rules 1 and 2 apply in the same calculation, follow Rules 1 and 2 in the order they are needed while also keeping Rule 3 in mind.

Example: $(3.22 - 3.034) \times 5.61 = 1.0659$ (calculator answer with premature rounding)

$= 1.1$ (incorrect answer due to preamture rounding)

$= 1.04346$ (calculator answer correctly determined)

$= 1.0$ (correct answer)

5. Conversion factors that are exact numbers have an infinite number of significant figures.

Example: There are exactly 3 feet per yard. How many feet are there in 2.7 yards?

$2.7 \times 3 = 8.1$ (two significant figures in correct answer, not one)

6. In cases in which the logarithm of a number needs to be determined, such as in converting [H$^+$] to pH or transmittance to absorbance, the number of digits in the mantissa of the logarithm (the series of digits to the right of the decimal point) must equal the number of significant figures in the original number.

Example: $[H^+] = 4.9 \times 10^{-6}$ M

$pH = -\log [H^+] = 5.31$

Appendix 4: Answers to Questions and Problems

CHAPTER 1

1. See Section 1.1.
2. An assay is an analysis in which a named material is analyzed for that named material. For example, the assay of a tablet of ibuprofen is an analysis of the tablet for ibuprofen content. An analysis is not an assay when the analyte is some material other than the named material. An analysis of a calcium supplement for a color additive is not an assay.
3. Qualitative analysis is identification—analysis for "what" a substance is or what is in a sample. Quantitative analysis is the analysis for quantity or "how much" of a given analyte is in a sample. Examples of qualitative analysis: (1) identify the contaminant giving an off-color to a manufactured product; (2) identify the byproduct of a chemical reaction; (3) identify an organic substance leaching into ground water from a hazardous waste site. Examples of quantitative analysis: (1) the determination of the nitrate concentration in a drinking water sample; (2) the determination of the concentration of the active ingredient in a pharmaceutical preparation; (3) the determination of the concentration of an additive in a solvent manufactured at a chemical plant.
4. It is a qualitative analysis because the task is to identify the material, not to determine quantity.
5. See Section 1.2.
6. The activities carried out in a "wet laboratory" would probably include sample preparation and wet chemical analysis procedures (for example, extractions, solution preps, and titrations)—activities that do not utilize sophisticated electronic instrumentation.
7. A wet chemical analysis would likely be chosen when a more precise result is needed or when the analyte is a major, rather than a minor, constituent.
8. A representative sample is one that has all the characteristics in exactly the same proportions as the bulk system from which it is taken.
9. (1) Obtain the sample, (2) prepare the sample, (3) carry out the analysis method, (4) work the data, and (5) calculate and report the results.
10. A sample is a portion of a larger bulk system that has all the characteristics of the bulk system and is acquired to test it in a laboratory. To "obtain a sample" means to carry out a process by which a sample is acquired from the bulk system such that the sample truly represents the bulk system and is not altered in any way before the analysis takes place in the laboratory.
11. To "prepare a sample" means to carry out laboratory procedure by which a sample is appropriately readied for the analysis method chosen. Such procedures include (but are not limited to) drying, dissolving, extracting, crushing, etc.
12. An analytical method is an operation (possibly involving many steps) by which data on a prepared sample and associated standard(s) are obtained and recorded so that the quality or quantity of the analyte in the sample can be ultimately determined. It is the heart of an analysis in the sense that it is the step following sample preparation in which the crucial data are acquired.

13. To "carry out the analytical method" means to perform the series of steps in which the crucial data for qualitative or quantitative analysis is directly obtained.

14. The raw data (measurements obtained directly from the laboratory equipment used without undergoing any additional process) usually needs to be used in calculations, graphing, etc. in order to be put into a form that is most useful. Once the data are "worked up" in this way, then it is ready for final calculation such that "results" are obtained and able to be reported.

15. A laboratory worker that has "good analytical technique" is one that takes special care in carrying out manual tasks to ensure that all data are obtained in a careful manner so that errors involved in handling samples and standards are eliminated or at least minimized.

16. The stirring rod is wet with your sample solution and rinsing it back into the beaker ensures that all of your sample is present in the beaker for subsequent reactions and/or measurement.

17. "GLP" stands for "good laboratory practices." These are federal regulations governing FDA- and EPA-affiliated laboratories and pertain to proper procedures in the laboratory to ensure that results are obtained in as trustworthy a manner as possible. "SOP" stands for "standard operating procedure." These are step-by-step written procedures that are specially approved by laboratory directors for carrying out certain specific tasks.

18. GLP regulations address such things as labeling, record keeping and storage, documentation and updating of SOPs and laboratory protocols, identifying who has authority to change SOPs, the processes by which they are changed, and audits.

19. There is a certain amount of error associated with all measurements regardless of the care with which the device was calibrated. This error is the uncertainty inherent in the measurement and the human error, either determinate or indeterminate, that can creep into an experiment. If the analyst and his/her client are to rely on the results, these must be taken into account.

20. Determinate errors are avoidable blunders that are known to have occurred. Indeterminate errors are errors that are assumed to have occurred but there is no direct knowledge of such or errors that are inherent in measurement in general. They are the errors that are dealt with by statistical treatment of the data and/or results.

21. A bias is an error that is known to occur each time a given procedure is carried out. Its effect is usually known such that a correction factor can be applied.

22. This is a determinate error because it is an error that is known to have occurred. Indeterminate errors are either errors inherent in a measurement or human errors that are not known to have occurred.

23. This is an indeterminate error because it is inherent in the use of the analytical balance.

24. Yes, if all determinate errors are eliminated and if indeterminate errors are taken into account by statistics.

25. "Accuracy" refers to how close a measurement or result is to being correct. "Precision" refers to the repeatability of a measurement. A precise measurement or result is not necessarily accurate.

26. They are very precise because they are within the uncertainty in the last significant figure expressed. It is not known with certainty if the results are accurate, although good precision usually indicates good accuracy.

27. The measurements are not precise because they differ widely. We cannot say anything about accuracy because we do not know the correct answer.

28. The results are still quite precise (see the answer to question 25 above), but given the additional information, we now know that they are not accurate.

29. Standard deviation is 0.0013. Relative standard deviation is 0.000075.

30. The percent standard deviation is 1.3%. The stated precision is exceeded.

31. Because it relates the value of the standard deviation to the value of the mean.

32. This means that if the experiment is carried out properly, that the precision of the data or result is such that a relative standard deviation of 1% or less can be achieved.

33. If the result of the analytical testing of this batch shows that it is out of statistical control (by being plotted outside the action limits on a control chart, for example), then this batch must be at least quarantined for further testing. If further testing indicates the same result, then the batch must be rejected.

34. A quality control chart is a visual aid for determining whether a given analytical result is outside the action limits determined for the results for that procedure. If it is outside the action limits, the cause may be a problem with the procedure, among other things.

35. (a) 5.0562, 5.066, none (b) 0.55984, 0.55995, 0.5615 (c) 43.476, 43.49, 43.49 (d) 12.540, 12.559, none (e) 0.1185, 0.1183, 0.1190 (f) 0.09745, 0.09745, none (g) 32.850, 32.85, 32.85

36. (a) 9.3%, 93 ppt, (b) 33.1%, 331 ppt, (c) 0.0144, 1.445, 14.4 ppt, (d) 0.0098, 0.98%, 9.8 ppt, (e) 0.3707, 0.0047, 0.0126, 1.26%, 12.6 ppt, (f) 0.1102, 0.00065, 0.0059, 0.59%, 5.9 ppt, (g) 1.4410, 0.0057, 0.0040, 0.40%, 4.0 ppt, (h) 56.37, 0.28, 0.0050, 0.50%, 5.0 ppt, (i) 0.423, 0.0082, 0.20, 2.0%, 2.0×10^1 ppt, (j) 23.67, 0.16, 0.0069, 0.69%, 6.9 ppt

37. (a) 2.015, 16.37 ± 0.23, (b) 2.447, 0.238 ± 0.018, (c) 4.604, 5.20 ± 0.21, (d) 2.365, 67.30 ± 0.08, (e) 2.015, 1.265 ± 0.037, (f) 3.707, 38.034 ± 0.146

38. (a) 0.047, no, (b) 0.046, no, (c) 0.0537, no, (d) 0.037, no, (e) 0.013, no, (f) 0.020, no.

39. (a) 0.28, 0.45, 0.62, 14.50, no, (b) 0.312, 0.333, 0,937, 0.015, yes, (c) 0.74, 1.09, 0.68, 47.51, no, (d) 0.106, 0.120, 0.883, 2.447, yes, (e) 0.06, 0.25, 0.24, 29.25, no, (f) 0.024, 0.031, 0.77, 0.259, yes.

CHAPTER 2

1. A representative sample must be obtained and transported to the laboratory safely without alteration, and the sample must be appropriately prepared (such as dried, dissolved, extracted, etc.) for the particular method to be used.

2. If a sample does not represent the bulk system in the manner intended, if the sample's integrity is not maintained during transportation, or if the preparation schemes go awry and do not provide the intended product, then the analytical results to be reported will be incorrect regardless of whether the analytical method was performed without error.

3. A representative sample is a sample that has all the characteristics in exactly the same proportions as the bulk system from which it came.

4. If a particular part of the bulk system under investigation is known to have an analyte concentration very different from the rest of the system, then taking a selective sample taken from that part of the system and analyzing it independently would make sense.

5. (a) At a given site, sample at different lateral positions, flowing water and semistagnant water, different depths, etc., then combine all samples into one composite sample. (b) The film is likely produced in different lots and on different production lines. Take random samples manufactured on a given day on a given production line (from a given lot) and combine. (c) Randomly sample several tablets from the bottle and combine. (d) Sample at different locations, at different depths, etc. and combine. (e) Take samples from all sides of the building, at different heights, under the eaves, close to the ground, etc. and combine. (f) Tissue from a particular organ may be more advantageous than others. Obtain samples from whatever organ is targeted and combine.

6. A bulk sample is the original undivided sample that was taken directly from the bulk system being characterized. A primary sample is the same as a bulk sample. A secondary sample (also subsample) is a part of the primary sample taken to the next step. A laboratory sample is a sample taken to a laboratory for analysis. The laboratory sample could be a primary sample or a subsample. A test sample is that part of a laboratory sample actually measured out for the method used.

7. The act of obtaining samples from a bulk system is subject to errors that can neither be detected nor compensated due to the bulk system often being nonhomogeneous and the sample therefore possibly not being exactly representative. Such errors are "indeterminate" and must be dealt with by statistics.

8. If the integrity of a sample is called into question, a paper trail, or chain of custody, must be examined in order to discover errors. This chain of custody documents who had custody of the sample and what actions were performed so that the sample integrity can be verified.

9. Glass containers can leach trace levels of metals and contaminate the sample.

10. It can be refrigerated to slow down the bacterial action.

11. Answer must be discovered in a reference book or Web site.

12. Particle size reduction makes dissolving procedures more efficient and extracting procedures more accurate because of the improved contact of the solvent with the sample. If accompanied by thorough mixing, particle size reduction also results in more homogeneous samples, which are more representative.

13. If the sample is not homogeneous, its division into smaller samples may result in a test sample that has a different overall composition than the original sample.

14. An extraction is the removal of the analyte from a sample through contact with a solvent in which the analyte is soluble.

15. Extraction is an incomplete dissolution. As such, certain parts of the sample are dissolved while other parts are not. The parts that are dissolved are thought of as having been "extracted" or removed" from the sample.

16. A solid–liquid extraction can be as simple as mixing the solid material in a flask with solvent followed by filtration, or a Soxhlet extraction. The extract in a Soxhlet extraction would no longer have the solid sample with the extract, so it would not require a filtration.

17. A supercritical fluid is a state of matter achieved by high temperature and extremely high pressure, exceeding the so-called critical temperature and pressure for that substance. The solvent properties of a supercritical fluid are much improved over the normal solvent properties of that fluid.

18. Vapor pressure is a measure of the tendency of a substance to be in the gas phase at a given temperature. At the boiling point of a liquid mixture, the component with the higher vapor pressure will have a higher concentration in the condensing vapors and thus will be purer than it was before.

19. A high vapor pressure compared with the hardness minerals means that water has a much stronger tendency to evaporate such that upon distillation, water will vaporize free of contamination from the minerals and, when the vapors are condensed, the water will be much purer than before.

20. The difference in vapor pressure and boiling point of water and dissolved hardness minerals is substantial, such that one simple distillation will result in significant purification. In the case of two liquids, however, these differences are much narrower, such that many distillations, or a fractionating column, are required.

21. Distilled water is water that is obtained by boiling tap water and recondensing the vapors. Deionized water is water that has been passed through a bed of resin that exchanges the ions of dissolved minerals with hydrogen and hydroxide ions, the combination of which is water.

22. (a) water, (b) HNO_3, (c) HF, $HClO_4$, or aqua regia, (d) HCl, (e) aqua regia, (f) HF, (g) H_2SO_4

23. (a), (g), (l) sulfuric acid
    (b), (h), (i) hydrofluoric acid
    (c), (e), (f) nitric acid
    (d), (j) hydrochloric acid
    (k) perchloric acid

24. A mixture of concentrated HNO_3 and concentrated HCl in the ratio 1:3 by volume.

25. Refer to Table 2.2 and accompanying discussion.

26. (a) nitric acid, (b) hydrochloric acid, (c) hydrochloride acid, (d) sulfuric acid, (e) hydrofluoric acid, (f) sulfuric acid, (g) sulfuric acid
27. HCl—(b)
    HNO_3—(e)
28. Fusion is the dissolving of a sample with the use of a molten inorganic salt called a "flux."
29. Refer to the answer to question 23 above.
30. Such a material must be insoluble in the flux. Examples are platinum, gold, nickel, and porcelain.
31. A solid–liquid extraction is an extraction in which an analyte in a solid sample is extracted into a liquid solvent. A liquid–liquid extraction is an extraction in which an analyte dissolved in a liquid sample is extracted into a second liquid solvent that is immiscible with the first. A solid-phase extraction is an extraction in which an analyte or its liquid solvent is extracted via contact with a solid material (sorbent) as the solution passes through a cartridge containing the sorbent.
32. (1) Analysis of water for pesticide residue; (2) analysis of soil for metals.
33. They are nonpolar and, as such, do not mix with water and will extract nonpolar solutes from water.
34. Diethyl ether is highly flammable and also highly volatile, a combination that can result in an explosion due to a flame source igniting the vapors in a wet laboratory in which it is being used. In addition, decomposition over time results in the formation of highly unstable and explosive peroxides in the stored containers. Precautions include avoiding open flames, working in a fume hood, storing the containers in explosion-proof refrigerators, the use of metal containers, which slow the peroxide formation, and disposal after about 9 months of storage.
35. *n*-Hexane, benzene, toluene, and diethyl ether are less dense than water, and thus would be the top layer in an extraction experiment. Liquids with a higher density than water would sink to the bottom of the separatory funnel.
36. (a) All the organic liquids mentioned are toxic to a certain extent, and certainly they should never be ingested. However, chloroform is more toxic than the others and requires special handling.
    (b) *n*-Hexane, toluene, and diethyl ether are flammable.
37. A purge and trap procedure is one in which a volatile analyte is purged from solvent by helium sparging and trapped on a sorbent held in a cartridge through which the helium then passes.
38. See Section 2.9.2.
39. Solvent exchange is a process in which the analyte solution is evaporated to dryness and the residue reconstituted with a different solvent.
40. In a liquid–liquid extraction experiment, two immiscible liquids, one pure and one with the analyte dissolved in it, are brought into intimate contact by vigorous shaking in a common container (separatory funnel). If the analyte is more soluble in the pure liquid than in the original liquid, its molecules move from the original to the pure during the shaking. Following the shaking, the immiscible liquids separate into two liquid layers in the container and most of the analyte is found in the extracting solvent.
41. The two liquid phases must be immiscible and the analyte must be more soluble in the extracting solvent.
42. The separatory funnel is the glass or plastic container that is used for liquid–liquid extractions. It is shaped like an inverted teardrop and has a stopper at the top and a stopcock at the bottom (see Figure 2.5). It is used as described in question 40 above in a repeated shaking and venting procedure as pictured in Figure 2.6a. Following the extraction, the liquid layers are allowed to separate and the bottom layer drawn off through the stopcock.

43. If the water containing the iodine is placed in the same container as pure hexane and the container shaken, extraction will be demonstrated because when the iodine transfers to the hexane layer, this layer will turn pink.

44. (a) 4.6 M, (b) 0.0447 M

45. If we assume that the concentration of analyte in the original solvent is 0.060 M before extraction, then, after rounding the answer of the calculation according to significant figure rules, the concentration in the extraction solvent after extraction is also 0.060 M (A very large distribution coefficient as in this problem would indicate that virtually all the analyte is extracted.).

46. Distribution coefficient = 0.731, % extracted = 19.6%

47. 0.046 g

48. (a) 35.8, (b) 2.17, (c) 0.321 M, (d) 0.00770 M, (e) 0.241, 0.0198 M, (f) 5.6, 0.059 M, (g) 1.10, 0.131 M, (h) 0.77, 0.0222 M, (i) 0.00782 M, 0.00536 M, (j) 0.0110 M, 0.0496 M

49. (a) 88%, (b) 0.0064 g, 45%, (c) 0.0051 g, 3.6, 78%, (d) 0.0047 g, 16, 77%, (e) 0.0006 g, 13, 84%, (f) 0.00026 g, 0.0022 g, 88%, (g) 0.0030 g, 0.0015 g, 0.15

50. These two quantities are the same, except that distribution ratio takes into account all dissolved forms of the analyte, whereas the distribution coefficient takes into account only one form. If only one form exists, then the two are identical.

51. An evaporator is a device that eliminates excess solvent from a liquid solution and concentrates the solutes by vaporizing a portion or all of the solvent. The modern way to accomplish this is by using a stream of inert gas blowing over the surface of the solution while applying gentle heat.

52. A solid–liquid extraction is the selective dissolution of a component of a solid material through contact with a liquid solvent. Two methods of accomplishing this are (1) shaking the solid with the solvent in the same container for a specified period and (2) using a Soxhlet extractor, in which fresh solvent is continually cycled through the solid for a specified period.

53. No. Besides the convenient use of the separatory funnel for the actual extraction, they are also designed for easy separation of two immiscible liquids after the extraction through the stopcock. It would not be easy to separate a liquid from a solid through the stopcock.

54. Potassium, iron, phosphorus, and other elements in soil; formaldehyde residue in insulation, cellophane, and other materials; pesticide and herbicide residue in plants.

55. Two liquids in a mixture typically have a significant vapor pressure and similar boiling points. Thus, a clean separation does not occur with only a simple distillation. A fractional distillation is usually required.

56. A theoretical plate consists of one evaporation/condensation step in the distillation process. The height equivalent to a theoretical plate is the length of fractionating column in which is contained one theoretical plate.

57. The quality of the reagent must be assured in order for the analyst to have the confidence that the sample and prepared reagents will not be contaminated and give inaccurate results.

58. "ACS certified" means that a reagent meets or exceeds the specifications of purity set by the American Chemical Society.

59. No, because they generally are not pure enough.

60. "Spectro Grade" or "Spectranalyzed" for spectrophotometric analysis. "HPLC grade" for liquid chromatographic analysis.

61. See Section 2.14.

CHAPTER 3

1. Gravimetric analysis is a wet chemical method of analysis in which the measurement of weight is the primary measurement and most of the time the only measurement that is made on the analyte and its matrices.

2. Gravimetric analyses do not usually require standard solutions of any kind and do not require the calibration of any equipment beyond a balance.

3. Mass is a measure of the "amount" of a sample. Weight is the measure of the gravitational pull on this amount of sample. On the surface of the earth, a weight measurement is taken to be the same as a mass measurement, utilizing the same units, although technically, they are not the same.

4. A weighing device is called a balance because a weight is often determined by balancing the object to be weighed with a series of known weights across a fulcrum. Modern "torsion" balances, however, do not operate this way, but are still called "balances."

5. A single-pan balance is a balance with a single pan, which is used only for the object to be weighed. In the older styles, a constant counterbalancing weight is used while a number of removable weights on the same side as the object are added or removed in order to obtain the weight of the object. The more modern single-pan balances are of the torsion variety.

6. To say that a weight measurement was made to the nearest 0.01 g means that the second digit past the decimal point is the last digit available on the balance used. To say that the precision of a balance is ±0.1 mg (±0.0001 g) means that the fourth place past the decimal point is the last digit that can be obtained by that balance.

7. A top-loading balance is an electronic balance with the sample pan on the top. It is not enclosed, which means that it is capable of measuring only to the nearest hundredth of a gram (see Figure 3.3a).

8. To say a balance has the "tare" feature means that the balance can be zeroed with an object, such as a piece of weighing paper, on the pan. This helps to obtain the weight of a chemical directly without having to obtain and subtract the weight of the weighing paper.

9. Analytical balances are balances that measure to 0.1 or 0.01 mg.

10. A number of considerations are important. For example, the analytical balance must be level, the pan must be protected from air currents, the object to be weighed must be at room temperature, the object must be protected from fingerprints, etc.

11. A desiccator is a storage container used to either dry samples or, more commonly, to keep samples and crucibles dry and protected from the laboratory environment once they have been dried by other means. An indicating desiccant is a desiccant that changes color when saturated with adsorbed water. Drierite™ is an example of a desiccant. It is anhydrous calcium sulfate.

12. To check the calibration of a balance is to weigh a known weight to verify that the balance gives the correct weight.

13. It depends on how heavy the sample must be. If it is less than a gram, then an analytical balance must be used because only such a balance would give four significant figures. However, if the sample is greater than 10 g, then an ordinary balance is satisfactory since it would give 4 significant figures.

14. A physical separation is a separation that results from some nonchemical operation, such as evaporating a solvent, filtering, etc. A chemical separation uses a chemical reaction to effect the separation.

15. (a) 29.031%, (b) 1.6265 g, 18.81%, (c) 0.4642, (d) 8.5219 g, 0.8823 g, 10.35%, (e) 8.450%, (f) 32.16%, (g) 68.93%

16. (a) Loss on drying is the determination of the percentage weight loss that occurs on a sample as a result of heating at or below the boiling point of water or as a result of desiccation.

    (b) 50.594%

17. 27.239%

18. "Heat to constant weight" refers to repeating heating, cooling, and weighing operations performed until there is no longer any loss in weight.

19. "Loss on ignition" is the determination of the percentage weight loss that occurs on a sample as a result of heating to ignition temperatures. "Residue on ignition" is the determination of the percentage residue weight remaining after heating to ignition temperatures.
20. Volatile organics: 48.301%, residue: 51.699%
21. Volatile organics: 55.992%, residue: 44.008%
22. (a) 12.81%, (b) 0.706 g, 5.67%, (c) 0.4005 g, 0.623 g, 15.6%, (d) 1.3237 g, 0.2141 g, 16.17%, (e) 67.53%, (f) 41.54%, (g) 1.4815 g, 0.4079 g, 27.53%, (h) 43.78%
23. (a) 92.447%, (b) 94.120%, (c) 2.7864 g, 85.311%, (d) 0.4548 g, 49.28%, (e) 1.5195 g, 1.2631 g, 83.126%, (f) 79.012%, (g) 83.433%, (h) 1.6145 g, 1.2298 g, 76.172%, (i) 86.37%
24. Insoluble matter in a general analysis report on the label of a chemical container is the percentage of the material that does not dissolve in water.
25. Sand, 78.565% Salt, 21.435%
26. 57.6%
27. (a) 72.05%, (b) 83.9 g, 63.2%, (c) 147.7 g, 61.0%, (d) 53.19%, (e) 46.66%, (f) 44.55 g, 8.72 g, 19.6%, (g) 48.9%
28. (a) 0.5744, (b) 0.3430, (c) 0.8084, (d) 1.473, (e) 0.7071, (f) 0.6065, (g) 0.5501, (h) 1.103
29. (a) 1.15730, (b) 1.29963, (c) 1.1768
30. (a) 0.3622, (b) 0.9309, (c) 0.96618
31. 0.0984 g
32. (a) 1.1 g, (b) 2.33 g, (c) 1.8 g, (d) 72.2 g, (e) 2.20×10^2 g, (f) 201 g
33. 0.8131 g
34. 1.513 g
35. 49.17%
36. 0.1840 g
37. 34.38%
38. 18.18%
39. 43.30%
40. 26.905%
41. 26.31%
42. There is no chance for the weighed sample to be anywhere but in your beaker. Also eliminates the chance of contamination.
43. To get the precipitate particles to "clump" together to make them more filterable.
44. From the time you weigh the item the first time until you weigh it the second time. It is during that time that the weight of a sample would be inaccurate due to fingerprints, since such added weight would alter the calculated weight of the sample.
45. The precipitate and filter paper are sopping wet at this point and would saturate the desiccant. Desiccators are mostly used to keep things dry once they are already dry, not to dry things that are wet.
46. The percent would be lower than the true percent because the weight of the precipitate (the numerator in the percent calculation) would be lower.

CHAPTER 4

1. Gravimetric analysis utilizes primarily weight measurements and may or may not involve chemical reactions. Titrimetric analysis utilizes both weight and volume measurements and always involves solution chemistry and stoichiometry.
2. Titrimetric analysis is sometimes called volumetric analysis because it is characterized by the frequent measurement of solution volume utilizing precision glassware.
3. For gravimetric analysis, a solution may be needed to react with the analyte. Otherwise, it consists of just physical separation operations and usually initial and final weight

measurements. For titrimetric analysis, solutions are always needed to react with the analyte, and these solutions must be standardized. Also, a critical measurement is a volume measurement (buret reading).

4. See Section 4.2 for the definition of terms.

5. (a) 0.195 M, (b) 3.13 M, (c) 0.01041 M, (d) 0.01607 M, (e) 1.8 M, (f) 2.40 M

6. (a) NaOH: 40.00, HCl: 36.461, (b) NaOH: 40.00, H_2SO_4: 49.04, (c) HCl: 36.46, $Ba(OH)_2$: 85.6710, (d) NaOH: 40.00, H_3PO_4: 32.67, (e) HCl: 36.46, $Mg(OH)_2$: 29.160, (f) NaOH: 40.00, H_3PO_4: 49.00, (g) NaOH: 40.00, Na_2HPO_4: 141.98, (h) NaOH: 40.00, H_3PO_4: 98.00, (i) Na_2CO_3: 52.9945, HCl: 36.46

7. (a) 0.159 N, (b) 5.16 N, (c) 10.8 N, (d) 9.92 N, (e) 0.0492 N, (f) 0.0121 N, (g) 0.0428 N

8. 0.2411 M

9. (a) Dissolve 2.8 g of KOH in water, dilute to 500.0 mL, and shake.
   (b) Dissolve 2.2 g of NaCl in water, dilute to 250.0 mL, and shake.
   (c) Dissolve 36 g of glucose in water, dilute to 100.0 mL, and shake.
   (d) Dilute 4.2 mL of 12.0 M HCl with water to 500.0 mL and shake.
   (e) Dilute 12 mL of 2.0 M NaOH with water to 100.0 mL and shake.
   (f) Dilute 56 mL of 18.0 M H_2SO_4 with water to 2.0 L and shake.

10. (a) Dissolve 6.8 g of KH_2PO_4 in water, dilute to 500.0 mL, and shake.
    (b) Dilute 1.5 mL of 18.0 M H_2SO_4 with water to 500.0 mL and shake.
    (c) Dissolve 7.1 g of $Ba(OH)_2$ in water, dilute to 750.0 mL and shake.
    (d) Dissolve 1.6 g of Na_2CO_3 in water, dilute to 200.0 mL and shake.
    (e) Dissolve 15 g of $NaHCO_3$ in water, dilute to 700 mL and shake.
    (f) Dilute 14 mL of 15.0 N $Ba(OH)_2$ with water to 700.0 mL and shake.
    (g) Dilute 1.0 mL of 15 M H_3PO_4 with water to 300.0 mL and shake.

11. (a) Dissolve 3.5 g of NaOH in water, dilute to 250.0 mL, and shake.
    (b) Dilute 15 mL of 6.0 M NaOH with water to 250.0 mL and shake.

12. 29 mL

13. 93 mL

14. (a) Dissolve 18 g of KNO_3 in water, dilute to 500.0 mL and shake.
    (b) Dilute 39 mL of 4.5 M KNO_3 with water to 500.0 mL and shake.

15. Use 5.0 mL of 17 M acetic acid and 12 g of sodium acetate. Dissolve the sodium acetate in water, add the 17 M acetic acid, dilute to 500.0 mL, and shake.

16. 1.56×10^3 mL

17. The ST/T ratio is the ratio of the balancing coefficients for the substance titrated (ST) and the titrant (T) in the balanced chemical equation. The units of these coefficients are moles ST and moles of T, respectively. Multiplying the moles$_T$ by this ST/T ratio then causes the moles$_T$ unit to cancel, such that we are left with moles$_{ST}$.

18. Multiplying L_T by M_T gives us moles$_T$, i.e., the liters unit cancels as you can see in Equation 4.17. Then, multiplying moles$_{ST}$ by the mole ratio ST/T gives us moles$_{ST}$ as explained in the answer to question 17 above.

19. When dealing with equivalents in a reaction, the ST/T ratio is always 1/1. This means that, at the end point of a titration, you have added a number of equivalents of titrant equal to the number of equivalents of substance titrated.

20. Multiplying L_T by N_T gives us equiv$_T$ because the liter unit cancels as you can see in Equation 4.21. Equation 4.23 then follows because of the facts stated in the answer to question 19 above.

21. See Section 4.6.1.

22. See Section 4.6.1.

23. 0.05586 M

24. 0.2671 M

25. 0.1056 M

26. 0.2250 N
27. 0.08798 N
28. 0.1733 N
29. No. The equivalent weight of sulfuric acid is different from that of hydrochloric acid, but it doesn't enter into the calculation.
30. See Section 4.6.2
31. See Section 4.6.2
32. SRM stands for "standard reference material." It is a standard chemical manufactured and certified by the National Institute for Standardization and Technology (NIST) as being exactly as labeled. It is the ultimate standard. CRM stands for "certified reference material." It is a standard chemical manufactured and certified by a vendor as being exactly as labeled. It is "traceable" to a SRM, meaning it has been compared with and certified with the use of an ultimate standard.
33. Titer is an alternate concentration expression for a titrant that is specific to a particular substance titrated. It is the weight of the substance titrated that is consumed by 1 mL of the titrant.
34. 2.22
35. See Section 4.7.
36. See Section 4.7.
37. 75.87%
38. 21.22%
39. 38.37%
40. 36.94%
41. (a) false, (b) false, (c) true, (d) true, (e) false, (f) true, (g) false, (h) false
42. (a) measuring pipette
    (b) the last bit of solution in the tip should NOT be blown out into the receiving vessel
    (c) otherwise, the solution would be diluted with the film of water on the inside surface.
43. A volumetric pipette, because the diameter of the tube where the calibration line is located is narrower.
44. No. A volumetric flask is calibrated TC, not TD.
45. A frosted ring near the top means the pipette is calibrated for blow-out.
46. A volumetric flask is a precision piece of glassware with a single calibration line in a narrow neck. An Erlenmeyer is not a precision piece of glassware and has graduation lines on it only as a rough indication of volume.
47. Rinsing a pipette with the solution to be transferred assures that the liquid film adhering to the inside surface is the solution to be transferred and not water, which would contaminate/dilute the solution to be transferred.
48. The serological pipette is calibrated through the tip and may be used by adjusting the meniscus to the line that would deliver the desired volume by letting it drain completely (and blowing it out). The Mohr pipette is only calibrated to the lowest calibration line and cannot be drained below this line.
49. (a) A volumetric flask is not calibrated to deliver a volume, only to contain. There is a difference because the thin liquid film adhering to the inner wall is contained but not delivered.
    (b) I would tell him to use a 50-mL volumetric pipette.
50. Because the liquid film adhering to the inner wall would contaminate/dilute the solution being transferred if it were not rinsed first with this solution.
51. By dilution: measure out 6.22 mL of the 4.021 M solution as precisely as possible (measuring pipette) and dilute it to 100.0 mL (volumetric flask) and shake. By weight, the pure solid: Weigh out 2.650 g on an analytical balance, place in a 100-mL volumetric flask, dissolve in water, dilute to 100.0 mL, and shake.

52. If I chose a Mohr pipette, it would not be calibrated for blow-out. If I chose a serological pipette, it would be calibrated for blow-out. A blow-out pipette would have one or two frosted rings completely circumscribing the top of the pipette (see Figure 4.20 for details on how to make the delivery).

53. (a) Use a volumetric flask and an analytical balance.
    (b) Standardization is an experiment by which a solution concentration is determined with good precision.

54. (a) Volumetric pipette: one calibration line. Serological pipette: many graduation lines.
    (b) Volumetric pipette: not calibrated for blow-out. Serological pipette: calibrated for blow-out.
    (c) Serological pipette, because as volumetric pipettes have just one calibration line, they are not manufactured for odd volumes such as 3.72 mL.
    (d) Both are calibrated TD.

55. Pipette-calibrated TC are used for viscous solutions that don't drain well. For these, a pipette contains the solution and then the contained volume is rinsed out into the receiving vessel.

56. Pipetters: a positive displacement device, meaning that 100% of the liquid drawn in is forced out with a plunger rather than allowed to flow out by gravity. The typical pipetter employs a bulb concealed within a plastic fabricated body, a spring-loaded push button at the top, and a nozzle at the bottom for accepting a plastic disposable tip. They may be fabricated for either single or variable volumes. In the latter case, a ratchet-like device with a digital volume scale is used to "dial in" the desired volume. Examples are shown in Figures 4.24, 4.25, and Figure 4.26.
    Micropipette and micropipetter: pipetters that are used for microliter volumes.
    Repipet: a bottle-top dispenser.
    Digital buret: a bottle-top dispenser that delivers 0.01 mL increments from a reagent bottle containing the titrant.
    Automatic titrators: an electronic titrator that draws the titrant from a reagent bottle and stores it in a built-in reservoir fitted with a plunger. The titration is performed by pressing a key on a keypad. One tap of the key delivers 0.01 mL to the titration flask.

57. Most chemists and technicians agree that Class A glassware items do not need to be checked for proper calibration.

58. Checking the calibration of a pipette or volumetric flask involves weighing a delivered volume of water or a contained volume of water (both at the temperature indicated on the label) to determine if the delivered or contained volume weighs what it is supposed to weigh. Of course, the balance used must have been properly calibrated as well.

59. A graduated cylinder is sufficient because this volume does not affect the result and is not entered into the calculation.

CHAPTER 5

1. The equivalence point must be easily and accurately detected, the reaction involved must be fast, and the reaction must be quantitative.

2. Monoprotic acid: an acid that has just one hydrogen ion to donate per molecule.
   Polyprotic acid: an acid that has more than one hydrogen ion to donate per molecule.
   Monobasic base: a base that will accept just one hydrogen ion per formula unit.
   Polybasic base: a base that will accept more than one hydrogen per formula unit.
   Titration curve: a plot of pH vs. mL of titrant showing the manner in which pH changes vs. mL of titrant during an acid–base titration.
   Inflection point: a point in a titration curve at which the slope of the curve is near zero.

3. These curves are shown in Figures 5.1a and 5.1b. Both curves show the same pattern after the inflection point because in both cases, the acids have been neutralized at that point and the pH depends only on the added NaOH. The acetic acid curve shows a higher pH level leading up to the equivalence point than the HCl curve. Also, the inflection point of the HCl curve covers a broader pH range than that of the acetic acid. The reason for these differences is that acetic acid is a weak acid, meaning fewer hydrogen ions in solution for the same total concentration. Fewer indicators will work for the acetic acid titration because of this narrower range.

4. These curves are shown in Figure 5.3a and 5.3b. Both curves show the same pattern after the inflection point because in both cases, the bases have been neutralized at that point, and the pH depends only on the added HCl. The ammonium hydroxide curve shows a lower pH level leading up to the equivalence point than the NaOH curve. Also, the inflection point of the NaOH curve covers a broader pH range than that of the ammonium hydroxide. The reason for these differences is that ammonium hydroxide is a weak base, meaning more hydrogen ions in solution for the same total concentration. Fewer indicators will work for the ammonium hydroxide titration because of this narrower range.

5. These curves are shown in Figures 5.7 and 5.8. The only similarity is the final pH level near the end of the titration. At that point, both acids have been neutralized and in each case, the pH depends only on the added NaOH. The phosphoric curve shows a higher pH level at the beginning. This is because it is a weaker acid and there are fewer hydrogen ions in solution. The phosphoric acid curve shows three inflection points, whereas the sulfuric acid curve shows essentially one inflection point. This is because phosphoric acid has three weakly acidic hydrogen ions to be neutralized per molecule, whereas sulfuric has just two hydrogen ions that are strongly acidic and are neutralized together. The inflection point of the sulfuric acid covers a broader pH range and many indicators would work. With phosphoric acid, the analyst has a choice of which inflection point to use and the indicator choice depends on which is chosen. Once a range is chosen, there would be fewer choices for the indicator because of the narrower pH range of each inflection point.

6. (a) See Figure 5.1b.
   (b) See Figure 5.3b.

7. (a) See Figure 5.1b.
   (b) See Figure 5.3a.
   (c) See Figure 5.3b.

8. No (see Figure 5.12). For Experiment 10, the second inflection point was chosen and is at too low a pH range to utilize the phenolphthalein end point.

9. No (see Figure 5.1b). The inflection point is at too high a pH range for the bromocresol green end point.

10. The point where the titration curve is undergoing the greatest change is the equivalent point. In the first derivative, this is shown as the maximum point. In the second derivative, this is the point where the curve crosses the x-axis.

11. The indicator must change color in the range of approximately 6.5 to 9.0. Phenolphthalein would be chosen over bromocresol green.

12. The "P" in KHP refers to "phthalate." The phthalate ion is a benzene ring with two carboxyl groups on adjacent carbons (see Figure 5.9). Potassium hydrogen phthalate (KHP) is useful as a primary standard because it possesses all the qualities sought in a primary standard (see Chapter 4, Section 4.6.2).

13. THAM, or TRIS, is "tris-(hydroxymethyl)amino methane." Its structure is $(HOCH_2)_3CNH_2$. It is a base because it contains the $-NH_2$ group, which accepts a hydrogen ion to form $-NH_3^+$. It is useful as a primary standard because it possesses all the qualities sought in a primary standard as discussed in Chapter 4, Section 4.6.2.

14. Sodium carbonate accepts two hydrogen ions per formula unit and each is accepted separately during a titration, hence two inflection points. In the region leading up to the second inflection point, carbonic acid is a product of the reaction. Because carbonic acid is in equilibrium with CO_2 and water, the reaction is sluggish and does not go to completion unless the CO_2 is eliminated by boiling. The second inflection point is at too low a pH range for phenolphthalein to work, but is just right for bromocresol green.

15. "Total alkalinity" is the capacity for a volume of water to neutralize an added standard acid solution usually to obtain a pH of 4.5. It is expressed as the millimoles of H^+ per liter of water.

16. 6.14 mmol/L

17. A back titration is a titration in which the titrant is added in excess and the excess amount titrated with a second titrant, the so-called back titrant, so as to come back to the end point.

18. Calcium carbonate is not soluble in water. The addition of the standard acid to a tablet immersed in water causes it to dissolve, but rather slowly. A direct titration is not appropriate because the reaction is not fast and it would be difficult to detect the end point. A back titration is therefore useful because the standard acid can be added in excess so that the calcium carbonate completely dissolves. The excess acid can then be titrated with a standard base.

19. 85.53%

20. (a) It is used to dissolve/digest the sample.
    (b) It is one option for the acid in the receiving flask to react with the ammonia.

21. We must subtract the excess equivalents of titrant (which is equal to the equivalents of back titrant used) from the total equivalents of titrant added so that we have the equivalents of titrant that actually reacted with the analyte.

22. 4.832%

23. 5.052%

24. Soda pop, aspirin, cleaning products containing ammonia, etc. (see Table 5.1).

25. Buffer solution: a solution that resists changes in pH even when a strong acid or base is added or when it is diluted with water.
    Conjugate acid: the product of the neutralization of a base that is an acid because it can lose the hydrogen ion that it gained during the neutralization.
    Conjugate base: the product of the neutralization of an acid that is a base because it can gain back the hydrogen ion that it lost during the neutralization.
    Conjugate acid/base pair: a pair of compounds consisting of a base and its conjugate acid or an acid and its conjugate base.
    Buffer capacity: the capacity of a buffer solution to resist pH changes.
    Buffer region: the region of a titration curve leading up to the inflection point.

26. Sodium acetate is a conjugate base when it results from the neutralization of acetic acid. The acetate ion can gain back the hydrogen ion that it lost and become acetic acid again. Ammonium chloride is a conjugate acid when it results from the neutralization of ammonia. The ammonium ion can lose the hydrogen ion it gained and become ammonia again.

27. The Henderson–Hasselbalch equation is an equation expressing the relationship among pH, pK_a, and the log of the ratio of the concentrations of the base to its conjugate acid, or an acid to its conjugate base. It is derived from the K_a or K_b expression (see Equations 5.25 through 5.28 in the text). They are each a form of this equation.

28. pH = 2.77

29. pH = 3.25

30. pH = 7.65

31. 5.7

32. A certain combination of chloroacetic acid and sodium chloroacetate would give a pH of 3.00 since the pH range for this combination is 1.8–3.8. From question 28, the K_a of chloroacetic acid is 1.36×10^{-3}. The ratio is 0.74.

33. This is the same question as question 15 in Chapter 4. Use 5.0 mL of 17 M acetic acid and 12 g of sodium acetate. Dissolve the sodium acetate in water, add the 17 M acetic acid, dilute to 500.0 mL, and shake.

34. Weigh 9.1 g of THAM and 32 g of THAM hydrochloride. Place both in the same 500 mL container. Add water to dissolve. Add water to the 500-mL mark and shake.

35. pH = 10.06

36. Monodentate, bidentate, hexadentate: these are the adjectives that describe a ligand in terms of the number of bonding sites (pairs of electrons) available for bonding to the metal ion. Mono = 1, bi = 2, hexa = 6.

    Ligand: the charged or uncharged chemical species that reacts with a metal ion forming a complex ion.

    Complex ion: a charged aggregate consisting of a metal ion in combination with one or more ligands.

    Coordinate covalent bond: a covalent bond in which the two shared electrons are contributed to the bond by only one of the two atoms involved.

    Water hardness: a term used to denote the Ca^{+2}, Mg^{+2}, and Fe^{+3} content of water both quantitatively and qualitatively.

    Aliquot: a portion of a larger volume of a solution, usually a transferred or pipetted volume.

37. Ligand: the charged or uncharged chemical species that reacts with a metal ion forming a complex ion. Examples: See Table 5.4.

    Complex ion: a charged aggregate consisting of a metal ion in combination with one or more ligands. Examples: See Table 5.4.

38. $CoCl_4^{2-}$ = complex ion; Cl^- = ligand

39. Ammonia is an example of a monodentate ligand (see Table 5.4 for other examples). EDTA is a hexadentate ligand. A monodentate ligand is one that has just one bonding site for bonding to a metal ion.

40. Bidentate, because there are two sites (hence, "bi"–) at which a metal ion will bond.

41. (a) The ligand is the reactant (left side of the equation) consisting of the three aromatic rings and two nitrogens.

    (b) The complex ion is the product of the reaction (right side of the equation).

    (c) The ligand is bidentate because it has two bonding sites that bond to the metal ion.

    (d) The complex ion is a chelate because the ligand involved has two more bonding sites.

42. (a) 6, (b) 1, (c) hexadentate

43. A basic pH is needed in order to expose all bonding sites on the EDTA (i.e., remove all acidic hydrogens) so that it can react completely with the metal ions. A pH of 8 is not basic enough for this (see Figure 5.21) and at a pH of 12, magnesium ions precipitate as the hydroxide and are lost to the analysis. A pH of 10 is a pH that will work.

44. (a) A complex ion formed between calcium ions and the eriochrome black T.

    (b) The free eriochrome black T ligand.

    (c) As a masking agent, cyanide acts as a ligand and forms a complex ion with a metal ion, in effect preventing the metal ion from interfering in an analysis. We say the metal ion is "masked" and that the cyanide is a "masking agent."

45. See answer to question 43.

46. (a) 6.2 mg of Mg are dissolved and the solution diluted to 250.0 mL.

    (b) 22.5 mg of Ag are dissolved and the solution diluted to 750.0 mL.

    (c) 24.0 mg of Al are dissolved and the solution diluted to 600.0 mL.

    (d) 7.5 mg of Mg are dissolved and the solution diluted to 500.0 mL.

(e) 7.50 mg of Fe are dissolved and the solution diluted to 250.0 mL.

(f) 12.5 mg of Cu are dissolved and the solution diluted to 100.0 mL.

47. (a) 12.5 mL
    (b) 9.0 mL
    (c) 6.2 mL
48. (a) 63.5 mg NaCl dissolved and diluted to 500.0 mL
    (b) 25.0 mL of the 1000.0 ppm diluted to 500 mL
49. (a) 7.50 mg of Fe dissolved and diluted to 250.0 mL
    (b) 54.3 mg dissolved and diluted to 250.0 mL
    (c) 7.50 mL of the 1000.0 ppm diluted to 250 mL
50. (a) 19.6 mg dissolved and diluted to 100.0 mL
    (b) 5.00 mg of copper metal dissolved and diluted to 100.0 mL
    (c) 5.00 mL of the 1000.0 ppm diluted to 100.0 mL
51. (a) 13 mL
    (b) 0.013 g
    (c) 0.022 g
52. (a) 35.7 mg
    (b) 25.3 mg
    (c) 36.1 mg
53. 4.70 g of disodium dihydrogen EDTA dihydrate are dissolved and diluted to 500.0 mL
54. 3.7 g
55. (a) 0.0103 M
    (b) 0.0196 M
    (c) 0.03688 M
    (d) 0.05755 M
    (e) 0.00775 M
    (f) 0.006418 M
    (g) 0.006578 M
    (h) 0.008402 M
    (i) 0.009371 M
56. (a) 404.2 ppm $CaCO_3$
    (b) 536.4 ppm $CaCO_3$
    (c) 271.7 ppm $CaCO_3$
    (d) 160.7 ppm $CaCO_3$
    (e) 110.2 ppm $CaCO_3$
    (f) 314.5 ppm $CaCO_3$
    (g) 283.5 ppm $CaCO_3$
57. Oxidation: the loss of electrons or the increase in oxidation number.
    Reduction: the gain of electrons or the decrease in oxidation number.
    Oxidation number: a number indicating the state an element is in with respect to bonding.
    Oxidizing agent: a chemical that causes something to be oxidized while being reduced itself.
    Reducing agent: a chemical that causes something to be reduced while being oxidized itself.
58. (a) +5, (b) +3, (c) +6, (d) +5, (e) +7, (f) +1, (g) +4, (h) +6, (i) +3, (j) +3
59. (a) +1, (b) −1, (c) +5, (d) 0, (e) +3, (f) +7
60. (a) +3, (b) 0, (c) +6, (d) +6, (e) +6
61. (a) +7, (b) −1, (c) 0, (d) +3, (e) +1
62. (a) +4, (b) −2, (c) 0, (d) +6, (e) +4, (f) −2, (g) +6, (h) +4, (i) +4, (j) +6
63. (a) +5, (b) +3, (c) +5, (d) +3, (e) +4, (f) +5, (g) −3, (h) +6, (i) 0, (j) +5
64. (a) CuO (or Cu) was reduced (+2 to 0) and NH_3 (or N) was oxidized (−3 to 0).
    (b) Cl_2 was reduced (0 to −1) and KBr (Br) was oxidized (−1 to 0).

65. (a) Mg is the reducing agent because it was oxidized from 0 to +2. HBr is the oxidizing agent because H is reduced from +1 to 0.
    (b) Fe is the reducing agent because it was oxidized from 0 to +3. O_2 is the oxidizing agent because it was reduced from 0 to −2.

66. (a) No, this is neutralization; no change in oxidation number.
    (b) Yes, S and N have changed oxidation number.
    (c) Yes, both Na and H have changed oxidation number.
    (d) No, this is neutralization; no change in oxidation number.
    (e) No, there were no oxidation number changes.
    (f) Yes, both K and Br have changed oxidation number.
    (g) Yes, both Cl and O have changed oxidation number.
    (h) No, this is neutralization; no change in oxidation number.

67. (b) is redox; Cu is oxidized, HNO_3 (N) is reduced. Cu is reducing agent, HNO_3 the oxidizing agent.

68. (a) (2) is redox
    (b) HCl (or H)
    (c) Zn
    (d) lose

69. (a) $3 H_2O + 4 Cl^- + 3 NO_3^- \rightarrow 4 ClO_3^- + 6 H^+ + 3 N^{3-}$
    (b) $2 H^+ + 2 Cl^- + 2 NO_3^- \rightarrow 2 ClO_2^- + N_2O + H_2O$
    (c) $ClO^- + 2 NO_3^- \rightarrow ClO_3^- + 2 NO_2^-$
    (d) $3 ClO^- + 2 H^+ + 2 NO_3^- \rightarrow 3 ClO_2^- + 2 NO + H_2O$
    (e) $4 ClO_3^- + SO_4^{2-} \rightarrow 4 ClO_4^- + S^{2-}$
    (f) Balanced as is.
    (g) $H_2O + 4 IO_3^- + SO_3 \rightarrow 4 IO_4^- + 2 H^+ + S^{2-}$
    (h) $Cl^- + 2 H^+ + SO_4^{2-} \rightarrow ClO^- + SO_2 + H_2O$
    (i) $3 Cl^- + 8 H^+ 4 SO_4^{2-} \rightarrow 3 ClO_4^- + 4 S + 4 H_2O$
    (j) $6 Br^- + 6 H^+ + SO_3 \rightarrow 3 Br_2 + S + 3 H_2O$
    (k) $5 I^- + 4 H^+ + 4 NO_3^- \rightarrow 5 IO_2^- + 2 N_2 + 2 H_2O$
    (l) $12 H_2O + 8 P + 5 IO_4^- \rightarrow 8 PO_4^{3-} + 24 H^+ + 5 I^-$
    (m) $3 H_2O + 3 SO_2 + BrO_3^- \rightarrow 3 SO_4^{2-} + 6 H^+ + Br^-$
    (n) $10 Fe + 30 H^+ + 3 P_2O_5 \rightarrow 10 Fe^{3+} + 6 P + 15 H_2O$
    (o) $2 Cr + 6 H^+ + 3 PO_4^{3-} \rightarrow 2 Cr^{3+} + 3 PO_3^- + 3 H_2O$
    (p) $5 Ni + 16 H^+ + 2 PO_4^{3-} \rightarrow 5 Ni^{2+} + 2 P + 8 H_2O$
    (q) $6 H^+ + 2 MnO_4^- + 5 H_2C_2O_4 \rightarrow 2 Mn^{2+} + 8 H_2O + 10 CO_2$
    (r) $6 I^- + 14 H^+ + Cr_2O_7^{2-} \rightarrow 3 I_2 + 2 Cr^{3+} + 7 H_2O$
    (s) $Cl_2 + H_2O + NO_2^- \rightarrow 2 Cr^- + NO_3^- + 2 H^+$
    (t) $3 S^{2-} + 8 H^+ + 2 NO_3^- \rightarrow 3 S + 2 NO + 4 H_2O$
    (u) Balanced as is.

70. (a) $6 S_2O_3^{2-} + 14 H^+ + Cr_2O_7^{2-} \rightarrow 3 S_4O_6^{2-} + 2 Cr^{3+} + 7 H_2O$
    (b) 0.4475 M

71. 0.03438 M

72. 0.02497 M

73. 0.007842 M

74. 19.76%

75. 70.37%

76. 55.76%

77. 42.08%

78. The intensely purple colored MnO_4^- solution becomes easily visible when there is no longer any "ST" available to react.

79. (a) It takes on electrons readily.
    (b) If potassium permanganate contacts oxidizable substances, a reaction will take place and the concentration of the permanganate will change to an unknown value.
80. An indirect titration is one in which the "ST" is determined indirectly by titrating a second species that is proportional to the "ST." A back titration is one in which the end point is intentionally overshot and the excess back titrated.
81. Iodometry is titrimetric method involving iodine. It is an indirect method because the product of the reaction of "ST" with I$^-$ (I$_2$) is titrated.
82. KI: the "titrant" from which the I$_2$ is liberated.
    Na$_2$S$_2$O$_3$: the titrant for the liberated I$_2$.
    K$_2$Cr$_2$O$_7$: the primary standard for the Na$_2$S$_2$O$_3$.
83. An indicator may be added that reacts with the titrant once the precipitation reaction is complete. The color change may be difficult to discern due to the presence of the precipitate, but an excess of the titrant may be added and then a blank correction applied. It may also be possible to detect the end point with an electrode.

CHAPTER 6

1. (1) Obtain the sample, (2) prepare the sample, (3) carry out the analysis method, (4) work the data, (5) calculate and report results.
2. "Wet" methods are those that involve physical separation and classical chemical reaction stoichiometry, but no instrumentation beyond an analytical balance. "Instrumental" methods are those that involve additional high-tech electronic instrumentation, often complex hardware and software. Common analytical strategy operations include sampling, sampling preparation, data analysis, and calculations. Also, weight or volume data are required for almost all methods as part of the analysis method itself.
3. Sampling activities are important because if the sample does not represent what it is intended to represent, all operations that follow, whether they are wet methods or instrumental methods, will not give a reliable result. Sample preparation schemes are important because if the sample is not prepared properly for the method chosen, the method, whether a wet method or an instrumental method, again will not give a reliable result.
4. In an instrumental method, a series of reference solutions are usually prepared to calibrate the equipment. In addition, a chemical reaction, often at the heart of a wet method, is not necessarily required for instrumental methods.
5. The general principle of analysis with analytical instrumentation is depicted in Figure 6.2. Some property of the standards and sample solution is detected and measured by the instrument. An electronic signal is generated proportional to this property and read on the readout device.
6. The three major classifications, with instrument names in parentheses, are spectroscopy (spectrometer), chromatography (chromatograph), and electroanalytical chemistry (no specific name).
7. Spectroscopic methods involve the use of light and measure either the amount of light absorbed (absorbance) or the amount of light emitted by solutions of the analyte under certain conditions. Chromatographic methods involve more complex samples in which the analyte is separated from interfering substances using specific instrument components and electronically detected, with the electrical signal generated by any one of a number of detection devices.
8. Calibration is a procedure by which any instrument or measuring device is tested with a standard in order to determine its response for an analyte in a sample for which the true response is either already known or needs to be established.

9. In the case of an analytical balance, the standard is a known weight. This weight is measured, and if this result and the known value are the same, the balance is calibrated. If they are not the same, the balance is taken out of service and repaired. In the case of a pH meter, the standard is a buffer solution. The pH of this buffer solution is measured, and if the known pH and the measured pH are the same, the meter is calibrated. If they are not the same, the readout is electronically tweaked until it gives the correct result. It is then said to be calibrated.

10. The "standard curve" is the plot of an instrument's readout vs. concentration, the data for which are the results of measuring a series of standard solutions prepared for the experiment.

11. Preparing and measuring a series of standard solutions, and plotting the standard curve is the usual process of calibration when the response of a device to a standard is not known in advance (see answer to question 8). In other words, a series of standards establishes the result that is not known in advance. With such a calibration in effect, the operator can then measure samples by comparing the measured result for the sample to the standards results.

12. Samples solutions can be held in place in a small container inside the instrument, injected into the instrument with the use of a syringe, aspirated into an instrument with a sucking device, or not placed inside the instrument at all, but externally tested by dipping a probe into it.

13. A sensor is a kind of translator. It receives specific information about the system under investigation and transmits this information in the form of an electrical signal.

14. Sensor, signal processor, power supply, and readout device.

15. The advantages of a single standard are that it takes less time and is less involved. The series of standards is preferred because we do not rely on just one data point (for which an error may go undetected) and we can establish the response over a range of concentrations rather than at just one concentration.

16. See Section 11.5.3.

17. 0.514

18. 7.3 ppm

19. 0.655 ppm

20. C_u = 6.55 ppm

21. The linearity (or lack of linearity) of the readings is only known for the range of concentrations of the standards prepared. Without testing other standards, it is not known if the linearity extends beyond this range, and so the answer to the unknown cannot be reliably determined.

22. 0.54

23. 0.140

24. 4.84

25. The interpolation of the unknown results relies on the establishment of the linear relationship between the readout and the concentration.

26. The method of least squares is a procedure by which the best straight line through a series of data points is mathematically determined. More details are given in Section 6.3.4. It is useful because it eliminates guesswork as to the exact placement of the line and provides the slope and y-intercept of the line.

27. The best straight line fit is obtained when the sum of the squares of the individual y-axis value deviations (deviations between the plotted y values and the values on the proposed line) are at a minimum.

28. Linear regression is another name for the process of determining the straight line for a series of data points via the method of least squares.

29. Slope, y-intercept, correlation coefficient, and concentrations of samples.

30. Perfectly linear data refers to data in which the instrument readout is exactly the same multiple of the concentration at all concentrations measured and all the points lie exactly on the line.

31. Values that have at least two nines (0.99) are satisfactory for some devices, but values with up to three nines, and sometimes four nines, are attainable in some cases.

32. Serial dilution is the preparation of a series of solutions by always diluting the solution just prepared to make the next one. For example, to make solutions with concentrations 10, 20, 30, 40, and 50 ppm, serial dilution would mean to prepare the 50 first, then to prepare the 40 from the 50, the 30 from the 40, the 20 from the 30, etc.

33. The series of standard solutions does not work well when the instrument readout is dependent on some other variable factor in addition to the concentration, such as variable injection volume in gas chromatography or the variable solution viscosity in atomic absorption.

34.

Concentration	Volume (mL) of 1000 ppm Needed
1.00	0.0500
2.00	0.100
3.00	0.150
4.00	0.200
5.00	0.250

A volume of 1000 ppm needed is pipetted into separate 50 mL volumetric flasks and water is added to each to the 50 mL mark. Each flask is then shaken to make the solutions homogeneous. The pipette needed would be a small serological pipette, perhaps 0.50 mL capacity. Alternatively, a pipetter, such as described in Chapter 4, can be used.

35. Assuming the need to achieve three significant figures in the concentrations, the following volumes of the 100.0 ppm stock would need to be pipetted for 25.00 mL each of 2.00, 4.00, 6.00, and 8.00 ppm solutions respectively: 0.500 mL, 1.00 mL, 1.50 mL, and 2.00 mL.

36. The blank (reagent blank) is a solution that contains all the substances present in the standards and the unknown (if possible) except for the analyte. A sample blank takes into account any chemical changes that may take place as the sample is taken and/or prepared (see Section 6.5.2 for an example).

37. See Section 6.5.3.

38. Most of the time, the unknown sample requires some pretreatment, such as dilution, extraction, or, if it is a solid, dissolving. The analytical concentration of the analyte in the untreated sample must usually be reported, rather than the concentration in the sample solution. Thus, a calculation is usually required to obtain the final answer.

39. (a) 1.5 ppm, (b) 1.2 ppm, (c) 3.81 ppm, (d) 1.50×10^3 ppm, (e) 159 ppm, (f) 3.34×10^3 ppm

40. (a) 7.22 mg, (b) 0.620 mg, (c) 0.482 mg, 1.93 ppm, (d) 0.408 mg, 0.815 ppm, (e) 1.36 mg, 0.272 ppm, (f) 7.06 mg, 1.41 ppm, (g) 1.40 mg, 5.60×10^2 ppm, (h) 0.232 mg, 155 ppm, (i) 0.229 mg, 76.3 ppm, (j) 0.0320 mg, 32.0 ppm, (k) 1.36 ppm, (l) 129 ppm.

41. 0.0131 mg

42. 0.462 mg

43. 1.87×10^3 ppm

44. 4.58×10^4 ppm

45. Assuming three significant figures in the volume measurements, 535 ppm.

46. 0.002562 g, 52.06 ppm

47. 7.50×10^2 ppm

48. (a) 2.74 ppm, (b) 11.0 ppm, (c) 1.92 ppm, (d) 435 ppm, (e) 189 ppm, (f) 256 ppm

49. Assuming three significant figures in the volume measurements, 2.09×10^3 ppm.

50. Assuming three significant figures in the volume measurements, 0.650 ppm.

51. Data acquisition by computer refers to the use of a computer to obtain data (instrument readout values) directly by interfacing to the instrument. These data are then found in

the computer's memory, or on disk, and is not necessarily recorded independently on a recorder or in a notebook.

52. The volume of data acquired in modern laboratories is such that computer storage is most efficient and eliminates the need to provide the space for hard copies.

53. For plotting standard curves.

54. Laboratory Information Management System (see Section 6.7.2).

CHAPTER 7

1. See Section 7.1

2. See Section 7.1

3. Some qualities of light are best explained if we describe the light as consisting of moving particles, often called photons or quanta (called the particle theory of light). Other qualities are best explained if we describe the light as consisting of moving electromagnetic disturbances referred to as electromagnetic waves (called the wave theory of light). Thus, light has a dual nature.

4. Mechanical waves require matter to exist. Electromagnetic waves do not. As outer space is relatively free of matter, electromagnetic waves can exist there while mechanical waves cannot.

5. See Section 7.2.1

6. Light with wavelength 627 Å

7. Light with frequency 7.84×10^{13} sec^{-1}

8. Light with wavelength 591 nm

9. Light with energy 5.23×10^{-14} J

10. Light with frequency 7.14×10^{13} sec^{-1}

11. (a) A, (b) A, (c) A

12. (a) decreased, (b) decreased, (c) decreased

13. IR light has a longer wavelength and lower frequency and wavenumber than UV light.

14. Energy and frequency: radio waves < infrared light < visible light < UV light < X-rays. Wavelength: X-rays < UV light < visible light < infrared light < radio waves.

15. Lower limit: approximately 350 nm. Upper limit: approximately 750 nm

16. (a) UV has more energy than IR.

    (b) UV causes electronic transitions; IR causes vibrational and rotational transitions.

17. $\eta = c/v$, and $\eta = \sin i/\sin r$

18. A refractometer does not measure the speed of light, though the definition of refractive index shown in Equation 7.10 indicates that it is proportional to the speed of light. A refractometer measures the angle of incidence and the angle of refraction and so utilizes the definition shown in Equation 7.11.

19. A refractometer measures the two angles and calculates the refractive index according to Equation 7.11.

20. Refractive index is useful for both qualitative analysis and quantitative analysis. The refractive index of a pure liquid at a given temperature is unique to that liquid. Hence, it can be used to help identify a liquid. The refractive index of an analyte solution depends on the concentration of the analyte, and thus, it can also be used for quantitative analysis.

21. A special sensor consisting of an array of diodes (a diode is a small device that converts light intensity into an electrical signal) is used, and when the angle of reflection changes, a different diode receives the light and generates the signal as shown in the figure. This results in a measurement of the angle of reflection and then, by calculation, the refractive index of the sample.

22. Yellow-colored objects appear to be yellow because yellow wavelengths are reflected and not absorbed like the other wavelengths.

23. See Figure 7.16 and accompanying discussion.

24. If the energy of light striking the atoms or molecules exactly matches an energy transition possible within the atoms or molecules, i.e., the transition of an electron from a lower state to a higher state, the transition will occur, and the energy that once was light is now possessed by the atoms or molecules, i.e., absorption.

25. An electronic transition is one in which an electron in an atom or molecule is moved from one electronic state to another with the absorption of the equivalent energy, such as from UV or visible light. A vibrational transition is one in which a molecular bond's vibrational state changes due to the movement from one vibrational energy level to another because of the absorption of the equivalent energy, such as from short wavelength infrared light. A rotational transition is one in which a molecule's rotational state changes due to the movement from one rotational energy level to another because of the absorption of the equivalent energy, such as from longer wavelength infrared light. Electronic transitions require the most energy—the energy of visible or UV light. Rotational transitions require the least energy—longer wavelength IR light.

26. See Figure 7.21.

27. UV/VIS spectrophotometry is a technique that measures molecules and complex ions using light in the UV regions of the electromagnetic spectrum. IR spectrometry is a technique that measures molecules using the infrared region of the electromagnetic spectrum. Atomic spectroscopy is a technique that measures atoms using light in the UV and visible regions of the electromagnetic spectrum.

28. An absorption spectrum is the plot of absorbance vs. wavelength—the unique pattern of absorption useful for qualitative analysis. A molecular absorption spectrum is of the "continuous" variety, whereas an atomic absorption spectrum is a "line" spectrum. Only specific wavelengths get absorbed by atoms because only specific energy transitions are possible (no vibrational transitions—only electronic). Both vibrational and electronic are possible with molecules, and thus all wavelengths get absorbed to some degree.

29. A line spectrum is an absorption or emission spectrum that displays a series of vertical lines indicating that only certain narrow wavelength bands (lines) are absorbed or emitted. A line spectrum results when atoms are measured. This is the case because there are no vibrational levels in atoms and therefore only very few transitions are allowed.

30. It is because of the presence or absence of vibrational energy levels in the species in question. Atoms do not have vibrational levels and so very few transitions are possible. Molecules have vibrational levels superimposed on the electronic levels, so many transitions are allowed.

31. Because no two chemical substances display identical spectra.

32. A transmission spectrum is a plot of T or %T vs. wavelength. An emission spectrum is a plot of emission intensity vs. wavelength.

33. An energy level diagram meant to depict atomic absorption will have arrows pointing upward to indicate that energy is being absorbed resulting in a transition from lower to higher energy levels. An energy level diagram meant to depict atomic emission will have arrows pointing downward to indicate that energy is being emitted resulting in a transition from higher energy levels to lower levels.

34. $T = I/I_o$, where I_o = intensity of light striking the detector with the blank in the path of the light. I = intensity of light striking the detector with a sample in the path of the light.

35. (a) 0.0857
    (b) 0.308
    (c) 0.613

36. (a) 0.331
    (b) 0.539
    (c) 0.166

37. (a) 0.239
    (b) 0.465
    (c) 0.315
38. (a) 40.6%
    (b) 13.1%
    (c) 32.4%
39. Beer's law is A = abc, where A is absorbance, a is absorptivity, b is pathlength, and c is concentration.
40. 1.04
41. 3.72×10^{-5} M
42. 9.27×10^{-6} M
43. (a) 0.118 (b) 0.981 L/mole cm (c) 7.87×10^{-6} M
44. T = 0.400. To calculate the molar absorptivity, the molarity of the solution would be needed.
45. 1.46×10^{5} L/mole cm
46. 1.83×10^{-5} M
47. 1.8×10^{-5} M
48. 0.404 cm
49. 0.0234 cm
50. 0.886
51. 1.67×10^{3} L mol^{-1} cm^{-1}
52. (a) 0.259
    (b) 8.38×10^{3} L mol^{-1} cm^{-1}
    (c) 0.505
    (d) 0.428
    (e) 90.2%
53. (a) 0.186
    (b) 5.98×10^{3} L mol^{-1} cm^{-1}
    (c) 0.635
    (d) 0.625
    (e) 89.6%
54. 3.79 ppm
55. 3.31 ppm
56. In order, from top to bottom: e, f, a, g, h, d, b, i, c
57. See Figure 7.29
58. The wavelength giving the most absorbance is the wavelength giving the best sensitivity.
59. It does not change the pattern in the sense that it is still the same fingerprint of the compound. It does change the level of the absorption, as shown in Figure 7.32.

CHAPTER 8

1. See Figure 7.21.
2. See Section 8.2.1.
3. The tungsten filament source is a visible light source. It is a light bulb with a tungsten filament and emits the visible wavelengths with significant intensity, but the intensity varies with wavelength, as it does with all light sources. The deuterium lamp is a UV light source. It contains deuterium gas at a low pressure and emits UV light when electricity is applied across a pair of electrodes. The xenon arc lamp is a source for both visible and UV light. It contains xenon at a high pressure and UV and visible light are generated via a discharge across a pair of electrodes.

4. No, the intensity varies with wavelength (see Figure 8.1 for the tungsten filament intensity profile).

5. Either a radiation filter (glass) or monochromator is used. A light filter absorbs all wavelengths except for a somewhat narrow band of wavelengths (see Figure 8.2). A monochromator utilizes a dispersing element in combination with a slit to select a very narrow wavelength band (see Figure 8.4).

6. Bandwidth is the width of the wavelength band that is allowed to exit a monochromator. The narrowness of this band is called the resolution. High resolution corresponds to a very narrow bandwidth and vice-versa.

7. An absorption filter is a wavelength selector consisting of a piece of glass that transmits only a certain rather narrow band of wavelengths from a light source and absorbs the rest. A monochromator is a wavelength selector consisting of a dispersing element/slit combination that selects a very narrow band of wavelengths by sliding the spray of wavelength coming from the dispersing element across the exit slit as shown in Figure 8.4.

8. A monochromator is a wavelength selector. It consists of two slits and a dispersing element in combination. The dispersing element splits the light from a source into the wavelengths of which it is composed. Upon rotating this element, different wavelengths pass through the "exit slit," and thus the position of the dispersing element dictates what wavelength is selected.

9. Turning the knob on the exterior of the instrument rotates the dispersing element of a monochromator in the interior of the instrument such that the spray of wavelengths coming from the dispersing element slides across the exit slit. At 728 nm, only a very narrow wavelength band at 728 nm emerges from the exit slit.

10. A diffraction grating is a dispersing element consisting of a highly polished mirror with a large number of regular narrowly spaced lines inscribed on the surface.

11. The sample compartment must consist of a light-tight box so that no stray light—only light from the instrument's light source—reaches the light sensor.

12. A single-beam spectrophotometer is one in which a single, continuous light beam shines from the light source through the monochromator and sample compartment to the detector. A double-beam spectrophotometer is one in which a single beam of light from the source is split into two beams in order to provide certain advantages. A single-beam instrument would be used in situations in which the expense of the double-beam is not warranted, such as in less precise work.

13. The precalibration of a single-beam spectrophotometer consists of tweaking the readout to read 100% T when the blank is in the sample compartment.

14. It is slow and tedious for an experiment in which a molecular absorption spectrum is measured. It is slow and tedious work because the wavelengths are manually scanned in small increments, and each time the wavelength is changed, the calibration step with the blank needs to be performed due to the variability of light intensity from the light source at the different wavelengths.

15. Rapid-scanning single-beam instruments exist. The absorption spectrum of the blank is first obtained followed by that of the sample. The sample scan is then adjusted to its proper measurement by using the spectrum of the blank.

16. Since the time between reading the blank and reading the sample can be significant, the instrument may lose its calibration due to minor electrical fluctuations either in the source or detector.

17. A light chopper is a rotating partial mirror used to split a light beam into two beams. A beamsplitter is a mirror with slots also used to split a light beam into two beams (see Figures 8.7, 8.8, and 8.9). A chopper creates a double beam in time, whereas a beamsplitter creates a double beam in space.

18. A double-beam spectrophotometer is one in which either a beamsplitter or light chopper is used to create two beams of light in order to deal with the problem of variable light intensity of the different wavelengths emitted by the source.

19. The two designs are those shown in Figures 8.10 and 8.11. In one case, the second beam passes through the blank, whereas in the other case, the second beam goes directly to a detector. See Section 8.2.3.3.

20. (a) Do not have to continually replace sample with blank when obtaining a molecular absorption spectrum.
    (b) Errors due to light source and detector fluctuations are minimized.
    (c) Accurate rapid scanning of wavelengths is possible.

21. A double-beam instrument is preferred for rapid scanning because adjustments for intensity changes after each wavelength can be made immediately before a sample is read. With a single-beam instrument, there is a delay.

22. A double-beam instrument is preferred for the reasons expressed in the answer to question 20.

23. A photomultiplier tube is a light sensor combined with a signal amplifier (see Section 8.2.4).

24. See Section 8.2.4.

25. A diode array spectrophotometer is one that utilizes a series of photodiodes to detect the light intensity of all wavelengths after the light has passed through the sample (see Figure 8.12). The advantage is that an absorption spectrum can be measured in a matter of seconds.

26. Cuvettes for visible spectrophotometry can be made of clear, colorless plastic. The only requirement is that none of the visible light wavelengths be absorbed, which is the reason that they must be colorless.

27. Cuvettes are matched if they are identical in terms of pathlength and reflective/refractive properties. Cuvettes used for calibration and the analysis of samples after calibration must be matched so that absorption readings are due solely to concentration effects and not cuvette differences.

28. Things that may be different include pathlength and the reflective/refractive properties at the interior and exterior interfaces.

29. An interfering substance absorbs the same wavelength as the analyte or otherwise inhibits the accurate reading of the analyte.

30. A deviation from Beer's law refers to the linear relationship between absorbance and concentration becoming nonlinear as in Figure 8.15.

31. The optimum working range for percent transmittance (to avoid instrumental deviations from Beer's law) is between 15% and 80%, which corresponds to an absorbance range of 0.10 to 0.82.

32. See Section 8.4.3.

33. The wavelength calibration can be checked by comparing the maximum absorbance wavelength for a known substance as measured by the instrument to what it is reported to be otherwise.

34. See Section 8.4.4.

35. The energy of fluorescence is less than the energy of absorption due to the vibrational relaxation that occurs both in the excited state and in the ground state when the fluorescing molecule seeks to return to the ground electronic state from the excited state. The wavelength of fluorescence is longer than the wavelength of absorption due to the fact that wavelength is inversely proportional to energy (see Figure 8.16).

36. See the answer to question 35.

37. Need one to select the wavelength of absorption and one to select the wavelength of fluorescence.

38. The fluorescence is measured (with a monochromator and phototube/readout) at right angles to the incoming light beam.

39. (a) Fluorometers have two monochromators and a right-angle configuration.

   (b) Two monochromators are needed, one to select the wavelength of absorption and one to select the wavelength of fluorescence. A right-angle configuration is important in order to avoid measuring the light from the source while trying to measure the fluorescence.

40. The differences are as mentioned in the answer to question 39 (a) (see Figure 8.17 for the instrument diagram).

41. Fluorescence intensity.

42. The electrons in benzene ring systems are the kinds of electrons that undergo the energy changes required for fluorescence.

43. To say fluorometry is more selective means that there are fewer interferences. To say that it is more sensitive means that it can detect smaller concentrations.

44. A fluorometric procedure would be used when analyzing for a substance that is able to fluoresce and when the absorption procedure is prone to interferences and does not give a satisfactory sensitivity.

45. More sensitive, more selective.

46. Advantage—virtually no interferences exist.

   Disadvantage—not useful for very many things.

47. (a) Fluorometers have two monochromators and a right-angle configuration. Absorption spectrophotometers do not.

   (b) Fluorometry is more sensitive

   (c) Absorption spectrophotometry is more highly applicable.

48. Infrared absorption causes vibrational transitions. UV-Vis absorption causes electronic transitions.

49. UV, quartz glass. IR, inorganic salt crystals.

50. Similar: both are molecular fingerprints and therefore useful for qualitative analysis and both display absorption behavior over a wavelength range.

   Different: IR displays sharper absorption bands that have greater specificity. IR spectra are usually displayed as transmission spectra rather than absorption spectra. IR spectra utilize wavenumber, rather than wavelength, on the x-axis.

51. Infrared absorption patterns can be more directly assigned to more specific structural features.

52. Fourier transform infrared spectrometry.

53. (a) FTIR

   (b) FTIR

   (c) Double-beam dispersive

   (d) FTIR

   (e) Double-beam dispersive

   (f) FTIR

54. An interferometer is a device that utilizes a moveable and a fixed mirror to manipulate the wave patterns of a split light beam so as to create constructive and destructive interference in this beam.

55. (a) FTIR

   (b) FTIR

   (c) Double-beam dispersive

   (d) FTIR

56. Inorganic compounds consist of ionic bonds, which do not absorb infrared light and therefore would present no interfering absorption bands.

57. Describing a liquid as "neat" means that it is pure and not the solute in a solution.

58. (1) sealed cell, (2) demountable cell, (3) sealed demountable cell

59. See Section 8.8.1.

60. The spacer between the NaCl windows.
61. Refer to Figure 8.25. When filling a cell with a liquid that has an unusually high viscosity, two syringes may be used—one to "push" the liquid into the cell by pushing down on the plunger of the syringe containing the liquid in the inlet port and one to "pull" the liquid into the cell by pulling up on the plunger of an empty syringe in the outlet port.
62. (a) The windows may fog or become disfigured since water dissolves them.
    (b) Water will interfere with the detection of alcohols in the spectrum.
63. See (a) in the answer to question 62.
64. The solvent will exhibit absorption bands that may interfere with those of the analyte. Carbon tetrachloride has only one kind of bond, the C–Cl bond, and this bond will not absorb IR light at wavelengths that are usually important.
65. (1) Dissolve in a solvent and then use a liquid sampling cell to measure the solution, (2) dissolve in a solvent, place several drops of the solution on a salt plate, evaporate the solvent and measure the residue, (3) KBr pellet, (4) Nujol mull, (5) reflectance.
66. A KBr pellet is a thin wafer of a mixture of KBr and sample made by first thoroughly mixing the sample with dry, powdered KBr, and then compressing a quantity of this mixture in a laboratory press.
67. (1) The best pellets are made from dry KBr. (2) The presence of water will result in absorption patterns that may cause the analyst to make erroneous conclusions.
68. A hydraulic press may be used to help make a higher quality KBr pellet.
69. A Nujol mull is a mixture of a solid sample with Nujol, or mineral oil, to more conveniently obtain an infrared spectrum of the solid sample.
70. Neither utilize solvents or other materials that would present possibly interfering absorption bands in the spectrum.
71. The mineral oil has a rather simple spectrum (only carbon–hydrogen bonds). Additionally, as with solvents, a computer maybe used to subtract the mineral oil bands from the spectrum of the solid.
72. See Section 8.9.5.
73. The diffuse reflectance method is a method for solids in which the powdered solid, held in a cup, is irradiated with the IR beam. The scattered reflected light is captured by the detector and the spectrum displayed.
74. Yes, gases may be measured by IR spectrometry (see Section 8.9.6).
75. Figure 8.39: a broad, fairly strong absorption band at 3300 cm^{-1} indicating an alcohol. No benzene ring absorptions or carbonyl group. It is an aliphatic alcohol.
    Figure 8.40: strong, sharp band at 1700 cm^{-1} indicating a carbonyl group. Absorption bands on the high side of 3000 cm^{-1} and a series of weak bands between 1700 and 2000 cm^{-1} indicating a benzene ring. Possibly benzaldehyde or a similar compound.
    Figure 8.41: strong, sharp absorption at 1700 cm^{-1} indicating a carbonyl group. No other significant patterns except the C–H pattern on the low side of 3000 cm^{-1}. It is an aliphatic aldehyde or ketone.
    Figure 8.42: a benzene ring is indicated because of the band on the high side of 3000 cm^{-1} and the series of weak peaks between 1700 and 2000 cm^{-1}. Aliphatic C–H bonds are also indicated (absorption bands on the low side of 3000 cm^{-1}). Possibly ethylbenzene or a similar structure.
76. (a) It appears that Figure 8.43 is the spectrum of an alcohol because of the broad absorption in the 3500 cm^{-1} range.
    (b) Yes, Figures 8.43 and 8.44 are both spectra of compounds that have a benzene ring. The three patterns to observe for benzene rings are sharp bands on the high side of 3000 cm^{-1}, a series of weak peaks between 1600 and 2000 cm^{-1} and two peaks between 1500 and 1600 cm^{-1}.

(c) Yes, Figure 8.45 is probably the spectrum of a compound with just C–H bonds. It does not have a benzene ring, however, because the spectrum does not display the absorption patterns of a benzene ring (see answer to question 76 (b)).

(d) Yes, Figure 8.43 is the spectrum of a compound that has a carbonyl group because it displays the strong sharp peak at 1700 cm^{-1}.

(e) Figure 8.43: benzophenone, Figure 8.44: 3-methylphenol, Figure 8.45: n-pentane.

77. Calibrate the instrument by preparing a series of standard solutions of the analyte in a solvent, measuring the absorbance of each at an appropriate wavelength and plotting absorbance vs. concentration. Obtain the concentration of the unknown from this graph (see Section 8.11 for details).

CHAPTER 9

1. Spectral lines for atoms are absorption or emission bands that are so narrow, they appear as lines rather than bands.

2. (a) See Figure 9.1. The sources of the lines are the very specific energy transitions that are allowed in atoms.

   (b) The two most common sources of atomic emission spectra are flames and the ICP torch.

3. A given emission line is caused by a transition between the same two energy levels as the corresponding absorption line. Since both therefore represent the same energy difference, they occur at the same wavelength.

4. See Section 9.1.

5. An atomizer is a device that forms atoms from ions. Examples include flames, graphite furnace, ICP source, vapor generators, etc.

6. Most common atomic absorption techniques: flame atomic absorption and graphite furnace atomic absorption. Most common atomic emission techniques: ICP and flame emission.

7. (1) solvent evaporates, (2) ions atomize, (3) atoms are raised to excited states, (4) excited atoms drop back to ground state and emit light.

8. (a) Flame AA is therefore useful because there is a large percentage of unexcited atoms present that can absorb light from the light source.

   (b) Resonance is the continuous movement of atoms back and forth between the ground state and excited states.

9. (a) 1800 K, (b) 2300 K, (c) 2900 K, (d) 3100 K

10. Air/acetylene and N$_2$O/acetylene flames are the most commonly used because their temperatures are high enough to provide sufficient atomization for most metals while not burning at too fast a rate.

11. An oxygen/acetylene flame has a high burning velocity, which decreases the completeness of atomization and thus lowers the sensitivity.

12. The element to be analyzed is contained in the cathode of the hollow cathode lamp, since its atoms become excited and emit light and this light is what is needed for absorption in the flame. No monochromator is needed because the wavelength is already specific for the atoms in the flame.

13. Unless the analyte metal is contained in the cathode, the lamp will not emit the required wavelength.

14. Many hollow cathode lamps are needed because each element, since it must be contained in the cathode, requires a different lamp. Some lamps, however, are "multielement." The element analyzed must be contained in the cathode so that its line spectrum will be generated and absorbed by the same element in the flame.

15. EDL stands for "electrodeless discharge lamp." It is an alternative to the hollow cathode lamp as a light source in atomic absorption spectroscopy.
16. A nebulizer is a device that converts a flowing liquid to a fine mist or cloud. A nebulizer is needed in conjunction with a premix burner so that the analyte solution can be sufficiently mixed with the fuel and oxidant gases prior to reaching the flame.
17. Refer to Figure 9.9 and accompanying discussion.
18. Since the solution, air, and fuel are premixed, some solution droplets will not make it all the way to the flame. These collect in the bottom of the mixing chamber unless they are allowed to drain out.
19. The light chopper serves to allow the detector to differentiate between the light emitted by the flame and the light originating from the light source. This, in turn, allows the detector to measure an absorbance that is free of the interfering light emitted by the flame.
20. The single-beam instrument uses a single light beam, albeit modulated, from the light source through the flame and monochromator to the detector. In a double-beam instrument, a second beam is created to bypass the flame to be rejoined after the first beam has passed through the flame (see Figures 9.11 and 9.12). The advantage of the single beam is a less expensive and less complicated instrument. The advantage of the double beam is that it eliminates problems due to source drift and noise.
21. The reference beam in the AA instrument does not pass through the blank, but merely bypasses the flame. Thus, the fluctuations in light intensity are accounted for, but the blank adjustment must be made at a separate time.
22. Room light is mostly eliminated by the monochromator positioned between the flame and the detector. Room light that is the same wavelength as the light being measured is eliminated along with flame emissions by modulating or chopping the light from the source such that the detector electronics is able to generate a signal based only on the light from the source.
23. The primary line is the line in the line spectrum that is used most often for analyzing for that element because it is the most sensitive and useful. Secondary lines are other lines that are sometimes used for various reasons.
24. Controls that need to be optimized are: wavelength, slit width, lamp current, lamp alignment, aspiration rate, burner head position, and fuel and oxidant flow rates (see Section 9.3.5 for details).
25. Yes, Beer's law applies. The width of the flame is the pathlength and each analyte has an absorptivity for the conditions chosen.
26. A chemical interference is one in which the sample matrix affects the chemical behavior of the analyte. A spectral interference is one that interferes with accurate measurement of the desired spectral line.
27. Matrix matching refers to the preparation of standards in such a way that their matrices match that of the sample as closely as possible. It is important because there may be a component in the sample that affects the reading in some way and unless that component is present in the standards at the same concentration level, it may affect the results in a negative way.
28. The standard additions method is one in which the standards prepared for the standard curve consist of the sample to which varying amounts of a standard solution have been added. It helps with the problem of chemical interferences because the sample is the matrix for the standards and the interference occurs in all measurements such that its effect is negated.
29. (1) It is not possible to prepare a blank with the same matrix and (2) the sample concentration is outside the range of the standards and the linearity of the standard curve cannot be verified.
30. To release the calcium from the sample matrix so that it can be atomized in the flame.

31. Background absorption is light absorption due to molecular substances or particles in the flame. Background correction refers to the technique in which the background absorption is isolated and subtracted out. The absorption of a continuum light beam passing through the flame along with the light from the source allows the required subtraction to take place.

32. See Section 9.3.7.

33. The graphite furnace method of atomization utilizes a small graphite tube furnace to electrically heat a small volume of the analyte solution contained inside rapidly to a temperature that eventually causes atomization.

34. The analyte solution is placed (injected) into the furnace with a micropipette or autosampler. Following this, a temperature program is initiated in which the furnace heats rapidly to (1) evaporate the solvent, (2) char the solid residue, and finally (3) atomize the analyte, creating the atomic vapor.

35. Compare Figure 9.11a with Figure 9.18. The difference is that the light beam passes through the furnace tube rather than the flame.

36. Argon gas is needed to provide an inert atmosphere for the graphite surfaces so that it is not oxidized (and damaged) by exposure to air at the high temperature. Cold water is needed to provide rapid cool-down between measurements. A source of high voltage is needed to accommodate the rapid heating to the very high temperature required.

37. See Figure 9.21.

38. It must be protected from air because air, at the high temperature achieved during the experiment, can oxidize, damage, and disintegrate the graphite surface.

39. The absorbance signal originates from a very small volume of solution placed in the furnace, and since the furnace is continuously flushed with an inert gas, the vapors from this volume are swept out of the furnace after a short time.

40. Advantages: (1) high sensitivity, (2) only small volumes of sample are needed, (3) better detection limit. Disadvantages: (1) matrix effects, (2) poor precision.

41. The Zeeman background correction is a background correction procedure for graphite furnace AA in which a powerful pulsing magnetic field is used to shift the energy levels of atoms and molecules and thereby shift the wavelengths that are absorbed and allow subtraction of the background.

42. Inductively coupled plasma. FP, because emission is measured and not absorption.

43. See Section 9.5.

44. Advantages: (1) more sensitive, (2) broader concentration range measurable, (3) multielement analysis possible. Disadvantages: (1) cost

45. (a) See answer to question 40.
    (b) See answer to question 44.

46. Mercury is the only metal that is a liquid at room temperature and is therefore the only metal that has a significant vapor pressure such that an atomic vapor can be created without heat.

47. The hydride generation technique is a technique in which volatile metal hydrides are formed by chemical reaction of the analyte solutions with sodium borohydride. The hydrides are guided to the path of the light, heated to relatively low temperatures, and atomized. It is useful because it provides an improved method for arsenic, bismuth, germanium, lead, antimony, selenium, tin, and tellurium.

48. See Table 9.3 and discussions in the text.

49. (a) ICP
    (b) Atomic fluorescence
    (c) Flame AA, graphite furnace AA, atomic fluorescence
    (d) Spark emission
    (e) Flame photometry, atomic fluorescence, ICP, spark emission
    (f) Graphite furnace AA
    (g) ICP

50. (a) T, (b) T, (c) F, (d) T, (e) F
    (f) F, (g) T, (h) F, (i) T, (j) T
    (k) T, (l) F, (m) T, (n) F, (o) F
    (p) T, (q) F, (r) T, (s) T, (t) T
51. Sensitivity is the concentration of analyte that will produce an absorption of 1%. Detection limit is the concentration that gives a readout level that is double the noise level in the baseline (see Table 9.2).

CHAPTER 10

1. It is not uncommon for real-world analysis samples to be very complex in terms of the number of chemical substances present. The study of modern separation science is thus very important in analytical chemistry from the standpoint that many potentially interfering substances must be identified and eliminated.
2. Chromatography is the separation of mixture components as a result of the varying degrees of interaction that the mixture components have with a mobile phase and a stationary phase.
3. GC, gas chromatography; LC, liquid chromatography; GSC, gas–solid chromatography; LSC, liquid–solid chromatography; GLC, gas–liquid chromatography; LLC, liquid–liquid chromatography
4. (1) Adsorption: mixture components are separated based on varying degrees of adsorption forces between them and the stationary phase.
   (2) Partition: mixture components are separated based on solubility differences in the liquid stationary phase.
   (3) Ion exchange: ionic mixture components are separated based on the varying strength of ionic bonds formed with sites on the stationary phase.
   (4) Size exclusion: mixture components are separated based on the varying abilities to penetrate the pores on the stationary phase.
5. Partition chromatography utilizes the varying solubilities of the mixture components in a liquid stationary phase. Absorption chromatography utilizes the varying tendencies for mixture components to adhere to the surface of a solid stationary phase.
6. (a) "B" would emerge first because polar mixture components tend to dissolve more in the polar mobile phase and thus will come through the column with the mobile phase and emerge first. "A," being nonpolar, will tend to remain behind in the stationary phase.
   (b) "A," being nonpolar, will emerge first since it will tend to dissolve more in the nonpolar mobile phase.
7. (a) no, (b) yes, (c) yes, (d) no, (e) yes, (f) no
8. The four types are partition, adsorption, ion exchange, and size exclusion
   (a) Partition only
   (b) None
   (c) All four
9. Size-exclusion chromatography. The separation occurs because the stationary phase particles are porous and the small molecules enter the pores and are slowed from passing through the column, whereas the large molecules pass through more quickly since they do not enter the pores.
10. With a cation exchange resin, the bonding sites are negatively charged and cations are exchanged. With an anion exchange resin, the bonding sites are positively charged and anions are exchanged.
11. IEC, ion-exchange chromatography; IC, ion chromatography; SEC, size-exclusion chromatography; GPC, gel-permeation chromatography; GFC, gel-filtration chromatography

12. The five configurations studied in this chapter are (1) paper chromatography, (2) thin-layer chromatography, (3) open-column chromatography, (4) gas chromatography, (5) high-performance liquid chromatography.

13. (a) In open-column chromatography, the stationary phase consists of solid particles, or a thin film of liquid bonded to solid particles, and is contained in a vertical tube. In HPLC, the stationary phase also consists of solid particles or a thin layer of liquid bonded to solid particles, but it is contained in a metal tube that doesn't have to be in any particular spatial orientation.

    (b) In open-column chromatography, the column is held vertically, so the force moving the mobile phase is gravity. In HPLC, the force that moves the mobile phase is the pressure from a high-pressure pump.

    (c) In the open column procedure, fractions of eluate are collected in tubes and analyzed later by some specified method. In HPLC, an electronic sensor (detector) on the eluate end of the column senses the mixture components as they elute and displays peaks, the sizes of which are related to quantity.

14. (a) In thin-layer chromatography, the mobile phase moves by capillary action. In GC, the mobile phase moves by gas pressure.

    (b) Thin-layer chromatography is a planar procedure. GC is a column procedure.

    (c) In thin-layer chromatography, R_f factors are used. In GC, retention times are used.

15. LSC is liquid–solid chromatography. Strictly speaking, any type that utilizes a solid stationary phase, namely adsorption, ion exchange, and size exclusion, can be referred to as LSC. However, adsorption chromatography is the only one of this group that is routinely referred to as LSC.

16. The sensor sends an electronic signal to the data system, and this signal increases and then decreases again as the as the individual mixtures elute, thus creating the appearance of a peak on the screen.

17. (a) j, (b) l, (c) m, (d) n, (e) l, (f) h, (g) k

18. (a) stationary phase, partition
    (b) size exclusion
    (c) electrophoresis
    (d) HPLC
    (e) adsorption
    (f) thin layer

19. Blank lines, left to right, starting upper left: water, liquids, or dissolved solids, size exclusion, porous polymer beads, any liquid type, polymer beads with ionic "site," ions, gas, thin liquid film, thin layer, liquids, or dissolved solids.

20. (a) false, (b) true, (c) true, (d) false, (e) true
    (f) false, (g) true, (h) true, (i) false, (j) true
    (k) true, (l) true, (m) false, (n) false
    (o) false, (p) true, (q) true

21. (a) Thin-layer chromatography
    (b) Partition chromatography
    (c) Ion-exchange chromatography
    (d) High-performance liquid chromatography
    (e) Partition chromatography and adsorption chromatography
    (f) Open-column chromatography
    (g) Partition chromatography
    (h) Paper chromatography and thin-layer chromatography

22. The retention time of a mixture component is the time from when the sample is first injected until the component's peak is at its apex. Adjusted retention time is the difference between the retention time of the mixture component and the retention time of air (see Figure 10.14).

23. See Figure 10.14. x-axis is time; y-axis is the electrical signal.
24. See Figure 10.17.

CHAPTER 11

1. Most often, the column is a capillary column, meaning that it has a very small diameter; it has the stationary phase liquid bonded to the capillary's internal wall, and it can be up to 300 ft long.
2. A liquid mixture is drawn into the syringe and injected through a rubber septum into the injection port where it is quickly evaporated. The mobile phase gas, usually helium, flows through the injection port and sweeps this mixture into the column where it encounters the stationary phase. It then separates into individual mixture components through its back and forth movement between the two phases. After completely traversing the column, these components elute from the column and move into the detector. In the detector, a particular sensing operation occurs (there are a number of different detector designs) and each individual component's electronic signal are sent to the computer. In the meantime, the mixture components is either destroyed in the detection process, or they emerge from the detector and evaporate into the air in the room.
3. Vapor pressure can be defined as the tendency of liquid substance to escape the liquid phase and become a gas or the pressure exerted by the gas molecules of a substance above a liquid containing that substance.
4. Vapor pressure is a measure of the tendency of a substance to be in the gas phase at a given temperature. Since different mixture components will have different vapor pressures, the separation occurs in part due to the different tendencies of the components to be in the mobile gas phase.
5. A nonpolar mixture component with a high vapor pressure would have the shorter retention time because both the high vapor pressure and the fact that it has a polarity different from the stationary phase means that it would likely be found in the mobile phase most of the time.
6. Such a mixture component would have a low vapor pressure and a high solubility in the stationary phase.
7. The three major components of a GC are the injection port, the column (with oven), and the detector.
8. The injection port, to flash vaporize the sample; the column, since the mixture components must remain gaseous and since vapor pressure depends on temperature; and the detector in order to keep the mixture components from condensing.
9. The size of the liquid sample injected into the column must be small because after it is converted to a gas, its volume must still be small so as to fit onto the column all at once. If it were too large, it would not fit onto the column all at once and would bleed onto the column gradually over a period of time. This would mean that, at best, the eluting peaks would exhibit fronting and tailing or, at worst, would be broad and quite likely overlap with each other to the point of exhibiting poor or no resolution.
10. A split injection is one in which the volume injected is internally split and a much smaller volume actually enters the column while most of it is vented to the air through a split valve. A splitless injection is one in which the injector is capable of splitting the volume, but the split valve is closed such that the volume is not split.
11. The four modes of injection are (1) direct injection, (2) split injection, (3) splitless injection, and (4) on-column injection.
12. GC columns are coiled up in order to fit in the small space (see Figure 11.2 for an example).
13. An open-tubular capillary column is a very long (30–300 ft), narrow diameter tube in which the stationary phase is held in place by adsorption on the inside wall. Such a column

is useful because it allows the use of a very long column (for better resolution) with minimal gas pressure required.

14. The packed column can be from 2 to 20 ft in length, typically has a diameter of 1/8 or 1/4 inches, and has small particles, often coated with a thin layer of liquid stationary phase, packed in the tube. The open-tubular capillary column can be up to 300 ft in length, has an extraordinarily small diameter (capillary), and has the liquid stationary adsorbed on the inside surface of the tube. In terms of separation ability, the open-tubular capillary column is better because the mixture components contact more stationary phase (column is longer) while passing through the column. The amount injected for the open-tubular capillary column must be much less (0.1 mL maximum as opposed to 20 mL for the 1/8 inch packed column) because the column diameter is much less and a greater volume would overload it.

15. "Analytical GC" is the use of GC solely for analysis, qualitative or quantitative. "Preparative GC" is the use of GC for preparing pure samples for use in another experiment.

16. See Section 11.5.3.

17. Chromosorb is the trade name given to diatomaceous earth, the decayed silica skeletons of algae. It is the substrate material on which the liquid stationary phase is adsorbed in packed columns.

18. A changing of the column temperature in the middle of the run.

19. The temperature of the column affects both the vapor pressures of the mixture components and the solubilities of the mixture components in the stationary phase. Sometimes a low temperature separates some mixture components that are not separated at higher temperatures. However, the lower temperature may result in inordinately long retention times for other mixture components. Temperature programming allows the best of both worlds—complete separation of short retention time components at the low temperature while shortening retention times of the others when the column temperature is raised later.

20. A temperature program should be used that begins at 80°C to facilitate separation of A and B, then increases to 150°C to decrease the retention times of C and D. This would ensure resolution of all components in a reasonable time.

21. Higher carrier gas flow rates would result in shorter retention times because when the mixture components are present in the mobile phase, they would progress to the opposite end of the column at a faster rate.

22. A universal detector is one that generates an electronic signal for all mixture components regardless of their identity. More sensitive means that smaller quantities can be detected. More selective means that only some mixture components can be detected and that there are fewer interferences for these components.

23. (a) FID, (b) ECD, (c) TCD, (d) TCD, (e) FID
    (f) MS, (g) MS, (h) ECD, (i) NPD, (j) IR
    (k) FPD, (l) PID, (m) Hall

24. (a) FID, (b) TCD, (c) TCD, MS, (d) ECD, NPD, MS, IR
    (e) TCD, (f) ECD, (g) FID, (h) TCD, ECD, IR, PID
    (i) Hall, (j) MS, IR

25. (a) flame ionization detector, (b) electron capture detector
    (c) flame ionization detector, (d) electron capture detector
    (e) mass spectrometer detector, (f) mass spectrometer detector
    (g) thermal conductivity detector, (h) flame ionization detector
    (i) electron capture detector, (j) mass spectrometer detector
    (k) thermal conductivity detector

26. (a) Flame ionization
    (b) Thermal conductivity
    (c) Thermal conductivity

(d) Electron capture

(e) Flame ionization

27. (a) More universal, does not destroy sample, safer to use.

(b) More sensitive

28. GC-MS stands for gas chromatography–mass spectrometry. The detector is a mass spectrometer. This detector determines the mass to charge ratio of fragments resulting from the destruction of component molecules by a high energy electron beam and displays the mass spectrum, a graph of signal intensity vs. mass to charge ratio. This mass spectrum is a molecule fingerprint and can be fully recorded as the peak is being traced, adding a definitive dimension to qualitative analysis by GC.

29. If an unknown is totally unknown, retention time data are limited because there is uncertainty about whether untested compounds have the same retention times as those tested. However, if the unknown is not totally unknown and only a limited number of compounds is known to possibly be in the sample, retention time data can be quite valuable if all the known compounds have different retention times. In that case, it is a matter of matching up retention times of known compounds with the peaks in the unknown's chromatogram.

30. (a) Yes. Benzene, ethylbenzene, and isopropylbenzene

(b) Toluene, n-propylbenzene

(c) There may be other compounds, whose retention times have not been measured, which may exhibit the same retention times as toluene and n-propylbenzene. Also, there is no match for compounds B and D among the known liquids measured.

31. Retention time data by itself are never sufficient to identify mixture components unless it is known that the standards tested were the only possibilities.

32. (a) carrier gas, mobile phase

(b) retention time

(c) thermal conductivity

(d) electron capture

(e) injection port, column, detector

(f) peak size

(g) theoretical plates, resolution

(h) temperature programming

(i) preparative GC

(j) FID

(k) open-tubular column

33. Response factor = 0.939 cm²/mg. % methylene chloride = 31%.

34. An internal standard is a compound that is added to the standard solutions in consistent amounts in order to eliminate the problem of variable injection volume. The inability to consistently inject a given volume accurately means that the analyte peak size is not reproducible. However, the ratio of the analyte peak size to the internal standard peak size is reproducible, and so this ratio is plotted on the y-axis of the standard curve vs. the analyte concentration.

35. The internal standard method uses an internal standard substance added in a constant amount to all standards and the sample. Area ratio of analyte peak to internal standard peak is plotted vs. concentration of analyte. The standard additions method uses the addition of the analyte in increasing amounts to the sample. Peak area is plotted vs. concentration added, and the line is extrapolated to zero peak area to get the sample concentration.

36. (a) IS: series of standard solutions are prepared in which the analyte is present at increasing known concentrations and a second substance, the internal standard, is present at a constant concentration. This amount of internal standard is also added to the unknown. SA: standards are prepared by adding a constant known amount of the analyte to the

unknown sample with the chromatogram measured after each addition. Alternatively, a series of standards could be prepared with the unknown as the diluent.

(b) IS: the analyte peak and the internal standard peak are well resolved from each other and the solvent peak(s). SA: the analyte peak grows proportional to the amount of analyte added.

(c) IS: area ratio vs. concentration. SA: peak size vs. concentration added (extrapolation required).

37. (a) Pipet increasing amounts of ethyl alcohol into the flasks. Pipet a constant amount of isopropyl alcohol into each of the flasks. Dilute standards with an appropriate solvent. Unknown must have isopropyl added.

(b) Ethyl alcohol and isopropyl alcohols peaks are well resolved. All other peaks need not be resolved (there would be many in the gasoline sample.)

(c) Area ratio vs. concentration.

38. Response factor method:

(a) No sample or solution preparation required. One chromatogram of unknown sample required. Considerable experimentation required to determine response factors, however (see text.)

(b) One chromatogram of unknown needed. Also, one chromatogram of each compound in the unknown in the pure state, in order to determine response factors.

(c) No plotting required.
    Serial dilution method:

(a) Prepare series of standards with analyte in increasing concentration. Unknown measured as is.

(b) Only analyte peak need be resolved.

(c) Peak size vs. concentration: internal standard method and standard additions method—see question 35.

39. Peak tailing and fronting.

40. A leaky or plugged syringe, a worn septum, a leak in the pre- or post-column connections, or a contaminated detector.

41. An insufficiently conditioned column, a detector that has not achieved a stable temperature, a faulty, or contaminated detector.

42. Peaks from both injections will appear on the chromatogram and confuse the interpretation of the chromatogram.

43. Perform a rapid "bakeout" via temperature programming after the analyte peaks have eluted; use pure reagents, or replace or clean septa, carrier, or column.

CHAPTER 12

1. High-performance (or pressure) liquid chromatography.

2. LC, liquid chromatography; LLC, liquid–liquid chromatography; LSC, liquid solid chromatography; BPC, bonded phase chromatography; IEC, ion-exchange chromatography; IC, ion chromatography; SEC, size exclusion chromatography; GPC, gel permeation chromatography; GFC, gel filtration chromatography

3. "High performance" refers to the fact that complex mixtures can be resolved and quantitated accurately in a very short time.

4. HPLC (high-performance liquid chromatography) is an instrumental chromatography technique in which the mobile phase is a liquid. Refer to Chapter 12, Figure 12.1, and accompanying discussion, for details of the HPLC system.

5. Refer to Figure 12.1. The mobile phase path begins in the solvent reservoir where a tube with a metal sinker/filter is immersed. A high-pressure pump draws the solvent out of the

reservoir, through the pump, and past the sample injection device where the sample is introduced. It then proceeds to the guard column, where undesirable sample components are removed, and from there to the analytical column where the components of interest are separated. The detector then detects mixture components as they elute. After passing through the detector, the mobile phase is channeled to waste.

6. Speed and overall performance.
7. (a) HPLC, high-pressure pump; GC, pressure from compressed he cylinder
   (b) HPLC, liquid (composition varies); GC, gas (usually helium)
   (c) HPLC, inside a short (about 1 ft) metal tube; GC, inside a glass or metal tube that can be about 1/4 in. in diameter and be up to 300 ft in length and can be either a capillary tube, 1/8 in or 1/4 in. in diameter
   (d) HPLC, all types; GC, partition and adsorption
   (e) HPLC, not applicable; GC, separation of mixture components assisted by differences in vapor pressures
   (f) HPLC, special injection valve; GC, mixture syringe-injected into flowing carrier gas in heated injection port
   (g) HPLC, solubility in liquid mobile phase vs. interaction with stationary phase; GC, vapor pressure and interaction with stationary phase
   (h) HPLC, often standard analytical instruments but with flow cell; GC, many different designs that detect mixture components in the gas phase
   (i) HPLC, computerized data system; GC, computerized data system
   (j) HPLC, chromatogram with peaks; GC, chromatogram with peaks
8. Mixture components move back and forth between the mobile phase and stationary phase as they pass through the column. In GC, the mobile phase is a gas, so the vapor pressure, the tendency to be in the gas phase, is important.
9. The pump operates by creating a vacuum to draw the mobile phase from the reservoir. If there are dissolved gases in the mobile phase, they can withdraw from solution when under this vacuum.
10. Mobile phases and samples contain small particulates that can damage the column and other hardware unless they are removed by filtration.
11. Degassing is the removal of dissolved gases from mobile phases and samples. It can be accomplished by (1) reducing the pressure in the air space above the liquid in the container, (2) placing the container holding the liquid in an ultrasonic bath, or (3) both (1) and (2) at the same time.
12. Helium sparging is a method for degassing a solvent consisting of a vigorous bubbling of helium gas through the solvent.
13. Mobile phases and samples contain small particulates that can damage the column unless they are removed by filtration.
14. Whether the samples or mobile phase contain an organic solvent dictates what filter material to use.
15. Paper may not be chemically inert to organic solvents. Nylon, on the other hand, is inert to organic solvents.
16. For those samples and mobile phases that contain an organic solvent.
17. A guard column is a short, less-expensive LC column that is placed ahead of the analytical column in an HPLC system. The purpose of a guard column is to adsorb and retain mixture components that would contaminate the more expensive analytical column. In-line filters are relatively coarse filters (compared with prefilters) placed in the mobile phase line to filter out particulates that may be introduced on-line, such as from sample injection.
18. The in-line filter ahead of the column is a last second filtering precaution, just in case particulates were introduced with the sample.
19. The HPLC pump must be capable of providing extremely high pressure and pulsation-free flow.

20. A reciprocating piston pump is a pump that utilizes a piston in a cylinder to pull and push the liquid mobile phase from the mobile phase reservoir through the HPLC system. Two check valves (back-flow preventers) are in-line to help force the liquid in only one direction (see Figure 12.3 of the text).

21. A check valve is a device used in a flow stream to allow flow in only one direction. It is needed in the HPLC system to ensure that mobile phase flows in only one direction through the pump.

22. Isocratic elution—one mobile phase composition for entire run. Gradient elution—mobile phase composition is altered during the run in some preprogrammed manner.

23. A gradient programmer is that part of an HPLC instrument that provides for the gradual changing of the mobile composition in the middle of the run.

24. The gradient elution method for HPLC is the method in which the mobile phase composition is changed in some preprogrammed way in the middle of the run. The device that accomplishes this is called the gradient programmer and is placed between the mobile phase reservoir and the pump. It is useful in experiments in which altering the mobile phase composition assists with the resolution of the mixture.

25. A parameter that alters retention time (and hence resolution) is altered in the middle of the run in each case.

26. Individual mixture components vary as to their solubilities in different mobile phases and thus will display characteristic retention times with each different mobile phase composition. Thus, changing the mobile phase composition in the middle of the run will slow down some components and speed up others and change the resolution of all peaks.

27. "Solvent strength" refers to the ability of a particular mobile phase (solvent) to elute mixture components. For example, if use of a particular solvent results in short retention times for mixture components, it is said that the solvent is strong. If use of such solvent increases retention time, it is a weak solvent.

28. (1) The septum material may not be compatible with all mobile phases, thus creating the possibility for contamination. (2) The system is under too high a pressure to make the septum-piercing method a viable possibility.

29. See Figures 12.6, 12.7 and accompanying discussion.

30. See Figures 12.6, 12.7 and accompanying discussion.

31. The analyst may either (1) install a smaller loop on the injector or (2) partially fill the current loop, using the syringe to measure the volume.

32. Normal phase: stationary phase is polar, mobile phase is nonpolar. Reverse phase is just the opposite.

33. Reverse phase, since the mobile phase is polar and the stationary phase is nonpolar.

34. For reverse phase, common mobile phases are water, methanol, acetonitrile, and mixtures of these. Common stationary phases are phenyl, C_8, and C_{18}. For normal phase, common mobile phases are hexane, cyclohexane, and carbon tetrachloride. Common stationary phases are structures that include cyano, amino, and diol groups.

35. Bonded-phase chromatography is a type of liquid–liquid chromatography in which the liquid stationary phase is chemically bonded to the support material (as opposed to being simply adsorbed). The stationary phase can be either polar or nonpolar and thus both normal phase and reverse phase are possible.

36. Both nonpolar.

37. C18: reverse phase. Adsorption chromatography: normal phase.

38. Both cation exchange resins.

39. Cation exchange resins have negatively charged bonding sites for exchanging cations, while anion exchange resins have positively charged sites for exchanging anions.

40. Ion chromatography is the modern name for ion-exchange chromatography. The mobile is always a pH-buffered water solution, such as an acetic acid/sodium acetate water solution.

41. GPC utilizes nonpolar organic mobile phases, such as THF, trichlorobenzene, toluene, and chloroform, to analyze for organic polymers such as polystyrene. GFC utilizes mobile phases that are water-based solutions and is used to analyze for naturally occurring polymers, such as proteins, and nucleic acids.
42. (a) Normal and reverse-phase HPLC
    (b) Size-exclusion HPLC
    (c) Normal and reverse-phase HPLC
43. In reverse-phase chromatography, the polar mixture components would elute first since they would be attracted by the polar mobile phase and repelled by the nonpolar stationary phase. In normal phase chromatography, nonpolar mixture components would elute first since they would be attracted by the nonpolar mobile phase and repelled by the polar stationary phase.
44. Mobile phase composition, stationary phase composition, flow rate, temperature of stationary phase.
45. (a) Isocratic elution refers to the same mobile phase composition being used for the entire run, changed only by shutting down the pump and restarting it. Gradient elution refers to changing the mobile phase composition in the middle of the run.
    (b) Normal-phase chromatography refers to a procedure in which a nonpolar mobile phase is used in combination with a polar stationary phase. Reverse-phase chromatography refers to a procedure in which a polar mobile phase is used in combination with a nonpolar stationary phase.
    (c) The UV absorbance detector is sensitive, but not universal. The refractive index detector is universal, but not sensitive, cannot be used with gradient elution, and is subject to temperature effects.
46. The fixed-wavelength detector utilizes a glass filter as the monochromator. The variable-wavelength utilizes a slit-dispersing element-slit monochromator. The latter design is used to maximize sensitivity by setting the monochromator to the wavelength of maximum absorbance. If this is not important, then the fixed-wavelength design may be preferred because it is less expensive.
47. A diode array detector is a UV absorbance detector in which the light is not dispersed until after it has passed through the flow cell. The dispersed wavelengths spray across an array of photodiodes that detect the absorbed wavelengths all at once, allowing for simultaneous readings, which translate into the absorption spectrum. Thus, comprehensive qualitative analysis is possible, as well as rapid changeover of wavelengths between peaks to maximize peak size and to delete interferences.
48. detector, advantage, disadvantage
    (a) UV, sensitive, not universal
    (b) RI, universal, not sensitive
    (c) F, very sensitive, not universal nor highly applicable.
    (d) Cond, selective for ions, not universal—only for ions.
    (e) Amp, broad applicability, frequently needs servicing
    (f) LC-MS, excellent for qualitative analysis, expensive
    (g) LCIR, sensitive and fast, not water compatible
49. Not all potential mixture components will absorb UV light.
50. See the answer to question 48 above for some examples.
51. (a) fluorescence, (b) refractive index, (c) conductivity, (d) UV absorbance, (e) refractive index, (f) fluorescence, (g) UV absorbance, (h) conductivity
52. Yes, because the absorptivity of mixture components varies.
53. A suppressor is a device that selectively removes ions from a flowing solution. In an ion-exchange HPLC experiment in which the mobile phase contains ions, a suppressor must be used to remove these ions so that the ions of the mixture can be detected. A conductivity

detector measures ions by the conductivity that they induce in the mobile phase, so the mobile phase ions constitute an interference.

54. FTIR, like UV absorbance, refractive index, etc., is a technique for liquid solutions. However, it has the advantage in that it is fast, thus allowing a complete spectrum to be obtained as a given mixture component elutes, and making it an extremely powerful tool for qualitative analysis.

55. See Section 12.7.

56. Plugged in-line filter is a typical cause of unusually high pressure. The solution is to back-flush the filter, or otherwise clean it.

57. See Section 12.8.

58. This is a symptom of "channeling" in the column. The problem is solved by replacing the column.

59. (1) Slow elution of chemicals adsorbed on the column, (2) temperature effects, such as with the refractive index detector, (3) a contaminated detector.

60. 4.51 mg

61. See Section 12.9 in the textbook for these descriptions.

62. See "Introduction" section under Section 12.9 in the text.

63. Capillary electrophoresis is an electrophoresis technique in which the mixture components are separated in a capillary tube and detected with an "on-line" detector after the separation occurs. The advantages include smaller quantity of sample and qualitative and quantitative analysis in a much shorter time.

64. See Section 12.9.2.1.

65. See Section 12.9.2.3.

CHAPTER 13

1. A mass spectrometer is an instrument that analyzes a sample by detecting and measuring the ions generated from the analyte atoms or molecules in an "ion source."

   A mass spectrum is the data obtained in a mass spectrometry experiment. It is a plot of the signal intensity from each ion vs. the mass to charge ratio of these ions.

2. See Section 13.1.

3. Notice that the less intense signal is labeled as the isotope of carbon with mass number 13. The stronger signal is due to the isotope of carbon with mass number 12. So the two signals are observed because of the existence of these different isotopes and the Carbon-12 signal is more intense because this isotope is more abundant.

4. The three largest peaks: $43 - CH_3CO^+$, $57 - CH_3COCH_2^+$, $15 - CH_3^+$.

5. Qualitative analysis is possible because the mass spectrum is a unique fingerprint of the compound. Quantitative analysis is possible because a given ion signal, or summation of two or more signals, is proportional to the quantity of the compound.

6. The five components are the sample inlet, the ion source, the mass analyzer, the ion detector, and the vacuum system. The sample inlet introduces the sample to the system and passes it along to the ion source. The ion source generates ions from the sample. The mass analyzer separates the ions from each other on the basis of the mass to charge ratio. The ion detector detects each ion individually as each emerges from the mass analyzer and generates an electrical signal for each. The vacuum system evacuates the ion source, mass analyzer and ion detector so that no ions other than those from the sample are measured.

7. The sample is fed into the ion source from a chromatography instrument or other source. They then pass through the ion source where the ions are formed. The ions then move into the mass analyzer where they are separated from each other. Finally, the electronic signal is generated and amplified as the ions pass one at a time into the ion detector.

8. If the components of a mass spectrometer were not evacuated (1) air molecules in the ion source would form ions along with sample molecules and would thus contaminate the

sample ions and (2) air molecules in the mass analyzer would collide with the sample ions as they move and interfere with their detection.

9. The mixture components can first be separated by instrumental chromatography before being fed into the mass spectrometer. In fact, the mass spectrometer is the chromatography detector in such a scenario.

10. The ion source is where the atoms or molecules get transformed into the ions and molecular fragments that are subsequently detected and measured.

11. The thermal ionization (TI) ion source is used for the analysis of a radioactive metal for isotopic distribution. The radioactive metal is electroplated onto a rhenium filament prior to being "desorbed" and vaporized as a result of the heating in the ion source.

12. In the EI ion source, the molecules of the analyte move through a shower of high-energy electrons and break apart into positively charged fragments (ions) as a result. These fragments are the ions that are separated and measured by the subsequent path through the mass analyzer and ion detector.

13. The five ion sources mentioned are (1) the thermal ionization (TI) ion source, (2) the electron impact (EI) ion source, (3) the chemical ionization (CI) ion source, (4) the atmospheric pressure chemical ionization (APCI) ion source, and (5) the electrospray (ESP) ion source.

14. In the mass analyzer component of a mass spectrometer, the ions originating in the ion source are separated as they move so that they can be detected one at a time by the ion detector.

15. The three designs mentioned in this chapter are (1) quadrupole mass analyzer, (2) time-of-flight mass analyzer, and (3) magnetic sector mass analyzer.

16. See the detailed information given for each in Section 13.3.

17. The ion detector collects ions as they emerge from the mass analyzer and generates an amplified electrical signal for each.

18. A dynode is an electrode that is designed such that each electron impinging on its surface causes two or more electrons to be emitted from this surface.

19. Both make use of dynodes to amplify an electrical signal. In a photomultiplier tube, the source of the electrons is a photocathode at which light causes the initial release of electrons. In an ion detector, it is ions impacting a dynode that cause the initial release of electrons.

20. A mass spectrum is a plot of signal intensity vs. mass-to-charge ratio, m/z. "Relative abundance" is often used as the label for the y-axis because it is proportional to the signal intensity.

21. The base peak is the peak with the greatest intensity.

22. The molecular ion peak is the peak that is due to the ion that has the entire analyte molecule intact—the ionized analyte molecule.

23. Clusters of mass peaks around certain m/z values are due to the presence of isotopes.

24. A computer can scan libraries of mass spectra to match a measured mass spectrum with one in the library so as to identify the analyte.

25. Inductively coupled plasma–mass spectrometry.

26. Besides being an atomizer and a source of atomic emissions in atomic spectroscopy, the ICP source also is an ion source for mass spectroscopy. It provides the mass spectrometer with a source of charged monatomic metal ions.

27. As an atomic spectroscopy technique, the ICP is an atomizer and an excitation/emission source for analyzing samples for metals. As a mass spectroscopy technique, the ICP is an ion source, providing a mass spectrometer with a source of charged monatomic metal ions.

28. ICP-MS is a much more sensitive technique compared with the atomic spectroscopy techniques studied in Chapter 8, including the ICP emission technique. Compare the sensitivity data in Table 9.1 with those in Table 13.1.

29. It is a natural fit because the mobile phase flow rate is very slow through a capillary column, and this mobile phase is then not a significant detriment to the detection of mixture components by the mass spectrometer.
30. We say that a GC-MS is two-dimensional because while the chromatography peak is being traced, the mass spectrum may be measured. Thus, not only does the operator obtain the chromatogram, but he/she also obtains the mass spectrum at the same time without further effort.
31. TIC refers to "total ion chromatogram." This means that the signal measured for the chromatography is the sum total of the all the signals representing all the ions seen in the mass spectrum. The alternative is referred to as "selective ion monitoring," or SIM. With SIM, only selected ion signals are measured and summed to give the chromatography signal.
32. See the answer to question 31 above.
33. SIM provides two important advantages: greater sensitivity and greater selectivity. By selecting only one or a few ion signals to measure and sum, there can be more scans and therefore a greater total signal for the same concentration, thus increasing sensitivity. Selectivity also is improved since the instrument is zeroing in on only one or a few signals due to the analyte and not the sum total of all mass spectrum peaks, which may include other mixture components if the chromatographic resolution was poor.
34. Qualitative analysis is improved since the MS provides the mass spectrum fingerprint that can be compared with the library of spectra to identify the analyte. However, quantitative analysis is also improved due to the greater sensitivity and selectivity that can be achieved.
35. See the discussion describing the APCI and ESI ion sources used for LC-MS in Section 13.8.
36. See the discussion in Section 13.8.
37. LC-MS ion sources are referred to as "soft" ion sources because the energy used to create the ions is much less than with the EI source. Thus, only the ions forming as a result of the chemical reaction in the APCI source or only the ions forming as a result of the high voltage and rapid evaporation of the mobile phase described for the ESI ion source enter directly into the mass analyzer and are detected. The data are therefore not the same mass spectrum measured as a result of the fragmenting of the molecules in the EI source.
38. Tandem mass spectrometry is the use of two mass analyzers back to back with a "collision cell" in between. The collision cell utilizes ionized gas molecules from an external source to collide with the ions generated in the first mass analyzer. The result is a fragmentation similar to what occurs in the EI source, and this results in the measurement of the usual mass spectrum.

CHAPTER 14

1. Potentiometry is a technique in which electrode potential is measured and related to concentration, whereas voltammetry/amperometry is a technique in which the current flowing at an electrode is what is measured and related to concentration.
2. Oxidation-reduction reactions involve electron transfer typically by direct collision between chemical species in solution. In electroanalytical chemistry, electron transfer occurs, but through electrical conductors rather than by direct collision.
3. A cell is a complete electroanalytical system consisting of an electrode at which reduction occurs as well as an electrode at which oxidation occurs, and including the connections between the two. A half-cell is half of a cell in the sense that it is either one of the two electrodes (and associated chemistry) in the system, termed either the reduction half-cell

or the oxidation half-cell. The anode is the electrode at which oxidation takes place. The cathode is the electrode at which reduction takes place. An electrolytic cell is one in which the current that flows is not spontaneous, but rather due to the presence of an external power source. A galvanic cell is a cell in which the current that flows is spontaneous.

4. A galvanic cell operates of its own accord as a result of a spontaneous redox reaction. An electrolytic cell operates as a result of an external power source (e.g., a battery) in the circuit.

5. A salt bridge is a connection between half-cells that allows ions to diffuse between them, a requirement for a complete circuit so that a current can flow in the external circuit.

6. Potentiometry uses galvanic cells; amperometry/voltammetry uses electrolytic cells.

7. The standard reduction potential, symbolized by $E°$, is a number reflecting the relative tendency of a reduction half-reaction to occur. In a table of standard reduction potentials, such as Table 14.1, those half-reactions at the top of the table have positive numerical values and have a strong tendency to occur. Those near the bottom have negative values and have a tendency to go in the reverse direction.

8. (a) no, (b) no, (c) yes

9. -1.38 V.

10. To the left because $E°_{cell}$ is negative.

11. (a) $+1.57$ V.
    (b) yes, because $E°_{cell}$ is (+).

12. $+1.10$ V. Reaction proceeds to the right.

13. $+0.60$ V.

14. -1.57 V.

15. The answers differ only in that the signs of the calculated $E°$'s are different. The reason is that the reactions in the two problems are the same except reversed.

16. $E°_{cell} = +0.51$ V; yes, $E°_{cell}$ is (+).

17. $+2.11$ V.

18. -0.74 V.

19. (a) -0.92 V, (b) $+3.56$ V, (c) $+3.143$ V, (d) -0.297 V, (e) $+1.0164$ V

20. (a) No. $E°_{cell}$ is negative. (b) Yes. $E°_{cell}$ is positive. (c) Yes. $E°_{cell}$ is positive. (d) No. $E°_{cell}$ is negative. (e) Yes. $E°_{cell}$ is positive.

21. See Section 14.4.

22. $+0.42$ V.

23. 3.15 V.

24. 0.21 V. (for the reduction reaction.)

25. $+1.78$ V. (for the reduction reaction.)

26. (a) $+0.830$ V, (b) -0.764 V, (c) $+0.733$ V, (d) $+0.181$ V, (e) $+0.7131$ V

27. -0.72 V.

28. $+2.20$ V.

29. -0.52 V (for $2 \, Cr^{3+} + 3 \, Ni \leftrightarrows 2 \, Cr + 3 \, Ni^{2+}$)

30. -1.13 V

31. -0.94 V

32. (a) $+0.379$ V, 6, $+0.330$ V; (b) $+1.108$ V, 2, $+0.330$ V; (c) 0.029 V, 3, 0.029 V; (d) -0.4794 V, 2, -0.4498 V; (e) $+1.5614$ V, 2, $+1.4726$ V; (f) -2.67 V, 3, -2.8901 V.

33. Potentiometry is the measurement of electrode potential in chemical analysis procedures to obtain qualitative and quantitative information about an analyte. The reference electrode is a half-cell, which is designed such that its potential is a constant, making it useful as a reference point for potential measurements. Ground is the ultimate reference point in electronic measurements. It is a common wire threading throughout a circuit, often connected to the frame of the instrument and to the ground prong on the electrical outlet. An indicator electrode is the electrode that gives the information sought—either the potential or current that leads to the answer to the analysis.

34. Equations 14.6 and 14.7 clearly show that the potential of the SCE is directly proportional to the log[Cl⁻] inside the electrode. Since the [Cl⁻] is a constant because the solution is saturated with KCl, then the E must be constant, a requirement of a reference electrode.

35. The potential of this electrode depends only on the chloride ion concentration inside. If the KCl is saturated, it is assured that the chloride ion concentration is constant, which in turn means that the potential of this electrode never changes, a requirement of a reference electrode.

36. The salt bridge for reference electrodes consists of porous fiber tips that provide for the diffusion of ions in and out of this half cell.

37. The Nernst equation defining the potential of the silver/silver chloride electrode is Equation 14.9. Since the [Cl⁻] in such an electrode is a constant the potential also must be a constant (the requirement of a reference electrode) because the [Cl⁻] is the only variable on which the potential depends.

38. See Equations 14.10. 14.11, 14.12, and accompanying discussion.

39. The result of the derivation is Equation 14.12. The explanation leading up to this equation clearly shows that the potential, E, is directly proportional to the pH of the solution into which the solution is dipped.

40. Either one buffer solution, with a pH near that of the solution being tested, or two buffer solutions, with pH values that "bracket" that of the solution being tested, are usually used. When the pH electrode is dipped into a buffer solution during the standardization step, the pH reading on the meter is adjusted to the given pH using the "standardize" knob, or other control, on the meter.

41. A combination pH electrode consists of a pH electrode and a reference electrode in a single probe.

42. Both provide an electrolyte solution for the electrical contact between two half-cells so that ions can freely diffuse between the two half-cells and a current can flow externally allowing the desired measurement to be made.

43. An ion-selective electrode is a half-cell that is sensitive to a particular ion like the pH electrode is sensitive to the hydrogen ion.

44. (b) Nitrate concentration = 2.36×10^{-4} M

45. In gel-filled electrodes, the interior reference solution is gelatinized. As such, there is no loss of solution through the salt bridge and no contaminating solution can enter. It cannot be refilled with reference solution. In double-junction electrodes, the usual electrode is inside another glass or plastic encasement and there are two junctions for contact to the external solution. The purpose of this design is to prevent contamination in either direction.

46. (1) So that ions can diffuse freely through the salt bridge and (2) so that the electrode can be refilled with reference solution.

47. Over the short term: tip, include the porous salt bridge, should be immersed in a soak solution (pH = 7 usually recommended). Over the long term: the protective plastic sleeve with which the electrode was shipped should be repositioned over the tip.

48. A platinum electrode is needed in some half-cells to provide a surface at which electrons can be exchanged. Such a surface is lacking where there is no solid metal as part of the half-reaction, such as with the ferrous/ferric half-reaction.

49. A potentiometric titration is one in which the end point is detected by measuring the potential of an electrode dipped into the titrated solution.

50. Potentiometry utilizes galvanic cells, whereas amperometry uses electrolytic cells. Potentiometric methods measure a potential in a galvanic cell arrangement. Amperometric and voltammetric methods measure a current in an electrolytic cell.

51. See Section 14.6.

52. (1) Its surface is continuously renewed—thus avoiding problems with surface contamination.
    (2) The solution is automatically stirred when the drop falls.
53. Polarography is the measurement of the current flowing at a dropping mercury electrode as the potential applied to this electrode is changed. Voltammetry is the measurement of the current flowing at a stationary electrode as the potential applied to this electrode is changed.
54. An amperometric titration is one in which the end point is determined by monitoring the current flowing at an electrode
55. The moisture (water) in a sample
56. See Section 14.7
57. The prefix "bi-" is used because it uses two platinum "prongs" rather than one platinum and a reference electrode as in a "normal" method.
58. To prevent water contamination from humid laboratory air.
59. Conditioning the solvent means to eliminate all water in the solvent by titrating it with the Karl Fischer reagent. This is done to keep from measuring this water rather than the water in the sample.
60. In the volumetric method, the titrant is added from an external reservoir. In the coulometric method, the titrant (iodine) is generated internally via an electrochemical reaction.
61. The titrant is standardized by titrating a known amount of water with the titrant and calculating the titer of the titrant, the weight of water consumed by 1 mL of titrant.
62. Titer is the weight of water consumed by 1 mL of the Karl Fischer titrant. 0.543 mg/mL
63. 22%
64. The anode/cathode assembly generates the iodine. The dual pin platinum electrode detects the end point.
65. The coulombs of current used to generate the iodine is proportional to the amount of iodine needed to consume the water. Titrant is not added from a buret.

CHAPTER 15

1. X-ray diffraction: measurement of atomic and ionic spacings in crystal structures and thicknesses of thin metal films.
   X-ray absorption: quantitative analysis of heavy metals in matrices of lighter metals.
   X-ray fluorescence: quantitative analysis for elements in a wide variety of materials.
2. X-ray tube: a metal anode is bombarded with high-energy electrons causing inner shell electrons to be ejected and replaced by higher shell electrons. The loss in energy of these electrons as they drop to the lower levels is on the order of the energy of X-rays and X-rays are emitted.
   X-ray fluorescence: X-rays emitted by an X-ray tube irradiate an elemental material causing inner shell electrons to be ejected. X-ray emission then follows as with the X-ray tube.
3. K_α emission: an X-ray emission resulting from an electron dropping from the L shell to the K shell.
   L_β emission: an X-ray emission resulting from an electron dropping from the N shell to the L shell.
   $K_{\alpha 2}$ emission: an X-ray emission resulting from an electron dropping from the next to lowest sublevel in the L shell to the K shell.
4. Diffraction is the scattering of light that occurs when the light is directed at a structure that has regularly spaced lines or points in it, and these spacings are similar to the wavelength of the light. The scattering takes place because of internal reflections that occur and the destructive and constructive interferences that then also occur.

5. Bragg's law is: $n\lambda = 2d \sin\theta$.

"n" is an integer, the order of the diffraction, "λ" is the wavelength in angstroms, "d" is the interplanar spacing, and "θ" is the diffraction angle in degrees (see Figure 15.3.)

6. 2.04 Å

7. 65.75°

8. It is fluorescence because a longer wavelength (lower energy) is emitted than the wavelength absorbed.

9. It is atomic fluorescence because it occurs with unbound atoms rather than with molecules.

10. Use of the analyzer crystal (wavelength dispersive system) means better resolution, but it is slower and less sensitive.

11. See Section 15.1.5.

12. Nuclear magnetic resonance

13. The nuclear energy transitions that occur with the absorption of radio frequency light occur only in a strong magnetic field.

14. Radiofrequency wavelengths, because the energy required to cause the nuclear energy transitions in a magnetic field are on the order of radio frequency energy.

15. The two states are: When the small magnetic field due to the spinning/processing nucleus (1) aligns with the applied magnetic field and (2) when it opposes the applied magnetic field. The latter is a slightly higher energy state.

16. Most of the time, the nucleus measured is the ^{1}H nucleus, which is essentially a proton. Thus, the "P" in "PMR" stands for "proton."

17. Fourier transform nuclear magnetic resonance.

18. Hertz: cycles per second; megahertz: million cycles per second.

19. A 60-MHz instrument requires a field strength of 14,092 gauss, whereas a 100-MHz instrument requires 23,486 gauss.

20. "Gauss" is a unit of magnetic field strength.

21. Seven components: magnet, sample holder, RF generator, RF detector, sweep generator, sweep coils, data system (see Section 15.2.2 for the function of each).

22. In FTNMR experiments, very brief repeating pulses of Rf energy are applied to the sample while holding the magnetic field constant. The resulting detector signal contains the absorption information for all the nuclear energy transitions of all ^{1}H nuclei. The Fourier transformation sorts out and plots the absorption as a function of frequency, giving the same result, the NMR spectrum, as in a traditional experiment but in much less time.

23. y-axis: absorption, x-axis: chemical shift (in ppm)

24. The chemical shift is the effect that the environment of a nucleus has on the position of an absorption peak in a NMR spectrum. Such peaks are "shifted" to an extent-dependent on this environment.

25. The structural features near each hydrogen in the molecule, since they display a small and variable magnetic field of their own, impact the magnetic field that these hydrogens see and thus the apparent magnetic field values at which the molecule absorbs.

26. (a) 1, (b) 2, (c) 3, (d) 1, (e) 2

27. There are two kinds of hydrogen, the methyl group hydrogen and the hydrogen of the alcohol functional group, thus two peaks.

28. TMS provides a reference for displaying peaks of each type of hydrogen. The chemical shifts observed for each type of hydrogen are so slight that they would be difficult to measure without this reference.

29. Since there are three separate peaks, there are probably three different kinds of hydrogens in the structure.

30. The integrator trace tells us that the number of hydrogens represented by the peak at 5 ppm is one-half the number at 4.2 ppm and one third the number at 1.8 ppm.

31. Three peaks means three different kinds of hydrogens. There are three different kinds of hydrogens in ethyl alcohol, but only two different kinds in diethyl ether.

32. The integration trace shows that the three peaks are in the ratio 1:2:3, left to right. This is what is expected with ethyl alcohol because there are one of one kind of hydrogen, two of another, and three of another. For ethyl methyl ether, a ratio of 3:2:3 would be expected.

33. One quartet is expected because the carbon with two hydrogens is adjacent to a carbon with three (n) equivalent hydrogens (n + 1 = 4). One triplet is expected because the carbon with three hydrogens is adjacent to a carbon with two (n) equivalent hydrogens (n + 1 = 3).

34. The spectrum is that of isopropyl alcohol. The septet is due to the splitting of the peak for the hydrogen on the center carbon because of the six (n) equivalent hydrogens on either side. The doublet is due to the splitting of the peak for the six hydrogens on the two methyl groups because of the one hydrogen on the adjacent carbon. The singlet is due to the hydrogen on the OH group.

35. The integrator trace would show that the singlet is due to one hydrogen, the septet due to one hydrogen, and the doublet due to six hydrogens.

36. A given peak will be split into a number of peaks equal to the number of equivalent hydrogens on the adjacent carbons plus 1 (the n + 1 rule). This represents a clue about the structure that can help lead to identification of the compound.

37. Food products: soup, hot chocolate. Hygiene products: toothpaste, shaving gel.

38. See Section 15.3.2.

39. Capillary viscometry measures the time it takes for the test fluid to flow through a capillary tube. The viscosity is calculated from this time and the calibration constant of the tube.

40. Fluids of different viscosities flow through the capillary tube at different rates depending on their viscosities.

41. The Ostwald viscometer is the general name given to a capillary viscometer.

42. A Cannon–Fenske viscometer is a capillary type of viscometer. It utilizes the flow through a capillary tube as a means of measuring viscosity.

43. The Cannon–Fenske viscometer is pictured in Figure 15.16. The U-tube has a bend in the center. There are several "bulbs" in the "U," which allows a greater volume of fluid to be tested, and this also means a long and precise time measurement. The Ubbelohde viscometer is pictured in Figure 15.17. In this design, the "U" is completely vertical and, like the Cannon–Fenske viscometer, also has several bulbs for containing the fluid. The Cannon–Fenske viscometer is used for measuring the kinematic viscosity of transparent Newtonian liquids, especially petroleum products and lubricants. The Ubbelohde viscometer is also used for the measurement of kinematic viscosity of transparent Newtonian liquids, but by the "suspended level" principle.

44. 0.117 cS s^{-1}

45. 13 cS

46. Rotational viscosity methods measure viscosity by measuring the torque required to rotate a spindle immersed in the fluid sample.

47. See Section 15.4.1.

48. DTA, differential thermal analysis; DSC, differential scanning calorimetry; TGA, thermogravimetric analysis.

49. All the heat being added to the sample is being used to melt the sample rather than to raise its temperature.

50. If there is no physical or chemical change occurring in the sample as the temperature of the surrounding increases, there will be no difference in the two temperatures. If there is an endothermic physical or chemical change occurring, the temperature of the surroundings increases at a faster rate than that of the sample. If there is an exothermic physical or chemical change occurring, the temperature of the sample increases at a faster rate.

51. The negative peak means that the temperature of the surroundings is increasing at a faster rate than the temperature of a sample. This occurs, for example, when the sample melts.

52. The difference is $T_S - T_R$, in which T_S is the temperature of the sample and T_R is the temperature of a reference material. This difference is negative when an endothermic process occurs because T_S lags behind T_R due to the fact that the sample absorbs heat from the environment while its temperature remains the same. The opposite is true for an exothermic process.

53. In DTA, temperature is monitored. In DSC, heat flow is monitored. The data obtained are similar. DTA can be considered a more global term covering all differential thermal techniques, whereas DSC is a DTA technique that gives calorimetric (heat transfer) information.

54. "S" refers to the sample. "R" refers to the reference material. There must be two locations because the heat flow to each must be monitored individually so as to produce the T data.

55. Indium is a common reference material for DSC.

56. Endothermic and exothermic physical and chemical changes produce results in DSC. Melting is just one such change. Other examples would include chemical reactions that are either endothermic or exothermic. Physical changes would include deformations that do not involve melting.

57. See Section 15.4.4.

58. See Section 15.5.

59. "Optical rotation" refers to the rotation of the plane of polarized light as it passes through a sample.

60. $CH_3CHOHCH_2CH_3$, $CH_3CHClCH_2CH_3$, $CH_3CHBrCH_2CH_3$, $CH_3CH(OCH_3)CH_2CH_3$

61. Samples of 2-butanol are not optically active because they are actually equal mixtures of two enantiomers (racemic mixtures), so their optical rotations cancel out.

62. Each will rotate the plane of polarized light equally in opposite directions. Such a mixture is not optically active because the rotations cancel.

63. A polarimeter measures the degree of rotation of the plane polarized light as this light passes through the sample.

Index

A

Absorbance, 166, 198, 203
Absorption filters, 217, 218
Absorption of light, 193–201
   by atoms, 194–196
   by molecules and complex ions, 196, 197
Absorptivity, 204
Absorption spectra/spectrum, 197–201
   effect of concentration on, 207
Accuracy, 4, 13, 447
Acetic acid
   properties of, 28, 29
   titration of, with sodium hydroxide, 115, 116
Acids; *see also* titrations, acid-base
   conjugate, 128
   monoprotic, diprotic, triprotic, polyprotic; *see also*
      specific acids, 113
Action limits, 447
Activity, 132
Activity coefficient, 132
Adjusted retention time, 301, 303
Adsorption chromatography, *see* Chromatography,
   adsorption
Alkalinity, 124
Ammonium hydroxide
   properties of, 29
   titration of, with hydrochloric acid, 116
Amperometric detector, *see* High performance liquid
   chromatography, detectors
Amperometric/voltammetric methods; *see also*
   Amperometry/voltammetry, 388, 405
   amperometric detector, *see* High performance liquid
      chromatography, detectors
   amperometric titration, 406
   polarography, 406, 407
   stripping voltammetry, 407
Amperometry/voltammetry, 405, 406
   electrode terminology, 405
Analysis, 1
Analyte, 1
Analytical balance, *see* Balance, analytical
Analytical Chemistry, 1
Analytical process, 3, 165, 166
Analytical Science, 1
Anode, 388
Antacid tablet, titration of, 125
Aqua regia
   properties of, 28, 29
   composition of, 28, 29
Arc and spark emission spectroscopy, 260, 282
Aspiration rate, 271
Assay, 1
Atmospheric pressure chemical ionization I(APCI), *see*
   Mass spectrometry, ion sources
Atomic fluorescence spectroscopy, 282, 283

Atomic spectroscopy; *see also* specific techniques, 198,
   259–283
Atomization, 259
Atomizer, 259
   flame, 261
   graphite furnace, 261
   inductively coupled plasma, 261
Automatic titrator, 75, 102, 103
Auto-sampler, 167

B

Back titration, *see* titration, back
Balances, 49–51
   care of, 52
   choice of, 52, 53
   electronic, 49
   top-loading, 50
Bandwidth, 216, 218
Bases; *see also* titrations, acid-base
   conjugate, 128
   monobasic, dibasic, tribasic, and polybasic; *see also*
      specific bases, 113
Beamsplitter, 221, 222
Beer's Law, 169, 204
   deviations from, 229, 230
Beer's Law plot, 205, 206
Bias, 5
Blank, 173
   reagent, 173, 174
   sample, 174
Bonded phase chromatography, *see* Chromatography,
   bonded phase
Bottle-top dispensers, 102
Bromcresol green, 117, 118
Buffer solutions, 127
   recipes for, 133
   theory of, 129–133
Buffer capacity, 129
Buffer region, 129
Bulk sample, *see* Samples, various types
Bulk system, 3
Buret, 74, 75
   digital, 75
Burner, *see* Premix burner

C

Calibration, 14, 447
   of balances, 52
   of glassware and other devices, 103, 104
   of instruments, 167, 168–170
Calibration curve, *see* Standard curve
Capacity factor, 305
Capillary columns, *see* Gas chromatography, columns
Capillary electrophoresis, 306, 359–362

detectors, 361, 362
electroosmotic flow, 361
free solution, 360
zone, 360
Cathode, 388
Certified Reference Material (CRM), 447
Chain of custody, 21, 447
Check valve, 344
Chelate, 134
Chemical analysis, 1
Chemical ionization (CI), *see* Mass spectrometry, ion
    sources
Chopper, 221, 223, 224
Chromatogram, 301, 302
Chromatographic methods, 166
Chromatographs, 166
Chromatography, 291
    adsorption, 293, 294, 296, 349, 350
    ascending, 295, 296
    bonded phase (BPC), 292
    descending, 295, 296
    gas (GC); *see also* Gas chromatography, 291
    gel filtration (GFC), 295
    gel permeation (GPC), 295
    general discussion, 291–306
    high performance liquid (HPLC); *see also* High
        performance liquid chromatography, 291
    instrumental; *see also* Gas Chromatography and
        High performance liquid chromatography,
        301–306
    ion-exchange (IEC or IC), 294–296, 349, 350
    liquid (LC); *see also* Liquid chromatography, 291
    open-column, 298–301
    paper, 296–298
    partition, 292, 293, 296, 348
    radial, 295, 296
    size exclusion (SEC), 295, 296, 349, 350
    thin-layer (TLC), 296–298
Cold vapor mercury atomic absorption, 282, 283
Colorimeter, 215
Complex ion formation reactions, 133–144
Complex ions, 133
    Terminology, 133–135
Composite Sample, *see* Samples, various types
Concentration of solutions, 75–79, 443, 444
    molarity, 75–77, 443, 444
    normality, 77–79, 444
    percent, 443
    ppm, 138, 139, 444
Conductivity detector, *see* High performance liquid
    chromatography, detectors
Confidence level, 10
Confidence limits, 10
Conjugate acid-base combinations; *see also* Acid,
    conjugate and Base, conjugate, 132
Continuous and line spectra, 199
    spectral lines, 259
Control, 174, 447
Control chart, 174, 447
Coordinate covalent bond, 133
Correlation coefficient, 172
Current Good Manufacturing Practices (cGMP), 447
Cuvette, 205
    selection and handling, 228, 229

D

Data acquisition, 178, 179
Degassing of HPLC mobile phases, *see* High performance
    liquid chromatography, mobile phase handling
Degrees of freedom, 6
Desiccator, 51, 52
Detection limit, 283, 447
Detectors
    for gas chromatography, *see* Gas chromatography,
        detectors
    for spectrophotometers, 227, 228
Deviation, 6
Diffraction grating, 220
Diffuse reflectance, 244
Diode array detector, *see* High performance liquid
    chromatography, detectors
Diode array spectrophotometer, *see* Spectrophotometer,
    diode array
Dispersing element, 218, 219
Distribution coefficient, 34
Distillation
    of water, *see* Water, distillation of
    of mixtures of liquids, 39–41
    fractional, 40
DSC, *see* Thermal analysis, differential scanning
    calorimetry
DTA, *see* Thermal analysis, differential thermal
    analysis
Dynode, 376

E

EDTA, 135–138
    standardization of solutions of, 142, 143
Electric current, 387
Electroanalytical methods; *see also* Amperometric/
    voltammetric methods and Potentiometric
    methods, 166, 387–409
    terminology, 388–391
Electrodeless discharge lamp, *see* Sources of light,
    Electrodeless discharge lamp
Electrolytic cell, 389
Electrolytic conductivity detector, *see* Gas
    Chromatography, detectors
Electromagnetic spectrum, *see* Spectrum,
    electromagnetic
Electromagnetic waves, *see* Light, wave theory
Electron capture detector, *see* Gas chromatography,
    detectors
Electron impact ion source, *see* Mass spectrometry, ion
    sources
Electronic transition, *see* Energy, electronic transition
Electroosmotic flow, *see* Capillary electrophoresis,
    electroosmotic flow
Electrophoresis, 306, 357–362, 387
    capillary, *see* Capillary electrophoresis
    gel, 359
    zone, 359
Electrospray ionization, *see* Mass spectrometry, ion
    sources
Electrothermal atomic spectroscopy; *see also* Graphite
    furnace atomic absorption, 260
Emission of light; *see also* Fluorescence, 201, 202

Emission spectrum, 201
End point, 75
Energy,
    electronic transition, 195
    vibrational transition, 197
Energy level diagram, 195
    depicting fluorescence, 231
Energy of light, *see* Light, energy of
Equivalence point, 75, 113
    detection of, 116–118
Equivalent weight of acids and bases, 78
Eriochrome black T, *see* Indicator, EDTA titrations
Errors, 5
    determinate, 5
    indeterminate, 5
    random 5
    systematic, 5
Ethylenediaminetetraacetic acid, *see* EDTA
Excited state, 195
Extinction coefficient, 204
Extraction
    from liquid solutions, *see* Liquid-liquid
        extraction
    purge and trap, 32, 316
    solid-liquid, 24, 38, 39
    solid phase, 30
    soxhlet, 24, 38, 39
    supercritical fluid, 24
    thermal, 25
Extraction efficiency, *see* Liquid-liquid extraction
Evaporators, 38

F

Filtering of HPLC mobile phases, *see* High
        performance liquid chromatography, mobile
        phase handling
Filters, *see* absorption filters
Flame atomic absorption, 260, 262–274, 283
    burner, *see* Premix burner
    flames, 262, 263
    interferences, *see* Interferences
    optical path, 267–269
    safety and maintenance, 274
Flame emission; *see also* Flame photometry, 260
Flame ionization detector, *see* Gas Chromatography,
        detectors
Flame photometric detector, *see* Gas Chromatography,
        detectors
Flame photometry, 281, 283
Flow spoiler, 266, 267
Fluorescence, 202, 231
    detector, *see* High performance liquid chromatography,
        detectors
    spectrum, 232
Fluorometer, 186, 232
Fluorometry, 231–233
Formation constant, 135
Fourier transform IR spectrometry; *see also* IR molecular
        spectrometry, 234, 235
Fraction collector, 298, 299
Frequency, *see* Light, frequency of
FTIR, *see* Fourier transform IR spectrometry
Fusion, 30

G

Galvanic cell, 389, 390
Gas chromatography (GC); *see also* Chromatography,
        instrumental, 311–330
    column temperature, 319–320
    columns, 311–318
    detectors, 321–325
    importance of solubility in, 311–313
    instrument components, 312–314
    importance of vapor pressure in, 311–313
    qualitative analysis with, 325, 326
    quantitative analysis with; see also Chromatography,
        instrumental, 326–329
    sample injection, 314–316
    stationary phase, 318, 319
    troubleshooting, 329, 330
Gas chromatography – mass spectrometry, 324, 325,
        378–380
    selective ion monitoring (SIM), 380
    total ion chromatogram (TIC), 379
gel electrophoresis, *see* Electrophoresis, gel
Gel filtration chromatography, *see* Chromatography, gel
        filtration
Gel permeation chromatography, *see* Chromatography, gel
        permeation
GC-MS, *see* Gas chromatography – mass spectrometry
Good Laboratory Practices (GLP), 447, 448
Grades of chemicals, 41
Gradient elution (in HPLC), *see* High performance liquid
        chromatography, elution methods
Graphite furnace atomic absorption, 260, 262, 275–278, 283
    advantages and disadvantages, 277, 278
    optical path, 275
    temperature program, 277
Gravimetric analysis, 49–64
    chemical separation of analyte, 58–63
    details of methods, 53–63
    physical separation of analyte, 53–58
Gravimetric factors, 59–61
    use of, 62, 139, 140
Ground state, 194
Guard column, 344

H

Hall detector, *see* Gas chromatography, detectors
Height equivalent to a theoretical plate (HETP),
    in instrumental chromatography, 304
    in distillation, 40
Helium sparging, *see* High performance liquid
        chromatography, mobile phase handling
Henderson-Hasselbalch equation, 129–130
High Performance Liquid Chromatography (HPLC);
        see also Chromatography, instrumental,
        341–357
    column selection, 348–350
    comparisons with GC, 341, 342
    detectors, 350–355, 387
    elution methods, 345, 346
    mobile phase handling, 342–344
    normal phase, 348, 350
    pumps, 344, 345
    qualitative analysis with, 355, 356

quantitative analysis with; *see also* Chromatography, instrumental methods, 355, 356
reverse phase, 348, 350
sample injection, 346, 347
summary of method, 341, 342
troubleshooting, 356, 357
Histogram, 9, 10
Hollow Cathode Lamp, *see* Sources of light, hollow cathode lamp
Hydride generation atomic absorption, 282, 283
Hydrochloric acid
properties of, 27, 29
titration of, with sodium hydroxide, 114
Hydrofluoric acid
properties of, 28, 29

I

ICP, *see* Inductively Coupled Plasma
ICP-MS, *see* Inductively Coupled Plasma – mass spectrometry
Indicator, 75, *see also* specific indicators
for EDTA titrations, 138
Indirect titration, *see* titration, indirect
Inductive coupled plasma, 261, 278–281, 283
advantages and disadvantages, 281
coupled with mass spectrometry, 281, 378
optical path, 280
standard curve, 281
Inductively coupled plasma – mass spectrometry, 378
detection limits, 379
Inflection point, 114
Infrared light, 189, 190, 233
Injection port (in gas chromatography), *see* Gas chromatography, sample injection
Insoluble matter in reagents, 54, 56
Instrumental analysis, 2, 165–179
Interferences, 291
in flame atomic absorption, 271–274
in spectrochemical analysis, 229
Interferometer, 233–235
Internal reflectance, 242, 243
Internal standard method, *see* Gas chromatography, quantitative analysis with
International Organization for Standardization (ISO), 448
Iodometry, 150, 151
Ion-electron method (for balancing oxidation-reduction reactions), 147, 148
Ion chromatography, *see* Chromatography, ion-exchange
Ion-exchange chromatography, *see* Chromatography, ion exchange
Ion-selective electrode, *see* Potentiometry, indicator electrodes
IR liquid sampling cells, 205, 235–240
pathlength of, 205, 233
IR molecular spectrometry, 198, 233–248
IR spectra, 233, 239
interpretation of, 244–247
ISO, *see* International Organization for Standardization
ISO 9000, 448
Isocratic elution (in HPLC), *see* High performance liquid chromatography, elution methods

K

Karl Fischer titration, 406–409
KBr pellet, 240–242
Kjeldahl titration, 125–127

L

LIMS, *see* Laboratory Information Management System
Laboratory Information Management System, 179
Laboratory sample, *see* Samples, various types
Ligand, 133, 134
Light,
characteristics of, 185–189
energy of, 187, 188
frequency of, 187, 188
particle theory of, 185
wave theory of, 185–189
wavelength of, 186
wavenumber of, 186–189
Line spectrum, *see* Continuous and line spectra
Liquid chromatography–mass spectrometry (LC–MS), 380, 381
atmospheric pressure chemical ionization (APCI), 373, 380
electrospray ionization (ESI), 373, 380–M
Liquid chromatography–mass spectrometry/mass spectrometry (LC–MS/MS), 381, 382
Liquid-liquid extraction, 30–38
use of a separatory funnel in, 30, 31
theory of, 34, 35
calculations for, 35–38
extraction efficiency, 37
Loop injector, *see* High performance liquid chromatography, sample injection
Loss on drying, 54–56
Loss on ignition, 54, 55
Luminescence, 231

M

Magnetic sector mass analyzer, *see* Mass spectrometry, mass analyzers
Masking, 135
Mass, 49
Mass spectra, 371, 372, 377, 378
Mass spectrometer detector, *see* Gas chromatography, detectors and also Mass spectrometry
Mass spectrometry, 371, 382
basic principles, 371, 372
ion detector, 371, 376
ion sources, 371–373, 378–380
mass analyzers, 371, 373–376
sample inlet systems, 372, 373
tandem, 381, 382
vacuum system, 371
Matrix, 1
Mean, 6
Median, 6
Measurement system, 448
Method of least squares, 171, 172
Microwaves, 189
Mobile phase, 291, 292
Mode, 6

Moisture; *see also* Loss on drying
    determination of, 55
    in samples, 51
    Karl Fischer method for, 406–409
Molar absorptivity, 204
Molarity, *see* Concentration of solutions
Molecular spectroscopy; *see also* UV/VIS molecular
        spectrophotometry and IR molecular
        spectrometry, 185
Monochromators, 217–220

N

Nebulizer, 266
Nernst equation, 394–396
Nitric acid,
    properties of, 28, 29
Nitrogen/Phosphorus detector, *see* Gas Chromatography,
        detectors
NMR, *see* Nuclear Magnetic Resonance spectrometry
Nonclinical laboratory study, 448
Normal distribution curve, 8
Normal phase HPLC, *see* High performance liquid
        chromatography, normal phase
Normality, *see* Concentration of solutions, normality
Nuclear magnetic resonance spectrometry (NMR),
        422–428
    analytical uses, 428
    instrumentation, 423–425
    NMR spectra, 425–428
Nujol mull, 242

O

Optical path
    In double-beam and single-beam
Optical rotation, 437–439
Oxidation, 144
Oxidation number, 144–146
Oxidation-reduction reactions, 144–152, 387, 388
Oxidizing agent, 146

P

Paper chromatography, *see* Chromatography, paper
Particle size,
    analysis, 54, 57, 58
    reduction, 23
Partition chromatography, *see* Chromatography, partition
Partition coefficient, 34
Parts per million, *see* Concentration of solutions, ppm
Pathlength, 204, 205
    in flame AA, 205
    in IR spectrometry, 205, 236, 237
Percent transmittance, 200
Percentage analyte calculations,
    gravimetric analysis, 54, 55, 62
    titrimetric analysis, 88–91
Perchloric acid
    properties of, 28
Person, 448
pH meter, 166
pH electrode, *see* Potentiometry, indicator electrodes, and
        Potentiometric methods, pH measurement

Phenolphthalein, 117, 118
Phosphorescence, 231
Phosphoric acid,
    titration of, with sodium hydroxide, 120, 121
Photodiodes, 228
Photoionization detector, *see* Gas chromatography, detectors
Photomultiplier tube, 227
Phthalic acid, molecular structure of, 121
Pipetters, 100–102
Polarimeter; *see also* Optical rotation, 438
Population standard deviation, *see* Standard deviation
Potassium biphthalate
    molecular structure of, 121
    titration of, with sodium hydroxide, 121
Potassium permanganate, 150
Potentiometric methods; *see also* potentiometry, 388
    ion-selective measurement, 401–403
    pH measurement, 401
    potentiometric titration, 404, 405
ppm, *see* Concentration of solutions, ppm
Post-run calculations in instrumental analysis, 174–178
Potentiometry; *see also* Potentiometric methods, 396–405
    care and maintenance of electrodes, 403, 404
    indicator electrodes, 396, 399–403
    reference electrodes, 396–399
Power supply, 168
Precipitates
    in gravimetric analysis, 64
Precision, 4, 13, 448
Premix burner, 256–267, 271
Preparation of solutions, 79–82, 444, 445
    dilution, 80–82, 444, 445
    molarity, solid solute, 80, 81, 445
    normality, solid solute, 81, 82, 445
    percent, 444, 445
    ppm, 139, 140, 445
Primary line, *see* Spectral lines
Primary sample, *see* Samples, various types
Prism, 220
Purge and trap, *see* Extraction, purge and trap

Q

Q test, 12, 13
Quadrupole mass analyzer, *see* Mass spectrometry, mass
        analyzers
Qualitative analysis, 2
Quality assurance, 448
Quality assurance unit, 448
Quality control, 448
Quality system, 448
Quantitative analysis, 2
    in gas chromatography, *see* Gas chromatography,
        quantitative analysis with
    in high performance liquid chromatography, *see*
        High performance liquid chromatography,
        quantitative analysis with
    in IR spectrometry, 247
    in UV-VIS spectrophotometry, 205, 206

R

Radio and TV light, 189
Random sample, *see* Samples, various types

Raw data, 448
Reagent, 41
Reducing agent, 146
Reduction, 144
Reference material (RM), 448
Reference substance, 448
Reflectance methods for IR analysis of solids, 242–244
Refractive Index, 190–193
  Detector, *see* High performance liquid
    chromatography, detectors
Refractometer, 185, 186, 191
Refractometry, 185, 190–193
Relative retention, *see* Selectivity
Relative standard deviation (RSD), *see* Standard deviation
Repipet, *see* Bottle-top dispensers
Representative sample; *see also* Samples, various types,
    3, 19
Residue on Ignition, 54, 56
Resolution
  of wavelengths, 217
  of chromatography peaks, 302
Resonance, 262
Response factor method, *see* Gas chromatography,
    quantitative analysis with
Retention time, 301, 303
Reverse phase HPLC, *see* High performance liquid
    chromatography, reverse phase
Right-angle configuration, 232

S

Salt bridge, 389, 390
Sample, 3
Sample compartment, 220
Sample preparation, 3, 23–25, 26–41, 165, 166
Sample standard deviation, *see* Standard deviation
Samples
  dissolution of, 26–30
  extraction of, *see* Extraction
  handling of, 21–23
  homogenization and division, 23
  integrity of, 22
  obtaining, 19–23, 165
  preservation of, 22
  stability of, 32
  various types, 20
Sampling, in IR spectrometry
  liquid, 235–240
  solid, 240–244
Secondary line, *see* Spectral lines
Sampling Statistics, 20, 21
Saturated calomel electrode, *see* Potentiometry, reference
    electrodes
Secondary sample, *see* Samples, various types
Selective sample, *see* Samples, various types
Selective ion monitoring (SIM), *see* Gas chromatography –
    mass spectrometry
Selectivity (in chromatography), 304, 305
Sensitivity
  In atomic spectroscopy, 282, 283, 379
Sensor, 168
Separatory funnel, *see* Liquid-liquid extraction
Serial dilution, 173
Shelf life, 41

Signal processor, 168, 193
Significant figure rules, 451, 452
Size exclusion chromatography, *see* Chromatography, size
    exclusion
Slits, 218, 219
Sodium carbonate
  in fusion, 30
  titration of, with hydrochloric acid, 122, 123
Sodium hydroxide
  titration of, with hydrochloric acid, 116
Solid phase extraction, *see* Extraction, solid phase
Solid-liquid extraction, *see* Extraction, solid-liquid
Solids in water and wastewater, 54, 56
Solvent exchange, 32
Solvent strength, 346
Solution concentration, *see* Concentration of solutions
Solution preparation, *see* Preparation of solutions
Solvent extraction, *see* Liquid-liquid extraction
Sources of light, 215, 216
  deuterium lamp, 216
  spectral line, 259, 263–265
  tungsten filament lamp, 216, 217
  xenon arc lamp, 216
  hollow cathode lamp, 264, 265, 271
  electrodeless discharge lamp, 265
Spark emission spectroscopy, *see* Arc and spark emission
    spectroscopy
Spectral lines; *see also* Sources of light, 259, 260, 268, 269
  primary line, 268
  secondary lines, 268
Spectrometer, 166, 185, 186
Spectrochemical methods, 185–207
Spectrometry, 185; *see also* spectrochemical methods
  atomic, *see* Atomic spectroscopy
  IR, *see* IR molecular spectrometry
  mass, *see* Mass spectrometry
Spectrophotofluorometer, 232
Spectrophotometer, 185
  diode array, 223–227
  double-beam molecular, 222–226
  UV-VIS Single-beam molecular, 220, 221, 226
  UV, 215
  UV-VIS, 215
Spectrophotometry, *see* spectrometry and spectroscopy
  UV-VIS, *see* UV-VIS molecular spectrophotometry
Spectroscopic methods; *see also* spectrochemical
    methods, 166
Spectroscopy; *see also* spectroscopic methods, 185
Spectrum, absorption, *see* Absorption spectrum
Spectrum, electromagnetic, 189
  visible region, 189, 190
Spectrum, transmission, *see* Transmission spectrum
Sponsor, 448
Sputtering, 265
Standard additions method, 272, 273
Standard cell potential (Eocell), 393, 394
Standard curve, 167, 169, 170
  in flame atomic absorption, 269, 270
  in instrumental chromatography, 306
Standard deviation, 6–8
Standard operating procedure (SOP), 448, 449
Standard reduction potential (E°), 391–393
Standard reference material (SRM), 449
Standard solution, 73, 172, 173

Standardization, 73, 84
  using a primary standard, 86–88
  using a standard solution, 84–86
Standardize, 73
Stationary phase, 291, 292
Stoichiometry of titration reactions, 82–84
Stopcock, 74
Student's t, 10, 11
Study, 449
Study director, 449
Subsample, *see* Samples, various types
Substance titrated, 74
Sulfuric acid
  properties of, 27, 29
  titration of, with sodium hydroxide, 118–120
System suitability, 449

T

Temperature programming (in gas chromatography), *see* Gas chromatography, column temperature
Test article, 449
Test sample, *see* Samples, various types
Test substance, 449
Test system, 449
Testing facility, 449
Total Quality Management (TQM), 449
TGA, *see* Thermal analysis, thermogravimetric analysis
Theoretical plates
  in chromatography, 302, 304
  in distillation, 40
Thermal analysis, 434–437
  differential thermal analysis (DTA), 434–436
  differential scanning calorimetry (DSC), 434–436
  instrumentation, 436, 437
  thermogravimetric analysis (TGA), 56, 434
Thermal conductivity detector, *see* Gas chromatography, detectors
Thermal ionization source, *see* Mass spectrometry, ion sources
Thin-layer chromatography (TLC), *see* Chromatography
Time of flight mass analyzer (TOF), *see* Mass spectrometry, mass analyzers
Titer, 88
Titrant, 74
Titration, 74
  acid-base, 113–127
  back, 124–126
  curves, 114–123
  indirect, 126, 127, 150, 151
  midpoint of, 131
  requirements for a successful, 113
Titrimetric analysis, 73–104
Total ion chromatogram (TIC), *see* Gas chromatography – mass spectrometry, total ion chromatogram
Traceability, 449
Transmission spectrum, 200
Transmittance, 200, 202
Tris-(hydroxymethyl)amino methane (tris, THAM), 122
  buffer solution made with, 130, 132
  titration of, with hydrochloric acid, 122
Troubleshooting,
  of gas chromatographs, *see* Gas chromatography, troubleshooting
  of liquid chromatographs, *see* High performance liquid chromatography, troubleshooting
  of spectrophotometers, 230, 231

U

Ultra-performance liquid chromatography (UPLC), 349
Ultraviolet light, 190
UPLC, *see* Ultra-performance liquid chromatography
UV absorption detector, *see* High Performance Liquid Chromatography, detectors
UV/VIS molecular spectrophotometry, 198, 215–231

V

Validation, 449
Vapor pressure, 26, 39, 40, 311
  in gas chromatography, 311, 312
Viscometers, 431–433
  Cannon-Fenske, 431
  rotational, 433
  Ubbelohde, 432
Viscosity, 428–433
  capillary, 430–433
  rotational, 433
  temperature dependence, 430
  terminology, 429–430
Visible light, 189–190
Volumetric analysis, *see* Titrimetric analysis
Volumetric glassware, 32, 91–100
  buret, 74, 75, 98–100
  cleaning and storing, 100
  flask, 73, 74, 91–94
  pipets, 73, 74, 94–98
Voltammetric methods, *see* Amperometric/voltammetric methods

W

Warning limits, 449
Water
  purification of, 25, 26
  distillation of, 25, 26
  deionization of, 26
  hardness of, 137, 138, 141–143
Wavelength; *see also* Light, wavelength of
  of maximum absorbance, 205, 206
  selection of, 216–220
Wavenumber, *see* light, wavenumber of
Weighing
  bottles, 63, 64
  by difference, 63, 64
Weight, 49
Wet chemical analysis, 2

X

X-ray light, 190
X-ray methods, 417–422
  X-ray diffraction, 418–420
  X-ray fluorescence spectroscopy, 421, 422

Z

Zone electrophoresis, *see* Electrophoresis, zone

Formula Weights

AgBr	187.772	Co₂O₃	165.864	KClO₃	122.549	NaCN	49.008

Formula	FW	Formula	FW	Formula	FW	Formula	FW
AgBr	187.772	Co$_2$O$_3$	165.864	KClO$_3$	122.549	NaCN	49.008
AgCl	143.321	Co$_3$O$_4$	240.798	KCN	65.116	Na$_2$CO$_3$	105.989
Ag$_2$CrO$_4$	331.730	Cr$_2$O$_3$	151.990	K$_2$CO$_3$	138.206	NaF	41.988
AgI	234.772	Cu$_2$O	143.091	K$_2$CrO$_4$	194.191	NaHCO$_3$	84.007
AgNO$_3$	169.873	CuO	79.545	K$_2$Cr$_2$O$_7$	294.185	Na$_2$H$_2$EDTA·2 H$_2$O	372.23
Ag$_2$O	231.735	CuSO$_4$	159.610	KHC$_2$O$_4$	128.13	NaH$_2$PO$_4$	119.977
AgSCN	165.952	CuSO$_4$·5 H$_2$O	249.686	KHC$_8$H$_4$O$_4$ (KHP)	204.23	Na$_2$HPO$_4$	141.959
Al$_2$O$_3$	101.961	Fe(NH$_4$)$_2$(SO$_4$)$_2$·6 H$_2$O	392.141	K$_2$HPO$_4$	174.176	NaOH	39.997
Al(OH)$_3$	78.004	FeO	71.844	KH$_2$PO$_4$	136.085	Na$_3$PO$_4$	163.944
Al$_2$(SO$_4$)$_3$	342.154	Fe$_2$O$_3$	159.688	KHSO$_4$	136.170	NaSCN	81.074
As$_2$O$_3$	197.841	Fe$_3$O$_4$	231.533	KI	166.003	Na$_2$SO$_4$	142.044
BaCO$_3$	197.336	HBr	80.912	KIO$_3$	214.001	Na$_2$S$_2$O$_3$·5 H$_2$O	248.186
BaCl$_2$	208.232	HC$_2$H$_3$O$_2$ (acetic acid)	60.05	KIO$_4$	230.001	NH$_3$	17.031
BaCl$_2$·2 H$_2$O	244.263	HCO$_2$C$_6$H$_5$ (benzoic acid)	122.12	KMnO$_4$	158.034	NH$_4$Cl	53.492
BaCrO$_4$	253.321	HCl	36.461	KNO$_3$	101.103	NH$_2$(HOCH$_2$)$_3$ (THAM)	121.136
BaO	153.326	HClO$_4$	100.459	KOH	56.105	(NH$_4$)$_2$C$_2$O$_4$·H$_2$O	142.110
Ba(OH)$_2$	171.342	H$_2$C$_2$O$_4$	90.04	K$_3$PO$_4$	212.266	NH$_4$NO$_3$	80.043
BaSO$_4$	233.391	H$_2$C$_2$O$_4$·2 H$_2$O	126.07	KSCN	97.182	(NH$_4$)$_2$SO$_4$	32.141
Bi$_2$O$_3$	465.959	HNO$_3$	63.013	K$_2$SO$_4$	174.261	(NH$_4$)$_2$S$_2$O$_8$	228.204
Bi$_2$S$_3$	514.159	H$_2$O	18.015	MgCl$_2$	95.210	PbCrO$_4$	323.2
C$_6$H$_{12}$O$_6$ (glucose)	180.16	H$_2$O$_2$	34.015	MgO	40.304	Pb$_3$O$_4$	685.6
C$_{12}$H$_{22}$O$_{11}$ (sucrose)	342.30	H$_3$PO$_4$	97.995	Mg(OH)$_2$	58.320	PbSO$_4$	303.3
CHCl$_3$	119.38	H$_2$S	34.082	Mg$_2$P$_2$O$_7$	222.555	P$_2$O$_5$	141.945
CO$_2$	44.010	H$_2$SO$_3$	82.080	MgSO$_4$	120.369	Sb$_2$O$_3$	291.518
CaCl$_2$	110.983	H$_2$SO$_4$	98.080	MnO$_2$	86.937	SiF$_4$	104.080
CaCO$_3$	100.087	HSO$_3$NH$_2$ (sulfamic acid)	97.095	Mn$_2$O$_3$	157.874	SiO$_2$	60.085
CaC$_2$O$_4$	128.097	HgO	216.59	Mn$_3$O$_4$	228.812	SnCl$_2$	189.615
CaF$_2$	78.075	Hg$_2$Cl$_2$	472.09	Na$_2$B$_4$O$_7$·10 H$_2$O	381.373	SnO$_2$	150.709
CaO	56.077	HgCl$_2$	271.50	NaBr	102.894	SrSO$_4$	183.68
Ca(OH)$_2$	74.093	Hg(NO$_3$)$_2$	324.60	NaC$_2$H$_3$O$_2$	82.034	SO$_2$	64.065
CaSO$_4$	136.142	KBr	119.002	Na$_2$C$_2$O$_4$	133.999	SO$_3$	80.064
CeO$_2$	172.115	KBrO$_3$	167.000	NaCl	58.443	TiO$_2$	79.866
Ce(SO$_4$)$_2$	332.245	KCl	74.551	NaClO	74.442		

Concentration Data for Commercial Concentrated Acids and Base

Acid or Base	Molarity	Density	% Composition (w/w)
Acetic acid (HC$_2$H$_3$O$_2$)	17	1.05	99.5
Ammonium hydroxide (NH$_4$OH)	15	0.90	58
Hydrobromic acid (HBr)	9	1.52	48
Hydrochloric acid (HCl)	12	1.18	36
Hydrofluoric acid (HF)	26	1.14	45
Nitric acid (HNO$_3$)	16	1.42	72
Perchloric acid (HClO$_4$)	12	1.67	70
Phosphoric acid (H$_3$PO$_4$)	15	1.69	85
Sulfuric acid (H$_2$SO$_4$)	18	1.84	96